# Introductory Solid State Physics

# Introductory Solid State Physics

Second Edition

**H. P. MYERS**

Chalmers University of Technology, Sweden

**UK** Taylor & Francis Ltd, 1 Gunpowder Square, London EC4A 3DE
**USA** Taylor & Francis Inc., 1900 Frost Road, Suite 101, Bristol, PA 19007


First published 1990
Reprinted 1991, 1994 (revised)
Second edition published 1997



**British Library Cataloguing in Publication Data**

A catalogue record of this book is available from the British Library.

ISBN 0-7484-0659-X (Cased)

ISBN 0-7484-0660-3 (Paperback)

**Library of Congress Cataloguing Publication data** are available

Cover design by Jim Wilkie

Typeset in Times 10/12pt by Santype International Ltd., Salisbury, Wiltshire

Printed in Great Britain by T.J. International Ltd, Padstow, Cornwall.

# Contents

Some Relevant Physical Constants and Conversion Factors

| | |
|---|---|
| Avogadro's number $N$ | $6.022 \times 10^{23}$ mol$^{-1}$ |
| Velocity of light $c$ | $2.998 \times 10^{8}$ m s$^{-1}$ |
| Boltzmann's constant $k_B$ | $1.381 \times 10^{-23}$ J K$^{-1}$ |
| Planck's constant $h$ | $6.626 \times 10^{-34}$ J s |
| $\hbar = h/2\pi$ | $1.055 \times 10^{-34}$ J s |
| Electron charge $\lvert e \rvert$ | $1.602 \times 10^{-19}$ C |
| Electron rest mass $m_e$ | $9.110 \times 10^{-31}$ kg |
| Proton rest mass $m_p$ | $1.673 \times 10^{-27}$ kg |
| Bohr magneton $\mu_B$ | $9.274 \times 10^{-24}$ J T$^{-1}$ |
| Electron volt equivalents | |
| 1 eV | $1.60 \times 10^{-19}$ J |
| 1 eV atom$^{-1}$ | $9.65 \times 10^{4}$ J mol$^{-1}$ |
| | $23.05 \times 10^{3}$ cal mol$^{-1}$ |
| $k_B T = 1$ eV | $T = 1.16 \times 10^{4}$ K |
| $\hbar\omega = 1$ eV | $\omega = 15.2 \times 10^{14}$ rad s$^{-1}$ |
| $h\nu = 1$ eV | $\nu = 2.42 \times 10^{14}$ Hz |
| $hc/\lambda = 1$ eV | $\lambda = 1.24 \times 10^{-6}$ m |
| $\mu_B B_0 = 1$ eV | $B_0 = 6.7 \times 10^{3}$ T |
| $\frac{1}{2}mv^2 = 1$ eV | $v = 5.93 \times 10^{5}$ m s$^{-1}$ |

# Preface to First Edition

There are several good texts directed at a general treatment of solid state physics, but as a teacher of a first undergraduate course in the subject I have not found them to be altogether satisfactory, partly because in some cases they are too all-embracing, and in others they argue from the general to the particular. They are excellent for the budding solid state scientist who will eventually take additional courses (probably reading the same text a second time in greater detail). However, the needs of most undergraduate students of physics, electrical engineering or materials science, who will take only one course in the subject, are not, I think, well met by these texts. At the other end of the scale are the short elementary texts that seldom satisfy the serious student or the teacher.

During some thirty years of teaching introductory solid state physics to engineering physics students, my own choice and presentation of subject matter have evolved, a process made easier by access to the earlier editions of C. Kittel's *Introduction to Solid State Physics.* My debt to that author will be obvious and is gratefully acknowledged. The notes on which the present text is based were developed during the past sixteen years, but the transition from notes to text has resulted in a book with much more material than can be included in a conventional undergraduate course.

I think that solid state physics is perhaps the branch of physics which, on macroscopic, microscopic and atomic scales, offers the widest range of phenomena for the application of both classical and quantum physics. The student must have a good background in general physics as well as atomic, quantum and statistical physics. Any attempt to incorporate such background material is out of the question and I have therefore taken much for granted. On the other hand, I believe that at the undergraduate level the subject poses little mathematical difficulty, but that there are considerable conceptual barriers to be overcome. There can also be problems in grasping the wide range of phenomena to be understood, which must be placed in a coherent physical framework.

My objectives have been (a) to limit the theoretical demands placed upon the student, yet cover the basic concepts; (b) to use simple models that still allow contact with the behaviour of real substances; (c) to provide a reasonably good coverage of subject matter and (d) to make contact with aspects of present-day research.

Tastes and needs differ. I suggest that the major portions of the first twelve chapters could form a core course. The remaining chapters might then serve as a basis for group or project study.

All formulae are presented in SI form and should, of course, be used with SI units; but this has not deterred me from the use of eV, cm, Å or gauss. Boltzmann's constant is denoted by $k_B$ in order to avoid confusion with $k$ for wave vector. $\mathbf{E}$ always denotes electrical field strength whereas $E$ denotes energy. I have also favoured the use of $\mathbf{B}$ rather than $\mathbf{H}$, so that the external magnetic field is usually denoted by $\mathbf{B}_0$ and measured in tesla or gauss. Integers are denoted by $n$, $m$, $p$ or $s$, but these symbols may also have different meanings in different contexts. Some illustrative physical data are provided, but no attempt has been made at completeness since student handbooks of physical data and formulae are now available.

I am grateful to the following persons who have kindly furnished me with illustrations: Y. Baer, O. Beckman, R. Behm, G. Dunlop, C. Nyberg, C. G. Tengstål, P. Toennies and C. Trapp. I should also like to thank all those authors and their publishers who have allowed the reproduction of diagrams based on those appearing in their publications.

Ö. Rapp and O. Beckman provided me with certain problems used in the text; C. Miller gave valuable comments on Miller indices; T. Claeson advised me on superconductivity and J. Gollub gave me much valuable criticism of an earlier manuscript. I extend my thanks to them all.

Finally I should like to thank the Chairman of Taylor and Francis, Bryan Coles, for his encouragement and many suggestions for improvement of the manuscript and the referees for much constructive criticism. Special thanks go to Julie Lancashire and John Haynes for their willing and expert editorial help.

H. P. Myers

# Preface to Second Edition

Taking into account previous restricted revisions of the first edition the present version presents an overhaul of the original text together with the addition of new features. Two major objectives have been the elimination of errors and an improved clarity; to these ends, in addition to textual changes, certain diagrams have been redrawn others retouched and new ones introduced.

The following major changes or additions to the text have been made. The original description of quasicrystals has been supplemented with a section on the structure of atomic clusters, the common feature being icosahedral symmetry. The description of dislocations has been rewritten in better conformance to current conventions whereas that of screening in the electron gas has been improved by the introduction of the Thomas-Fermi approximation to the dielectric function, together with mention of the Linhard version. The significance of the dielectric function for the cooperative interaction of electrons in metals is emphasized. In a book that pays particular attention to the electronic structure of the elemental metals a natural question that arises concerns the effects of alloying. The treatment of metallic behaviour has therefore been augmented with a qualitative description of the electronic structure of alloys, including 'pseudo-atom' alloys as well as those base on Cu, Ag or Au where the effect of alloying on the d states is especially important. This allows the magnetic and electrical properties of dilute alloys that display local magnetic moments to be treated in some detail in terms of the Anderson model of the resonant bound d state. A description of the associated Kondo problem is also included.

The original intention that the book should serve as a first introduction to solid state physics remains, but the ever-growing breadth of the subject demands that the student or teacher pick and choose.

*H. P. Myers*

# Introduction

## What is Solid State Physics?

'Solid state physics' is, as the name suggests, the physics of solid substances; nowadays one often uses the term 'physics of condensed matter', since this includes the study of liquids, especially liquid metals, but also molten salts, solutions, so-called liquid crystals and even substances like glycerine. Liquified gases are also important materials, and liquid helium is a subject so large and specialized that it provides more than enough material for separate books. From the present point of view, solid state physics is mainly the physics of crystalline solids. So liquids and amorphous substances make only limited appearances in this book. Most of the inorganic solids we encounter in our daily lives are crystalline, the notable exceptions being glass, which is a supercooled liquid, and soot, which is amorphous; there are, however, much more important amorphous materials than soot!

In crystalline solids the atoms are arranged in a regular manner which is significant for their behaviour and makes the theoretical treatment easier, but this does not mean that crystalline solids are easy to understand in detailed fashion. Consider the pure solid chemical elements (and they can all be obtained in solid form); we know that the properties of free atoms and their ions are governed by their electronic structures. We might very well expect that the properties of the solid elements should reflect the electronic structures of the separate atoms. In general terms this is true, and a good knowledge of the periodic table and the underlying electronic theory is just as valuable for solid state physicists as it is for chemists.

We shall have a lot to say about the properties of metals, which is not so surprising since of the 92 natural elements more than 60 are metals and of these 40 are transition metals (the rare earths and the actinides are included in the general class of transition metal). Although we shall discuss metallic properties in some detail, there is seldom mention of stainless steel or other commercial alloys; these lie outside our present sphere of interest. Furthermore, although we may mention or discuss a particular pure metal, we shall for the most part be concerned with metallic phenomena. Thus at the present level it is more important to understand why some substances are metals, others insulators and still others semiconductors than to have a detailed knowledge of the properties of individual elements. The pure metallic elements are the simplest metals. Alloys are much more difficult to understand. On the other hand, the simplest insulating materials (excluding the solid rare gases) are not necessarily elements such as sulphur, S, or iodine, I, but compounds like the alkali halides, for example NaCl; we shall therefore have something to say about these substances. All solids are difficult subjects for theoretical study, but some are more difficult than others, and the number of possible solid substances is vast even excluding organic and biologically significant materials.

A solid may be considered as a collection of ions (e.g. NaCl) or a collection of ions and loosely bound electrons (e.g. Al), not forgetting the special case of covalent crystals such as diamond. Ions are very heavy objects and electrons are small and insignificant except for their charge. Thus the positions of the ions determine the crystal structure. We shall often find it convenient to separate a solid into ions and valence electrons, although the latter may be difficult to define for a solid transition metal. Sometimes we choose to consider only the behaviour of the ions, as when calculating the heat capacity of a solid. In other instances we pretend the ions are spread out to form a uniform background of positive charge through which the outer valence electrons move; we call this an ideal free electron gas and it helps us to understand some aspects of metallic behaviour. In the end, however, we are obliged to consider how electrons behave in a periodic electrostatic potential produced by the regularly spaced ions. Only then can we understand why sodium, say, is a metal and germanium a semiconductor. However, the movement of electrons through real solids is a very difficult process to analyse because a solid metal, for example, contains millions upon millions of interacting particles. All the electrons and all the ions interact with one another owing to their electric charge. However, the ions are massive, and to start with we can assume that they remain fixed at their equilibrium positions. But this is not the case for the electrons. We have an extreme example of a 'many-body' problem where the behaviour of any particular electron is dependent upon the motion of all the other electrons. In principle it is not possible to separate the motion of individual particles; the whole system of interacting electrons is the true entity. One way to reduce the complexity is to assume that the many-body interactions are small and that it is practicable to consider the behaviour of the whole to be the sum of the behaviours of individual electrons. We assume that when discussing the motion of a particular electron the effect of all the other electrons gives rise to an average potential, so that the energy of our particular electron is a function only of its own coordinates. This is called the 'one-electron approximation' or the 'independent-particle approximation' and is the basis for almost all present-day calculations. In fact we shall never concern ourselves with exact solutions appropriate to particular substances, but consider simple models that show us the patterns of behaviour to expect in real life and that provide us with a good physical understanding of what goes on.

Now there is another feature of solids that is not at all apparent at first sight, namely that they are never perfect. What is a perfect solid? A possible answer could well include the following demands:

(a) perfect purity in an element or perfect stoichiometry in a compound;

(b) the solid must occur as an infinite single crystal; there should be no external surfaces and no grain boundaries;

(c) there must be no holes – no internal surfaces;

(d) every atom must occupy a lattice site – no atom should be out of place;

(e) there must be no empty lattice site – no missing atoms;

(f) the temperature must be absolute zero – no excitation of the system.

Neglecting the fact that it is impossible to obtain complete chemical purity, we shall find that it is thermodynamically impossible to avoid defects in crystals. The surface of a real crystal is unavoidable and is the source of the simplest defect, which is an empty lattice site or more simply an 'atomic hole'. There may be an extra atom in a 'wrong' place or a boundary surface between two crystals.

Such features are always present and often have a marked effect upon the physical properties of solids. Many solid state physicists confine themselves not to the study of defects but to a much more restricted field such as the study of particular types of defect in a narrow range of solids. The use of modern electron microscopes has enabled solid state physicists to see and follow the movement of crystal defects and correlate their properties with the behaviour of bulk matter. Defects are as integral a part of solids as are ions and electrons (although not so numerous of course). Electrons, ions and defects are the constituents of all solids and the source of solid state phenomena.

Which properties are of interest in solid state physics? Clearly if solids are crystalline we must be able to determine and describe the spatial arrangement of atoms characterizing a crystal. Ultimately we expect to predict the crystal structure of an element from a knowledge of the properties of the constituent atoms. It is clear, however, that a knowledge of crystallography is essential. When considering the geometry of crystals, we always assume them to be perfect and we do not need to know anything in detail about the atoms – we can represent them by hard spheres. Similarly, with defects and the mechanical properties of solids, our level of understanding, whilst relatively advanced, does not take into account the electrical character of matter but assumes that atoms are like billiard balls, and, if we do not look too closely, that we can often think of matter as an elastic continuum. With regard to the thermal properties of solids, particularly for insulators, we find that usually the heat capacity and thermal conductivity can be discussed without reference to the detailed structure of the atom. Furthermore, the elementary formulae of the kinetic theory of gases are often of use.

The electrical and magnetic properties of matter have always interested physicists. These properties have commercial applications and this gives added incentive to research, particularly in large corporations. Our purpose is to try and understand the origin of the electrical and magnetic properties of solids, and this requires a quantum-mechanical approach.

The spectroscopic study of atoms and molecules is an important branch of physics, which has received additional impetus with the advent of lasers and synchrotron light sources. Spectroscopy is also important in solid state physics: not only ordinary light is used, but also electromagnetic radiation covering the whole range from X rays to radio waves. The object is to obtain detailed information about the distribution of energy levels for electrons in solids. This is a very active research field at the present time.

Normally in our attempts to understand solid state phenomena we direct attention to the behaviour of atoms within bulk material; we assume that our samples are sufficiently large that their finite size does not affect their behaviour in a significant fashion. The number of surface atoms having properties different from those within the bulk of the sample is a very small fraction ($N^{-1/3}$) of the total number of atoms $N$ ($\approx 10^{18}$–$10^{23}$) comprising the sample. In theoretical models we also try to simulate the 'surface-free crystal', and to this end the sample is usually thought of as a finite element of crystal, large compared with atomic dimensions, but deeply embedded within a larger parent crystal. We assume that opposite faces of our embedded crystal element behave in an exactly similar fashion, and in this way surface effects are eliminated.

In practice the surface is unavoidable and has its own distinct features; furthermore, the surface is very important in technology. Friction and wear arise between moving contacting surfaces, and adhesion and chemical processes (catalysis) are also surface phenomena. In the past twenty five years surface physics has developed rapidly and now forms a major subdivision of solid state physics.

Nuclear phenomena are also valuable in studying solids: nuclear magnetic resonance and the Mössbauer effect are two examples. The present text attempts to cover these different facets of solid state physics.

Clearly solid state physics is a very large subject; it contains well-defined fields of interest, and in any one field (e.g. magnetism) one can easily become absorbed in significant but special problems. There are more solid state physicists in the world than any other kind of physicist; the size of the subject is both an advantage and a disadvantage. The advantage is that it offers a wide range of interesting phenomena for study and research. The disadvantage is that the range of subjects is so large that it is difficult for any one person to encompass them all. Thus there are few, if any, experts in solid state physics; on the other hand there are many experts in magnetism or X-ray photoelectron spectroscopy or diffusion or dislocation theory or semiconductor physics, but I doubt whether there are many experts in both X-ray photoelectron spectroscopy and dislocation physics for example.

## Why Study Solid State Physics?

Since at least half the research physicists in the world are occupied with some aspect of solid state physics, we presume the subject to have a certain significance and value as an aspect of physics in general. Thus we first establish that *solid state physics is an integral part of physics*, and, to the extent that physics is a valuable cultural and scientific pursuit, so is solid state physics. There is, however, an added incentive to have had some contact with solid state physics, because the subject deals largely with the properties of matter in its natural bulk form. Thus the solid state physicist studies, say, copper as copper metal, exactly as it exists in ordinary electrical connections. Furthermore, the phenomena that interest the solid state physicist are often those that technology makes use of. There is no doubt that technology has benefitted immensely from the developments of solid state research. Perhaps the most obvious field of development is that of electronics, especially the applications of semiconductors. However, there are less obvious examples. Consider for example studies of the optical properties of, say, CsBr – of what possible use could this be? In fact, CsBr is transparent to infrared radiation and makes a very good prism in infrared spectrometers, which are important in chemical analysis. In general, instrumentation and measurement techniques lean heavily on methods and apparatus initially developed for pure solid state research. X-ray diffraction and nuclear magnetic resonance are two prominent examples. Much of present-day technology is based on the applications of solid state phenomena. If this is the case then engineers, engineering physicists and materials scientists should have some idea of the basic origins of these phenomena – only in this way can they expect to follow and direct the development of new materials and new techniques. We can specifically mention the factors controlling the mechanical strength and stability of solids, the electrical and magnetic properties of metals and semiconductors, the occurrence of superconductivity, the development of new dielectric materials for the control and modulation of laser light and so on.

This book is planned with the following sequence of topics. Certain properties of solids may be understood with little or no knowledge of the electronic structure of atoms; the crystal structure of solids is one example; we may consider the atoms as points or at the most as hard spheres like billiard balls. The same is true when we discuss mechanical and thermal behaviour. We therefore begin with structure and mechanical and thermal properties.

On the other hand, a characteristic feature of metals is their electrical conductivity, so we proceed by assuming that the electrons responsible for the metallic properties form a simple gas, the free electron gas; this elementary approach is surprisingly effective in describing many aspects of metallic behaviour. Even so, such a model cannot tell us why certain solids are conductors whereas others are insulators or semiconductors; this is only possible when we treat the properties of electrons moving in periodic potentials; this problem, and its solution in principle, together form a central part of the book. The qualitative significance of the band structure of solids is thereafter considered in detail for both metals and semiconductors. The above aspects are covered in the first ten chapters. We have already stressed that solid state physics comprises a vast range of subjects and materials, so that the choice of more specialized topics is broad and to a certain extent a matter of taste or personal experience. Nevertheless, the basic as well as technical importance of magnetic and dielectric phenomena are such that they demand their place. Similarly, the surprising phenomenon of superconductivity and its unique theoretical interpretation justify its inclusion. The final chapters treat surface physics and nuclear solid state physics, two subjects that have developed rapidly in the post-war years, and in which major advances are still being made.

1

# The Materials and Methods of Solid State Physics

## 1.1 Phenomena and Materials

It is customary to associate the separation of solid state physics as an integral subdivision of physics with the discovery of X-ray diffraction in 1912; this was a prerequisite for any proper understanding of the properties of solids. Another landmark, for physics as a whole, was the development of quantum mechanics and its application to model problems during the decade 1926–36. Since the end of World War II, solid state physics has grown continuously in size, diversity and sophistication in both theory and experiment. Perhaps the most significant developments have been: the availability of intense neutron beams from nuclear reactors and spallation sources; the widespread use of $He_4$ and $He_3$ to obtain temperatures down to a few mK; the large calculating capacity of modern computers; the development of molecular beam techniques for the preparation of semiconductor films and the use of lithographic methods to fabricate nanometre structures. Furthermore the development of ultra-high vacuum techniques has furthered the study of surfaces, adsorbed layers and low-dimensional structures.

In principle, of course, any solid or liquid material is a possible object for study by a physicist, but traditionally, say up to 1950, specimens were almost always inorganic substances, pure metals, alloys, metal compounds such as CuO or ZnS and salts like the alkali halides or $AgNO_3$. Today the situation is different. The enormous growth of research activity in solid state physics and the bordering subjects of solid state chemistry and physical metallurgy has opened many new fields and led to the preparation of new materials or the search for samples amongst substances that formerly were thought to be of interest only to inorganic or organic chemists.

However, whatever the historical development and irrespective of whether interest is directed towards a basic understanding or the practical application of materials, the central role is played by *solid state phenomena.* Thus the occurrence of, say, magnetism or superconductivity attracts the curiosity of the scientist and captures the imagination of the electrical engineer. Usually we find it very difficult to understand solid state phenomena, and there is therefore the need to simplify problems as much as possible. This leads the scientist to the study of pure metals, simple alloys or simple well-defined compounds. On the other hand, the engineer or materials scientist, wishing to make practical use of solid state phenomena, rarely finds the pure elements suitable in themselves and, either by design or by trial and error, prepares substances that by virtue of their performance and cost can fill a practical need. We demonstrate these two aspects in a simple block diagram:

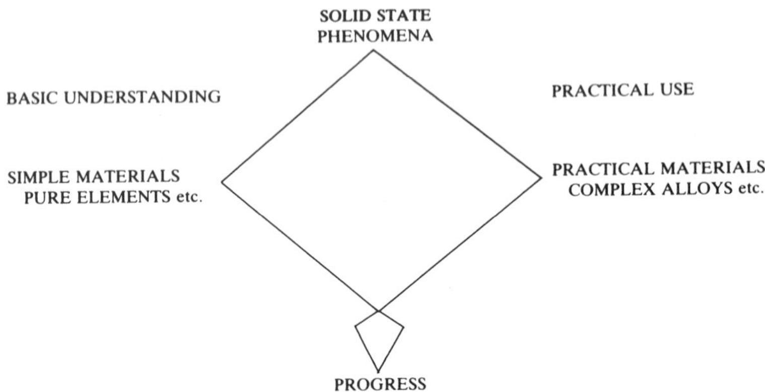


Progress in both the basic understanding and the practical utilization of phenomena and materials rests on mutual appreciation of purpose and interchange of information between those concerned primarily with the basic understanding and those occupied in the practical use of materials.

But what do we mean by a 'phenomenon'? We use the term to describe a particularly well-characterized form of physical behaviour, and in the first instance we concentrate on the behaviour *pattern* rather than on the substances giving rise to it. An example makes this easier to explain. In 1908 Kamerlingh Onnes liquefied $^4$He for the first time and found its boiling point to be 4.2 K. Earlier, in 1898, Dewar had succeeded in liquefying $H_2$, boiling point 20 K. These achievements made it possible to study the variation of the electrical resistance of a metal at low temperatures – a matter of considerable interest. After first trying Pt, Onnes in 1911 turned his attention to Hg, because this could be distilled to a high degree of purity. He found the resistance was not measurable below 4.15 K (Fig. 1.1). The resistance disappeared in an abrupt manner. We now know that this phenomenon of superconductivity arises in 28 pure metals in their ordinary bulk form and in several others under special conditions. There are hundreds of alloys and compounds that become superconducting. In this case the occurrence of superconductivity is the 'phenomenon'. Irrespective of the many different metals that may exhibit the effect, we believe that there must be some mechanism common to them all which is at work. Our first objective must therefore be to describe this mechanism that we presume common to all superconductors. Naturally we make use of all the experimental facts that are available, and by studying many different properties of superconductors we try to find out what other common features they possess – one such feature is that superconductors are usually not the best conductors of electricity at normal temperatures. We then attempt to define a model substance and, by using established knowledge and novel procedures, we try to calculate the physical behaviour of the model substance, hoping to show that below a certain temperature the electrical resistance disappears. If successful, we will have produced a theory of superconductivity. This

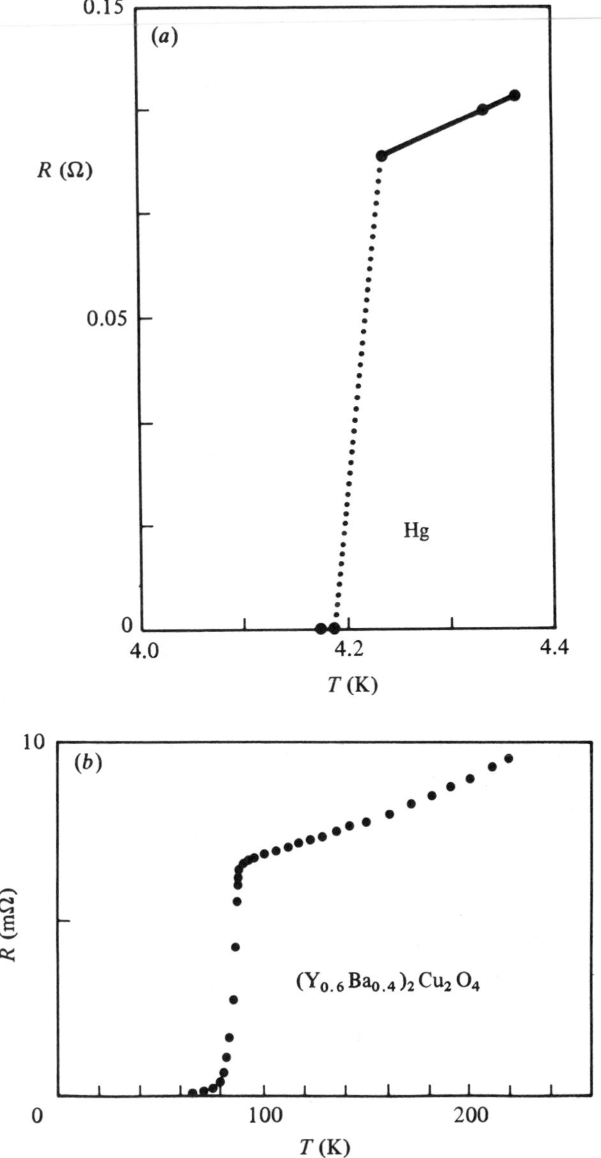


**Figure 1.1** Experimental data (Hg) marking the discovery of superconductivity (a) (after Kammerlingh Onnes 1911), together with those for a newly discovered high-temperature superconducting ceramic (b) (after Wu *et al.* 1987).

was first achieved by Bardeen, Cooper and Schrieffer in 1957. Attention may then be directed to the application or modification of the model theory to particular superconductors.

In this introductory text we must restrict attention to certain basic phenomena of the solid state, and in Table 1.1 we list them together with examples of associated measurable quantities. One of the great advantages of studying solid state physics is that we attempt to understand the properties of ordinary matter, or at least ordinary inanimate matter, as we meet it in our daily lives. We obtain an insight regard-

**Table 1.1**

| Phenomena | Measurables | Variables |
|---|---|---|
| Structural | Space symmetries<br>Interatomic spacing<br>Atomic and ionic volumes | $T$, $P$ |
| Mechanical | Elastic constants<br>Critical shear stress<br>Dislocation densities | $T$, $P$, structure |
| Thermal | Heat capacity<br>Heats of transformation<br>Thermal conductivity<br>Atomic diffusion<br>Defect densities | $T$, $P$, $V$, $\mathbf{B}$, structure |
| Electrical | Conductivity/resistivity<br>Optical reflectivity<br>Polarizability<br>Thermoelectric power | $T$, $P$, $\mathbf{E}$, $\mathbf{B}$, $\hbar\omega$, structure |
| Magnetic | Dipole moment<br>Susceptibility<br>Spectroscopic state<br>Magnetic structure | $T$, $P$, $\mathbf{B}$, structure |
| Dielectric | Polarizability<br>Dielectric constant<br>Optical absorption | $T$, $\mathbf{E}$, $\hbar\omega$, structure |

When we consider other than the pure elements it is clear that chemical composition, as for an alloy or compound, is a very important variable; in many cases the preparation of suitable specimens can demand great skill and considerable effort.

ing the origins of mechanical strength and plasticity in solids, the special properties of silicon that make it so important in semiconductor technology, the origins of metallic lustre and the colour of copper or gold, as well as more exceptional features like magnetism. The above are but a few randomly chosen examples and one could quote many more.

In actual experiments we use ordinary bulk matter; thus the magnetic susceptibility of a metal may be measured using a small cylindrical sample with dimensions 5 mm × 3 mm diameter, whereas the electrical resistance would be determined on, say, 20 cm wire drawn to a diameter of 0.1 mm. However, it is often the case, particularly for alloys, that the mechanical workability influences the form specimens can take. For the metallic elements, and for the measurements just mentioned, it would be sufficient that the samples were pure and well annealed so that internal stresses were absent. But what do we mean by pure metal? Normally metals like Al or Cu are available with purities in the region of 99.999%; in other words impurities occur to the extent of 10 parts per million (ppm). However, analytical data usually refer only to other metallic elements, and there may be considerable amounts of gaseous impurities $H_2$, $O_2$ or $N_2$ either dissolved in the sample or present in combined form, e.g. as the oxide. The transition elements are very reactive and particularly difficult to obtain better than 99.99% pure.

Another feature of ordinary bulk matter is that it is usually polycrystalline, i.e. it is composed of conglomerates of small crystals that are called 'grains'. The size, shape and orientation of grains or crystallites are very dependent on the previous thermal and mechanical history of the sample. In annealed pure metals the grains usually approximate regular polyhedra and have random orientations. The boundaries separating the grains have a finite thickness of order 10 Å and are often places where insoluble impurities collect. Normally the linear size of a grain may vary from the order of micrometres to millimetres, but it is also possible to obtain smaller grain sizes as well as very much larger ones – in the latter case we come into the realm of the single crystal. The polycrystalline character of bulk matter is often disadvantageous, and certain measurements can only be made on well-annealed single crystals, which are usually prepared artificially. The production of large single crystals of metals, alloys, compounds and salts has developed greatly in the past thirty five years. Several special techniques have been developed for their growth, and many substances are available commercially in single-crystal form. Some examples of the need for single crystals are

(a) when it is necessary to minimize the scattering of electrons, photons or phonons (quanta of lattice vibration);

(b) determination of the critical shear stress;

(c) studies of crystalline anisotropy in electrical or magnetic phenomena;

(d) studies of lattice vibrations by neutron scattering.

Single-crystal material is often needed in technical applications too; thus devices based on the semiconducting properties of Si use single-crystal material, and large optical prisms for infrared spectrometers are single crystals of NaCl, CsBr or the like. Quartz crystals are also found as optical components, but their piezoelectric properties give them wider application; solid state lasers are sometimes based on single crystals (glasses and liquids are also used as lasing media). Turbine blades of single-crystal material have been developed for jet engines. Shortly we shall emphasize the need for comparative studies in solid state physics; here we also need to change the conditions of experimentation, and the most important variables are perhaps the following.

(*a*) *Temperature*: $10^{-3}$ K–$3 \times 10^{3}$ K. At the present time it is within the reach of most laboratories to work in the range 1.5–2000 K, whereas rather special procedures are needed to reach extreme temperatures in a controlled fashion. Particular efforts are made to reach $< 10^{-3}$ K in the study of liquid He, but such work is confined to a handful of laboratories.

(*b*) *Energy* $\hbar\omega$: Many experiments make use of particle scattering; the particles and their energies may be

| | |
|---|---|
| photons | $10^{-3}$–$10^{6}$ eV, |
| neutrons | $10^{-1}$–$10^{6}$ eV, |
| electrons | 1–$10^{6}$ eV, |
| ions, α particles, protons etc. | $10^{4}$–$10^{6}$ eV. |

Low-energy photons, electrons and particularly neutrons are used to exchange energy with the solid, particularly to determine the possible excited states of the electrons and the ions.

High-energy radiation is used to cause damage, i.e. the lattice is disrupted and many defects created, with significant changes in physical behaviour; this kind of study has basic scientific importance, but is also intimately connected with the material problems of nuclear reactors. Ion beams may be used to cause radiation damage, but also for controlling the composition of thin layers of material as in semiconductor junctions; then one speaks of ion implantation.

(*c*) *Pressure*: $10^{-10}$–$10^{10}$ Pa.† The compressibilities of solids are so small that we can only influence the lattice spacing by applying hydrostatic pressures greatly in excess of that of the atmosphere. Specially designed high-pressure apparatus allows many physical properties to be determined as functions of pressure, leading to the observation of structural, electrical and magnetic changes of great interest. Using explosive techniques, one can obtain very high transient pressures $>100$ GPa. There is particular interest in the formation of a metallic phase of hydrogen with the aid of very high pressures, but so far the metallic state has not been found after subjection to pressures up to 200 GPa (Ashcroft 1995).

We reduce the pressure or create a vacuum around a specimen primarily to maintain a clean surface. In recent years the examination of solid surfaces using low-energy electron beams or photons has developed into what is known as surface science (see Chapter 14).

The preparation and maintenance of atomically smooth and impurity-free surfaces demands the use of special ultra-high vacuum techniques that keep the residual gas pressure in the range $10^{-8}$–$10^{-10}$ Pa. Under such conditions one can also provide controlled atmospheres of gases or metallic vapours and thereby study adsorbed layers.

(*d*) *Magnetic field strength* **B**: $10^{-15}$ to about 50 T. The lower limit is merely to demonstrate the sensitivity of a modern superconducting junction (squid) magnetometer. Normally we might expect to vary the magnetic field from say about $10^{-7}$ T to as high as we can arrange. The highest static fields that are at present available are around 20 T (present record 35 T) obtained in cooled coils driven by large DC generators. The projected high magnetic field laboratory in the USA plans to produce fields of 45 T. Lower field strengths up to 10 T are readily available from superconducting magnets.

Transient fields $>50$ T with a duration of about $10^{-4}$ s are obtained by discharging banks of condensers through small water-cooled silver coils. Why do we need such high magnetic fields? The general answer lies in the fact that the properties of solids are determined largely by the valence electrons, and these are subject to the Lorentz force

$$\mathbf{F} = -e(\mathbf{E} + \mathbf{v} \times \mathbf{B}).$$

High magnetic fields are essential if we are to determine the detailed dynamics of electrons in solids. The study of the properties of solids in very high magnetic fields $>50$ T is in its infancy. This is not difficult to appreciate when we consider the electromagnetic energies that are involved. For convenience we take a magnetic field with a flux density equal to 100 T (1 MG). The energy density associated with this

† 1 torr = 1 mmHg = 133.3 Pa; 1 bar = 750 torr = $10^5$ Pa.

field is

$$\frac{B^2}{2\mu_0} = \frac{10^4}{8\pi \times 10^{-7}} = 4 \times 10^{10}\ \mathrm{J\ m^{-3}}.$$

This is an enormous energy density, and the associated magnetic pressure is $4 \times 10^{10}$ Pa. The magnetic field is confined by the current sheath that produces it, and the material of the current-carrying coil must in turn support the pressure. The mechanical and thermal stresses involved are far greater than can be withstood by the silver or copper coils that are used and they literally explode. Very intense magnetic fields can only be obtained in pulsed form and are self-destructive, but since the coils explode (i.e. fly apart) the specimen may be preserved. Occasionally we may in an exercise put $\mathbf{B} = 100$ T, but it is important to remember that this is an exceptionally high value and not easily obtained: ordinary laboratory magnets seldom achieve fields in excess of 1.5 T. Similarly, when we discuss the magnetic properties of iron and other ferromagnetic materials we shall find that their behaviour may be described as though powerful molecular magnetic fields of order 100 T were active in the material. This cannot be the case, since the associated electromagnetic forces would cause the iron to disintegrate. The cooperative magnetic behaviour of solids arises from quantum-mechanical electrodynamic Coulomb interaction and is in no way of magnetostatic origin.

(*e*) *Electric field strength* $\mathbf{E}$: $<5 \times 10^8\ \mathrm{V\ m^{-1}}$. The electric field strength is a more important variable in technical applications because dielectric breakdown in insulating liquids and solids limits the performance of electrical equipment. Normally in solid state physics we are not so concerned with high uniform electrostatic fields because the atomic electrostatic fields experienced by the valence electrons are so very large, of order $10^{10}\ \mathrm{V\ m^{-1}}$, that all external perturbations are effectively very small, leading to linear behaviour, except in a special class of spontaneously polarized substances called ferroelectrics. However, in recent years the development of high-power lasers has enabled electric field strengths of about $5 \times 10^8\ \mathrm{V\ m^{-1}}$ to be achieved, and this has led to new research fields in optics. Here again these intense electric fields often cause mechanical damage on account of the associated mechanical and thermal stresses to which they give rise.

## 1.2 The Periodic Table

Condensed matter is formed by the aggregation of atoms, and the properties of the aggregate are dependent upon the electronic structure of the component atoms. It is therefore of the utmost importance that we appreciate the significance of the periodic table of the elements (Fig. 1.2), which is as vital for the classification of the physical properties of the solid elements as it is for their constituent atoms. We must always be prepared to make comparisons both with elements of slightly larger or smaller atomic numbers and with elements in the same column. Our present limited ability to understand fully the properties of any particular element, alloy or compound demands that we recognize similarities and differences in the behaviour of solids, and that we attempt to associate behaviour patterns with the positions of the constituent atoms in the periodic table. We demonstrate the above point with a particular example. The cohesive energies† for the series of elements K–Kr and

† The cohesive energy is the energy required to separate a solid into its component atoms at absolute zero.

| IA | IIA | IIIB | IVB | VB | VIB | VIIB | ← VIII | | → | IB | IIB | IIIA | IVA | VA | VIA | VIIA | NOBLE GASES |
|---|---|---|---|---|---|---|---|---|---|---|---|---|---|---|---|---|---|
| 1<br>**H**<br>1.008 | | | | | | | | | | | | | | | | 1<br>**H**<br>1.008 | 2<br>**He**<br>4.003 |
| 3<br>**Li**<br>6.940 | 4<br>**Be**<br>9.013 | | | | | | | | | | | 5<br>**B**<br>10.82 | 6<br>**C**<br>12.010 | 7<br>**N**<br>14.008 | 8<br>**O**<br>16.000 | 9<br>**F**<br>19.00 | 10<br>**Ne**<br>20.183 |
| 11<br>**Na**<br>22.997 | 12<br>**Mg**<br>24.32 | | | | | | | | | | | 13<br>**Al**<br>26.97 | 14<br>**Si**<br>28.06 | 15<br>**P**<br>30.98 | 16<br>**S**<br>32.066 | 17<br>**Cl**<br>35.457 | 18<br>**Ar**<br>39.944 |
| 19<br>**K**<br>39.096 | 20<br>**Ca**<br>40.08 | 21<br>**Sc**<br>45.10 | 22<br>**Ti**<br>47.90 | 23<br>**V**<br>50.95 | 24<br>**Cr**<br>52.01 | 25<br>**Mn**<br>54.93 | 26<br>**Fe**<br>55.85 | 27<br>**Co**<br>58.94 | 28<br>**Ni**<br>58.69 | 29<br>**Cu**<br>63.54 | 30<br>**Zn**<br>65.38 | 31<br>**Ga**<br>69.72 | 32<br>**Ge**<br>72.60 | 33<br>**As**<br>74.91 | 34<br>**Se**<br>78.96 | 35<br>**Br**<br>79.916 | 36<br>**Kr**<br>83.7 |
| 37<br>**Rb**<br>85.48 | 38<br>**Sr**<br>87.63 | 39<br>**Y**<br>88.92 | 40<br>**Zr**<br>91.22 | 41<br>**Nb**<br>92.91 | 42<br>**Mo**<br>95.95 | 43<br>**Tc**<br>(99) | 44<br>**Ru**<br>101.7 | 45<br>**Rh**<br>102.91 | 46<br>**Pd**<br>106.7 | 47<br>**Ag**<br>107.88 | 48<br>**Cd**<br>112.41 | 49<br>**In**<br>114.76 | 50<br>**Sn**<br>118.70 | 51<br>**Sb**<br>121.76 | 52<br>**Te**<br>127.61 | 53<br>**I**<br>126.92 | 54<br>**Xe**<br>131.3 |
| 55<br>**Cs**<br>132.91 | 56<br>**Ba**<br>137.36 | 57<br>**La**<br>138.92 | 72<br>**Hf**<br>178.6 | 73<br>**Ta**<br>180.88 | 74<br>**W**<br>183.92 | 75<br>**Re**<br>186.31 | 76<br>**Os**<br>190.2 | 77<br>**Ir**<br>193.1 | 78<br>**Pt**<br>195.23 | 79<br>**Au**<br>197.2 | 80<br>**Hg**<br>200.61 | 81<br>**Tl**<br>204.39 | 82<br>**Pb**<br>207.21 | 83<br>**Bi**<br>209.00 | 84<br>**Po**<br>210 | 85<br>**At**<br>(210) | 86<br>**Rn**<br>222 |
| 87<br>**Fr**<br>(223) | 88<br>**Ra**<br>226.05 | 89<br>**Ac**<br>227.0 | | | | | | | | | | | | | | | |
| **LANTHANIDE SERIES** | | | | 58<br>**Ce**<br>140.13 | 59<br>**Pr**<br>140.92 | 60<br>**Nd**<br>144.27 | 61<br>**Pm**<br>(147) | 62<br>**Sm**<br>150.43 | 63<br>**Eu**<br>152.0 | 64<br>**Gd**<br>156.9 | 65<br>**Tb**<br>159.2 | 66<br>**Dy**<br>162.46 | 67<br>**Ho**<br>164.94 | 68<br>**Er**<br>167.2 | 69<br>**Tm**<br>169.4 | 70<br>**Yb**<br>173.04 | 71<br>**Lu**<br>174.99 |
| **ACTINIDE SERIES** | | | | 90<br>**Th**<br>232.12 | 91<br>**Pa**<br>231 | 92<br>**U**<br>238.07 | 93<br>**Np**<br>(237) | 94<br>**Pu**<br>(239) | 95<br>**Am**<br>(241) | 96<br>**Cm**<br>(242) | 97<br>**Bk**<br>(243) | 98<br>**Cf**<br>(244) | | | | | |

**Figure 1.2** The periodic table of the elements.

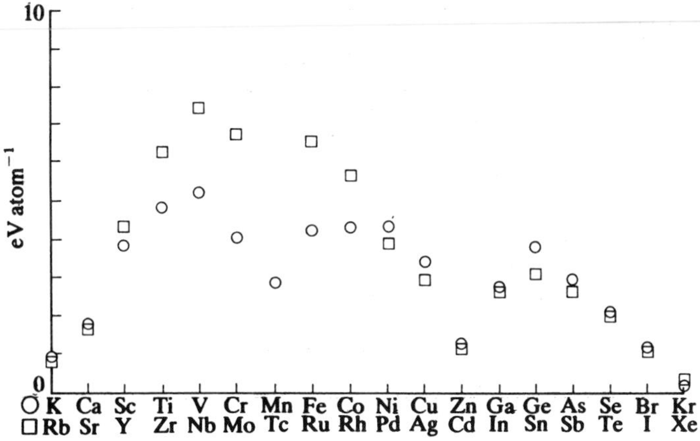


**Figure 1.3** The variation of cohesive energy for two parallel series of elements. (From Kittel 1986.)

Rb–Xe are plotted against atomic number in Fig. 1.3. The similarity in the patterns is striking but typical. We know that in the series K–Kr the 4s, 3d, 4p shells are successively filled with electrons, whereas in the series Rb–Xe the same is true for the 5s, 4d, 5p shells. The symmetry and occupancy of the shells are the same and it is not surprising that we find a parallel variation of the binding or cohesive energy. Let us do the same for the elements in vertical columns of the periodic table: we choose the solid rare gases Ne–Rn and the alkali metals Li–Cs; see Fig. 1.4. In both

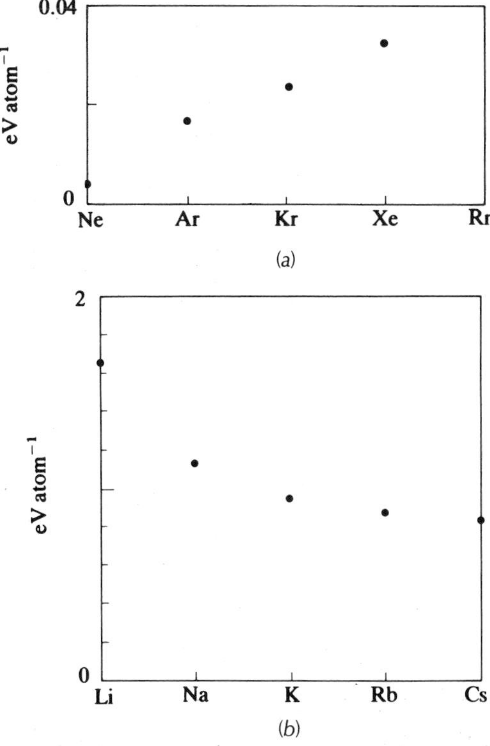


**Figure 1.4** The cohesive energy for certain solid rare gases (*a*) and for the alkali metals (*b*). (From Kittel 1986.)

cases there is a well-defined trend, but, whereas the binding energy for the alkali metals decreases with atomic number, that of the solid rare gases increases.

Perhaps this has something to do with the atomic size, but when we plot the average atomic diameter (Fig. 1.5) we find similar trends, both kinds of atom becoming larger as the atomic number increases. We shall return to this problem in a little while and for the moment it is sufficient that the chosen example illustrates the regular variation of a physical property of solid substances through the periodic system. It is always worthwhile plotting physical parameters along a series of elements or down a column of elements. Within any series (i.e. horizontal row of elements), the electron content and valence change as we proceed along the series, and this leads to pronounced variations in the physical properties of the solid elements. On the other hand, elements within any given column have different atomic size, on account of the different number of closed electron shells, but the same outer electron configuration of valence electrons: the free atoms and ions therefore have similar properties, and this is reflected in the behaviour of the solid elements too, although, as we have seen, this does not preclude regular changes within a given column (or group as it is often called).

In this book we devote most attention to the properties of pure elements, and although we cannot discuss the detailed behaviour of the individual elements, we separate them into categories to which we will often refer.

(*a*) *The simpler 'sp' elements*, as exemplified by the following:

| | | |
|---|---|---|
| monovalent alkali metals | Li, Na, K, Rb, Cs | Group 1A |
| divalent alkaline earth metals | Be, Mg, Ca, Sr, Ba | Group IIA |
| divalent metals | Zn, Cd, Hg | Group IIB |
| trivalent metals | Al, Ga, In, Tl | Group IIIB |
| quadrivalent elements | Si, Ge, Sn, Pb | Group IV |

The above elements, nearly all of which are metals, are characterized by closed shell 'inert gas' cores and valence electrons that occupy s or p states in the free atom. Their behaviour as solids is governed wholly by the outer s and p electrons and we shall often refer to them as the 'sp' elements or 'sp' metals. This is also true of the elements As, Sb and Bi, which we characterize as 'poor metals' but nevertheless metals. We have little space for these interesting but somewhat exceptional elements.

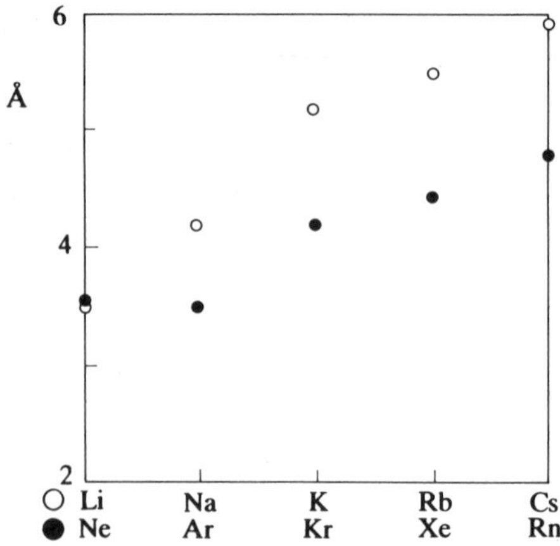


**Figure 1.5** Atomic diameter of the solid rare gases and alkali metals.

Similarly, we shall have little or nothing to say about the non-metallic elements such as S or Cl that have nearly filled sp shells. The reason is that these elements are often gaseous at room temperature or form molecular solids, i.e. they are aggregates of molecules and not atoms. The structures of these solid elements are often complicated and their physical properties are neither so diverse nor so interesting as for the metals. The rare gases form simple solids and are important as model systems, but again we must neglect their consideration in favour of more important materials. $^4$He and $^3$He are exceptional, a study of their behaviour demanding a special treatment.

(*b*) *The transition metals.* The term 'transition metal' is used to describe any element containing an incompletely filled d or f shell. We find the following series:

| | |
|---|---|
| First transition series (incomplete 3d shell) | Sc, Ti, V, Cr, Mn, Fe, Co, Ni |
| Second transition series (incomplete 4d shell) | Y, Zr, Nb, Mo, Tc, Ru, Rh, Pd |
| Third transition series (incomplete 5d shell) | La, Hf, Ta, W, Re, Os, Ir, Pt |

Thus there are 24 'd' transition metals. There is a complication, however, in that after $^{57}$La, which has outer electron configuration $5d^16s^2$, the 4f shell begins to fill and further electrons are not added to the 5d shell until the 4f shell is completed. Thus between La and Hf arise the 14 rare earth transition metals

(Ce, Pr, Nd, Pm, Sm, Eu, Gd, Tb, Dy, Ho, Er, Tm, Yb, Lu)

A similar series of 5f rare earth metals starts with Th, but these will not concern us.

Of the 92 naturally occurring elements, some 62 are metals and of these 40 are transition metals. So no excuse is needed if we choose to concentrate our attention primarily on metallic behaviour, and it may be noted that the highest strengths, the strongest magnetism and the highest superconducting temperature are associated with transition elements.

We have still to mention three very important metals that do not fall into either of the two above groups, namely Cu, Ag and Au. These are the coinage metals: they are monovalent, possessing a single s valence electron. They contain filled d shells and therefore we do not class them as transitional. Nevertheless the d electrons have energies very close to those of the outer s electron and are important for their physical behaviour; we call them 'post-transition' metals.

Consider the electronic structures of the following three metals:

| | |
|---|---|
| Al | (Ne core) $3s^23p^1$ |
| Mn | (Ar core) $3d^54s^2$ |
| Gd | (Xe core) $4f^75d^16s^2$ |

When aluminium atoms form the solid metal the outer valence electrons, $3s^23p^1$, on adjacent atoms interact strongly, lose their atomic character and become free from the atoms, otherwise aluminium would not conduct electricity. The same is true for the $4s^2$ electrons in Mn, and to a certain extent for the $3d^5$ electrons too. The d electrons, however, lie closer to the atom core and we say they are tightly (but not completely) bound to the atom: d electrons on adjacent atoms do interact, but to a very much smaller extent than the sp electrons.

On the other hand, in Gd, and the other rare earth metals, the 4f electrons lie deep within the atom and are shielded by the filled $5s^2 5p^6$ shells of electrons. Thus even in the solid metal the 4f electrons on adjacent atoms do not interact directly with one another and they maintain their atomic character. The 5d and 6s electrons provide the metallic properties of Gd.

When we recall that, within the limits set by the Pauli principle, electrons occupy states so as to maximize the total spin, and thereby the magnetic moment, we can expect that magnetic phenomena are a characteristic feature of transition metals, their alloys and compounds. Magnetic phenomena therefore form a significant part of solid state physics.

Assuming that we have some prior knowledge of the elements, a glance at the periodic table shows that we can make a reasonably clear division of the solid elements into metals (i.e. conductors), semiconductors and insulators. One of our major objectives will be to obtain a convincing explanation for the occurrence of these distinct forms of electrical behaviour.

## 1.3 The Potential Energy

An important question is why do atoms form the condensed phases that we call liquids and solids? In other words, what are the mechanisms of cohesion or crystal binding? To take a specific example, we must ask what is the difference in energy between $6 \times 10^{23}$ free independent atoms of Na and 23 g of Na metal at zero Kelvin. Experiment shows that solid Na at 0 K is more stable than the vapour by an amount 1.13 eV atom$^{-1}$, which is a very small quantity compared with the total energy of a single Na atom. This we can readily appreciate from the fact that the first and second ionization potentials of the Na atom are 5.14 and 52.43 eV respectively, and the neutral atom contains eleven electrons. The binding energy of a solid is in fact a small quantity, being the difference between two large and almost equal energies. Calculations are becoming more and more reliable, but any attempt at even an outline of this subject is more suitable for the end rather than the beginning of a course in solid state physics. Nevertheless, we can make some progress using simple qualitative ideas. Condensed phases are stable because *long-range attractive forces* caused by the electrostatic Coulomb attraction between the valence electrons and the ions overwhelm the *short-range repulsive forces* that arise when atoms, and in particular their inner closed shells of electrons, come into contact. The Pauli principle is very effective in preventing the overlap of closed electron shells because this can only arise if electrons in these shells are driven to high energies. This leads to a rapid increase in the total energy as closed shells on adjacent atoms are pressed into contact. This is the origin of the incompressible character of liquids and solids.

Diagrammatically, we can express the energy of the solid relative to the same mass of dilute vapour in terms of a pair potential that is the above difference in energy per atom described as a function of interatomic spacing; it contains attractive and repulsive terms and has the form shown in Fig. 1.6. In principle this pair potential is calculable, but in practice the difficulties are great, although significant progress has been made in recent years. In certain cases, e.g. ionic salts, it is usual to represent the repulsive component $U_R$ by a simple analytic function of interatomic separation involving two parameters that are to be determined by comparison with experimental data.

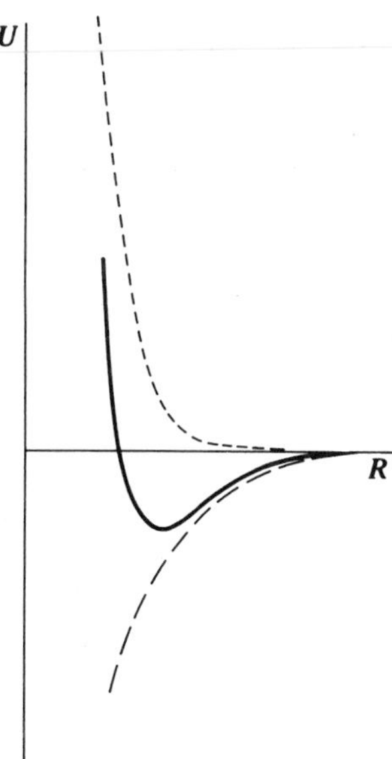


**Figure 1.6** The qualitative form of the variation of potential energy with atomic separation. The resultant energy is the difference between the attractive (— —) and repulsive (- - -) components.

Two expressions often assumed to represent repulsive forces are

$$U_{\mathrm{R}} = \frac{B}{R^{12}}$$

(which is called the Lennard-Jones potential) and

$$U_{\mathrm{R}} = \lambda \mathrm{e}^{-R/\rho}$$

(which is called the Born–Mayer potential). $B$, $\rho$ and $\lambda$ are parameters determined by comparison with experiment. The former expression is empirical, but the latter has a basis in quantum mechanics. In practice the problem is more complicated since we must also predict the crystal structure, and, just as the binding energy is not large, the *differences* in binding energy for different structures are very small indeed. However, we can readily appreciate that the average equilibrium interatomic spacing, and thereby the density, is governed by the position of the minimum in $U$, whereas the depth of the minimum is a direct measure of the binding energy at 0 K. Furthermore, if we calculate $U$ for different temperatures, we find that the minimum becomes shallower and moves to higher values of the interatom spacing, thereby demonstrating thermal expansion. The depth of the minimum also determines the surface energy and the surface tension, whereas the curvature of the potential in the neighbourhood of the minimum controls the elastic constants and the normal modes of atomic vibrations.

## 1.4 Crystal Binding and Valence Charge Distributions

The long-range attractive forces that hold atoms together in solids arise from the Coulomb interaction between valence electrons and the ions. Although the physical origin of the interaction is the same in all substances, we know from experience that it is convenient to divide the different solid types into characteristic groups that we define as 'ionic salts', 'covalent compounds', 'metals', etc., and we associate these groups with rather well-defined geometries for the valence charge distribution. These different classes of solid may also be associated with different variants of the long-range Coulomb attractive forces as the following discussion illustrates.

(*a*) *van der Waals interaction.* This arises between neutral spherically symmetrical charge distributions best exemplified by the inert gas atoms. It occurs because although the average charge distribution is spherically symmetrical the atom is a dynamical system and charge density fluctuations arise, leading to a temporary electrical dipole moment. Of course, the time average of this dipole moment is zero, but its instantaneous value is finite and the associated electric field induces a similar but oppositely directed electric dipole on a neighbouring atom, leading to a weak attraction.

The associated attractive potential may be expressed as

$$U_{\mathrm{A}} = -\frac{A}{R_{ij}^{6}}, \tag{1.1}$$

so the pair potential may be written

$$U_{ij} = \frac{B}{R_{ij}^{12}} - \frac{A}{R_{ij}^{6}}. \tag{1.2}$$

The total potential energy of the solid becomes

$$U = \frac{1}{2}\sum_{i \neq j}\left(\frac{B}{R_{ij}^{12}} - \frac{A}{R_{ij}^{6}}\right). \tag{1.3}$$

The constants $A$ and $B$ are established with the help of experimental data, e.g. the known equilibrium interatomic separation and bulk modulus. These van der Waals interactions are weak, but are the source of binding in the inert gas condensed phases. The closed electron shells of the inert gases cause the atoms to act as hard spheres, and, with the exception of He, the solid phase adopts the face-centred cubic structure, which, as we shall see, is a typical geometry for the close packing of hard spheres. The first ionization potentials of the rare gases are large, $>10$ eV, and this means that electronic excitation of the solid rare gas can only arise from high energy ultraviolet radiation.† The liquid and solid rare gases are therefore transparent. The weak attractive force produces weak binding, and we find that these gases have low melting points and small heats of vaporization. Although of very limited practical importance, the simplicity of the rare gas solids compared with other substances makes them attractive for detailed study.

We can now return to Fig. 1.4(*a*) and appreciate why the inert gas binding energy increases with atomic number and atomic size. The binding arises from electrostatically induced dipole moments. The polarizability of the rare gas atom is therefore a significant quantity. In Chapter 12 we show that the polarizability of an atom is directly proportional to the atomic volume, and this is the reason why the binding energy increases with atomic number of the rare gas atom.

(*b*) *Ionic interaction.* As we shall soon see, the ionic salts, as typified by the alkali halides, form crystals with highly ordered geometrical ionic arrangements, a

† Ordinary visible light is limited to the range $1.6 < \hbar\omega < 3.2$ eV.

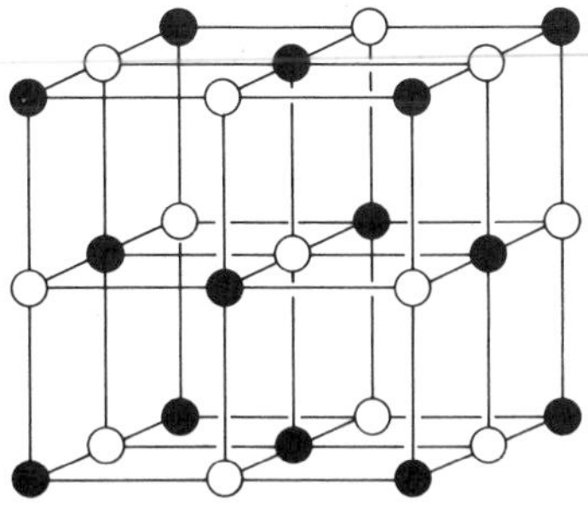

**Figure 1.7** Ionic arrangement in an alkali halide crystal. The anions are in fact much larger than the cations.

positive ion with charge $+Q$ being surrounded by negative ions with charge $-Q$ and vice versa (Fig. 1.7). We associate positive and negative point charges with particular lattice sites, and this situation is typical for compounds between elements that form spherical ions by charge transfer.

The archetype is the alkali halide such as NaCl or LiF. The alkali metal atom has a single valence electron outside an inert gas core; the ionization energy for this electron is relatively small, about 4–5 eV. The halogen atom, F, Cl, Br or I, immediately precedes the inert gas that terminates any given row or series of the periodic table and possesses seven sp electrons; it needs only one more to form an octet and become a spherical negative ion. In fact such an ion is a stable entity because the added electron polarizes the neutral atom and becomes attached to it with a considerable binding energy, which is called the electron affinity; for the halogens its value is about 3 eV, which is a significant fraction of the ionization energy of an alkali metal atom. It is therefore advantageous for neutral Na and Cl atoms to become ions by charge transfer, the resultant electrostatic attraction between these ions causing a reduction in potential energy far greater than the net energy needed to create the two ions. Thus the Coulomb forces that arise between the 'point' charges of the ions lead to the strong cohesion of this type of compound. For any given ion pair (not necessarily neighbours) in the solid there is a Coulomb potential energy

$$\frac{\pm Q^2}{4\pi\varepsilon_0 R_{ij}},$$

the sign being dependent upon whether the ions $i$ and $j$ have like or unlike charges. For the regular geometrical array of ions the total potential energy of any one ion $i$ in the presence of all other ions $j$ is

$$\sum_{j\neq i} \frac{\pm Q^2}{4\pi\varepsilon_0 R_{ij}}. \tag{1.4}$$

If the ions have a nearest-neighbour separation $R_0$ then the above expression may be written

$$\sum_{j\neq i} \frac{\pm Q^2}{4\pi\varepsilon_0 R_{ij}} = \frac{-\alpha Q^2}{4\pi\varepsilon_0 R_0}. \tag{1.5}$$

The parameter $\alpha$ takes different values for different crystal geometries; it is called the Madelung constant but we shall not concern ourselves with its calculation. Assuming that the repulsive interaction is confined to that between a given ion and its $z$ nearest neighbours, the total potential energy of one ion in the presence of all the other ions $j$ is then

$$U_i = z\lambda e^{-R_0/\rho} - \frac{\alpha Q^2}{4\pi\varepsilon_0 R_0}. \tag{1.6}$$

Here the repulsive forces are represented by the exponential expression in preference to the inverse power law adopted in the case of the rare gases. Again the constants $\lambda$ and $\rho$ must be determined with the aid of the known physical behaviour of the substance under consideration.

The Coulomb interaction arising in the ionic salts has long range and considerable strength, leading to some of the most chemically and mechanically stable structures; witness the mechanical rigidity and high melting points of so-called refractory materials like $Al_2O_3$ and $ZrO_2$ for example. Such substances find application as furnace linings in industry and as crucibles for the retention of liquid metals in laboratory work. They are called ceramic materials. Again, just as for the rare gas atoms, the ions are very stable and not readily excited by ordinary optical radiation so they are also transparent, unless they happen to contain transition metal ions whose incomplete d shells provide excitation possibilities leading to optical absorption and brilliant colours. The colours of glasses and many natural (or artificially produced) gem stones are often due to the presence of a small amount of transition metal impurity in what would otherwise have been a water-clear transparent substance. The best known example is perhaps ruby, where the characteristic red colour is a result of impurity $Cr^{3+}$ ions in an otherwise colourless $Al_2O_3$ crystal.

It is instructive at this point to emphasize the fact that metal ions are small and non-metal ions large. The reasons are clear; the metal ion (cation) has an excess positive charge and the remaining electrons are more strongly attracted to the nucleus. On the other hand, the non-metal negative ion (anion) has received extra electrons that, although stably attached to the ion, are not drawn into the core of the former neutral atom, but occupy orbitals of large diameter. The conventional way of representing the structure of such salts, as in Fig. 1.7, is therefore misleading and we should rather think of arrangements where the small metal ions are squeezed into the residual spaces left when the large spherical anions are packed together. Thus the oxide $Fe_3O_4$, which may be better appreciated as $FeOFe_2O_3$, can be thought of as a face-centred cubic array of contacting $O^{2-}$ ions into which the $Fe^{2+}$ and $Fe^{3+}$ ions are inserted (Fig. 1.8).

(c) *Covalent interaction.* This is a particularly important bonding mechanism in organic chemistry, but is significant in inorganic substances too. The elements C (diamond), Si and Ge all have four valence electrons and therefore a half-filled sp shell. They need four more electrons to form the particularly stable sp octets that characterize the rare gases. The conventional description of their stability (in the diamond cubic structure, Fig. 2.12) is that by the mutual sharing of their valence electrons they attempt to approximate the complete octet around each atom. The sharing of the electrons, between say Ge atoms, causes the valence electron charge to be distributed primarily between the atom sites, and we may think of it as a cement binding the ions together. The covalent bond, as it arises in the $H_2$ molecule, provides a distribution of valence electron charge between the $H^+$ ions in a manner completely complementary to that characteristic of the ionic bond as described for $Na^+Cl^-$. Although not so obvious as in the case of the ionic bond, it is still the electrostatic interactions between the valence electron charge clouds and the ions that produce the bond, but we cannot represent the associated potential energy in the simple manner that is applicable for the ionic point charges.

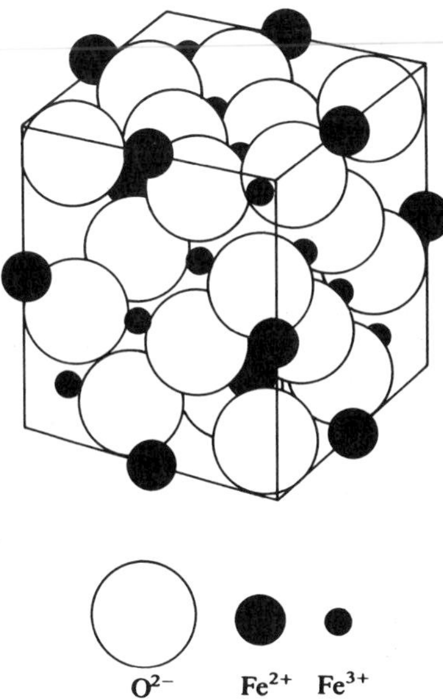


**Figure 1.8** Ionic arrangement in $Fe_3O_4$.

Electron sharing may also be described as the result of maximizing the constructive overlap of electron wave functions on adjacent atoms, and theoretical chemists have devised hybridized (mixtures of) wave functions to describe these overlapping or shared valence electrons. The $sp^3$ hybrid wave functions produce the tetrahedrally directed covalent bonds of the carbon atom and account for the occurrence of diamond, silicon and germanium in the 'diamond cubic' structure.

The ideal ionic and ideal covalent bonds correspond to well-defined and complementary valence charge distributions, but we also encounter substances where it makes sense to speak of a mixture of ionic and covalent bonding components. We use GaAs to illustrate this situation. GaAs has the same crystal structure as Ge, but Ga is trivalent and As pentavalent; there are, on average, four electrons per component atom, but if these are concentrated (shared) between the atoms in a symmetrical fashion we must necessarily find an element of ionic binding. This must be so because the atomic sphere associated with the Ga atom must contain somewhat more than three electrons, and that for the As atom somewhat less than five. The Ga will have fractional negative charge, the As fractional positive charge. Otherwise, if this situation is to be avoided, the valence charge distribution departs from the ideal symmetry as is found in say diamond or the hydrogen molecule. A consideration of BeO or ZnS leads to similar conclusions, and we expect the fractional ionic character to increase as the elements concerned lie closer to the opposite ends of a given sp series. The ideal covalent bond is, as we have said, formed between two electrons in similar eigenstates on similar atoms. We know from the theory of the hydrogen molecule that binding converts two similar eigenstates (appropriate to the two isolated components) into molecular orbitals of bonding (constructive overlap) and antibonding (destructive overlap) character as shown in Fig. 1.9, which shows the physical origin of the reduction of Coulomb energy in the extra charge density in the potential well between the nuclei for the bonding orbitals. These general results are of considerable qualitative significance in understanding the electronic structure of solids.

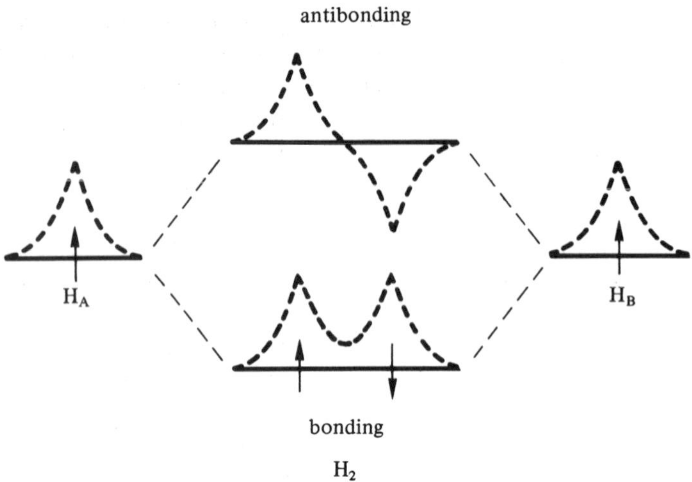


**Figure 1.9** The thick horizontal lines $H_A$ and $H_B$ represent the similar ground state energies of two separate hydrogen atoms, whereas those in the centre are the bonding and antibonding energy levels of the $H_2$ molecule associated with the linear combinations of the separate atomic wave functions. The dotted curves illustrate the qualitative form of the electron wave functions associated with the different levels. This pattern of bonding and antibonding levels is maintained in all molecular bonding interactions as well as between atoms in condensed matter (see also Box 11.1 on the $O_2$ molecule at the beginning of Chapter 11).

(*d*) *The metallic bond.* We use the word 'bond' when we can clearly recognize the forces that are at work, as in the dipole-dipole interactions of the van der Waals mechanism or the electrostatic forces between regular arrays of point charges. The covalent bond is more difficult to appreciate in this way, and we concentrate on the geometry of the charge distribution as being the recognizable trademark. The metallic bond is less discernible than the covalent bond, but we maintain that there is no real distinction between them. In simple metals we shall find that the sp electrons form what is for most purposes an electron gas of uniform density: thus it is as though the conventional covalent bond had been generalized and become uniformly spherically symmetrical, the directional character having disappeared. The electrical properties of metals demand that the valence electrons no longer be bound to the atoms; we can perhaps accept this situation if we think of a metal crystal as one very large molecule in which all the valence electrons are shared among all the atoms. It is important to recognize that there is no fundamental difference in character between the eigenstates occupied by electrons in aluminium and those in germanium.

If we are to obtain estimates of the binding energy of Ge or Al, we must realize that these elements are stable as solids at 0 K because the electrons, and principally the valence electrons, have lower total energy in the solids than in the corresponding vapours. The problem is not so transparent as for the ideal ionic salt.† We require a complete quantum-mechanical description of the dynamics of the valence electrons

† We must remember that a first-principles calculation for an ionic salt is just as difficult a problem as that for a metal. However, for the ionic salt we can make a simple representation in terms of attractive and repulsive components.

in the periodic potential of the regular array of ions. If we know the available eigenstates and the associated electron kinetic and potential energies then we can sum them to find the total energy of the solid, which is to be compared with that of the independent atoms. At our present state of knowledge we must be content with the somewhat unrigorous statement that the electrons 'cement' the ions together. It is not just that the calculation of the binding energy is a complicated problem, but, as we have seen, metals show considerable diversity and whereas the calculation of the binding energy of Na is tractable, the variation of the binding through a transition metal series (Fig. 1.4) is a much more difficult problem that has only recently been resolved.

We shall try to discuss this again in a qualitative fashion in Chapter 8. But we should return now to Fig. 1.4 and explain why the binding energy of the alkali metals decreases with atomic number. We have already seen that the atomic volume increases with atomic number, and since the alkali metals each have one valence electron, the uniform electron gas mentioned above must have a density that decreases with the atomic number of the metal: the cement, as it were, becomes thinner and the binding weaker.

Thus, although we have emphasized that chemical binding always arises from the electrostatic interactions between the valence electrons and the ions, it is the manner in which these interactions manifest themselves that characterizes the bond: that is why in the case of the rare gases the bond becomes stronger as the atomic number increases, whereas the opposite is true for the alkali metals.

## References

Ashcroft, N. W. (1995) *Phys. World* **8** Nr. 7, 43.

Kamerlingh Onnes, H. (1911) *Akad. Vetenschaffen* **14**, 113, 818.

Kittel, C. (1986) *Introduction to Solid State Physics*, 6th edn. Wiley, New York.

Wu, M. K., Ashburn, J. R., Torng, C. J., Hor, P. H., Meng, R. L., Gao, L., Huang, Z. J., Wang, Y. G. and Chu, C. W. (1987) *Phys. Rev. Lett.* **58**, 908.

## Further Reading

For a review of the progress in solid state physics between 1930 and 1980 consult The Beginnings of Solid State Physics, *Proc. Roy. Soc.* (1980) A **371** Nr. 1744, 1–177.

The following data reference handbooks may be found useful: *Physics Handbook*, The American Institute of Physics, McGraw Hill Inc., New York.

*'Kaye and Laby' Tables of Physical and Chemical Constants*, 15th Edition (1986), Longman, Harlow.

*The Elements*, Emsley, J. (1989), Clarendon Press, Oxford.

# 2

# Crystallography

## 2.1 Lattices

We are most familiar with condensed matter in the form of solid crystalline substances and, although interest in liquids and amorphous solids has grown considerably in the past few years, the physics of condensed matter is to a very great extent the physics of crystals. To begin, we therefore need to learn how to describe the regular geometrical arrangement of atoms in space that is the essence of crystallinity. Crystals are finite regular arrangements of atoms in space. In practice the atomic arrangement is never perfect, but in crystallography we neglect this aspect and describe crystals by reference to perfect infinite arrays of geometrical points called lattices. *A lattice is an infinite array of points in space so arranged that every point has identical surroundings.* All lattice points are geometrically equivalent. A lattice therefore exhibits *perfect translational symmetry* and, relative to an arbitrarily chosen origin, at a lattice point, any other lattice point has the position vector

$$\mathbf{r}_{123} = n_1\mathbf{a} + n_2\mathbf{b} + n_3\mathbf{c}. \tag{2.1}$$

The numbers $n$ are necessarily integral and the vectors **a**, **b** and **c** are fundamental units of the translational symmetry; the latter are arbitrary, but a sensible choice is usually that which gives the shortest vectors or the highest symmetry to the unit cell. On the other hand, by definition, the volume associated with a single lattice point is unique, but since there is a choice regarding the vectors **a**, **b** and **c**, it may take a variety of shapes as illustrated for a two-dimensional example in Figs 2.1 and 2.2.

The volume associated with a single lattice point is called the *primitive cell*, and this usually takes one of two forms. Either the lattice point is confined to the centre of the primitive cell, which is then determined by the planes bisecting the lines joining the particular point with its neighbours, or the primitive cell may be described as the unit of the mesh formed by the lines connecting lattice points; in the latter case points lie at the vertices of the cell and this is the primitive cell most often used in crystallography. The former case, where the lattice point is enclosed within the primitive cell, is called a *Wigner-Seitz cell* (Fig. 2.3) and is significant for

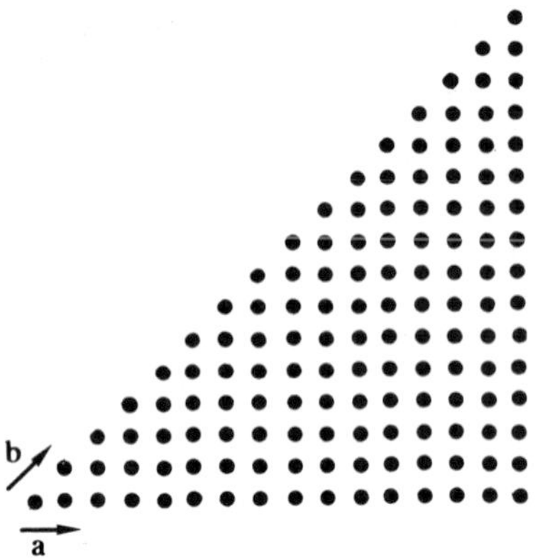


**Figure 2.1** A portion of a two-dimensional lattice: two possible primitive cell vectors are shown.

the description of the physical behaviour of crystals. It will be important in later work.

There are only fourteen different ways of arranging identical points in three-dimensional space so that they are in every way equivalent in their surroundings. These fourteen arrays are called *the Bravais lattices*; they are shown in Fig. 2.4 and listed in Table 2.1. It can be seen that the volume unit depicting the lattice is not always a primitive cell. This is because it is often convenient to use a larger volume called the *crystallographic unit cell*, the reason being that such a cell illustrates the symmetry in a more obvious manner and favours the use of orthogonal axes as is seen in Fig. 2.5.

Confining our attention to the pure elements, we find that many of them, particularly among the metals, have crystal structures that can be associated directly with the simpler Bravais lattices. This must not lead one to believe that there are only fourteen possible crystal structures. The fourteen Bravais lattices are determined by the allowable spatial symmetries of arrays of points. We can, however, associate a group of atoms with a lattice point, and this immediately opens new possibilities for structural arrangements. If to each lattice point we associate a group of atoms then the position vector of the *j*th atom in the group may be written

$$\mathbf{r}_{123j} = n_1\mathbf{a} + n_2\mathbf{b} + n_3\mathbf{c} + \mathbf{R}_j, \tag{2.2}$$

$\mathbf{R}_j$ being a position vector relative to the point $(n_1 n_2 n_3)$.

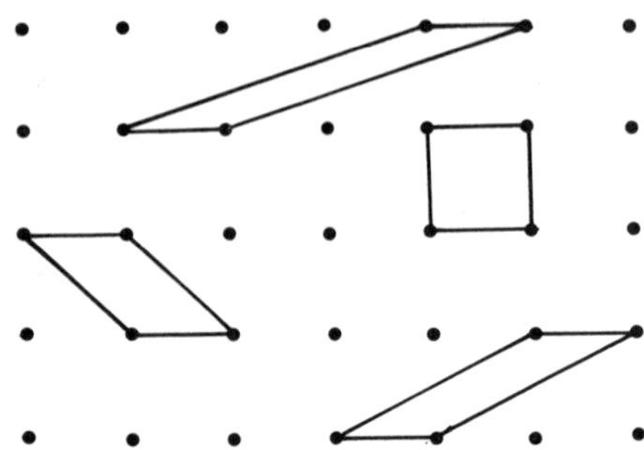

**Figure 2.2** The primitive cell has arbitrary shape but a fixed area or volume.

**Table 2.1** The seven crystal systems and fourteen Bravais lattices

| System | Conventional unit cell | Bravais lattice |
|---|---|---|
| Triclinic | $a \neq b \neq c$ | P (primitive) |
| | $\alpha \neq \beta \neq \gamma$ | |
| Monoclinic | $a \neq b \neq c$ | P |
| | $\alpha = \beta = 90° \neq \gamma$ | C (base-centred) |
| Orthorhombic | $a \neq b \neq c$ | P |
| | $\alpha = \beta = \gamma = 90°$ | C |
| | | I (body-centred) |
| | | F (face-centred) |
| Tetragonal | $a = b \neq c$ | P |
| | $\alpha = \beta = \gamma = 90°$ | I |
| Cubic | $a = b = c$ | P |
| | $\alpha = \beta = \gamma = 90°$ | I |
| | | F |
| Trigonal | $a = b = c$ | R (rhombohedral |
| | $120° > \alpha = \beta = \gamma \neq 90°$ | primitive) |
| Hexagonal | $a = b \neq c$ | P |
| | $\alpha = \beta = 90°$, $\gamma = 60°$ | |

The angles $\alpha$, $\beta$ and $\gamma$ are those between the base vectors, and are defined according to the usual geometrical convention so that $\alpha$ is the angle between **b** and **c** (cyclic).

The group of points or atoms that is associated with *every* lattice point is called a *basis*. A crystal structure is therefore the sum of two quantities: namely a lattice of points and a basis, which is a geometrical arrangement of atoms associated with every lattice point. Simple crystal structures have bases containing only a few atoms, but it is possible, particularly in biological materials, to find many hundreds of atoms in the basis. By way of illustration we consider the body-centred cubic array (Fig. 2.6). This is a true Bravais lattice, but the accepted unit cell may be considered as built up from a primitive simple cubic cell to which has been added a basis, namely one atom at a corner and one at the centre of the cube. Thus, although the arrangement of points is part of a true lattice, we treat it as a structure derived from the simple cubic lattice. An example of a more complicated structure built around the body-centred cubic lattice is that of $\alpha$-Mn. Although the lattice is body-centred cubic, the unit cell contains 58 atoms, the basis being a group of 29 Mn atoms.

Geometrically the introduction of the basis, being an assembly of points associated with each and every lattice point, introduces the possibility for new *symmetry elements* such as rotations and reflections of the basis about axes and planes

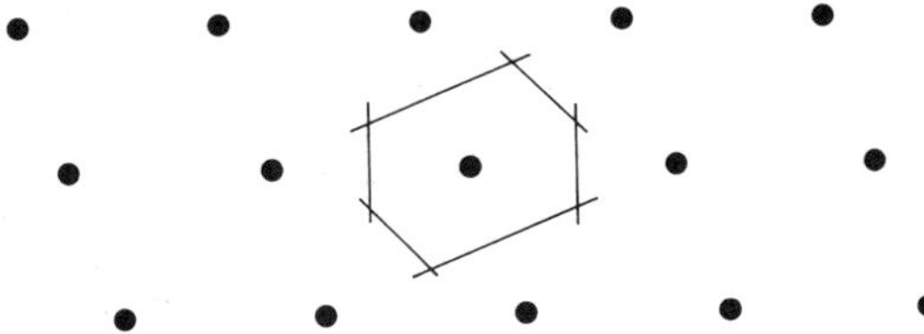

**Figure 2.3** The Wigner-Seitz cell that surrounds each lattice point.

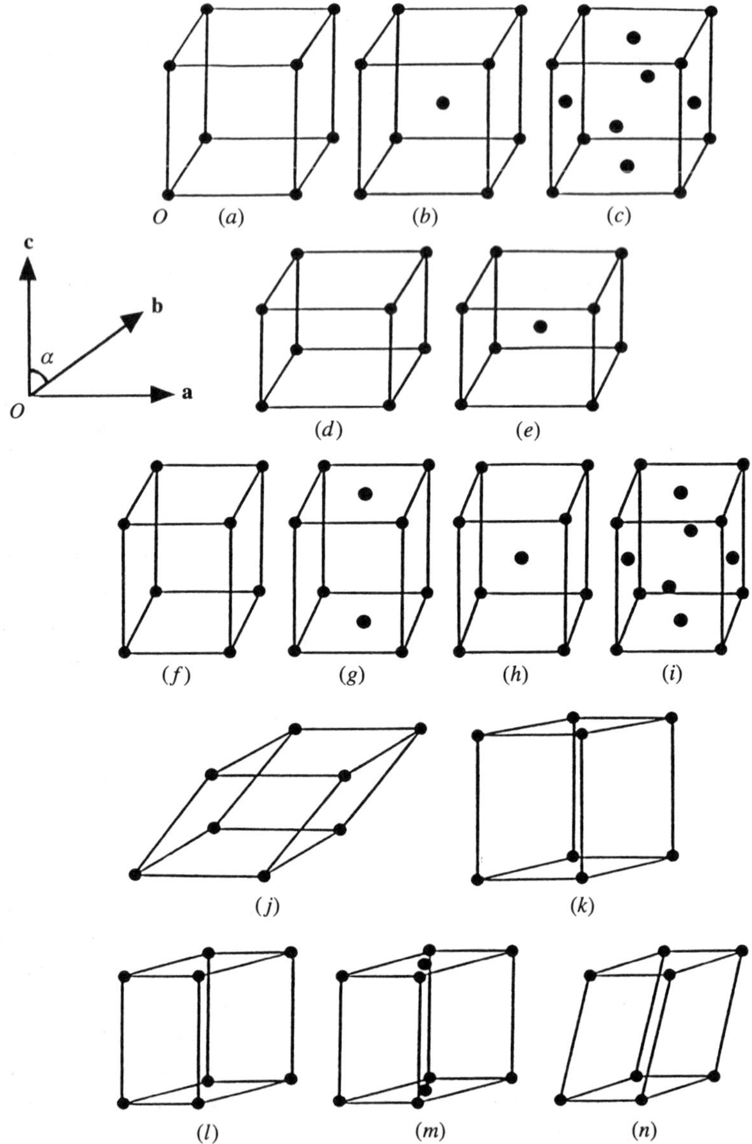


**Figure 2.4** The 14 Bravais space lattices; they may be described in terms of suitable primitive cells. However, it is more convenient and conventional to use a larger unit cell which often involves atoms in end, body-centre or face-centre positions. This procedure also has the advantage of allowing orthogonal axes. (a), (b), (c) cubic systems; (d), (e) tetragonal systems; (f), (g), (h), (i) orthorhombic systems; (j), (k) rhombohedral systems; (l), (m), (n) monoclinic and triclinic systems. Inspection shows that translational symmetry is incompatible with 5-fold rotational symmetry.

through the associated lattice point, each operation or combination of operations turning the arrangement into itself again.

In all, it is found that there are 230 different symmetry patterns available for three-dimensional structures by the combination of translational and point symmetry operations. Full details of these 230 space groups, as they are called, may be

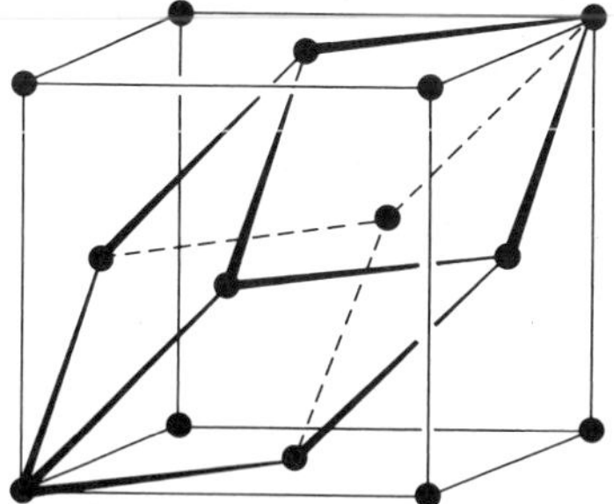

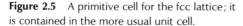

**Figure 2.5** A primitive cell for the fcc lattice; it is contained in the more usual unit cell.

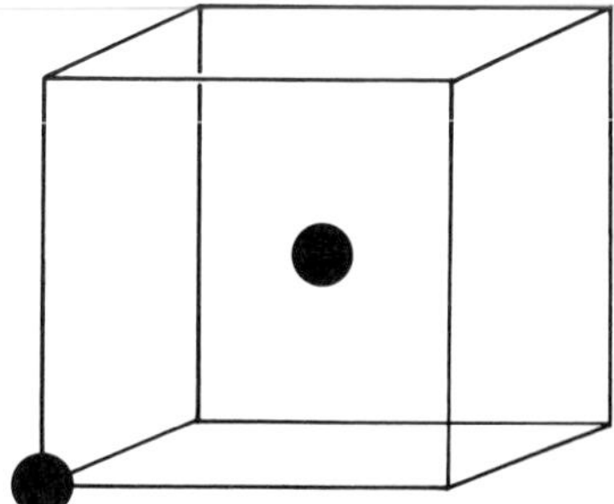

**Figure 2.6** The bcc lattice may be considered derived from the simple cubic lattice using the basis (000, $\frac{1}{2}\frac{1}{2}\frac{1}{2}$). There are eight such equivalent bases corresponding to the eight corner sites of the cube.

found in specialist texts. The number of different crystal structures, taking into account the separation of atoms and the composition of the basis, is, of course, immense. Nevertheless, each structure, however complicated, is, with regard to symmetry, limited to one of the 230 space groups. There is little to be gained at this stage from a detailed discussion of these space groups. The important aspect is not the complexity of the subject but how well one can manage with an acquaintance with its more elementary aspects.

## 2.2 Crystal Planes

A Bravais lattice is an ordered three-dimensional array of points in space, but it may also be considered as an assembly of two-dimensional arrays, i.e. planes. As Fig. 2.7 illustrates, there is an infinite number of different families of planes associated with any Bravais lattice, and each family of similar planes is characterized by the arrangement and density of points within each plane and a specific interplanar spacing. Figure 2.7 also clearly demonstrates that the larger the interplanar spacing the greater the density of lattice points within the plane; this follows directly from the exact similarity of all points in a Bravais lattice and the unique volume per lattice point.

Crystals are anisotropic, which is to say that the physical properties (e.g. electrical resistivity, magnetic susceptibility) are different when measured in different crystallographic directions, the more so the less symmetry the crystal possesses. We therefore need to describe the different lattice planes and directions in an exact manner, and for this purpose *Miller indices* are used. Since all points in a given Bravais lattice are equivalent, any choice of origin is completely arbitrary, as is the choice of coordinate axes. We choose our origin at a particular lattice point and whenever practicable use orthogonal axes, but the following discussion is not dependent on the latter choice. When describing a Bravais lattice as an assembly of similar planes it is important to remember that *all* the lattice points lie on the chosen assembly of planes. It is also the case that, whatever the composition of the plane chosen, we shall find that certain planes of the assembly (in very simple cases

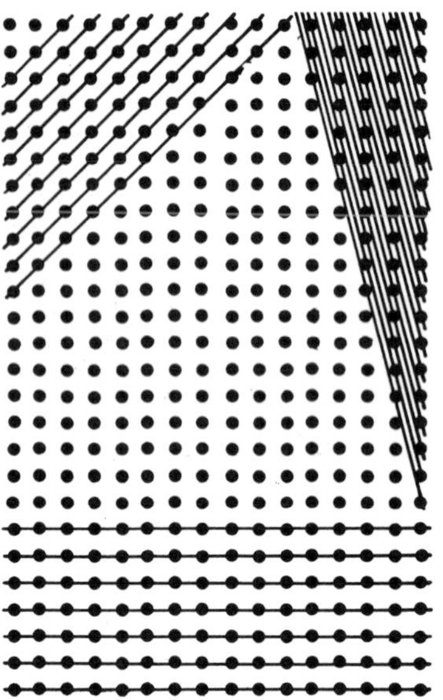

**Figure 2.7** A Bravais lattice may be considered as an assembly of identical lines in two dimensions or planes in three dimensions. There is an infinite number of ways of choosing such an assembly, three of which are depicted here.

sometimes all of them) always intersect the coordinate axes at lattice points. Since all planes in a given assembly are parallel, this means that every plane in this chosen family makes intercepts on the coordinate axes that stand in a definite rational ratio to one another. This is easily demonstrated for a two-dimensional net and applies equally to the three-dimensional lattice.

The Miller indices make use of this fact and allow all crystallographic planes to be described *within the unit cell*; they are derived by the following prescription.

(a) For the plane of interest, determine the intercepts $x$, $y$ and $z$ on the coordinate axes.

(b) Express the intercepts in terms of the base vectors of the unit cell,

$$\frac{x}{a}, \frac{y}{b}, \frac{z}{c};$$

these are not necessarily integers *but they do have rational ratios.*

(c) Form the reciprocals

$$\frac{a}{x}, \frac{b}{y}, \frac{c}{z};$$

(d) Express as the lowest triplet of integers *hkl*.

The triplet *hkl* written (*hkl*) describes that particular plane, of a family of similar planes, lying within the unit cell and nearest to the origin. Since we have chosen the origin at a lattice point, and since this point must itself lie on one of the planes of the family (*hkl*), the distance from the origin to the plane (*hkl*) is the interplanar spacing $d_{hkl}$. It follows that, for any family of planes indexable as *hkl*, the plane (*hkl*)

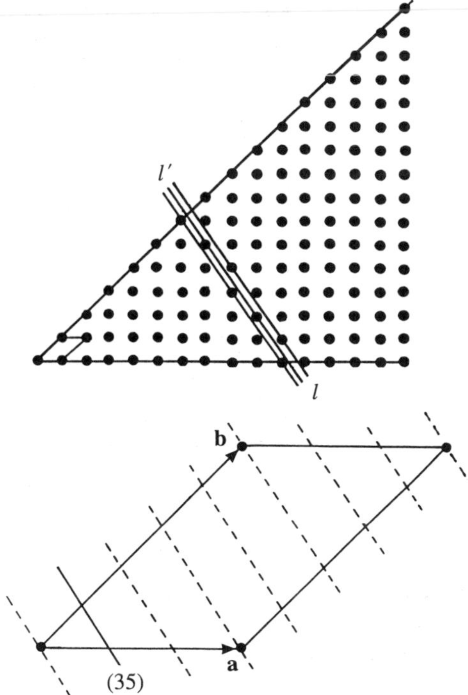


**Figure 2.8** The arbitrary line $ll'$ belongs to a family of similar lines that may be indexed as (35) in the chosen unit cell (in this case a primitive cell). The latter is shown enlarged in the lower diagram. The broken lines show the other planes in the (35) family that pass through the unit cell. The primitive cell has sides a and b as shown.

lying closest to the origin makes intercepts $a/h$, $b/k$, $c/l$ on the crystal axes.† *Thus all planes are describable within the unit cell.* Planes making negative intercepts on the axes are treated in similar fashion and the negative sign appears above the Miller index, e.g. $\bar{h}$ (pronounced bar $h$). The following notation is customary: a particular plane or set of parallel planes ($hkl$); a set of symmetrically equivalent planes $\{hkl\}$ (including negative indices). The use of Miller indices is best appreciated from Fig. 2.8. The positions of lattice or basis points are denoted by their coordinates expressed in terms of base vectors, thus the body-centre position is expressed as $\frac{1}{2}, \frac{1}{2}, \frac{1}{2}$. Directions in the lattice are specified by the coordinates of the lattice point that is nearest to the origin in the chosen direction. A direction is written $[uvz]$ and a set of equivalent directions $\langle uvz \rangle$. In cubic lattices (and only in these) one can define directions in terms of the normals to the lattice planes; thus $[hkl]$ indicates the direction normal to the plane ($hkl$). In structures that possess a centre of symmetry many of the planes containing the same indices are equivalent. Thus in the cubic systems the planes (123), (213), (321), (132), (231) and (312) have the same density of packing and the same interplanar spacing. They are therefore equivalent from these points of view, which are the relevant ones in X-ray diffraction. Remembering that each index may have a negative sign, we find that the indices 1, 2 and 3 may be arranged in 48 different ways. On the other hand the plane (111) has only one

† This may be inferred from Fig. 2.8.

distinguishable arrangement of three similar indices; each index, however, may independently take a plus or minus sign. So there are eight equivalent (111) planes in the cubic lattices. The total number of equivalent planes for given numerical values of the indices is called the multiplicity *p*.

The Bravais lattices are the scaffolding around which crystal structures are built. A crystal structure may be simple or very complex. Fortunately many of the metallic elements, their alloys and compounds have simple structures directly identifiable with the Bravais lattices. Thus sodium has a structure formed by placing a sodium atom at each point of the body-centred cubic (bcc) Bravais lattice; Cu, on the other hand, crystallizes with a face-centred cubic (fcc) arrangement of its atoms. Some metals crystallizing in these two basic structures are listed below together with their coordination number (i.e. the number of nearest neighbours) and the packing fraction (i.e. the fraction of space occupied by the atoms considered as contacting hard spheres):

*bcc coordination 8, packing fraction 0.680*

Li Na K Rb Cs
Ba

V Cr Fe (<910, >1390°C)

Nb Mo
Ta W
Eu

*fcc coordination 12, packing fraction 0.740*

Ca Sr Al Pb
Co Ni Cu Fe (910–1390°C)
Rh Pd Ag
Ir Pt Au
Ce Yb
Th

The other common structure among the metals is the *hexagonal close-packed arrangement* (hcp). If we imagine atoms as hard spheres then we can form close-packed planes with hexagonal symmetry (Fig. 2.9), which may then be stacked on

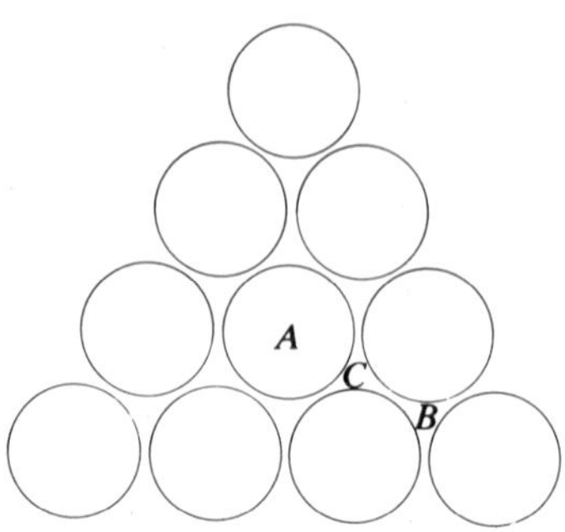


**Figure 2.9** Different sites in a close-packed plane of spheres are labelled *A*, *B* and *C*.

top of one another in a close-packed fashion, the spheres or atoms in one plane nesting perfectly in the depressions between atoms of adjacent planes. Suppose we start off with atoms in the positions *A*. We note that the other spaces between the atoms are of two kinds, *B* or *C*, and the second plane may be laid down in either position. Suppose we choose the *B* positions; then a third nesting layer may be placed over the original *A* positions or the sites *C* (both in the first layer). In fact we have two stacking possibilities, either . . . *ABCABCABC* . . . , which gives the fcc structure or . . . *ABABABAB* . . . , which gives the hcp structure. This is so because the fcc arrangement may be considered to be formed by nesting (111) planes, each plane being a close-packed hexagonal arrangement of spheres. The repetitive structural unit is then a group of three neighbouring planes in the sequence *ABC*. On the other hand the repetition of the sequence *AB* leads to the close-packed hexagonal arrangement, as Fig. 2.10 illustrates. For a given atomic diameter the density of packing is the same in both arrangements and the maximum that can be arranged, corresponding to a packing fraction 0.740. The bcc structure has packing fraction 0.680 and is a more open structure. Other sequences are found even in the elements. Thus Nd has the double hexagonal structure . . . *ABACAB* . . . .

The unit cell of hcp structure is shown in Fig. 2.10, but the basic unit is seen to be the rhombus based cell containing two atoms. The hcp structure is therefore built around this primitive cell using the basis 000, $\frac{1}{3}\frac{1}{3}\frac{1}{2}$. Crystallographic practice favours the complete hexagonal cell, and whereas one may apply the Miller system to index planes, its use is inconvenient because symmetrically equivalent planes do not have the same index numbers. To avoid this difficulty a four index system known as Miller-Bravais indices is used (Fig. 2.11). The geometry clearly shows that if we choose the base vectors $\mathbf{a}_1$, $\mathbf{a}_2$, $\mathbf{a}_3$, $\mathbf{c}$ then $\mathbf{a}_3 = -(\mathbf{a}_1 + \mathbf{a}_2)$, and similarly for the indices *hkil*, $i = -(h + k)$, so the third index is not usually written and planes are often labelled $(hk \cdot l)$. When indexing directions in the hexagonal structure, this relationship is preserved. If the atoms are truly spherical then the axial ratio of the hcp structure, $c/a$, is 1.633, and we have already seen each atom has 12 equidistant nearest neighbours. However, in real crystals deviations, both positive and negative, from the ideal value can occur.

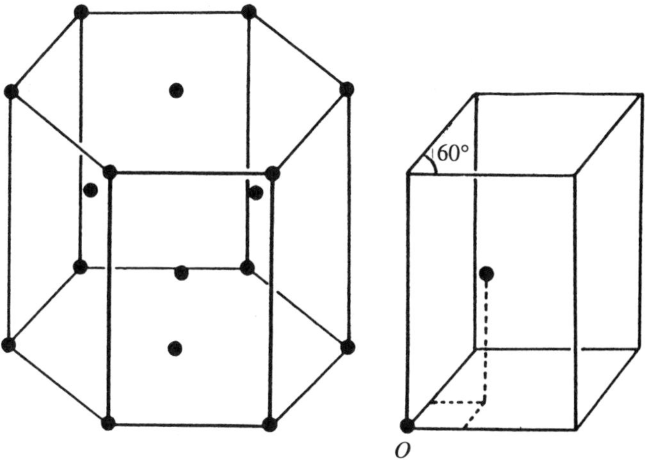


**Figure 2.10** The hexagonal close-packed structure. This arrangement, common among metals, is characterized by a rhombus based primitive cell and the basis (000, $\frac{1}{3}\frac{1}{3}\frac{1}{2}$), as shown on the right.

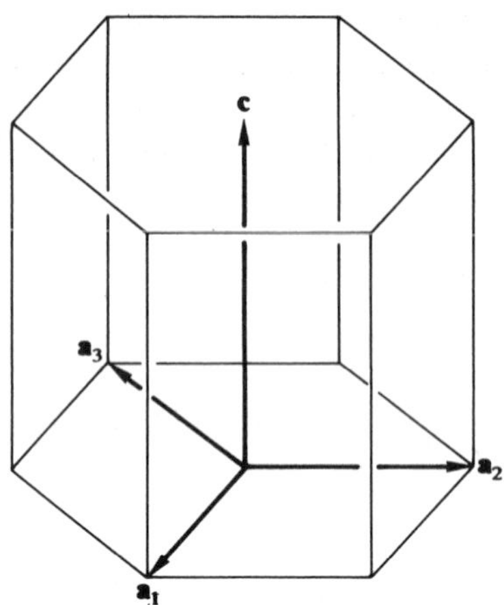


**Figure 2.11** Axes for the Miller-Bravais indices.

*hcp ideal coordination 12, packing fraction 0.74*

Li (<78 K) Na (<40 K)
Be Mg
Sc Ti Co (<400 K)–Zn
Y Zr Ru Cd
Hf Re Os Ti
Gd Tb Dy Ho Er Tm Lu

*Diamond cubic coordination 4, packing fraction 0.34*

The important semiconducting elements Si and Ge crystallize in the same structure as diamond, which is based on the fcc lattice with the basis 000, $\frac{1}{4}\frac{1}{4}\frac{1}{4}$. Thus to each of the four atoms making up the fcc cell we must add an extra atom giving eight atoms per cubic cell. As Fig. 2.12 shows, the atoms have the tetrahedral coordination characteristic of the $sp^3$ covalent bonds. The diamond cubic structure is much more open than the bcc and fcc structures, the packing fraction being 0.34.

Chemical compounds and alloys crystallize in a wide variety of structures. Compounds with the same structure are grouped together and the structure named after

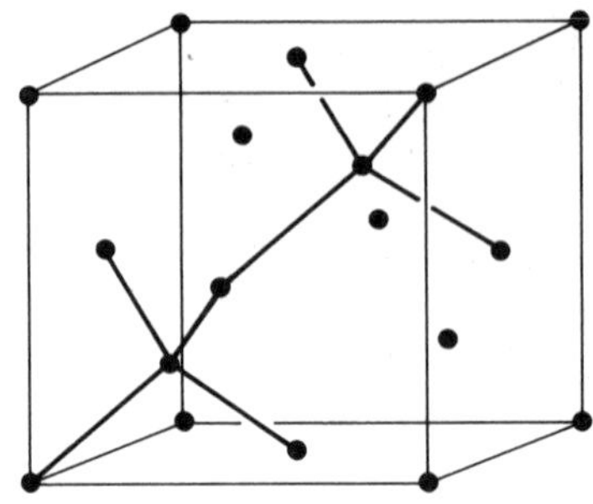

**Figure 2.12** The important elements Ge and Si crystallize in the diamond cubic structure. The arrangement is based on the fcc cell with the basis (000, $\frac{1}{4}\frac{1}{4}\frac{1}{4}$); the unit cell therefore contains 8 atoms.

an important member of the group. Thus, for example, the halides of the alkali metals, with the exception of Cs, crystallize in what is known as the NaCl structure (Fig. 2.13*a*); this structure is shared by a number of other compounds, including the oxides, sulphides, selenides and tellurides of the rare earth metals. CsCl has given its

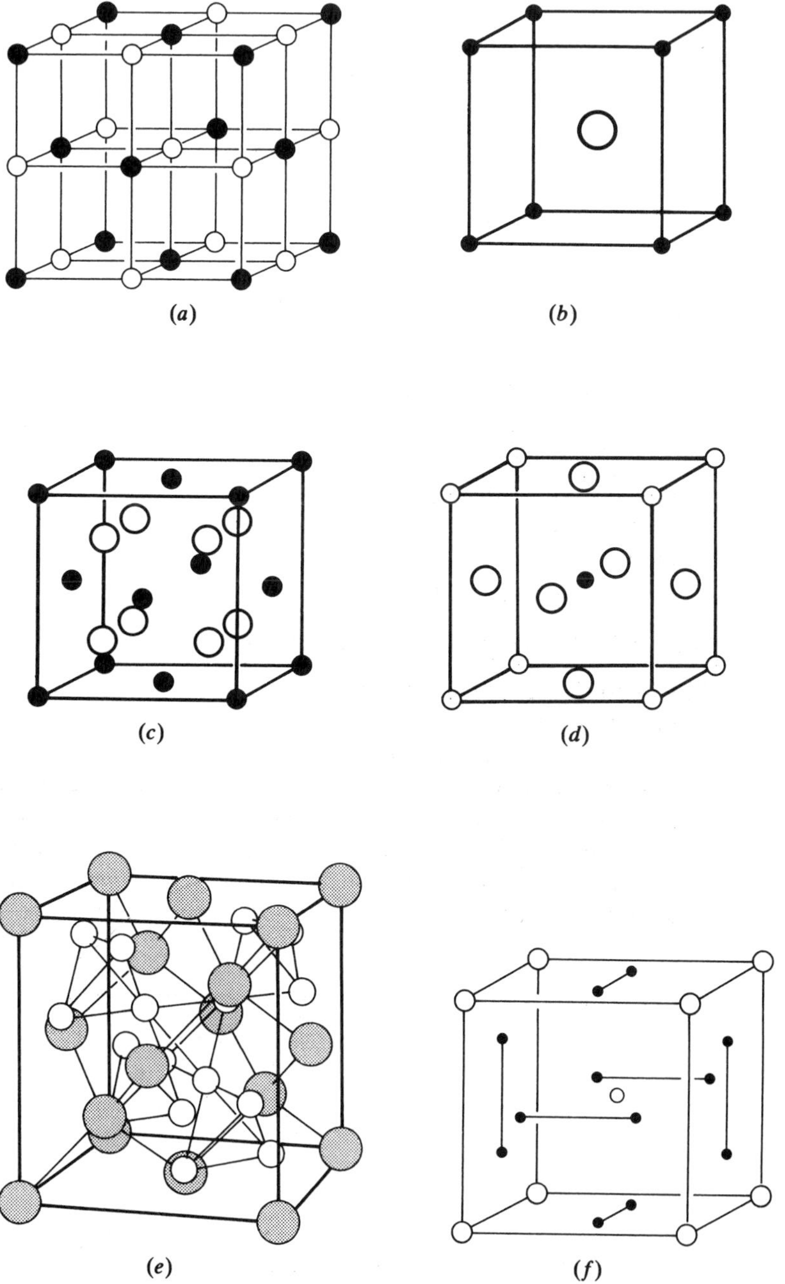

**Figure 2.13** Common crystal structures: (a) NaCl; (b) CsCl; (c) fluorite ($CaF_2$); (d) perovskite ($BaTiO_3$); (e) Laves phase (e.g. $Cu_2Mg$); (f) A15 or $\beta$-tungsten structure (e.g. $Nb_3Sn$). ((e) is after Wernick 1965.)

name to the structure shown in Fig. 2.13(*b*), which is common to many other substances, for example a particular form of brass with composition CuZn. Figures 2.13(*c*) and (*d*) show the fluorite ($CaF_2$) and perovskite ($BaTiO_3$) structures respectively. When metals form alloys, the different elements are usually distributed in random fashion on the crystallographic sites, but at certain compositions of the form $AB$, $AB_2$, $AB_3$ etc., the atoms $A$ and $B$ may prefer specific sites, and the alloy is said to develop an ordered structure. The tendency to order may be weak and the order may be destroyed by thermal agitation, as by heating to say 500°C; in other cases the ordered arrangement may be very stable and the alloy is similar to a chemical compound – one speaks of intermetallic compounds. One common structure for such compounds is the cubic Laves phase (Fig. 2.13*e*), which also exists in a related hexagonal form at the 1/2 ratio; here the geometrical sizes of the constituent atoms are important for the stability. Another important structure is the A15 or $\beta$-tungsten structure (Fig. 2.13*f*); certain superconductors with transition temperatures around 20 K adopt this structure.

We shall not describe the many different crystal structures that can occur: for full details see Wyckoff (1963), Pearson (1964) or Westbrook (1967).

## 2.3 Crystal Projections

We often need to study the relative orientations of crystal planes or their orientation with respect to some line of action in an external field. This is best done by mapping the main features of the crystal structure onto a two-dimensional surface. The principles are the same as those used in cartography, i.e. we project the crystal onto a flat surface. First the crystal is imagined to be placed at the centre of a sphere and the normals to all the major planes connect the crystal to the spherical surface, which is provided with an angular scale in the form of lines of longitude and latitude. The points of contact with the sphere define the *spherical projection.* To obtain a plane projection, we connect the south pole of the sphere to all the points of intersection lying in the northern hemisphere. The connecting lines cut the equatorial plane, giving rise to the *stereographic projection* (Fig. 2.14). The lines of longitude or latitude may also be projected onto the equatorial plane in this way, providing a scale of angular measurement called a Wulff net. The stereographic projection of the planes in the cubic structures are shown in Fig. 2.15. If the spherical projection is mapped onto the plane tangent to the north pole and from the centre of the projection sphere, we obtain the *gnomic projections.*

## 2.4 The Reciprocal Lattice

Clearly a family of crystal planes (*hkl*) is characterized by (a) the normal to the planes, which we can denote by the unit vector $\mathbf{n}_{hkl}$, and (b) the interplanar spacing $d_{hkl}$. Thus another way of describing a crystal structure might be to tabulate, in some suitable way, $\mathbf{n}_{hkl}$ and $d_{hkl}$. However, experience has shown that it is much more valuable and practical to define a new lattice in *reciprocal space* formed by the vectors

$$\mathbf{G}_{hkl} = 2\pi \mathbf{n}_{hkl}/d_{hkl}. \tag{2.3}$$

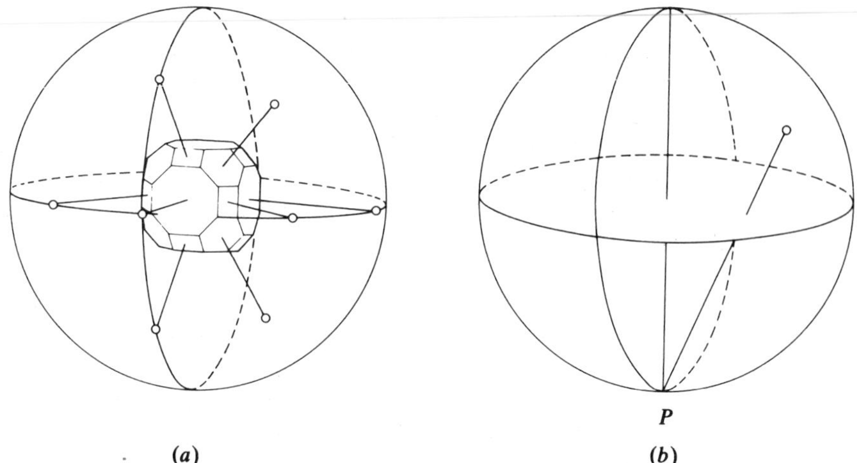


**Figure 2.14** (a) In the spherical projection we imagine the crystal placed at the centre of a sphere. Each crystal plane is connected to the sphere by its normal. The points of intersection of the normals with the sphere produce the spherical projection. (b) If we now imagine an equatorial plane, we may connect all points of the spherical projection in the 'northern hemisphere' to the 'south pole' P. These connecting lines cut the equatorial plane at points that form the stereographic projection of the upper half of the crystal. Only when the crystal lacks a centre of symmetry need we project the upper and lower halves separately. (After Phillips 1946.)

The factor $2\pi$ is included for later convenience. From a chosen origin our reciprocal lattice comprises all points $\mathbf{G}_{hkl}$, there being one point for every family of planes in the direct lattice. We define base vectors **A**, **B** and **C** for the reciprocal lattice so that

$$\mathbf{G}_{hkl} = h\mathbf{A} + k\mathbf{B} + l\mathbf{C}, \tag{2.4}$$

where

$$\mathbf{A} = \frac{2\pi\mathbf{b} \times \mathbf{c}}{\mathbf{a} \cdot (\mathbf{b} \times \mathbf{c})}, \quad \text{with } \mathbf{B} \text{ and } \mathbf{C} \text{ given by cyclic permutation.} \tag{2.5}$$

Clearly, with this definition of **A**, **B** and **C** we find that

$$\begin{aligned} \mathbf{A} \cdot \mathbf{a} &= 2\pi, \quad \mathbf{A} \cdot \mathbf{b} = \mathbf{A} \cdot \mathbf{c} = 0, \\ \mathbf{B} \cdot \mathbf{b} &= 2\pi, \quad \mathbf{B} \cdot \mathbf{c} = \mathbf{B} \cdot \mathbf{a} = 0, \\ \mathbf{C} \cdot \mathbf{c} &= 2\pi, \quad \mathbf{C} \cdot \mathbf{a} = \mathbf{C} \cdot \mathbf{b} = 0. \end{aligned}$$

We now show that the equations (2.4) and (2.5) are compatible with the definition (2.3).

In Fig. 2.16 let $O$ be the origin of both direct and reciprocal lattices. In keeping with our discussion of Miller indices, the representative plane $(hkl)$ makes intercepts on the base vectors $\mathbf{a}/h$, $\mathbf{b}/k$, $\mathbf{c}/l$ and the triangular portion of the plane has sides

$$\left(\frac{\mathbf{a}}{h} - \frac{\mathbf{b}}{k}\right), \quad \left(\frac{\mathbf{b}}{k} - \frac{\mathbf{c}}{l}\right), \quad \left(\frac{\mathbf{c}}{l} - \frac{\mathbf{a}}{h}\right).$$

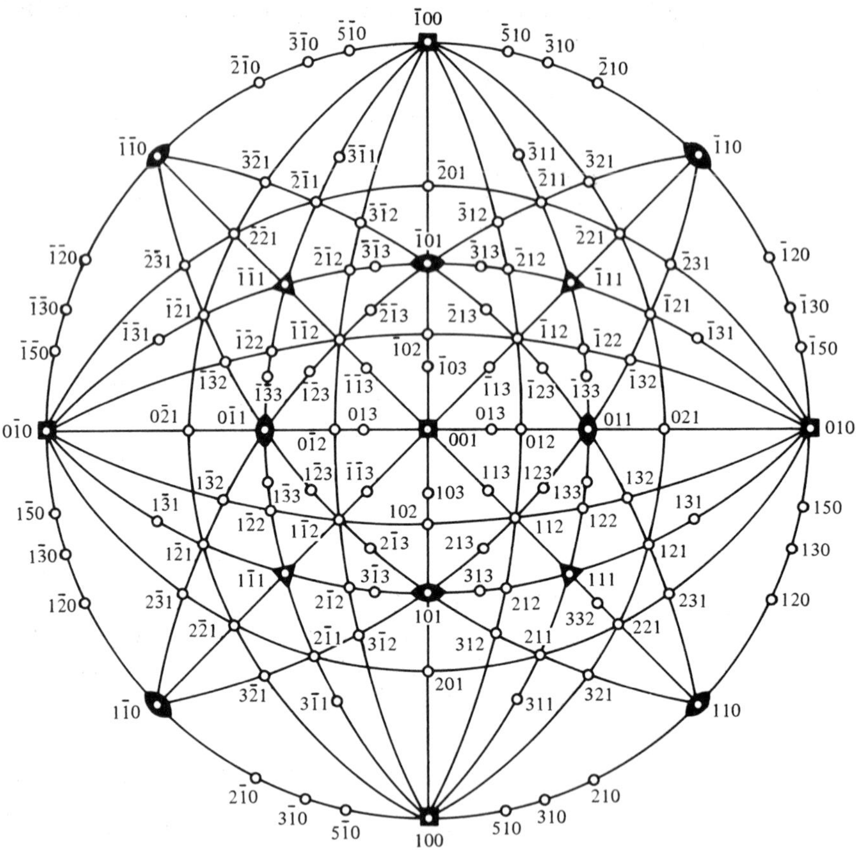

**Figure 2.15** Part of a stereogram for a cubic crystal. The various points shown correspond to the intersections on the equatorial plane of Fig. 2.14. The projection in question is made around the (001) plane, but any plane may be used. It is seen that certain groups of planes lie on 'great circles', i.e. diametral circles, and this facilitates the appreciation of angular relationships between crystal planes. (Reprinted with permission from C. S. Barrett and T. B. Massalski, *Structure of Metals*, 3rd edn, 1980, Pergamon, Oxford.)

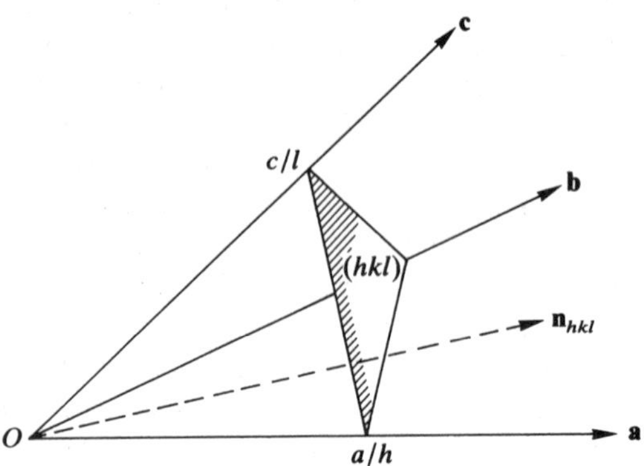


**Figure 2.16** An element of a three-dimensional lattice showing an arbitrary plane ($hkl$) referred to chosen axes and base vectors **a**, **b** and **c**.

The reciprocal vector $\mathbf{G}_{hkl}$ pierces the plane $(hkl)$, and, remembering the definitions (2.4) and (2.5), we readily see that

$$\left(\frac{\mathbf{a}}{h}-\frac{\mathbf{b}}{k}\right)\cdot\mathbf{G}_{hkl}=\left(\frac{\mathbf{b}}{k}-\frac{\mathbf{c}}{l}\right)\cdot\mathbf{G}_{hkl}=\left(\frac{\mathbf{c}}{l}-\frac{\mathbf{a}}{h}\right)\cdot\mathbf{G}_{hkl}=0.$$

Thus $\mathbf{G}_{hkl}$ is perpendicular to two vectors in the plane $(hkl)$ and is therefore perpendicular to the plane. What about the modulus of $\mathbf{G}_{hkl}$? We may write

$$\mathbf{n}_{hkl}=\frac{\mathbf{G}_{hkl}}{|\mathbf{G}_{hkl}|},$$

and furthermore

$$d_{hkl}=\frac{\mathbf{a}}{h}\cdot\mathbf{n}_{hkl}=\frac{\mathbf{a}}{h}\cdot\frac{\mathbf{G}_{hkl}}{|\mathbf{G}_{hkl}|}=\frac{2\pi}{|\mathbf{G}_{hkl}|}.$$

Thus

$$|\mathbf{G}_{hkl}|=\frac{2\pi}{d_{hkl}}.$$

Our equations (2.3), (2.4) and (2.5) are mutually consistent.

In constructing the reciprocal lattice we find all the points with position vectors $\mathbf{G}_{hkl}$ given by (2.4) and we take all possible values for $h$, $k$ and $l$. The reciprocal lattice is a true lattice and is infinite in extent. Thus we shall find a row of points of the form

$$\bar{\infty}00\ \ldots\ \bar{3}00,\ \bar{2}00,\ \bar{1}00,\ 000,\ 100,\ 200,\ 300,\ \ldots\ \infty00,$$

and the whole of reciprocal space is filled by such rows when $h$, $k$ and $l$ take all possible values. We have also demonstrated that each point in reciprocal space represents a family of identical planes in the direct lattice. For convenience let us consider a simple cubic direct lattice. It seems that our definition of the reciprocal lattice is not wholly in harmony with our concepts of Miller indices and crystal planes.

In our prescription for establishing the Miller index we have said that we must form the triplet of lowest indices. Thus the planes (200), (300), (400), ..., ($n$00) should all reduce to (100)! Furthermore, if we try to illustrate say the planes (300), what do we find? As Fig. 2.17 shows, these are planes parallel to (100) but with an interplanar spacing one third that of the (100) spacing. Most of the planes (300) do not pass through lattice points and such planes can hardly have physical significance. Surely only the planes (100) are 'real' and all planes of the form ($n$00), ($nn$0) and ($nnn$), $n > 1$ are redundant. It therefore seems that our definition (2.4) overdetermines the reciprocal lattice with regard to the number of physically significant planes in the direct lattice. However, in the next chapter we shall see that this problem is very simply resolved. For the moment we declare that it is physically justifiable to accept planes of the forms ($nh$ $nk$ $nl$) for all values of $n$, even though the majority of these planes do not pass through lattice points. In consequence we also declare that all the points of the infinite reciprocal lattice are physically significant.

The reciprocal lattice appears a very artificial concept. It was first advanced by Ewald in 1913 shortly after the discovery of X-ray diffraction by crystals. Later, in 1921, he used it to derive Bragg's law as described in the next chapter. The recipro-

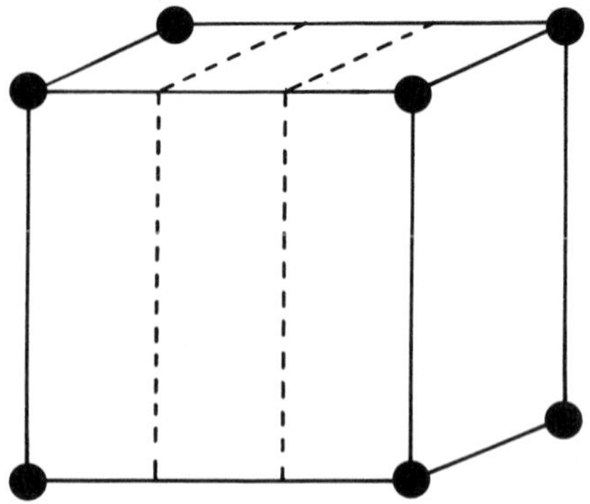

**Figure 2.17** In a simple cubic lattice only one third of the planes (300) pass through lattice points.

cal lattice, important as it is in X-ray and electron crystallography, has much wider significance. It is a space necessary for the description of the scattering of waves by a periodic array of scattering centres. In particular it is needed for the understanding of the properties of the valence electrons in crystalline solids. A travelling wave is often used to describe the valence electron wave function in a metal, i.e. $\psi(r) = A \exp i(\mathbf{k} \cdot \mathbf{r} - \omega t)$. It is found preferable to characterize the state by the wavevector $\mathbf{k}$ ($|\mathbf{k}| = 2\pi/\lambda$) rather than the wavelength $\lambda$. $\mathbf{k}$ has dimension reciprocal length. In due course we shall find that the space occupied by acceptable $\mathbf{k}$ values for valence electrons in solids is the reciprocal space as defined earlier, just as the ordinary space lattice is the space for denoting the positions of atoms. Of particular importance is the scalar product of a vector of the direct lattice $\mathbf{r}$ with a vector of the associated reciprocal lattice $\mathbf{G}$;

$$\mathbf{G}_{hkl} \cdot \mathbf{r}_{123} = 2\pi(n_1 h + n_2 k + n_3 l) = \text{integer} \times 2\pi \tag{2.6}$$

so

$$\exp i(\mathbf{G}_{hkl} \cdot \mathbf{r}_{123}) = 1. \tag{2.7}$$

## References

BARRETT, C. S. and MASSALSKI, T. B. (1980) *Structure of Metals*, 3rd edn. Pergamon, Oxford.

HENRY, N. F. M., LIPSON, H. and WOOSTER, W. A. (1961) *The Interpretation of X-ray Diffraction Photographs*, 2nd edn. Macmillan, London.

PEARSON, W. B. (1964) *Handbook of Lattice Spacings and Structure of Metals*. Pergamon, Oxford.

PHILLIPS, F. C. (1946) *An Introduction to Crystallography*. Longman, London.

WERNICK, J. H. (1965) *Physical Metallurgy* (ed. R. W. CAHN ), p. 213. North-Holland, Amsterdam.

WESTBROOK, J. (1967) *Intermetallic Compounds*. Wiley, New York.

WYCKOFF, R. (1963) *Crystal Structures*. Interscience, New York.

## Further Reading

LIMA-DE-FARIA, J. (Editor), *Historical Atlas of Crystallography*, (1990) Kluwer Academic Publishers, Kingston-upon-Thames.

## Problems

**2.1** If atoms are considered as contacting hard spheres show that:
(a) the bcc lattice has packing fraction 0.68;
(b) the fcc lattice has packing fraction 0.74;
(c) the hcp structure has $c/a = 1.633$.
(d) If the cube has side $a$ what is the 'atomic diameter' in the bcc and fcc cases?

**2.2** Given a bcc structure formed by the packing of hard spheres, what is the largest sphere that can be introduced (not at a lattice site) without distorting the original bcc arrangement? Where would these extra spheres be placed? Atoms introduced at other than proper lattice or structural sites are called 'interstitial atoms'.

**2.3** In the fcc lattice extra atoms may be introduced into what are called 'tetrahedral' and 'octahedral' sites, i.e. the surrounding lattice atoms lie at the vertices of a regular tetrahedron or octahedron. Can you determine where these interstitial sites lie?

**2.4** A bcc lattice may be simply deformed into the fcc lattice. Determine the necessary deformation. Referring the two lattices to the appropriate cubic axes, how will the planes (100), (110), (111) of the original bcc lattice be described in the resultant fcc lattice?

**2.5** In the tetragonal lattices we do not encounter the end-centred or face-centred forms. Using clear diagrams, show that if we attempt to form these lattices they may be described as simple tetragonal and body-centred tetragonal respectively.

**2.6** Determine the lattice and appropriate basis for the crystal structures shown in Figs 2.13(*a, b, d, f*).

**2.7** For the cubic systems show that

$$d_{hkl} = \frac{a}{(h^2 + k^2 + l^2)^{1/2}}.$$

What is '$d_{hkl}$' in the orthorhombic lattices?

**2.8** Show that, using the Miller-Bravais system of indices, the six prism planes of the hcp structure have similar indices, i.e. they are different combinations of the same digits. What would the indices be in the conventional Miller system?

**2.9** In Fig. 2.13(*c*) both primitive and unit cells of the fcc lattice are shown. Normally we use the unit cell and we describe important planes by the Miller indices, e.g. (100), (110), (111). If instead we were to use the primitive cell for our coordinate axes, what would the above-mentioned planes then have for Miller indices?

**2.10** There are only five two-dimensional Bravais lattices. An obvious example is the square net; what are the other four Bravais nets?

**2.11** Consider the conventional bcc unit cell, choose an origin at (000) and form the shortest translational vectors required to define the lattice. Then, using the definitions of the reciprocal lattice vectors, demonstrate that the associated reciprocal lattice has fcc form.

**2.12** Consider the two-dimensional net

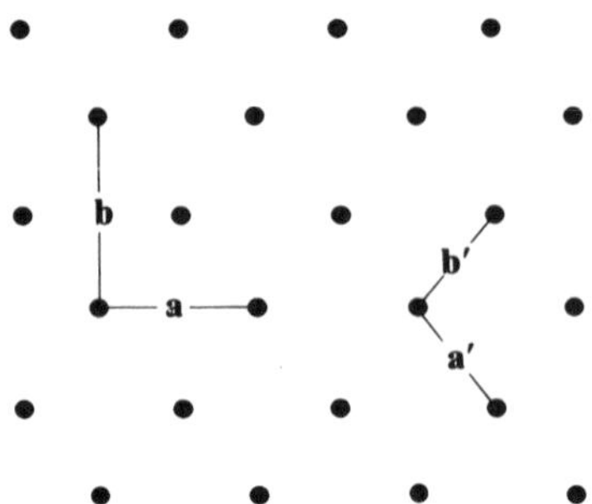


We may describe it in terms of a centred rectangular unit cell with base vectors **a** and **b**, or in terms of an arbitrary primitive cell with base vectors **a**′ and **b**′. Lattice lines may be given Miller indices $(h, k)$ based on the rectangular unit cell or indices $(H, K)$ when referred to the chosen primitive cell. Determine the relation between $(H, K)$ and $(h, k)$.

**2.13** Cobalt below approximately 400°C is stable in the hcp form with the stacking sequence . . . *ABABABAB* . . . . Sm is also formed by the stacking of closed-packed planes, but the sequence is . . . *ABABCBCACABABCBCACABABCBCAC* . . . . What is the repetitive grouping in Sm? (This structure also arises in Li metal below about 40 K and in some Au-Zn alloys quenched from high temperatures.)

# 3

# Diffraction

## 3.1 Theoretical Background

Crystal structures are determined by diffraction of X-ray, neutron or electron beams, and to proceed we need to know:

(a) the physical basis for crystal diffraction;

(b) the scattering power of an atom;

(c) the effect of lattice geometry.

We shall confine our discussion to X rays of wavelength about 1 Å, i.e. comparable to the distances separating atoms in crystals. The physical basis for diffraction lies in the interference effects produced by phase differences between rays elastically scattered from different atoms in the crystal. The scattering of X rays by atoms arises on account of their electron content. (Why don't we consider the scattering from the charged nuclei?) The electrons are accelerated by the electric field vector of the X-ray photon, and secondary X rays, with the same wavelength and phase as the incident radiation, are emitted. There are other inelastic scattering processes causing wavelength changes in the scattered radiation, but these do not give rise to diffraction. Thus, in the spirit of Huygens' principle, we may consider each atom in the sample to be a source of secondary spherical waves whose strength is controlled by the scattering power of the atom, a quantity known as the atomic form factor. It is proportional to the atomic number. For the moment we assume that the elastic scattering is isotropic.

In practice structure determination by X rays makes use of small crystals seldom greater than 1 mm in linear dimensions and the sample-to-detector distance is several centimetres. The detector may be a photographic film or a Geiger counter. The sample is irradiated by a collimated beam of monochromatic X rays with wave vector $\mathbf{k}$, and the scattered rays in the directions of constructive interference may be considered as parallel beams with typical wave vector $\mathbf{k}'$ and $|\mathbf{k}| = |\mathbf{k}'|$. Consider first two atoms of the sample; we label them $A$ and $B$. We choose $A$ as the zero of

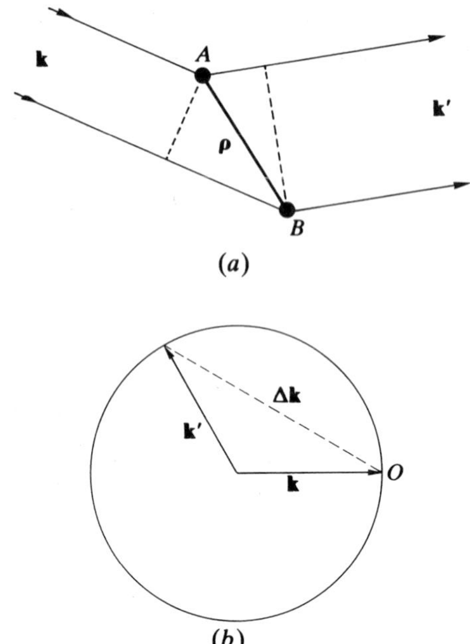


**Figure 3.1** In (a) the scattering of an incident plane wave, **k**, by two centres $A$ and $B$ is shown. The scattered wave has a component with wave vector $\mathbf{k}'$. The scattering of X rays by atoms may be elastic $|\mathbf{k}| = |\mathbf{k}'|$ or inelastic $|\mathbf{k}| \neq |\mathbf{k}'|$. The diffraction process is characterized by elastic scattering and $\Delta\mathbf{k} = \mathbf{G}$, a condition conveniently described in terms of Ewald's construction (b). A sphere of radius $2\pi/\lambda$ is constructed in reciprocal space with appropriate orientation of the incident **k** vector relative to the origin. The allowed diffracted rays are then determined by the reciprocal lattice points that lie on this sphere. The geometry also shows that the condition for Bragg reflection is $\Delta\mathbf{k} = -\mathbf{G}$. See also Fig. 3.2.

coordinates for the scattered rays and $B$ has position vector $\boldsymbol{\rho}$ relative to $A$ (Fig. 3.1*a*). The scattered spherical wave leaving the atom $A$ in the direction $\mathbf{k}'$ is written

$$\frac{\Phi f}{r'}\,\mathrm{e}^{\mathrm{i}(k'r' - \omega t)}\,, \tag{3.1}$$

$\Phi$ being the amplitude of the incident beam, $f$ the atomic form factor and $r'$ the distance to the detector measured from $A$.

The scattered amplitude from atom $B$ may be written

$$\frac{\Phi f}{r_{\mathrm{B}}}\,\mathrm{e}^{\mathrm{i}(k'r' - \omega t + \Delta)}\,,$$

where

$$\Delta = \boldsymbol{\rho} \cdot (\mathbf{k}' - \mathbf{k}) = \boldsymbol{\rho} \cdot \Delta\mathbf{k} \tag{3.2}$$

is a phase difference caused by the displaced position of atom $B$ relative to atom $A$ (Fig. 3.1*a*).† If we now consider any atom $j$ in the sample at $\boldsymbol{\rho}_j$ relative to atom $A$

† As drawn in Fig. 3.1(*a*), the phase of rays leaving atom $B$ is retarded relative to rays leaving atom $A$, i.e. $\Delta < 0$.

and distance $r_j$ from the detector then this atom produces a signal amplitude at the detector and in the direction $\mathbf{k}'$ equal to

$$\frac{\Phi f}{r_j}\,\mathrm{e}^{\mathrm{i}(k'r' - \omega t + \boldsymbol{\rho}_j \cdot \Delta\mathbf{k})}. \tag{3.3}$$

Thus the signal amplitude at the detector and in direction $\mathbf{k}'$ arising from the complete sample becomes

$$\sum_{\text{all atoms}} \frac{\Phi f}{r_j}\,\mathrm{e}^{\mathrm{i}(k'r' - \omega t)}\mathrm{e}^{\mathrm{i}\boldsymbol{\rho}_j \cdot \Delta\mathbf{k}}. \tag{3.4}$$

The fact that the $r_j$ in the denominator are all different is unimportant: this just means that each atom makes a slightly different contribution to the total signal strength, but the differences are insignificant because, as implied earlier, the maximum relative deviation in $r_j$ is only about 1%. The only term in (3.4) that can produce diffraction effects is

$$\sum_{\text{all atoms}} \mathrm{e}^{\mathrm{i}\boldsymbol{\rho}_j \cdot \Delta\mathbf{k}}, \tag{3.5}$$

which arises because atoms in different positions produce scattered waves with different phases.

Suppose the sample is a rectangular parallelepiped with dimensions $M\mathbf{a}$, $N\mathbf{b}$ and $P\mathbf{c}$ containing $MNP$ atoms. Then, expressing $\boldsymbol{\rho}_j$ in terms of the base vectors of the lattice, (3.5) becomes

$$\sum_{0}^{(M-1)(N-1)(P-1)} \mathrm{e}^{\mathrm{i}(m\mathbf{a} + n\mathbf{b} + p\mathbf{c}) \cdot \Delta\mathbf{k}}. \tag{3.6}$$

Let us consider just one of the three independent and similar terms in (3.6). We take

$$\sum_{0}^{M-1} \mathrm{e}^{\mathrm{i}m\mathbf{a} \cdot \Delta\mathbf{k}}.$$

This is a finite geometrical progression with ratio $\mathrm{e}^{\mathrm{i}\mathbf{a} \cdot \Delta\mathbf{k}}$, so the sum becomes

$$\sum_{m=0}^{M-1} \mathrm{e}^{\mathrm{i}m\mathbf{a} \cdot \Delta\mathbf{k}} = \frac{1 - \mathrm{e}^{\mathrm{i}M\mathbf{a} \cdot \Delta\mathbf{k}}}{1 - \mathrm{e}^{\mathrm{i}\mathbf{a} \cdot \Delta\mathbf{k}}},$$

and this may be rewritten as

$$\frac{\mathrm{e}^{\mathrm{i}M\mathbf{a} \cdot \Delta\mathbf{k}/2}(\mathrm{e}^{-\mathrm{i}M\mathbf{a} \cdot \Delta\mathbf{k}/2} - \mathrm{e}^{\mathrm{i}M\mathbf{a} \cdot \Delta\mathbf{k}/2})}{\mathrm{e}^{\mathrm{i}\mathbf{a} \cdot \Delta\mathbf{k}/2}(\mathrm{e}^{-\mathrm{i}\mathbf{a} \cdot \Delta\mathbf{k}/2} - \mathrm{e}^{\mathrm{i}\mathbf{a} \cdot \Delta\mathbf{k}/2})}. \tag{3.7}$$

This is one factor in the scattered amplitude, but since we can only measure X-ray intensity, we need to know the magnitude of this term, which, because the prefactors have modulus unity, becomes

$$\frac{\sin \tfrac{1}{2}M\mathbf{a} \cdot \Delta\mathbf{k}}{\sin \tfrac{1}{2}\mathbf{a} \cdot \Delta\mathbf{k}}.$$

This is just the expression governing the amplitude of light reflected from a linear grating. We find absolute maxima determined by

$$\sin \tfrac{1}{2}\mathbf{a} \cdot \Delta\mathbf{k} = 0, \quad \text{i.e. } \Delta\mathbf{k} = n_1\mathbf{A},\ n_1 \text{ integral,}$$

$$\mathbf{A} = 2\pi \frac{\mathbf{b} \times \mathbf{c}}{\mathbf{a} \cdot \mathbf{b} \times \mathbf{c}},$$

and subsidiary maxima, $M - 2$ of them, between adjacent principal maxima,† the latter are so overwhelmingly large that the subsidiary maxima are insignificant. When we consider all three terms in (3.6) our condition for a diffracted beam becomes

$$\Delta\mathbf{k} = n_1\mathbf{A} + n_2\mathbf{B} + n_3\mathbf{C}, \quad \text{the } ns \text{ all being integers.}$$

This is none other than a vector of the reciprocal lattice, which is always of the form

$$\mathbf{G}_{hkl} = h\mathbf{A} + k\mathbf{B} + l\mathbf{C}$$

and represents a family of planes ($hkl$) in the direct lattice. The condition for constructive interference in crystal diffraction becomes

$$\Delta\mathbf{k} = \mathbf{G}_{hkl} = h\mathbf{A} + k\mathbf{B} + l\mathbf{C}. \tag{3.8}$$

We see that each diffracted beam is to be associated with a particular family of crystal planes. This allows a simple description in terms of the *Ewald sphere* (Fig. 3.1*b*). We construct the reciprocal lattice according to the usual prescription. The incident radiation has wave vector **k**; this vector is arranged to end on the origin of the reciprocal lattice and is correctly oriented according to the incident geometry. Using the origin of **k** as centre, a sphere of radius $|\mathbf{k}| = 2\pi/\lambda$ is constructed. Wherever the surface of the sphere passes through a point of the reciprocal lattice the condition $\Delta\mathbf{k} = \mathbf{G}_{hkl}$ is fulfilled and one sees immediately the geometry of the resulting diffraction pattern.

It is also clear that if $2\theta$ is the angle separating incident and diffracted beams then

$$|\Delta\mathbf{k}| = 2|\mathbf{k}| \sin\theta = |\mathbf{G}_{hkl}| = \frac{2\pi}{d_{hkl}},$$

which leads to

$$2d_{hkl} \sin\theta = \lambda, \tag{3.9}$$

otherwise known as *Bragg's law*. Bragg's law is ordinarily discussed in terms of Bragg reflection as a result of constructive interference of rays diffracted from successive planes of a given family ($hkl$). Figure 3.2 shows that the condition for Bragg reflection is

$$n\lambda = 2d_{hkl} \sin\theta, \tag{3.10}$$

where $n$ is the order of diffraction.

† This is because between two adjacent principal maxima there arise $M - 1$ minima determined by the zeros of the numerator; the widths of principal maxima therefore vary as $M^{-1}$.

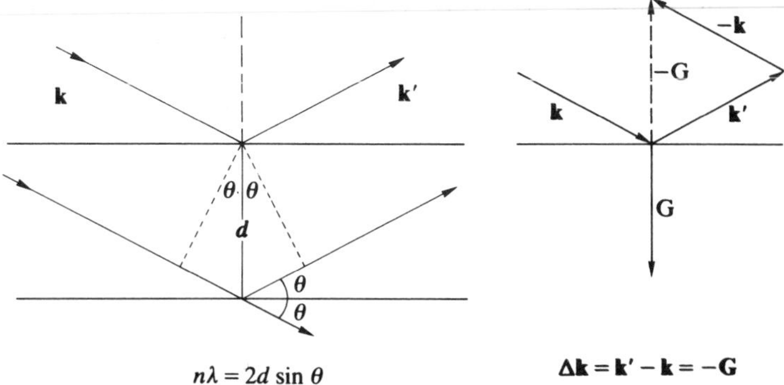


**Figure 3.2** Diffraction by an array of point scatterers may be considered as Bragg reflection from planar arrays.

How are we to take account of the different orders of diffraction? Let us rewrite Bragg's law in the form

$$\lambda = 2\,\frac{d_{hkl}}{n}\sin\theta.$$

Now the planes $(nh\ nk\ nl)$ have interplanar spacing $d_{hkl}/n$, so $n$th-order diffraction in planes $(hkl)$ is equivalent to first-order diffraction in planes $(nh\ nk\ nl)$. If we choose to describe diffraction by crystals as always arising in first order then we must introduce planes of the form $(nh\ nk\ nl)$. It was planes of just this kind that we found redundant at the end of the last chapter. Thus our redundancy problem is resolved if we assume that lattices (i.e crystals) always diffract X rays, neutrons, electrons or any other particle in first order.

## 3.2 The Atomic Form Factor

In the previous discussion we have assumed that the atom scatters X rays uniformly with a strength $f$; usually one compares the atom with a free electron and then the scattering factor is just $Z$, the atomic number. This means that the scattering power varies radically and monotonically with atomic number throughout the periodic table. An atom, however, does not scatter X rays in an isotropic fashion, the scattering being much weaker in the backward than the forward direction. The reason is that the atom has a size comparable to the wavelength of the X rays, and scattered rays leaving different parts of the atomic charge cloud are not exactly in phase. The difference in phase is zero at zero diffraction angle, but increases markedly as $\theta$ varies from 0 to 90°. The atomic form factor, as the scattering power is called, thus decreases rapidly with angle of diffraction as Fig. 3.3 demonstrates. The form factor and its angular dependence for the elements are usually provided in books concerned with X-ray crystallography.

## 3.3 The Structure Factor

The discussion so far has assumed that the diffracting system is a Bravais lattice occupied by one kind of atom. We have seen that real crystal structures are more

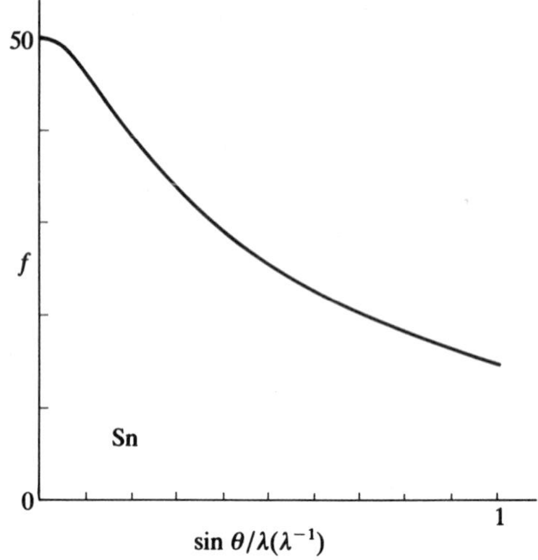


**Figure 3.3** The atom has linear extent similar to that of the X-ray wavelength; this causes a marked decrease of the atomic scattering power with angle of diffraction by the interference of scattered rays coming from different parts of the same atomic charge cloud.

complicated than this and that the repetitive unit is the basis. Our previous discussion holds, but *we must replace f by the scattering power of the basis*; this is calculated as already described by taking account of the phase differences between the rays scattered by different atoms. We need to calculate

$$\left(\sum_j f_j e^{i\Delta_j}\right)_{\text{basis}}. \tag{3.11}$$

The subscript $j$ now denotes the different atoms in the basis. We must reckon with the basis containing different atom sorts with different atomic form factors. We rewrite (3.11) as

$$\sum_j f_j e^{i\boldsymbol{\rho}_j \cdot \Delta\mathbf{k}}, \tag{3.12}$$

and using (3.8) this becomes

$$\sum_j f_j e^{i\boldsymbol{\rho}_j \cdot \mathbf{G}_{hkl}}. \tag{3.13}$$

The basis is

$$\boldsymbol{\rho}_j = u_j \mathbf{a} + v_j \mathbf{b} + w_j \mathbf{c}, \tag{3.14}$$

$u$, $v$ and $w$ being the coordinates of the atoms measured in units of the cell vectors.

Our condition for the formation of the diffracted beams, equation (3.8), remains unchanged, because it is a property of the lattice around which the crystal structure is built. The amplitudes of the diffracted beams are, however, determined by expression (3.13), which is called the *structure factor*:

$$\text{structure factor} = \sum_j f_j e^{i\boldsymbol{\rho}_j \cdot \mathbf{G}_{hkl}} = S_{hkl}.$$

Substituting (3.14) for $\boldsymbol{\rho}_j$ and using the usual expression for $\mathbf{G}_{hkl}$, we find

$$S_{hkl} = \sum_j f_j e^{i2\pi(hu_j + kv_j + lw_j)}. \tag{3.15}$$

To demonstrate the use of the structure factor, we compare the diffraction properties of the simple cubic and the body-centred cubic lattices.

For the *simple cubic lattice* with one atom at 000,

$$S_{hkl} = f e^{i2\pi 0} = f,$$

all values of $h$, $k$ and $l$ are permissable and the amplitudes of diffracted beams are governed solely by the angular dependence of the scattering factor $f$.

Using Bragg's law, we can determine the possible values of the Bragg angle $\theta$. We have seen that

$$\lambda = 2d_{hkl} \sin\theta, \tag{3.16}$$

so

$$\sin^2\theta = \frac{\lambda^2}{4d_{hkl}^2}.$$

For the cubic structures

$$d_{hkl}^2 = \frac{a^2}{h^2 + k^2 + l^2} \tag{3.17}$$

and

$$\sin^2\theta = \frac{\lambda^2}{4a^2}(h^2 + k^2 + l^2). \tag{3.18}$$

Each of the indices $h$, $k$ and $l$ may take any integral value and zero, leading to the following possible values of $h^2 + k^2 + l^2$:

$$0\ 1\ 2\ 3\ 4\ 5\ 6\ \cdot\ 8\ 9\ 10\ 11\ 12\ 13\ 14\ \cdot\ 16 \ldots .$$

Note the absence of the terms 7, 15, 23 etc.

We now consider the *body-centred cubic lattice*. Although the bcc structure is a true Bravais lattice, the use of the cubic unit cell containing two atoms demands that we describe it in terms of the simple cubic lattice using the basis 000, $\frac{1}{2}\frac{1}{2}\frac{1}{2}$; the atoms are assumed identical. Thus

$$\begin{aligned} S_{hkl} &= f(e^{i2\pi 0} + e^{i2\pi(h+k+l)/2}) \\ &= f(1 + e^{i\pi(h+k+l)}). \end{aligned}$$

Therefore

$$S_{hkl} = \begin{cases} 2f & \text{when } h + k + l \text{ is even,} \\ 0 & \text{when } h + k + l \text{ is odd.} \end{cases} \tag{3.19}$$

Diffracted beams exist only for Bragg angles $\theta$ appropriate to the following values of $h^2 + k^2 + l^2$:

$$0\ 2\ 4\ 6\ 8\ 10\ 12\ 14\ 16\ 18\ 20 \ldots .$$

On the other hand, if the basis contains two different atoms, as in the CsCl structure, all the diffracted beams of the simple cubic lattice arise, but the amplitudes of those with $h + k + l$ odd have

$$S_{hkl} = f_1 - f_2,$$

and those with $h + k + l$ even have

$$S_{hkl} = f_1 + f_2,$$

$f_1$ and $f_2$ of course being the atomic form factors of the two sorts of atom. Again the angular dependence of $f$ is superimposed on the above amplitude variations. Furthermore, when comparing the diffraction intensities from different families of planes, we must take account of the multiplicity. (See Problem 3.10).

The behaviour of the fcc and diamond cubic structures may be analysed similarly. The four cubic structures are compared in Fig. 3.4.

*Summary of cubic structures*

| | | |
|---|---|---|
| Simple cubic | all $h$, $k$, $l$ allowed | first reflection (100) |
| Body-centred cubic | only $h + k + l$ even allowed | first reflection (110) |
| Face-centred cubic | $h$, $k$, $l$ all odd or all even allowed | first reflection (111) |

## 3.4 Experimental Details

Unless one specializes in X-ray crystallography, it is often sufficient to have a knowledge of

(a) the Debye-Scherrer powder method for lattice parameter measurements (Box 3.1, p. 54), and

(b) the back-reflection Laue method for the orientation of massive single crystals (Box 3.2, pp. 55–6).

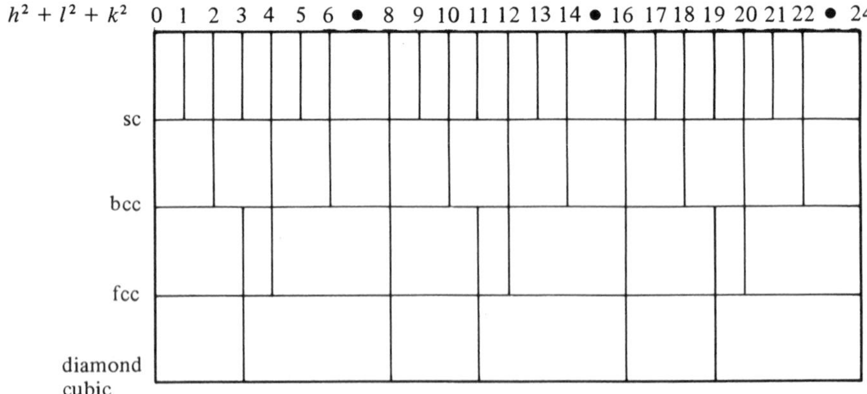


**Figure 3.4** Permitted diffracted beams in cubic systems are characterized by the values of $h^2 + k^2 + l^2$. The figure shows how the addition of a basis to the simple cubic primitive cell reduces the number of allowed beams, increasingly the more atoms in the basis. Note the regular sequence of each pattern. The correct angular separations are not reproduced in this diagram.

X rays are produced when energetic electrons collide with solids. The deceleration of the electron causes the emission of a broad band of X radiation with a short wavelength cut-off given by $hc/\lambda = eV$, where $V$ is the accelerating potential in volts. This is the *continuous spectrum.* Superimposed on this continuous background are sharp lines, the *characteristic spectrum* arising as a result of ionization of the inner shells of the atoms. Usually we are only concerned with the spectrum obtained when the $K$ shell is ionized (see Fig. 3.5). In this case one obtains the $K\alpha$ and $K\beta$ rays; the latter are filtered away in practice, leaving the $K\alpha_1$, $K\alpha_2$ lines, the $\alpha_1$ being twice as intense as the $\alpha_2$. They may be separated using a crystal monochromator, but this is not always necessary.

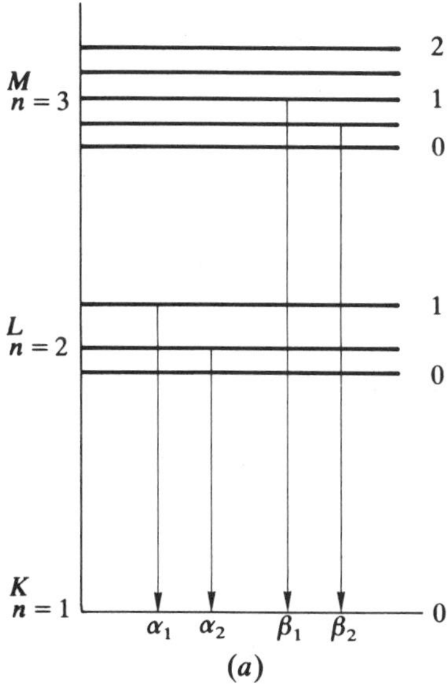


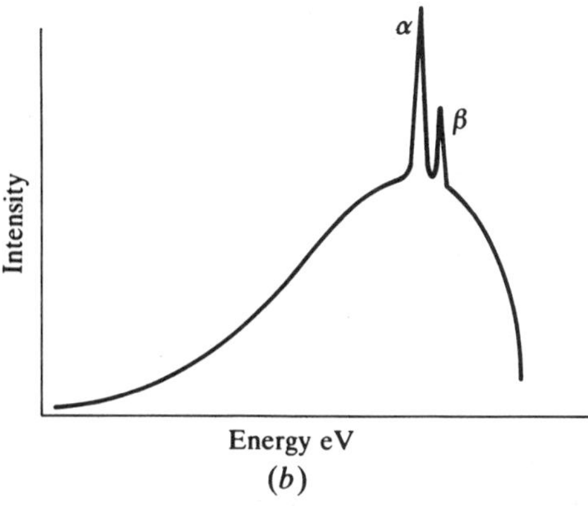


**Figure 3.5** (a) Origins of the $K\alpha$ and $K\beta$ spectra; the numbers associated with each level are the orbital quantum numbers. (b) is a qualitative picture of the spectra. Wavelengths for selected elements are given in Table 3.1.

**Table 3.1** X-ray wavelength (Å)

| | Cu | Ni | Fe | Mn |
|---|---|---|---|---|
| $K\alpha_1$ | 1.5404 | 1.655 | 1.932 | 2.098 |
| $K\alpha_2$ | 1.5444 | 1.658 | 1.936 | 2.101 |

**Box 3.1. The Debye-Scherrer Method**

In the Debye-Scherrer X-ray technique the specimen is a very fine powder contained in a thin glass capillary tube or, more conveniently, attached to a glass fibre, the latter having been coated with grease and then rolled in the powder. If the powder is sufficiently fine then there are very many crystals each in its own particular orientation, but in effect the number of crystals is so large that we may assume all possible orientations of the crystal structure are represented. This means that if monochromatic radiation is used then all the planes $\{hkl\}$ satisfying

$$\theta_{hkl} = \sin^{-1}(\lambda/2d_{hkl}) \leq \tfrac{1}{2}\pi$$

are capable of producing diffraction effects.

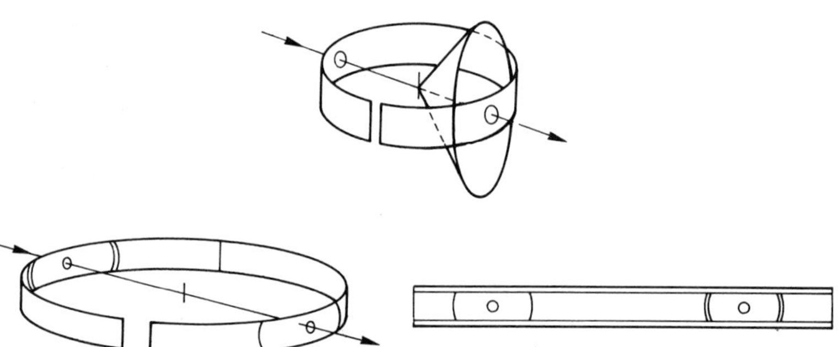

**Figure 3.6** (After Henry *et al.* 1961.)

Because the incident X-ray beam is an axis of symmetry, each set of planes produces a cone of diffracted rays, Fig. 3.6, the latter are recorded photographically and subsequently the Bragg angles are measured. We have

$$\sin^2\theta = \lambda^2/4d_{hkl}^2,$$

and, even though we may not always know the indices *hkl*, we can always determine the interplanar spacings. In cubic systems

$$\sin^2\theta = (h^2 + k^2 + l^2)\lambda^2/4a^2.$$

$h^2 + k^2 + l^2$ may, depending on the structure, take certain integral values apart from 7, 15, 23 etc. If we make a list of the measured $\sin^2\theta$ then a determination of the common factor $\lambda^2/4a^2$ allows a to be evaluated.

**Box 3.2. The Laue Method**

The back-reflection Laue technique is used to orient large crystals opaque to X rays (Fig. 3.7). In a crystal there often arise groups of planes whose intersections form parallel lines, e.g. the planes $\{10\bar{1}0\}$ in the hcp structure. Such a group of planes is called a zone and the common direction a zone axis. In the reciprocal lattice the points representing the several crystal

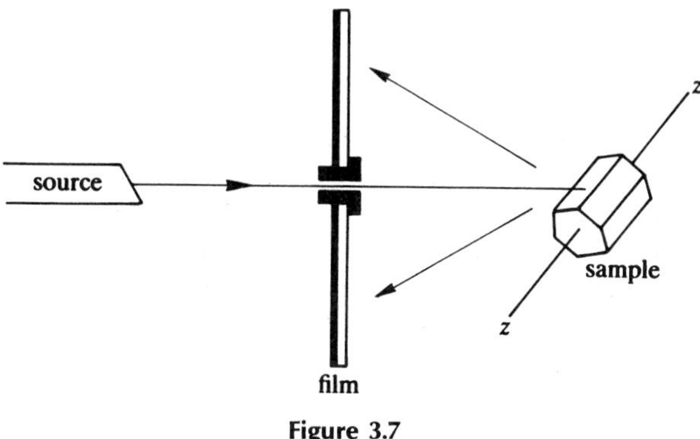


Figure 3.7

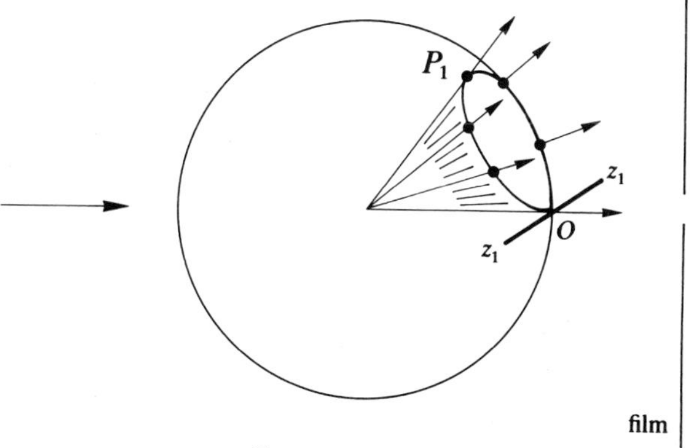


Figure 3.8

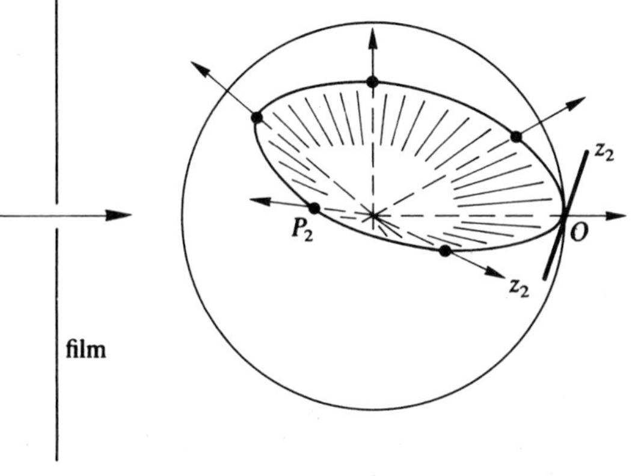


Figure 3.9

*continued*
planes of a zone lie on a plane perpendicular to the zone axis. Consider diffraction in the reciprocal lattice using the Ewald sphere (Fig. 3.8). If $P_1$ is a point of the reciprocal lattice lying on the surface of the Ewald sphere then we obtain immediately the direction of a diffracted ray. Now suppose the family of planes represented by $P_1$ belongs to a zone with zone axis $z_1$, as shown; then other planes belonging to this zone have their reciprocal lattice points on a plane perpendicular to $z_1$. If certain of these points also lie on the Ewald sphere, they will define a 'small' circle. The zone therefore produces diffracted rays describing a cone of angle $2\theta$ that includes the incident ray. It is clear that in transmission (i.e. the Laue method) the 'cone' of diffracted rays gives rise to an elliptical intersection with a photographic film placed perpendicular to the incident ray (Fig. 3.8). However, in back reflection, as demonstrated by the point $P_2$ associated with the zone $z_2$, the cone of diffracted rays produces a hyperbolic intersection with the film (Fig. 3.9). Thus, when orienting crystals using the back-reflection method, the diffraction spots are found to lie on hyperbolas. The angular relationships between the zones which these curves represent may be measured, and using the stereographic projection and the known structure we can orient the crystal.

The *Debye-Scherrer* method makes use of monochromatic X radiation ($K\alpha_1$, $K\alpha_2$) and a cylindrical powdered specimen with diameter of about 0.2 mm. The powder contains very many small crystals of all possible orientations, and furthermore it is rotated during exposure in order to simulate an infinitely fine distribution of orientations. Thus there are always crystals oriented to fulfil Bragg's law. The X rays impinge on the specimen at right angles to the cylinder axis and, by the symmetry about the incident direction, planes oriented for Bragg reflection produce cones of diffracted X rays. These are detected photographically and allow the measurement of Bragg angles $\theta$. The method has the following uses:

(a) determination of lattice spacing with a normal accuracy of about 1 in $10^4$;
(b) adaption for measurements at low and high temperatures;
(c) identification of alloy phases and the determination of phase boundaries;
(d) studies of phase transformations;
(e) structure determination (only in the simplest cases).

Usually, only qualitative measurement of the intensities of Debye-Scherrer lines is carried out. When intensity values are required it is more practical to use the modern version of the *Bragg spectrometer*. The incident radiation falls on a flat powdered specimen and the diffracted beam is detected with a Geiger counter, preceded if necessary by a crystal monochromator to isolate the $K\alpha_1$ radiation. The detector and specimen are rotated synchronously so that the detector moves through twice the angle rotated by the specimen. In this way, the conditions for Bragg reflection are satisfied and the many orientations provided by the powdered sample ensure that the required diffracted beams are obtained. Automatic scanning can be arranged and the method lends itself to use at high and low temperatures more readily than the photographic technique.

In solid state physics measurements are often made on single-crystal samples which are usually prepared in a particular orientation and cut from larger crystals grown artificially in special apparatus. The parent crystal may have a volume anywhere in the range 1–500 $cm^3$. The outer form of such artificial crystals does not necessarily provide information about the internal symmetry, and the orientation of the crystal is established using X-ray diffraction. Contrary to popular belief, X rays with $\lambda \approx 1$ Å are strongly attenuated by matter and thin metal foils are adequate as shutters. It is not possible to orient massive crystals by transmission experiments (although this is quite straightforward with neutron beams). One therefore uses the Laue back-reflection technique. Flat photographic film is fitted with a collimator. The X rays pass through the collimator, strike the sample and are diffracted backwards to produce a pattern of spots on the film. Since the specimen is a single crystal, one requires a continuous spectrum of X radiation to ensure that the particular orientation of the sample gives rise to diffraction effects. The film may be analysed and the orientation deduced, but in crystals with simple structures one can often detect the position of an axis of high symmetry by visual inspection (e.g. the $c$ axis in hexagonal metals such as Zn or Mg), and this allows a rapid assessment of the situation.

## 3.5 Electron and Neutron Diffraction

The physical basis for the diffraction of electron and neutron beams is the same as that for the diffraction of X rays, the only difference being in the mechanism of scattering. Electrons are scattered by the strong potential produced by the positive nucleus and the outer electrons of the atom: they are scattered much more strongly than X rays (by a factor of order $10^4$) and the diffracted beams are readily detected by photographic plates or observed 'live' by the images produced in fluorescent screens. To obtain $\lambda \approx 1$ Å, an electron needs energy of about 150 eV, but it is normally the case that accelerating potentials $(5\text{–}10) \times 10^4$ V are used. The wavelength is then much shorter than 1 Å and the Bragg angles $\theta$ correspondingly smaller. In principle the wavelength is continuously variable and the accelerating voltage must be carefully stabilized to provide monochromatic radiation. The electron beams demand the use of magnetic or electrostatic lenses and can only exist in a high vacuum. Furthermore, they are strongly absorbed by matter and specimens must be very thin for transmission experiments. Bulk specimens are studied by reflection at glancing incidence. Electron diffraction lends itself to the study of surfaces and may often be considered to arise from two-dimensional structures.

Neutrons of thermal energy (about 60°C) are concentrated in the moderators of research reactors, and in recent years reactors have been built solely to provide intense neutron beams $>10^{15}$ n cm$^{-2}$ s$^{-1}$ (as opposed to about $5 \times 10^{13}$ n cm$^{-2}$ s$^{-1}$ in an ordinary research reactor). The neutron energies are distributed, as are the energies of molecules in a gas, according to Maxwell's law. The equivalent wavelengths corresponding to the root-mean-square velocities of a neutron gas at 0°C and 100°C are 1.55 Å and 1.33 Å respectively – just right for diffraction by crystals. A suitable band of wavelength is selected by Bragg reflection from a chosen family of planes in a large crystal of, say, Cu. This monochromatic beam may then be used with a Bragg spectrometer and a powdered sample, or to study single crystals. The detector is based on $^{10}$B, the thermal neutron producing a nuclear reaction

$$^{10}\text{B} + \text{n} \rightarrow \text{Li} + {}^4\text{He}.$$

The neutrons are scattered by the nuclei of the sample. Since the nucleus has diameter of order $10^{-12}$ cm and is very much smaller than the wavelength, there is no angular dependence of the form factor, or scattering length as it is called in the case of neutrons. The neutron is momentarily captured, forming a compound nucleus and is then re-emitted. If the nuclei possess no spin then the scattering is the same for all nuclei and is said to be coherent. Interference and diffraction effects can arise just as for X rays. On the other hand, when the scattering nucleus has spin $I$ the compound nucleus may take two spin values $I \pm \frac{1}{2}$. The resultant scattering may be incoherent as well as coherent. Incoherent scattering events possess no phase correlation and interference effects do not arise. Occasionally the incoherent scattering is the dominant component, as in vanadium. Unlike the case of X-ray scattering by atoms, there is no regular variation of the neutron scattering length through the periodic table. Light atoms may scatter just as strongly as heavy ones. This makes neutron diffraction suitable for the study of compounds of light and heavy elements, and enables neutron scattering to reveal the ordered structures of such alloys as FeCo or CoZn where X-ray scattering would not. Since the scattering mechanism is a process involving the creation of an unstable compound nucleus, it is not surprising that different isotopes of the same element scatter differently. The neutron also has a magnetic moment and it can therefore interact with magnetic atoms, i.e. those possessing a net electron spin. This allows the determination of the magnetic structures of varying complexity found in the transition and rare earth metals, as well as compounds and alloys of these elements. The magnetic moment of an atom arises in the electrons, and magnetic scattering is therefore subject to the usual angular dependence as for X rays. (The electron also possesses a magnetic moment – why can we not study magnetic structure using electron diffraction?) The acquisition of information by inelastic neutron scattering about lattice vibrations is discussed in Section 5.11.

## 3.6 Non-periodic Structures

Solids are not necessarily crystalline. They may also be obtained as glasses and as amorphous aggregates that are not crystalline on any significant scale; i.e. there is no long-range correlation between atom positions and no translational symmetry. Short-range order over distances corresponding to one or two atomic spacings may arise, the degree of ordering depending on the type of bond that is formed. Amorphous metals are well described as random close-packed structures. Translational symmetry is also lacking in simple liquids, i.e. those composed of independent similar atoms. Absence of symmetry prevents a description in terms of geometrical concepts such as the unit cell. The liquid has a dynamic equilibrium in which the relative positions of the atoms are constantly changing. Nevertheless, we can imagine a time average of the liquid structure, which may be approximated by a radial distribution function (RDF).

Taking the origin as the momentary position of a given atom, we define the probability of finding another atom in an element of volume $d^3R$ at a distance $R$ from our chosen atom as

$$\frac{N}{V}\, g_2(R)\, d^3R. \tag{3.20}$$

$N/V$ is the average particle density and $g_2(R)$ is called the *pair distribution function.* Now when $R$ is large, say $>20$ Å, we do not expect any correlation between the positions of the two atoms, and the probability can only depend on the size of $d^3R$

and the average density, so $g_2(R)$ is unity for large $R$. At very small distances, $g_2(R)$ is zero because two atoms cannot occupy the same position. However, for a certain range of values of $R$, centred around the average particle separation, $g_2(R)$ varies in a pronounced manner (see Fig. 3.10). For an isotropic liquid, the radial distribution function is defined as

$$\mathrm{RDF} = 4\pi R^2 g_2(R). \tag{3.21}$$

Integration of the RDF between the limits 0 and $R$ tells us how many atoms are contained within a sphere of radius $R$ described around our chosen atom. One can define higher distribution functions, e.g. the three-particle distribution $g_3(R_{12}, R_{13})$, that are without doubt important in the liquid, but since there is no way of obtaining experimental knowledge about them, discussion of liquid and other disordered structures is usually confined to $g_2(R)$ and the RDF.

The diffraction of X rays or neutrons by a liquid, just as in the case of the crystal, may be described with the aid of the structure factor, but for the liquid or amorphous element we must define this quantity for the *complete sample* because there is no repetitive unit cell. The intensity of the diffracted ray is controlled by

$$\langle S_i(\mathbf{k}) S_j^*(\mathbf{k}) \rangle, \tag{3.22}$$

where the angle brackets signify a time average and

$$\left. \begin{aligned} S_i(\mathbf{k}) &= \sum_{i=1}^{N} \mathrm{e}^{\mathrm{i}\Delta\mathbf{k}\cdot\mathbf{r}_i} \\ S_j^*(\mathbf{k}) &= \sum_{j=1}^{N} \mathrm{e}^{-\mathrm{i}\Delta\mathbf{k}\cdot\mathbf{r}_j} \end{aligned} \right\} \tag{3.23}$$

In the liquid the atoms are in rapid motion and the intensity of diffracted rays depends upon the existence of an average correlation in the phase of rays scattered from different atoms. If within the lifetime of a photon there is no such correlation then each atom scatters as an independent unit and the phases are randomly distributed. The result is an isotropic scattering modified only by the atomic form factor and possible absorption effects. No Laue diffraction can arise.

For X rays with $\lambda \approx 1$ Å the photon has a lifetime of order $10^{-15}$ s and the root-mean-square velocity of an atom in a liquid is usually less than $10^5$ cm s$^{-1}$. In the lifetime of a photon, the atom moves about $10^{-2}$ Å, i.e. $10^{-2}\,\lambda$. Each photon is scattered and diffracted independently according to the instantaneous geometrical distribution of atoms in the liquid. Over the exposure time many millions of photons are diffracted and each photon contributes to the intensity of the final diffraction pattern which therefore carries information about the time averaged liquid structure. Experiment shows that the diffraction pattern consists of halos centred around the incident beam (Fig. 3.11). As is always the case, the diffraction pattern is the Fourier transform of the aperture function of the diffracting object (in this case the averaged geometrical structure and scattering power of the liquid). An interference function is defined as

$$a(k) = \frac{1}{N} \langle S_i(\mathbf{k}) S_j^*(\mathbf{k}) \rangle. \tag{3.24}$$

and it determines the intensity of the diffracted ray. Its importance lies in that $a(k) - 1$ and $g_2(R) - 1$ are a Fourier-transform pair, a fact that we state without

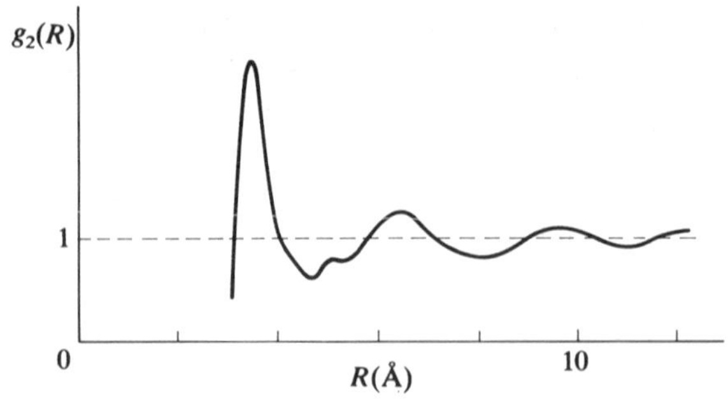


**Figure 3.10** The pair correlation function (normalized to unity for large $R$) for liquid Pb at 340°C as might be obtained from data like those shown in Fig. 3.11. (After North *et al.* 1968.)

proof. (For further details regarding the structure and properties of liquid metals see Faber 1972.)

Experiments with X rays and neutrons have been performed; on the whole they present many difficulties, but accurate data are becoming increasingly available. A typical plot of $g_2(R)$ is shown in Fig. 3.10. If the first maximum is well defined, as is usually the case, we can calculate an average number of nearest neighbours. The average separation of atoms in liquid metals just above the melting point is usually not very different from that in the solid, the difference arising in the coordination, i.e. number of nearest neighbours. In certain cases where strong directional binding occurs in the solid, as in Ge which has tetrahedral coordination, this may be lost when it becomes liquid, leading to a more isotropic and closer-packed arrangement of atoms. Such changes can produce significant alterations in behaviour, for example Ge which is a semiconductor in the solid state becomes a true metal when liquid.

In recent years amorphous solids have received more and more attention. Alloys and even some pure metals may be obtained in the amorphous state. When certain metals like Ga or Be are evaporated onto very cold substrates, at about 2 K, the atoms 'freeze' as they strike the substrate and form a disordered structure. Similarly, ion bombardment can remove crystalline order. Many covalently bonded substances are stable in the amorphous state even at room temperature. A relatively recent development is the production of metallic alloy glasses by extremely rapid

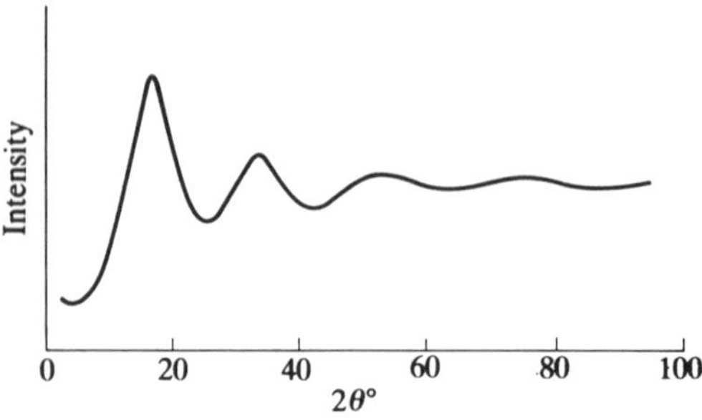


**Figure 3.11** A qualitative picture of the diffraction pattern from a liquid metal. $2\theta$ is the angle between incident and diffracted rays.

cooling ($>10^6$ K $s^{-1}$) of the melt. One cannot prepare amorphous pure metals this way, and elements such as Si, B or P appear to be necessary to stabilize the amorphous glassy structure. When formed, these alloys do not recrystallize until heated to 300–400°C. They have potential commercial uses, in particular they possess much better corrosion resistance than crystalline phases and certain magnetic alloys have potential application as transformer cores.

## 3.7 Icosahedral Structures

### 3.7.1 *Quasicrystals*

In 1984 two articles – one theoretical, the other experimental – were published almost simultaneously; both concerned icosahedral structures. In computer simulations of atomic motion in a liquid as it cools, it had earlier been found that atomic groupings with icosahedral symmetry were favoured. The theoretical paper suggested how icosahedral symmetry might arise in a solid, and furthermore demonstrated that the diffraction pattern from such a structure would be sharp and show characteristic five-fold symmetry never observed in diffraction from the Bravais lattices. The experimental paper, on the other hand, presented what appeared to be just this predicted icosahedral diffraction pattern in an alloy $Al_6Mn$ that had been produced by rapid, but not too rapid, cooling. These two papers have stimulated an increasing spate of activity in the study of what are now called quasi-crystals or quasi-periodic structures. We shall try to describe the essence of quasi-periodicity in two dimensions. An extensive account of the occurrence and structure of quasi-crystals is found in the book by Janot (1992).

First of all, we demonstrate why one never finds five-fold rotational symmetry in a Bravais lattice. Suppose, as in Fig. 3.12, that we have a two-dimensional Bravais lattice based on a pentagonal cell. The separations of the 'lattice' points are shown, and it is clear that the central point does not conform to our notion of a lattice because there is no translational symmetry. If there is a point $P$ at $\mathbf{r}_1$ then we must find an equivalent point at $-\mathbf{r}_1$, and this is not the case. Perhaps we should assume a decagonal cell as indicated by the dotted lines. However, this will not do either,

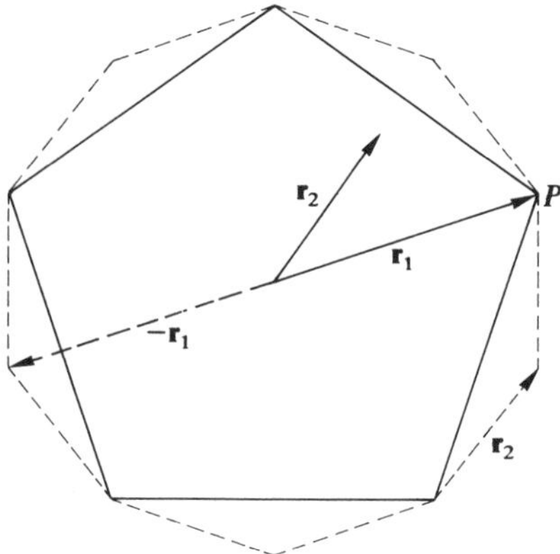


**Figure 3.12** Pentagonal symmetry is incompatible with translational symmetry.

because we can translate the edge vector $\mathbf{r}_2$ to our central point; there we expect to find another point of our supposed lattice, but this is not so since $|\mathbf{r}_2| < |\mathbf{r}_1|$, our presumed lattice spacing. Our assumptions of pentagonal or decagonal symmetry are not compatible with our definition of a lattice. How is it then that we can observe such rotational symmetries in experimental diffraction patterns from specimens that appear to be homogeneous over linear dimensions of mm or more?

To answer this question, we must digress to explain what is known as Penrose tiling. The question posed is whether it is possible to cover (i.e. tile) flat two-dimensional space in an irregular (i.e. aperiodic) manner with a finite number of different shapes of tile and if so what is the minimum number of shapes or tiles needed. Penrose showed that it is possible to use only two different tile shapes and to cover flat space perfectly with them in an infinite number of aperiodic ways. The Penrose tiles are rhombi with areas forming an irrational ratio; this ratio for the Penrose rhombi is the golden mean $\phi = 1.618\ldots$, which provides a direct connection with the pentagon (Fig. 3.13). The Penrose patterns are remarkable in that

(a) every finite region in any one pattern is contained within every other pattern;

(b) the different patterns differ in infinitely many ways, but are only distinguishable in the limit of infinite size;

(c) any region of linear extent $d$ is never further than a distance $2d$ from a similar region;

(d) the patterns contain local axes of five-fold symmetry.

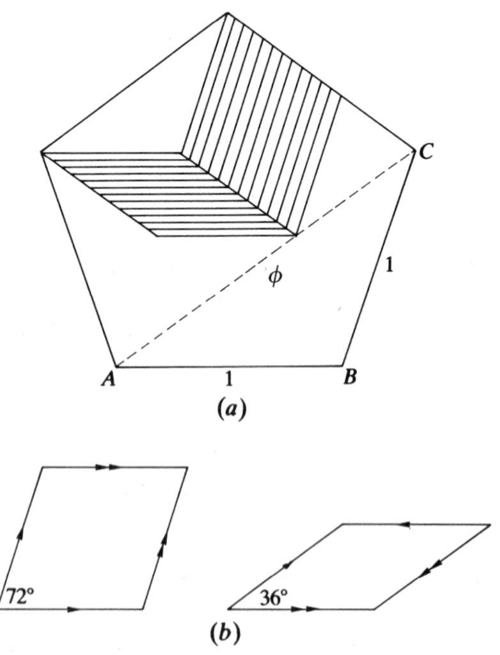


**Figure 3.13** (a) The unit pentagon and the Penrose tiles (shown shaded) derived from it. (b) The Penrose tiles; they may only be used if matching sides, as indicated by the arrows, are placed together. AC is the irrational number $\phi$.

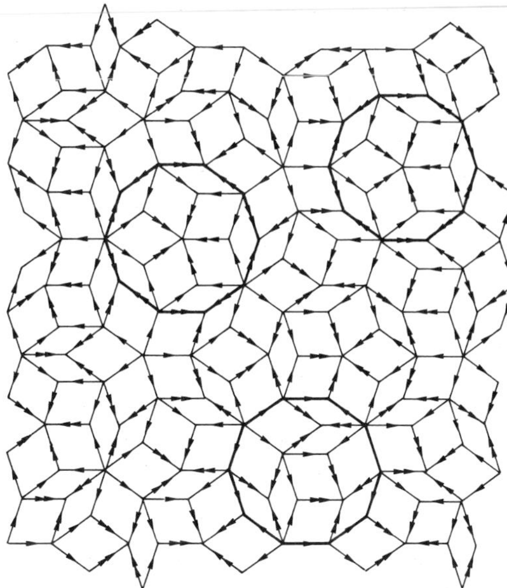

**Figure 3.14** A portion of a Penrose pattern showing local regions of five-fold symmetry. (From D. R. Nelson and B. I. Halperin, *Science*, **229**, 1985, 233. Copyright 1985 by the AAAS.)

Some of these features may be observed in Fig. 3.14. The key to appreciating the long-range quasiperiodicity that produces discrete diffracted rays lies in Fig. 3.15, which shows how we may discern one family of rows of tiles or cells; there are in fact five such families at intervals of 72°, i.e. in pentagonal symmetry. In each row of cells there is an irregular distribution of cells, but in a given family of rows one can identify a characteristic inter-row separation that will produce diffraction – as has been demonstrated optically. The pattern of Fig. 3.14 is that of a two-dimensional quasicrystal 'lattice' based on the irrational number $\phi$. This is the principal feature of quasicrystals – the structures are determined by two or more quantities (lengths, areas or volumes) that form irrational ratios with one another. Returning now to three dimensions, we show the icosahedron in Fig. 3.16. It is the most complicated of the five regular polyhedra known as the Platonic solids.† The icosahedron has twenty similar sides, each an equilateral triangle. It is associated with six five-fold, ten three-fold and fifteen two-fold axes of rotational symmetry. Just as pentagons cannot cover flat space perfectly, icosahedra cannot fill three-dimensional space. We can form an icosahedral cluster of spheres: thirteen are needed, a central sphere and twelve at the vertices. The separation of an outer sphere relative to the central sphere is some 5% less than the separation of neighbouring vertex spheres.

To explain the actual occurrence of symmetry elements characteristic of the icosahedron in macroscopic samples, we turn to Penrose tiling in three dimensions. Thus we may consider the existence of two rhombohedra (analogous to the rhombi of Fig. 3.14) that can together fill space perfectly without overlap or gaps arising. These rhombohedra produce an irregular but quasiperiodic arrangement of 'lattice points', and we find the elements of icosahedral symmetry in a manner analogous to

† The others are the regular tetrahedron, cube, octahedron and pentadodecahedron.

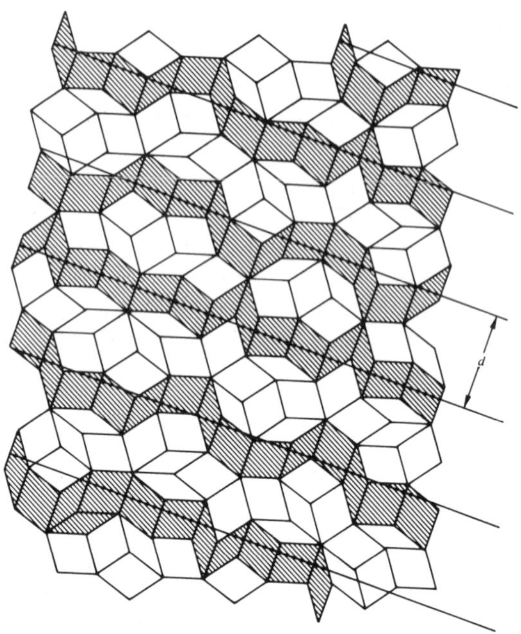

**Figure 3.15** One family of rows that may produce Bragg diffraction. The rows are defined by choosing cells with parallel bounding sides perpendicular to the row direction. (From D. R. Nelson and B. I. Halperin, *Science*, **229**, 1985, 233. Copyright 1985 by the AAAS.)

that whereby the pentagonal symmetry arises in the two-dimensional case. However, although one may understand how in principle quasiperiodic structures may occur, the question of deciding the positions of the atoms in these structures is exceedingly difficult. Since their discovery in 1984 quasicrystals have attracted much interest particularly with regard to structure and the origins of their stability. Many new compositions have been discovered and certain of them do not require rapid cooling to form the quasicrystalline phase (e.g. $Al_{65}Cu_{20}Fe_{12}$) so that samples with linear size $\sim 1$ mm are now obtainable.

Quasicrystals always contain two or more components and the quasiperiodicity may arise over a range of compositions; they may be considered a link between the perfectly periodic and the amorphous states; it is in fact possible to imagine a continuous variation from the quasiperiodic to the amorphous condition. (For further details see Nelson and Halperin 1985, Janot and Dubois 1988 and Janot 1992.)

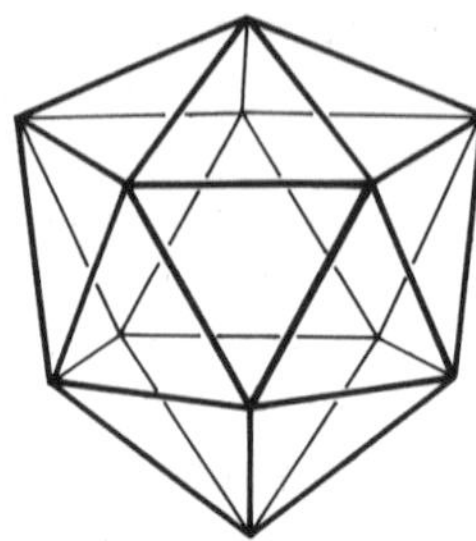

**Figure 3.16** The icosahedron, a polyhedron with twenty identical equilateral triangular sides.

### 3.7.2 Atomic clusters

The electronic structures of atoms and molecules are understood in great detail and, as the following chapters will show, so is the basic electronic structure of crystalline matter. In this respect there is no need to study ever larger clusters of atoms in order to obtain knowledge of bulk matter. Atomic clusters are interesting in their own right. A particular question one might ask concerns the size a cluster needs to have in order to exhibit bulk behaviour. Doubtless there is not one particular size; furthermore, such a size probably depends on the physical property in question. Nevertheless the finite size of any sample involves the presence of a surface, and surface atoms have an environment different from that of atoms in the bulk, leading to a surface energy.

It is straightforward to show that for a 1 cm cube of crystalline copper the ratio $r$ of the number of surface atoms to the total number of atoms in the sample is of order $10^{-7}$ – quite insignificant. The ratio $r$ scales as $1/l$, where $l$ is the cube edge size. A $(\mu\text{m})^3$ of Cu therefore has $r = 0.001$ and contains about $10^{11}$ atoms – quite big. On the other hand, a $(2.3\ \text{nm})^3$ cube of Cu contains some 1000 atoms and has $r = 0.45$. These figures would not be significantly different in spherical assemblies of Cu atoms. We now make an arbitrary decision and say that for a sample with $r > 0.1$ we might expect 'cluster' properties and for $r < 0.1$ the bulk behaviour probably prevails; $r = 0.1$ corresponds to some $10^5$ atoms and a linear size $l = 10$ nm.

In the past fifteen years the study of ultra-fine particles and atomic clusters of widely different chemical character has been in rapid development. Electronic, optical, magnetic and other properties have been studied, but, as might be imagined, the geometrical arrangement of the atoms in the cluster can only be inferred using theoretical predictions based on specific geometries and comparison with observed behaviour. A particular difficulty with small clusters or artificial molecules is that they may have several possible forms or isomers. Thus a cluster of metal atoms $M_6$ may arise as a planar molecule, a pentagon based pyramid or an octahedron; only calculation of the energies involved tells us which is the most stable and likely form and this is very expensive in computer time. Furthermore the standard theoretical procedures, e.g. the molecular orbital approach, are inapplicable to clusters containing more than ten or so atoms because the computational burden becomes too great. There are, however, other approaches.

Atomic clusters are formed from atomic vapours often produced in a laser vaporization source. A pulsed laser is focused onto a chosen target material causing local vaporization. The puff of vapour cools, partly by adiabatic expansion and partly by contact with a cold buffer gas, usually He or Ar. The clusters formed are carried by the buffer gas and ejected at supersonic speed from a nozzle to form a narrow beam. Dependent on circumstances the clusters have a temperature in the range 100–200 K. The cluster sizes present in the beam are determined by conventional mass spectrometry or, more customarily, by time of flight spectroscopy. The clusters are thermalized in the buffer gas then singly ionized by a special laser and thereafter accelerated by a fixed electric field. All clusters attain the same kinetic energy but have a velocity inversely proportional to the square root of their mass; different cluster masses therefore have different times of flight to the detector. The acceleration and detection are triggered to the vaporization laser pulse.

Some of the first studies of atomic clusters concerned Xe and Na. Xe is an inert gas and its electrons are tightly bound to the nucleus in stable closed shells. The

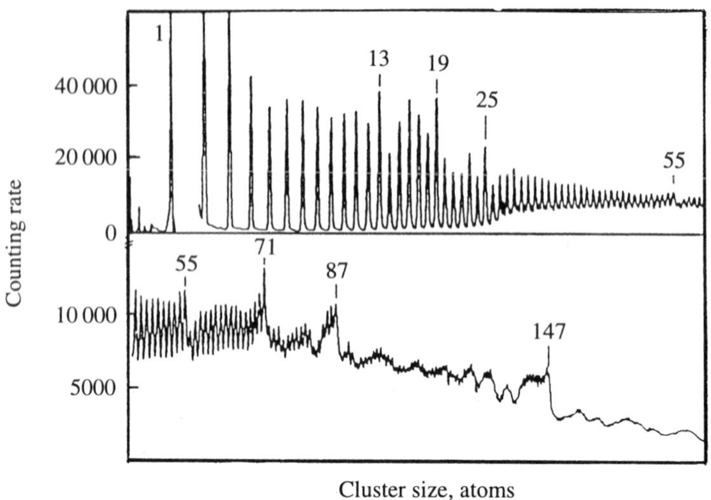


**Figure 3.17** The cluster spectrum for Xe atoms. The numbers indicate the cluster sizes based on icosahedral geometries. (After Echt *et al.* 1981.)

atoms act as hard spheres and interact only very weakly. The observed cluster mass spectrum, Fig. 3.17, shows characteristic variations and the observed peaks in abundance are associated with particularly stable structures. We expect clusters of different size to be produced with a uniform or smoothly varying abundance, but clusters may have different stabilities. Some may dissociate quickly and be poorly represented in the data, others may be particularly stable and therefore stand out above the rest.

The peaks in abundance at cluster sizes of 13, 19, 25, 55 . . . Xe atoms are associated with stable geometrical structures identical with or built around the icosahedron. Whereas structures based on the icosahedron are possible for individual clusters they cannot normally be used for conventional crystalline matter because they cannot fill space perfectly. Crystalline Xe has the face centred cubic structure.

The characteristic peaks in the mass spectrum of Na clusters, Fig. 3.18, arise at different mass numbers from Xe so they cannot have an icosahedral basis. In all probability the small Na clusters have spherical or spheroidal shapes and the stability is of electronic origin, a conclusion confirmed by measurement of the ionization energies of the clusters. The ionization energies are found to have maximum values at stability peaks and thereafter fall drastically for unit increase in cluster size. This is indicative of a shell structure for the valence electrons bound to the cluster, a conclusion supported by calculation. One very successful model for Na and similar metal clusters is an adaptation of the shell model used to describe the nucleus (see Krane 1988). In this approach it is assumed that a valence electron moves in a spherically symmetrical potential formed by a smoothed background of positive ionic charge and an average potential created by all the remaining electrons in the cluster including the spin dependent components (parallel spins giving rise to exchange interaction whereas antiparallel spins produce correlation corrections; these aspects are considered further in Section 6.8).

A self-consistent calculation in a potential of the form shown in Fig. 3.19 leads to the series of degenerate one electron energy levels shown: the degeneracy arises from

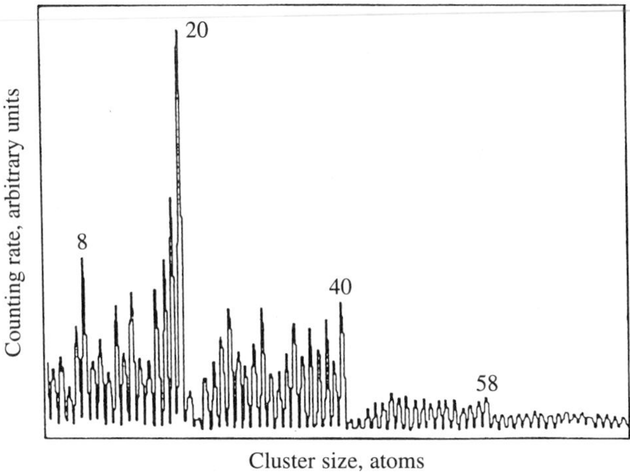


**Figure 3.18** The cluster spectrum for Na atoms. The particular stability of certain cluster sizes is associated with the occurrence of stable electron shell structures, see Fig. 3.19. (After Knight *et al.* 1984.)

the angular momentum associated with the level and nuclear terminology is used. Further development includes ellipsoidal cluster potentials. The valence electrons are then placed in these electron states in accordance with Pauli's principle. For Na such calculations predict closed electron shell stability at the following atom contents: 8, 20, 34, 58, 92, 138, 196, 268, 338, 440, 562, 704, 854, 1012, and larger clusters up to 2500 atoms size. The geometrical structures of these clusters are not known, but they are probably random close packed arrangements of atoms and may in fact be almost liquid-like. The basic input data are the chemical valence and effective diameter of the atom, i.e. the valence electron density.

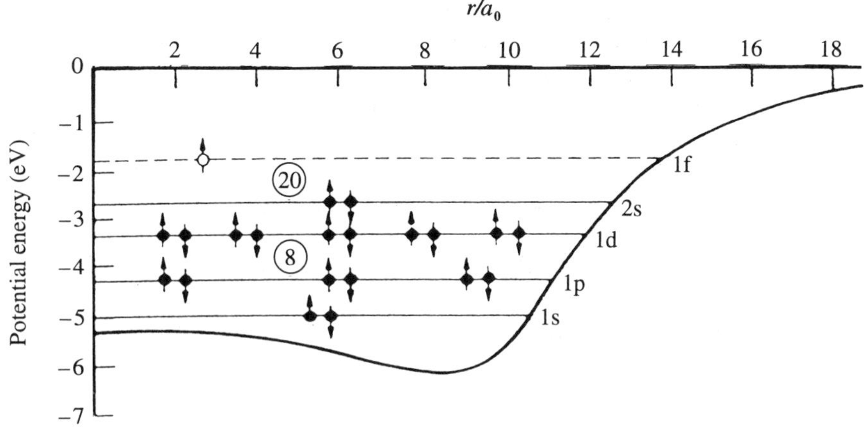


**Figure 3.19** The spherically symmetrical potential, assumed experienced by the valence electrons in sodium clusters, provides a series of degenerate electron levels which in turn produces the observed stabilities of particular cluster sizes (e.g. 8 and 20 atoms). The diagram shows the level structure for a 20 atom cluster. The radius $r$ of the cluster is given in units of the Bohr radius $a_0$. (After Eckardt 1984.)

For Na clusters containing more than about 2000 atoms the electronic shell structure is expected to be less significant for cluster stability, because the higher lying shells come closer and closer together in energy. It has in fact been demonstrated in experiment that for large clusters with more than 2000 atoms a new stability pattern develops which originates from a geometric layer structure based on completed icosahedra. This supershell structure has been observed up to about 25 000 atoms of Na; it is therefore concluded that these large clusters are different structurally from bulk Na which, as we have already seen, has bcc structure (so it seems that our earlier guess of $10^5$ atoms may not be far wrong). For further detail on the properties of atomic clusters of Na and other metals the reader is referred to de Heer (1993).

It may seem surprising that a rather featureless potential well like that of Fig. 3.19 together with one parameter, the electron density, should describe the cluster behaviour so well. This is primarily because of the smallness of the atomic core volume and the fact that the ionization energy is not so sensitive to the detailed geometrical distribution of the ions, in fact it is difficult to understand why the icosahedral supershell structure is observed in the ionization energies of large Na clusters. Later we shall see that a similar but simpler electron gas model can be successfully applied to certain bulk metals like Na (particularly in the liquid state).

### *3.7.3* $C_{60}$

Perhaps the most striking incident in the physics and chemistry of small particles was the discovery of the fullerenes and in particular $C_{60}$ in 1985. $C_{60}$ has a near spherical, hollow cage-like structure in which the carbon atoms form regular pentagons or hexagons with similar edge length. Twelve pentagons are disposed among twenty hexagons to produce, in a ball and stick model, the appearance of a soccer ball. The molecule has diameter 7.1 Å and the symmetry of the icosahedron. Its shape is actually that of a truncated icosahedron (Fig. 3.20).

Along with other carbon clusters of smaller and larger size, $C_{60}$ may be prepared by laser evaporation of graphite, but is obtained in greater amounts in the smoke

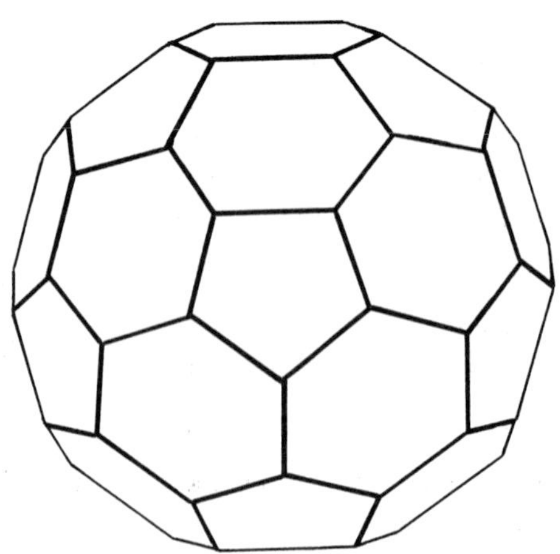

**Figure 3.20** The hollow cage structure of $C_{60}$.

from a DC carbon arc operating in He at a reduced pressure of about 13 kPa. $C_{60}$, unlike $C_{70}$ and other carbon particles that are formed together with it, is soluble in benzene, forming a rose coloured solution. Evaporation of the filtered solution then leaves a crystalline residue suitable for further study; it was first isolated in 1990. At room temperature the solid $C_{60}$ adopts a rather open face centred cubic structure with lattice parameter 14.2 Å, corresponding to a nearest neighbour separation of 10.04 Å. Each ball of carbon atoms is, however, constantly changing its orientation every $10^{-11}$ s. It therefore simulates spherical symmetry and there is no incompatibility between the icosahedral symmetry of the molecule and that of the crystal

**Table 3.2** Some data for the more common metallic elements

| Element | Structure | $a$ (Å) | Density ($10^3$ kg m$^{-3}$) | Mean atomic volume (Å$^3$) | Mean atomic radius (Å) | Ionic radius (Å)[a] |
|---|---|---|---|---|---|---|
| Li | bcc | 3.49 | 0.54 | 21.28 | 1.72 | 0.68 (+) |
| Na | bcc | 4.23 | 1.01 | 37.71 | 2.08 | 0.98 (+) |
| K | bcc | 5.23 | 1.91 | 71.43 | 2.57 | 1.33 (+) |
| Rb | bcc | 5.59 | 1.63 | 87.11 | 2.75 | 1.48 (+) |
| Cs | bcc | 6.05 | 2.00 | 110.5 | 2.98 | 1.67 (+) |
| Be | hcp | — | 1.82 | 8.26 | 1.25 | 0.30 (2+) |
| Mg | hcp | — | 1.74 | 23.26 | 1.77 | 0.65 (2+) |
| Ca | fcc | 5.58 | 1.53 | 43.45 | 2.18 | 0.94 (2+) |
| Sr | fcc | 6.08 | 2.58 | 56.18 | 2.38 | 1.10 (2+) |
| Ba | bcc | 5.02 | 3.59 | 62.5 | 2.46 | 1.29 (2+) |
| Al | fcc | 4.05 | 2.70 | 16.6 | 1.58 | 0.45 (3+) |
| Ga | complex | — | 5.91 | 19.61 | 1.67 | 0.62 (3+) |
| In | tetrag. | — | 7.29 | 26.11 | 1.84 | 0.92 (3+) |
| Si | diamond | 5.43 | 2.33 | 20.0 | 1.68 | 0.38 (4+) |
| Ge | diamond | 5.66 | 5.32 | 22.62 | 1.75 | 0.44 (4+) |
| Sn | tetrag. | — | 5.76 | 34.36 | 2.02 | 0.74 (4+) |
| Pb | fcc | 4.95 | 11.34 | 30.3 | 1.93 | 0.84 (4+) |
| Cu | fcc | 3.61 | 8.93 | 11.83 | 1.41 | 0.96 (+) |
| Ag | fcc | 4.09 | 10.50 | 17.09 | 1.60 | 1.13 (+) |
| Au | fcc | 4.08 | 19.28 | 16.95 | 1.59 | 1.37 (+) |
| Zn | hcp | — | 7.13 | 15.27 | 1.54 | 0.83 (2+) |
| Cd | hcp | — | 8.65 | 21.55 | 1.73 | 1.03 (2+) |
| Hg | rhomb. | — | 14.26 | 30.30 | 1.93 | 1.12 (2+) |
| Ti | hcp | — | 4.51 | 17.69 | 1.62 | 0.90 (2+) |
| V | bcc | 3.03 | 6.09 | 13.85 | 1.49 | 0.88 (2+) |
| Cr | bcc | 2.88 | 7.19 | 12.0 | 1.42 | 0.84 (2+) |
| Mn | complex | — | 7.47 | 12.23 | 1.43 | 0.80 (2+) |
| Fe | bcc | 2.87 | 7.87 | 11.77 | 1.41 | 0.76 (2+) |
| Co | hcp | — | 8.9 | 11.15 | 1.39 | 0.74 (2+) |
| Ni | fcc | 3.52 | 8.91 | 10.94 | 1.38 | 0.72 (2+) |

Ge and Si are semiconductors but become metals when in the liquid state.

1 nm = 10 Å

[a] Charge on cation in parentheses.

lattice. At about 260 K there arises a phase change to a simple cubic lattice structure. The carbon balls retain their former positions in a now contracted fcc arrangement, but they are no longer identical because the four $C_{60}$ molecules associated with the primitive cubic cell each have their own specific orientation about the different $\langle 111 \rangle$ directions. The situation is further complicated by the fact that the large near-spherical molecules hop between two distinct orientations with similar energies. However, at 86 K the molecules of $C_{60}$ stop rotating: they retain their fcc distribution but their orientations along the $\langle 111 \rangle$ directions become frozen in position.

$C_{60}$ is a semiconductor but may be alloyed with metallic elements, particularly the alkali metals, to become conducting and even superconducting at low temperatures. Its electronic structure is well understood and good agreement between calculated and observed infrared spectra has been obtained. The significance, however, is probably greatest for chemistry with prospects for new complex compounds based on the fullerenes. For further details consult David (1993), Huffman (1991), Prassides and Kroto (1992).

Atomic clusters may be collected and if large enough, $>1$ nm, may be observed in an electron microscope, in this way cylindrical versions of the fullerenes (nanotubes) have been discovered. The carbon atoms form single or concentric hexagonal sheets on a cylindrical surface. Several concentric cylinders may form a single nanotube with length of order $\mu$m and diameter 2–30 nm. The cylinders may have open or closed ends. The open tubes have been found to be excellent receptacles for metal atoms. It is thought that these developments will provide many new research areas and possibly commercial applications.

## References


Chow, M. Y., Cleland, A. and Cohen, M. L. (1984) *Solid State Commun.* **52**, 645.

David, W. I. F. (1993) *Europhys. News*, **24**, 71.

Echt, O., Sattler, K. and Rechnagel, E. (1981) *Phys. Rev. Lett.* **47**, 1121.

Eckardt, W. (1984) *Phys. Rev.* **B29**, 1558.

Faber, T. (1972) *Introduction to the Theory of Liquid Metals.* Cambridge University Press, Cambridge.

de Heer, W. A. (1993) *Rev. Mod. Phys.* **65**, 611.

Henry, N. F. M., Lipson, H. and Wooster, W. (1961) *The Interpretation of X-ray Diffraction Photographs*, Macmillan, Basingstoke.

Huffman, D. R. (1991) *Phys. Today* **44**, 11, 22.

Janot, C. (1992) *Quasicrystals, A Primer.* Oxford University Press, Oxford.

Janot, C. and Dubois, J. (1988) *J. Phys.* **F18**, 2303.

Knight, W. D., Clemenger, K., de Heer, W. A., Saunders, W. A., Chow, M. Y. and Cohen, M. L. (1984) *Phys. Rev. Lett.* **52**, 2141; **53**, 510.

Krane, K. S. (1988) *Introductory Nuclear Physics.* Wiley, New York.

Nelson, D. R. and Halperin, B. I. (1985) *Science*, **229**, 233.

North, D. M., Enderby, J. E. and Egelstaff, P. A. (1968) *Proc. Phys. Soc.* (Ser. 2) **1**, 1075.

Prassides, K. and Kroto, H. (1992) *Physics World*, **5**(4), 44.

## Further Reading

BAGGOT, J. (1994) *Perfect Symmetry*, Oxford University Press, Oxford.
ELLIOTT, S. R. (1983) *Physics of Amorphous Materials*, Longman, London.
LIMA-DE-FARIA, J. (Editor), (1990) *Historical Atlas of Crystallography*, Kluwer Academic Publishers, Kingston-upon-Thames.

## Problems

**3.1** In a Debye-Scherrer diffractogram, we obtain a measure of the Bragg angles $\theta$. In a particular experiment with Al powder, the following $\theta$ data were obtained when Cu $K\alpha$ radiation was used:

| | |
|---|---|
| 19.48° | 41.83° |
| 22.64° | 50.35° |
| 33.00° | 57.05° |
| 39.68° | 59.42° |

Aluminium has atomic weight 27 g and density 2.7 g cm$^{-3}$. Calculate Avogadro's number.

**3.2** An alkali halide is studied with the Debye-Scherrer technique and Cu $K\alpha$ radiation. The Bragg angles for the first five lines are 10.83°, 15.39°, 18.99°, 22.07° and 24.84°. Calculate

(a) the lattice parameter;
(b) the Miller indices for the planes producing the mentioned diffraction beams;
(c) the Miller index for the line(s) producing the largest allowable Bragg angle.

Then attempt to identify the alkali halide in question.

**3.3** Consider simple and close-packed hexagonal cells. Without resorting to the structure factor, determine in which structure(s) X-ray reflections arise from the planes (0001), (0002) and $(10\bar{1}0)$. Justify your answer.

**3.4** KCl and KBr are alkali halides with the NaCl structure. Reflections from the following planes are observed with X-ray diffraction:

| | | | | | | | | |
|---|---|---|---|---|---|---|---|---|
| KBr | (111), | (200), | (220), | (311), | (222), | (400), | (331), | (420) |
| KCl | | (200), | (220) | | (222), | (400) | | (420) |

Why is there this difference in two similar geometrical structures?

**3.5** In an examination of a certain element using the Debye-Scherrer technique the following diffraction angles were noted:

| | |
|---|---|
| 45.26° | 100.6° |
| 65.92° | 118.68° |
| 83.56° | 140.88° |

Cu $K\alpha$ radiation $\lambda = 1.540$ Å was used. Obtain two estimates of the atomic diameter of the element in question. (The two values are (a) that obtained assuming the atoms act as contacting hard spheres and (b) that obtained from the average volume per atom.)

**3.6** An alloy of 50% Au and 50% Zn crystallizes in the bcc structure. At high temperature the Au and Zn atoms are randomly distributed on the lattice sites, but after slow cooling an ordered structure wherein each atom is surrounded by atoms of the other sort arises. How will the Debye-Scherrer diffractograms of the two forms of this alloy differ?

**3.7** Describe Ewald's construction for X-ray diffraction. In a Debye-Scherrer experiment the following values of $\sin^2\theta$ were obtained with Fe $K\alpha$ radiation:

| | |
|---|---|
| 0.1843 | 0.6707 |
| 0.2450 | 0.6719 |
| 0.4887 | 0.7314 |
| | 0.7345 |
| | 0.9739 |
| | 0.9777 |

Fe $K\alpha_1$, 1.932 Å; Fe $K\alpha_2$, 1.936 Å. Determine the crystal structure and lattice parameter of the substance under study.

**3.8** A certain substance has cubic structure at room temperature. The (400) reflection arises at a diffraction angle 161.48° when Fe $K\alpha_1$ (1.932 Å) radiation is used. At 83 K this reflection is split in two, one line arising at 161.48° and the other, which is twice as strong as the first, at 163.38°. What is the structure at 83 K?

**3.9** Determine how the accuracy of the Debye-Scherrer method of lattice spacing measurement (here defined as $\Delta a/a$) depends upon the various parameters involved, e.g. wavelength, diffraction angle.

**3.10** In highly symmetrical structures, many families of planes are equivalent and possess the same interplanar spacing; these planes therefore contribute to the same diffraction line. Thus, in cubic systems, the planes (123), (321) and (213) are equivalent, but this is not the case in orthorhombic systems. The total number of equivalent planes for given $h$, $k$ and $l$ is called the multiplicity and together with the structure factor controls the intensities of diffraction lines. Remembering that $h$, $k$ and $l$ may take both positive and negative values, calculate the multiplicities of the following types of plane in both cubic and orthorhombic structures: $(hkl)$, $(hhl)$, $(hhh)$, $(0kl)$, $(0kk)$, $(00l)$.

**3.11** At 300 K aluminium has fcc structure with $a = 3.90$ Å; its coefficient of thermal expansion (linear) is 0.000025 $K^{-1}$. If Al is studied using the Debye-Scherrer technique with Cu $K\alpha_1$ radiation, what change in the Bragg angle for the (111) reflection will occur when the temperature is increased from 300 K to 600 K? How does this change in Bragg angle vary with the Bragg angle itself and which reflection would give the best accuracy if one were to measure the expansion coefficient in this manner?

**3.12** At room temperature a substance has fcc structure, but below −40°C it becomes tetragonal, the more so the lower the temperature: there is no other change in the distribution of the atoms on the lattice sites. Describe how each diffraction line up to $h^2 + k^2 + l^2 = 20$ changes when the structure becomes tetragonal. Assume $a = b > c$.

**3.13** Normally we use the pair distribution function when describing the structure of a liquid, but the concept is also applicable to a crystal. Illustrate the pair distribution function for the bcc lattice.

**3.14** Illustrate qualitatively the pair correlation function for a crystalline, a liquid and a gaseous substance.

**3.15** Consider the bcc structure with every lattice point occupied by the same kind of atom. How will the allowed Bragg reflections, $(hkl)$, be affected as the coordinates of the body-centred atom change from

$$\frac{1}{2}, \frac{1}{2}, \frac{1}{2} \quad \text{to} \quad \frac{1}{2n}, \frac{1}{2n}, \frac{1}{2n}$$

when $n$ is a positive integer? Calculate for $n = 2$ in detail.

**3.16** What is the significance of a complex structure factor?

**3.17** Which transitions give rise to the $K\alpha_1$, $K\alpha_2$, $K\beta_1$, $K\beta_2$ radiations from an element? What are the photon energies for these radiations in Cu, Ag and Au? In the case of Cu $K\alpha$

radiation, what is the weighted mean wavelength? Why are the component lines of different intensity?

**3.18** Consider the two-dimensional lattice of Problem 2.12. If each lattice point is occupied by the same kind of atom and if the lattice is studied by X-ray diffraction, what are the allowed reflections (*hk*) and (*HK*)? In general when a lattice is described by a primitive cell what are the allowed reflections (*hkl*)?

**3.19** Si and GaAs have the same crystal structure built around the fcc lattice with bases

Si (000), $(\frac{1}{4}\frac{1}{4}\frac{1}{4})$

GaAs Ga (000), As $(\frac{1}{4}\frac{1}{4}\frac{1}{4})$

Determine the Miller indices of the allowed Debye-Scherrer reflections for these two substances.

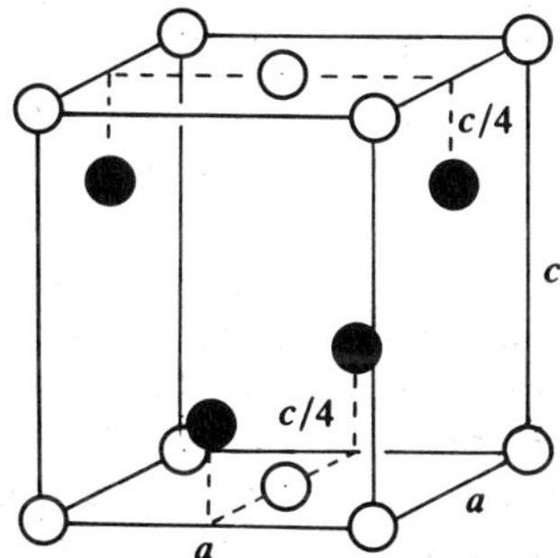


**3.20** A certain compound containing the elements A (○) and B (●) has a crystal structure with a unit cell as shown in the above diagram. Determine the following:

(a) The chemical composition of the compound.
(b) The underlying Bravais lattice and the structure basis.
(c) How the Debye-Scherrer X-ray diffraction pattern may be characterized in terms of the allowed values for *h*, *k* and *l*.

# 4

# Defects in Crystals

## 4.1 Mechanical Properties

The usefulness of a material for constructional purposes lies not in its strength but in a combination of strength and ductility. When loaded beyond a certain limit, called the yield stress, the material should not break, as does glass for example, but should yield plastically, becoming permanently deformed without rupturing. We say that glass is brittle whereas copper and aluminum are ductile. All pure metals have a greater or lesser degree of ductility, and the metallurgist endeavours to design alloys with greater and greater strength while maintaining an acceptable measure of ductility. We characterize the mechanical behaviour of a material with the aid of a stress-strain diagram (Fig. 4.1). Ideally this diagram shows the stress required to maintain a given rate of strain as a function of strain. In practice, the ability of the testing machine to respond to the behaviour of the sample can cause details of stress–strain curves to appear differently on different machines, but this will not concern us. The important variables are strain rate, temperature and composition of the specimen. We shall consider only the simplest principal features of mechanical behaviour as exemplified by a pure metal such as aluminium at room temperature under moderate rate of strain, say $10^{-3}\ s^{-1}$. Assuming our specimen to be polycrystalline with grain size about $10^{-2}$ mm, we should expect to find a curve similar to that of Fig. 4.1, which we can divide into three parts. There is an initial reversible elastic region where the stress and strain are linearly related. This enables us to define an elastic constant, Young's modulus $E$:

$$E = \frac{\text{stress}}{\text{strain}} = \frac{\text{force per unit area}}{\text{extension per unit length}}.$$

Young's modulus for the commoner metals lies within the range $2 \times 10^9$–$5 \times 10^{10}$ Pa. However, at a certain stress $\sigma_Y$, called the yield stress, behaviour becomes irreversible, and the material begins to be permanently deformed. The material yields and flows. Continued deformation demands ever increasing stresses to maintain the chosen deformation rate; this is called work hardening. Eventually the stress attains a maximum value, the fracture strength $\sigma_B$; thereafter the specimen deforms mark-

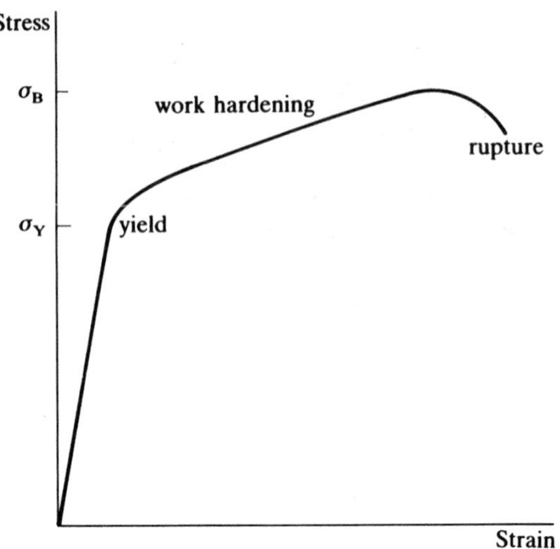


**Figure 4.1** A typical stress-strain diagram for a polycrystalline metal.

edly in a localized region under a rapidly decreasing stress. It then breaks into two. This pattern of behaviour is typical for polycrystalline materials although some, notably carbon steels, may exhibit special properties in the neighbourhood of the yield point. The important quantities are $E$, $\sigma_Y$, $\sigma_B$† and the total extension to rupture expressed as a percentage of the gauge length (i.e. length of specimen uniformly loaded).

Suppose now that we make similar tests on single crystals of aluminium; each specimen might well be of cylindrical shape with diameter of about 0.5 cm or a plate say 2 cm wide and 0.2 cm thick shaped as in Fig. 4.2. We should find the behaviour to be dependent on the orientation of the crystal relative to the stress axis. A typical result might well look like the curve in Fig. 4.3. In contrast with the polycrystalline case, yield occurs at a much lower stress and is followed by a region of easy glide where deformation proceeds at an almost constant stress little different from $\sigma_Y$. Eventually work hardening commences and the stress to maintain the deformation rate increases, finally leading to a rupture at a stress $\sigma_B$ similar to that for a polycrystalline specimen. The total extension may be much greater for the single crystal than for the polycrystalline case. If the crystal had been carefully electropolished to obtain a smooth surface then we would have seen that deformation produces a characteristic pattern of lines on the sample that are called slip bands, glide bands or deformation bands (Fig. 4.2); they provide us with an indication of how deformation proceeds. Detailed study of many single crystals clearly shows that deformation proceeds by shear, blocks of crystal apparently sliding over one another (Fig. 4.4). Careful experiment shows that one can associate the yield stress and the start of easy glide with the appearance of the first band, which can be used to define the plane on which slip arises and the direction of slip within the slip plane as well as their relation to the crystal and stress axes. In this manner Schmid found that for a given metal there is a preferred slip plane and slip direction. Furthermore, a definite critical shear stress $\tau_c$, resolved on the slip plane and in the slip direction, is needed

† In SI $\sigma_Y$ and $\sigma_B$ are replaced by $R_e$ (*résistance d'écoulement*) and $R_m$ (*résistance maximale*) respectively.

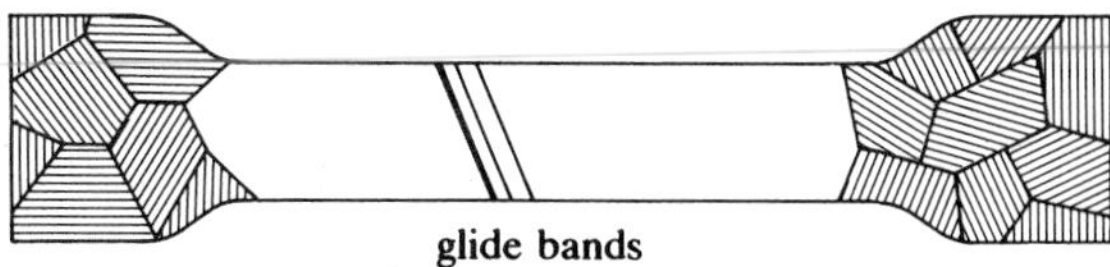


**Figure 4.2** A schematic picture of a sample, obtained as a single crystal over the gauge length, illustrating the occurrence of glide bands immediately after the onset of plastic deformation.

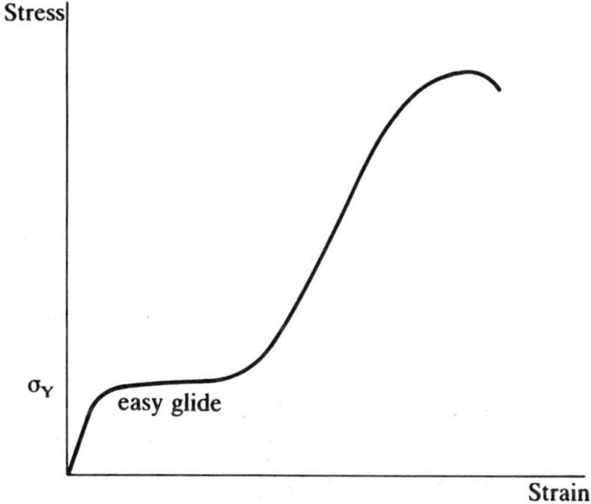


**Figure 4.3** The stress-strain diagram for a single crystal is very dependent on sample orientation. Here we show schematically what happens when a slip-plane system becomes active for small applied stresses, causing 'easy glide'. Eventually work hardening leads to rupture.

to promote glide. This critical resolved shear stress is independent of the stresses in the direction perpendicular to the slip plane. Using the notation of Fig. 4.5, we express Schmid's critical shear stress law as

$$\tau_c = \sigma_Y \cos \alpha \cos \beta. \tag{4.1}$$

Experiment has shown that *slip occurs most readily on close-packed planes and in directions of closest packing*; this has made Zn, Mg and similar metals favourite

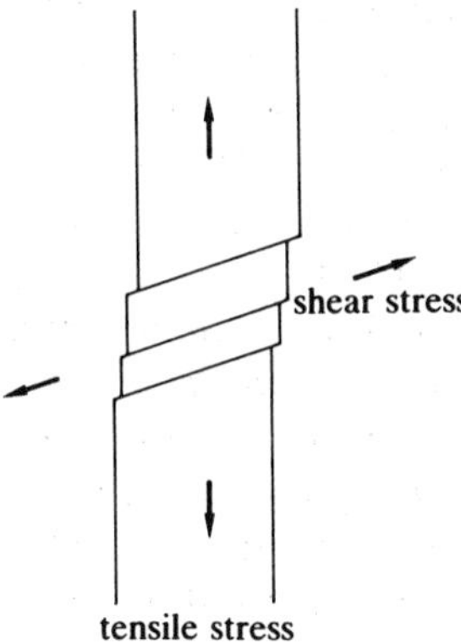


**Figure 4.4** Microscopic examination of single crystals shows that deformation proceeds by shear, blocks of crystal apparently sliding over one another.

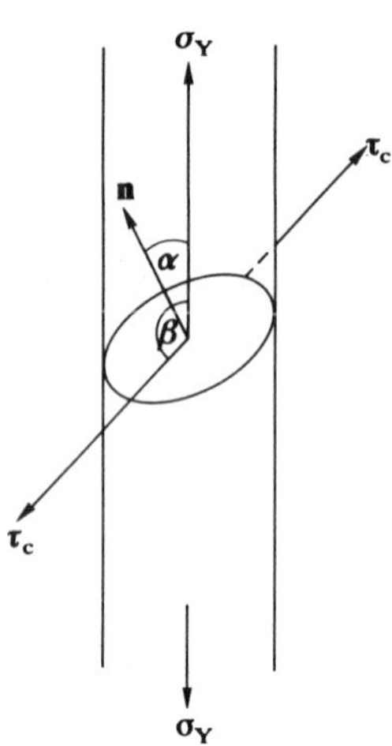


**Figure 4.5** The initial deformation arises when a critical shear stress $\tau_c$ is reached; $\tau_c$, $\sigma_Y$ and **n** are not usually in the same plane.

subjects for study. We recall that the close-packed planes are those with the largest interplanar spacing. When more than one family of slip planes exist, they may all operate at high temperatures or at high stresses, but normally one family of planes, the most suitably oriented, operates, and when this is exhausted new slip planes come into play. The observed critical shear stresses are small, of order $10^{-4}G$, where $G$ is the shear modulus.† Now if slip takes place by the rigid movement of one piece of crystal over another, we expect the critical shear stress to be of order $10^{-1}G$ (Fig. 4.6). The observed critical stresses are therefore some $10^3$ times smaller than theoretical expectation, and this must mean that rigid slip cannot occur.

## 4.2 Dislocations

Crystals are very soft because they contain line defects called dislocations. The concept was first introduced into crystal physics by Taylor (and independently by Orowan and by Polanyi) in 1934 in an attempt to explain the low values of critical shear stress; it was generalized by Burgers in 1938. The first direct experimental evidence for their existence came in 1948, when growth patterns on the surfaces of natural beryl crystals were found to be identical with those predicted by dislocation theory. Many techniques are now available for their study, the most important of which is, without doubt, transmission electron microscopy.

The simplest forms of dislocation are the 'edge' and the 'screw'. Formally we may think of them as being formed from blocks of perfect crystal, which are imagined cut, then deformed and thereafter rejoined (Fig. 4.7). In real crystals a dislocation causes a displacement equal to an interatomic separation as is clearly seen for both edge and screw types. The edge dislocation is a directed line defined by the extra 'half plane' inserted into the otherwise perfect crystal; Fig. 4.7*d* shows a positive

† In practice $G \approx E/2.5$; for an isotropic elastic solid $G = E/2(1 + \nu)$, $\nu$ being Poisson's ratio, the latter for a cylindrical rod is the fractional change in radius divided by the fractional change in length, $0 \leqslant \nu \leqslant \frac{1}{2}$.

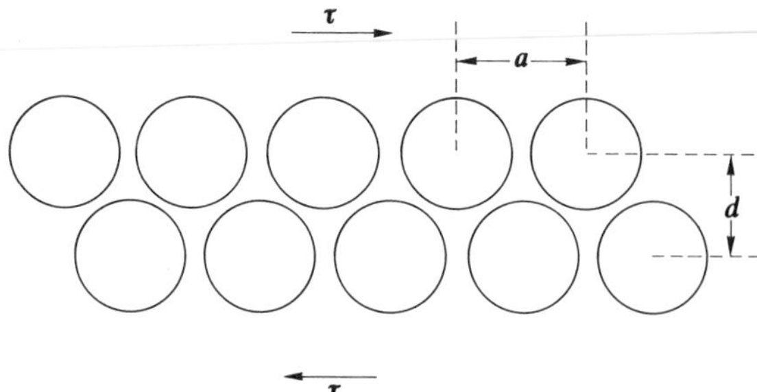


Two blocks of crystal move rigidly over one another. At the interface the upper plane of atoms must climb over those in the lower plane. The required shear stress will initially increase, go through a maximum and then become zero just before the upper plane falls into the next stable position one atomic spacing from the starting position. We assume that the required shear stress is

$$\tau = \tau_{\max} \sin \frac{2\pi x}{a},$$

and for small $x$

$$\tau \approx \tau_{\max} \frac{2\pi x}{a} \quad \text{and} \quad \tau = \frac{Gx}{d},$$

so

$$\tau_{\max} \approx G/2\pi, \qquad (d \approx a)$$

In rigid glide $\tau_c = \tau_{\max}$, so $\tau_c \approx G/10$.

**Figure 4.6** A simple calculation shows that shearing a crystal by rigid displacement of one portion past another would require $\tau_c \approx G/10$; a condition very different from that experimentally observed.

edge dislocation directed into the plane of the figure. The screw dislocation has right or left handed character.

We see that the dislocation line may move merely by a rearrangement of the atoms in its immediate vicinity; if the line runs out of the crystal then we obtain a step on the surface and effectively we have moved the upper portion of the crystal relative to the lower portion. An element of slip has occurred without our needing to move blocks of crystal rigidly. Thus, if dislocations exist in crystals, we might expect them to move under small shear stresses and their movement to produce slip. This implies that a crystal without any dislocations would be difficult to deform and ought to have the ideally large critical shear stress of order $10^{-1}G$. Such crystal can in fact be grown, often arising as very small objects called 'whiskers', and approximate the ideal behaviour. We can also appreciate why polycrystalline substances have greater yield strengths than single crystals. The grain boundaries prevent the dislocations from 'running out' of the grains and thereby prevent slip at low stresses. In fact, the yield strength of polycrystalline materials increases as the grain size decreases, being proportional to $d^{-1/2}$, where $d$ is the average grain 'diameter'.

How do dislocations arise in crystals? We do not know in detail, but generally speaking they are formed by the freezing in of lattice disorder when a metal solidifies. They are not thermodynamically stable, their configurational entropy being

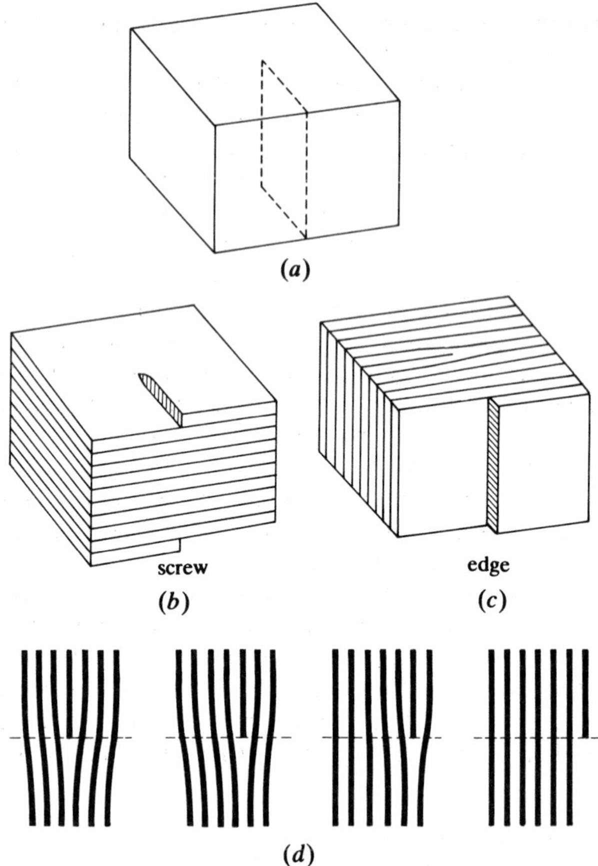


**Figure 4.7** A dislocation is a line defect which may be imagined as introduced into the perfect crystal by: (1) cutting the crystal along the slip plane (*a*); (2) partially displacing one portion with respect to the other by one interatomic spacing; (3) allowing the cut to heal. Depending on the direction of shear, either a screw (*b*) or an edge (*c*) dislocation is formed. The dislocation is clearly the boundary line separating the displaced (slipped) and undisturbed (unslipped) regions as is readily seen when we view the edge dislocation end-on in its slip plane, a sign and direction are attributed to the dislocation according to whether the half plane is inserted from above (+) or below (−) the slip plane, (*d*).

small, but once formed they are difficult to eliminate, although prolonged annealing can reduce their numbers. We shall see that they can be produced during deformation. Even a good well-annealed crystal contains of order $10^4$ dislocations $cm^{-2}$ whereas a heavily worked metal may have up to $10^{11}$ $cm^{-2}$. (These numbers are the average number of dislocation lines intersecting unit area within the crystal.)

Edge and screw dislocations are pure forms of dislocation and it is possible to have a continuous transition from one to the other, although this is not readily visualized on the atomic level (see Read 1953, p. 18). A dislocation line may be defined as the boundary between two regions of crystal, one of which has experienced slip while the other has not. Since the dislocation creates slip on an atomic scale its movement corresponds to a transfer of matter; this requires that the dislocation line end on the surface of the crystal, form a closed loop or be part of a network of intersecting dislocations. If this were not the case then excess atoms or holes would arise in the crystal.

The strength of a dislocation is defined by a vector, the Burgers vector **b**, which governs the direction and amount of slip that the dislocation produces. A description of the Burgers vector may be obtained in the following manner. In the real crystal, where the atoms are assumed normally to occupy the sites of a Bravais lattice, we may trace a closed Burgers circuit by starting at a particular atom site and moving in a series of interatomic steps finally to return to the starting point. It is not necessary to keep to a particular plane, but this simplifies the illustration. In Fig. 4.8*a* the Burgers circuit encloses an edge dislocation directed into the plane of the paper and with the extra half plane inserted from above the slip plane. The Burgers circuit is, by convention, traced in a right handed fashion relative to the sense of the dislocation line. There are however two choices regarding the direction of the dislocation line; one leads to a Burgers vector directed to the left, the other to a Burgers vector directed to the right. This is a direct result of the mirror symmetry about the extra half plane. The two choices are equally valid. Here a positive edge dislocation is defined as one with the half plane inserted from above and the line directed into the plane of the drawing.

We now assume the existence of a similar but ideal crystal free from all defects and retrace the same sequence of atomic steps that gave rise to the circuit in the real crystal. Now the circuit does not close: the end point F is not identical with the starting point S. The closure vector FS defines the Burgers vector for the dislocation, Fig. 4.8*b*. It is directed to the left. A negative edge dislocation has the extra half plane inserted from below the slip plane and is directed out of the plane of the

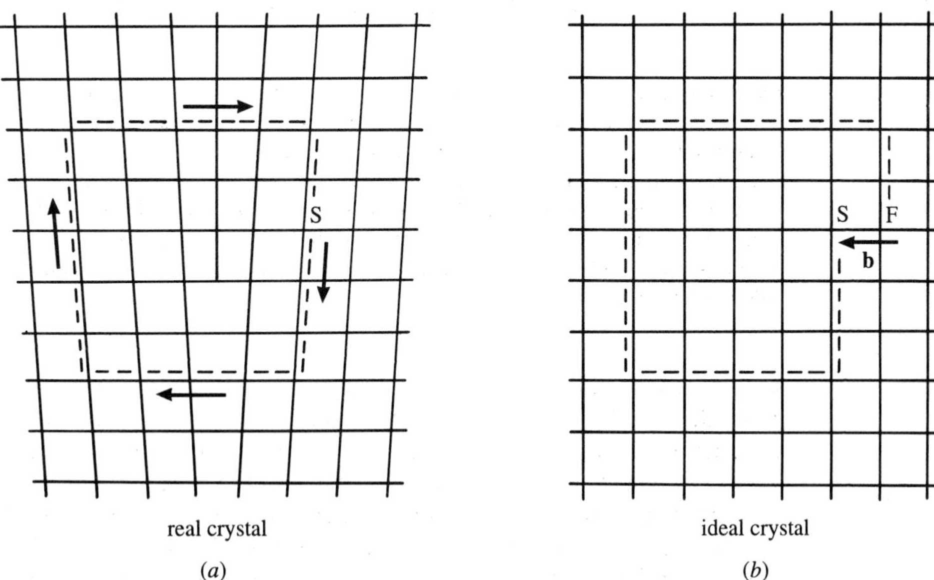


**Figure 4.8** To determine the Burgers vector **b** of a dislocation a closed circuit surrounding the dislocation is traced in the real crystal (*a*). This circuit is then mapped into an ideal crystal (*b*); the circuit does not close and the closure vector FS defines **b**. The dislocation is directed into the plane of the diagram and the circuit traced in the clockwise direction. This defines a positive edge dislocation.

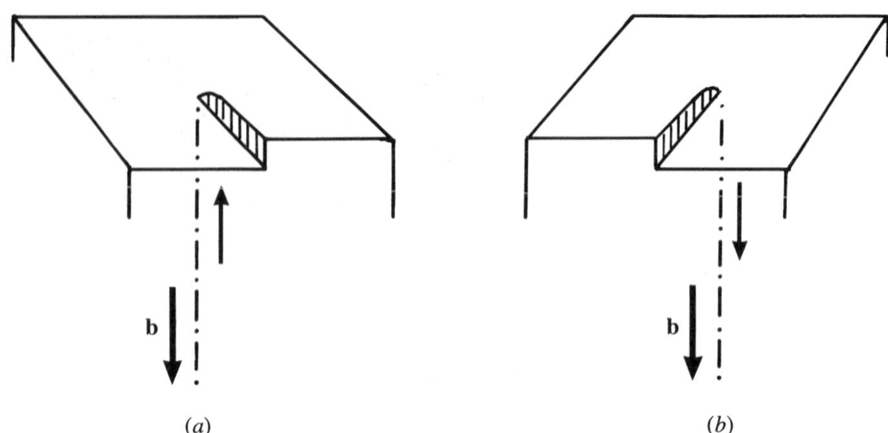


**Figure 4.9** The sense and Burgers vector of left handed (*a*) and right handed (*b*) screw dislocations.

paper. You may yourself verify that its Burgers vector is the same with regard to both strength and direction as for the positive edge dislocation. In similar fashion Burgers circuits may be drawn for screw dislocations to give the descriptions of Fig. 4.9.

The electron microscope has provided ample evidence for the existence of the various forms of dislocation, straight line segments, loops (that are not necessarily confined to one plane) and three dimensional networks, Fig. 4.10. On a given dislocation line or loop we must always find the same Burgers vector and in a network there must be continuity of the Burgers vector, so at a node (a point where two or more dislocation lines meet)

$$\sum \mathbf{b} = 0$$

The fact that dislocations (edge or screw) of opposite sign have the same Burgers vector is an important feature of a closed dislocation loop. A dislocation loop defines a boundary within which slip has occurred. The element of slip is the same throughout the whole area contained within the loop. The dislocation line has the same sense along its length and this demands that the signs of opposite portions of the loop be different, Fig. 4.11. Such a situation is also necessary to ensure that the dislocations annihilate one another when the loop shrinks to a point and the slip disappears.

### 4.2.1 Dislocation energy, jogs, sources

A dislocation possesses an energy, which resides in the elastic strain field surrounding it. This energy, which is readily calculated for the screw dislocation (Fig. 4.12), is proportional to $b^2$. This means that a dislocation always has the smallest possible Burgers vector, corresponding to the interatomic spacing in the direction of slip. A dislocation with larger strength would spontaneously dissociate into separate dislocations with the smallest Burgers vector; these dislocations would then separate owing to the repulsion of their stress fields. It is clear from Fig. 4.13 that dislocations with similar sign repel one another whereas those with opposite sign may

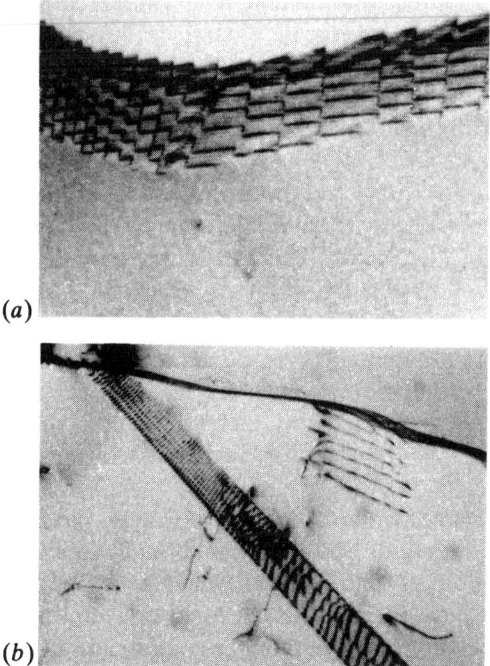


**Figure 4.10** Dislocations in crystals may arise as straight lines ending on the crystal surface or as complicated networks. (*a*) shows a network of dislocations forming a low-angle grain boundary in a ferritic steel; the sample is a very thin foil studied in an electron microscope. (*b*) is an electron microscope picture of a similar foil. Dislocations pass right through the thickness of the foil and appear as parallel lines in the photograph. They originate from the same source and move in the same slip plane; they are hindered in their motion by the large-angle grain boundary, which causes a dislocation 'pile up'. (Photographs by courtesy of G. Dunlop.)

attract and even annihilate one another. Alternatively, if they happen to lie on separate slip planes, they can give rise to a line of atomic holes, which are called lattice vacancies, or, if they overlap, to a line of extra interstitial atoms. These lattice holes or extra atoms not occupying lattice sites are called point defects. In certain

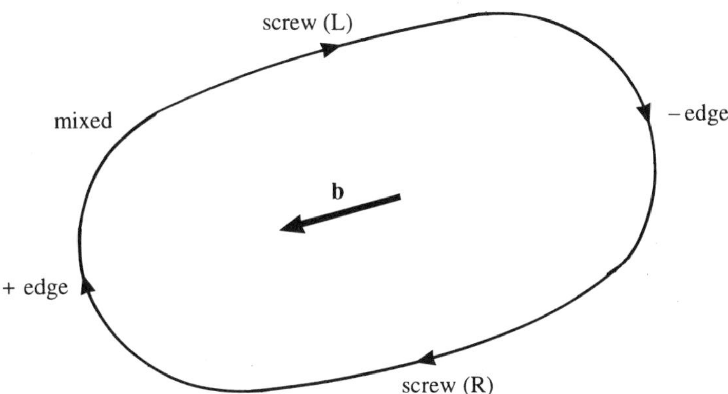


**Figure 4.11** A planar dislocation loop with clockwise sense illustrating the changing character and sign of the dislocation.

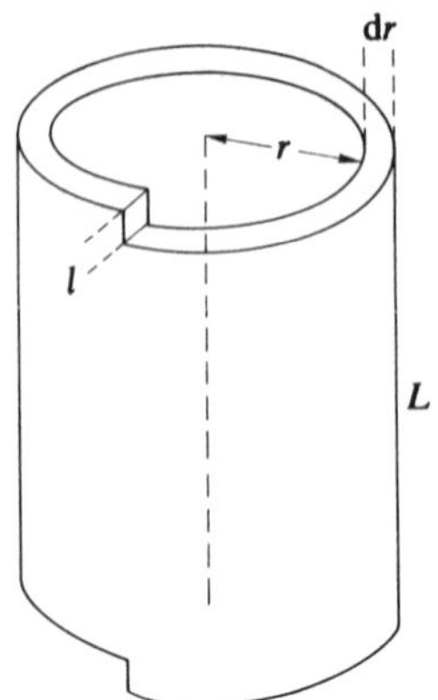


Consider a cylindrical element surrounding a screw dislocation, a displacement $l$ produces a shear strain $e = l/2\pi r$ and a shear stress $\tau = Ge$. An incremental displacement $\mathrm{d}l$ demands work $\mathrm{d}W = \tau L\, \mathrm{d}r\, \mathrm{d}l$, whence

$$\frac{W}{L} = \int_{r_0}^{R} \int_{0}^{b} \frac{G}{2\pi r}\, l\, \mathrm{d}r\, \mathrm{d}l = \frac{Gb^2}{4\pi} \ln \frac{R}{r_0}$$

is the work per unit length necessary to produce the dislocation. It is the elastic strain energy of the dislocation. $r_0$ is the radius of the dislocation core and $R$ an upper limit to the extent of the strain field determined by the dislocation density in the material.

**Figure 4.12** Calculation of the elastic strain energy associated with unit length of a screw dislocation (assuming insignificant core energy).

situations, the Burgers vector may be a fraction ($\frac{1}{2}$, $\frac{1}{4}$ or $\frac{1}{6}$) of an interatomic spacing – the dislocation is then called a partial dislocation; we shall describe a specific example later.

A dislocation line is not necessarily straight but may contain steps called jogs (Fig. 4.14). These steps are usually one or two atoms high and may be formed either by the diffusion of lattice vacancies to the dislocation, or when dislocations cross one another (Fig. 4.15). Usually an edge dislocation can only leave its primary slip plane by diffusion processes which become more probable the higher the temperature; this is called dislocation climb. When a dislocation contains many jogs,

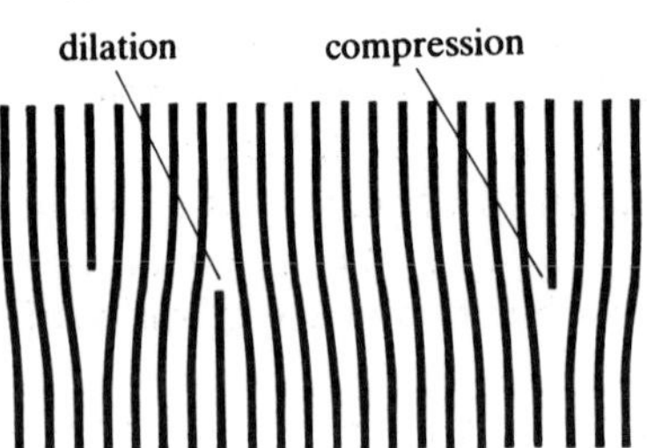


**Figure 4.13** In keeping with the definition of the Burgers vector, dislocations may have different signs, the extra half plane being above or below the slip plane as shown here. Each dislocation has regions of compression and dilation. Dislocations of different sign may wholly or partly annihilate one another, depending on whether or not they lie in the same slip plane.

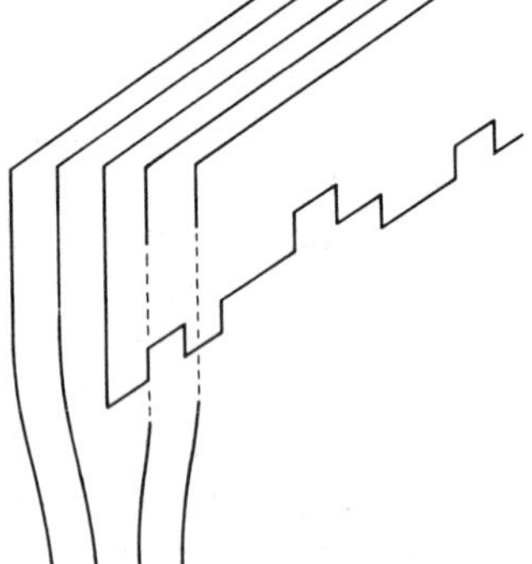

**Figure 4.14** A dislocation line is not necessarily straight on the atomic scale, but may present a 'ragged' appearance through the formation of 'jogs', which are steps of atomic size in the dislocation line.

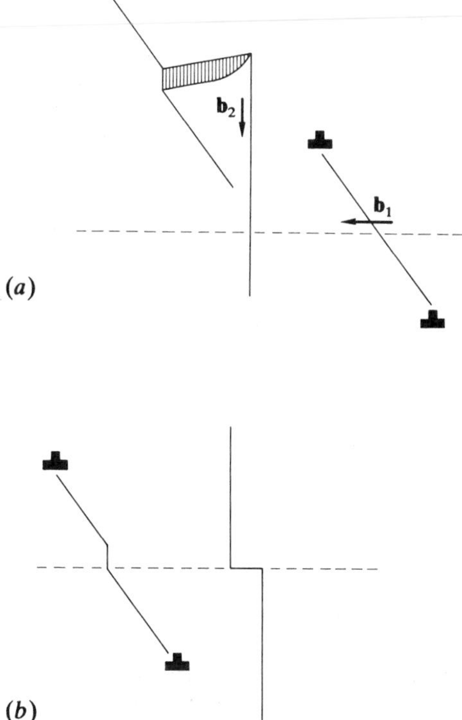


**Figure 4.15** Jogs are formed by the passage of dislocations through one another; (a) shows the separate dislocations before they cross and (b) the situation after crossing; jogs are introduced into both dislocations.

the vertical portions of the dislocation may not lie in close-packed slip planes and they will not glide so easily as the parent dislocation; the jogs therefore hinder the motion of the dislocation. This is particularly evident for a jog on a screw dislocation (Fig. 4.16); clearly such a jog has edge character (check the orientation of the Burgers vector, which is continuous throughout the complete dislocation) and its motion is incompatible with that of the screw dislocation. If by brute force we compel the arrangement to move then this produces either a row of extra atoms or one of lattice vacancies (Fig. 4.17). The jog in the screw dislocation acts as an anchor for the dislocation. It is also possible to form special dislocations that cannot move; these are called 'sessile.'

The important point is that although straight dislocations may move easily, they can also be anchored at certain positions either by jogs or by connection to sessile dislocations. This is an important aspect because it leads to one mechanism for the repeated production of dislocations. When slip occurs, one might think that the dislocations run out of the crystal, which rapidly becomes empty; it should therefore harden very quickly and approach ideal strength. Closer inspection of slip bands, however, shows that very large amounts of slip may arise on a given slip plane, and we are led to the conclusion that dislocations must be manufactured on these slip planes (Fig. 4.18).

One source of dislocation reproduction was proposed by Frank and Read, and

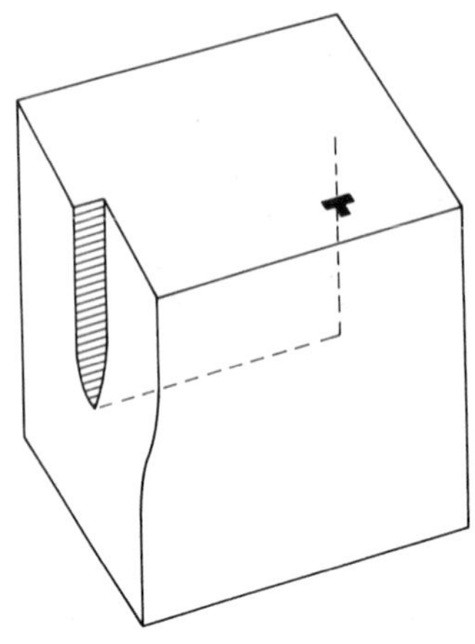

Figure 4.16 A dislocation cannot end in a crystal. Here a screw dislocation ends but creates an element of edge dislocation.

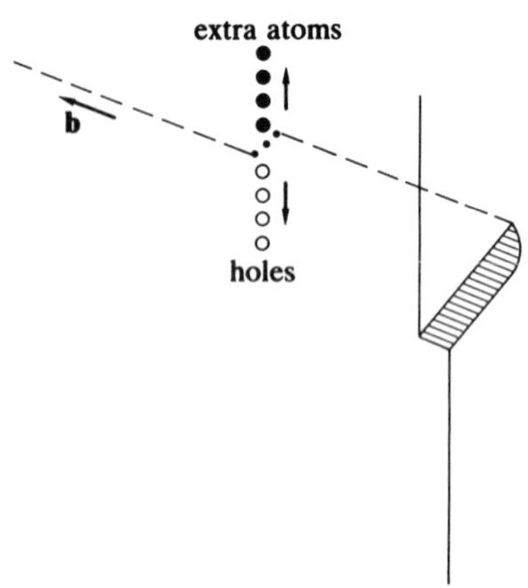


Figure 4.17 Matter must be conserved when dislocations move. Here, depending on the direction of movement, a jog creates lattice vacancies or interstitial atoms.

has been amply substantiated by experiment. We assume that a piece of straight edge dislocation is pinned at its ends by screw dislocations that cannot move (let us say that they contain many jogs). The sequence of lines in Fig. 4.19 shows how the line bows out under applied stress. Because it is fixed at the ends, it must eventually bow backwards; finally the portions meet to form a closed loop, but at the same time regenerate the original element. Under continued stress, this process can be repeated many times, producing a series of dislocation loops and multiple slip on a given plane.

### 4.2.2 Dislocation reactions, stacking faults

We have already mentioned that dislocations may build networks and that there must be conservation of the Burgers vector when dislocations meet at a node. This

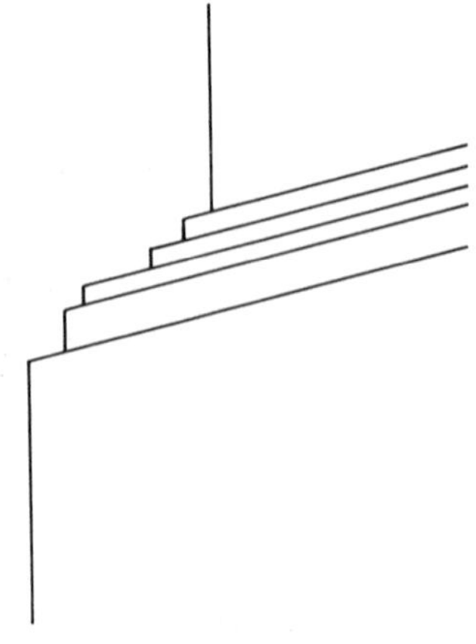

Figure 4.18 Electron microscope studies have revealed that multiple slip arises on the same slip plane showing that dislocations must be created on the slip plane (e.g. Fig. 4.10*b*).

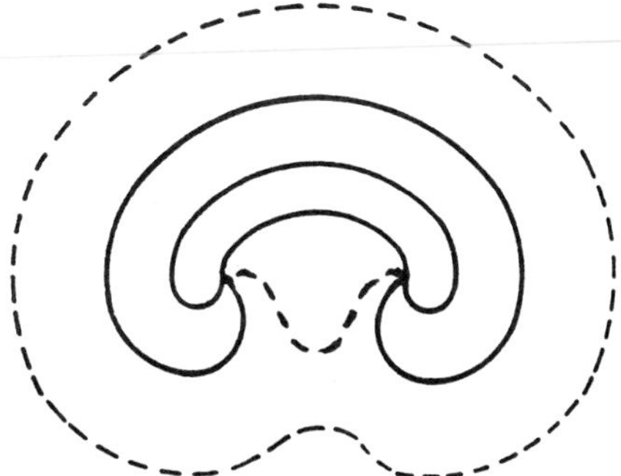

**Figure 4.19** The Frank-Read source: a straight portion of edge dislocation is anchored at its ends by screw dislocations of opposite sign. Under stress the edge dislocation successively bows out as shown; eventually it meets up with itself and generates a closed loop, but at the same time the original dislocation is reformed (broken line) which allows a continued production of loops.

means that a dislocation may dissociate into two different dislocations, so we can write

$$\mathbf{b} = \mathbf{b}_1 + \mathbf{b}_2 . \tag{4.2}$$

We call this a dislocation reaction, and a particularly important example arises in the fcc structure. In this structure, the slip planes are $\{111\}$ and the slip directions $\langle 110 \rangle$. Consider the (111) plane as drawn in Fig. 4.20. If we imagine such a plane to be built up of contacting spheres, we recognize one Burgers vector as $\frac{1}{2}a[10\bar{1}]$ (Fig. 4.20). However, the special hill-and-dale structure of the $\{111\}$ planes makes it very

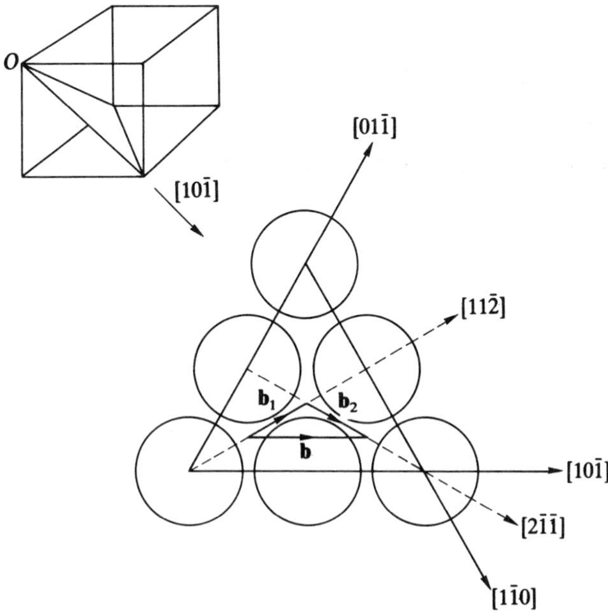


**Figure 4.20** Zig-zag slip within the $\{111\}$ plane of an fcc crystal. The Burgers vector **b** of the perfect dislocation may dissociate into the partial zig-zag dislocations with Burgers vectors $\mathbf{b}_1$ and $\mathbf{b}_2$.

unlikely that slip occurs by a single displacement of dislocation over this distance because atoms above the slip plane have to climb over those that are in this plane. It is more favourable for the displacement to follow the zig-zag path along the valleys between atoms in the slip plane. We describe this situation with the aid of Fig. 4.20, where we have labelled the relevant directions and vectors

$$\mathbf{b} = \tfrac{1}{2}a[10\bar{1}], \tag{4.3}$$

$$\mathbf{b}_1 = \tfrac{1}{3}(a[10\bar{1}] - \tfrac{1}{2}a[1\bar{1}0]) = \tfrac{1}{6}a[11\bar{2}]. \tag{4.4}$$

Similarly,

$$\mathbf{b}_2 = \tfrac{1}{6}a[2\bar{1}\bar{1}]. \tag{4.5}$$

Clearly $\mathbf{b} = \mathbf{b}_1 + \mathbf{b}_2$ since

$$\tfrac{1}{2}a[10\bar{1}] = \tfrac{1}{6}a[11\bar{2}] + \tfrac{1}{6}a[2\bar{1}\bar{1}].$$

$\mathbf{b}$ is the perfect dislocation, which may dissociate into the partial dislocations $\mathbf{b}_1$ and $\mathbf{b}_2$.

We decide whether this dissociation occurs by using the energy criterion. The energy is proportional to $b^2$, and

$$b^2 = \tfrac{1}{2}a^2 \quad \text{whereas} \quad b_1^2 + b_2^2 = \tfrac{1}{3}a^2,$$

so it appears energetically favourable for the dislocation to split into partial dislocations; we might then expect these to separate under mutual repulsion, leading to the situation described in Fig. 4.21. If the partial dislocations separate then they contain within their boundaries a region of partially slipped crystal that is called a stacking fault for the following reason. We know that the fcc structure is built by stacking close-packed planes in the sequence . . . *ABCABCABC* . . . (see Fig. 2.9).

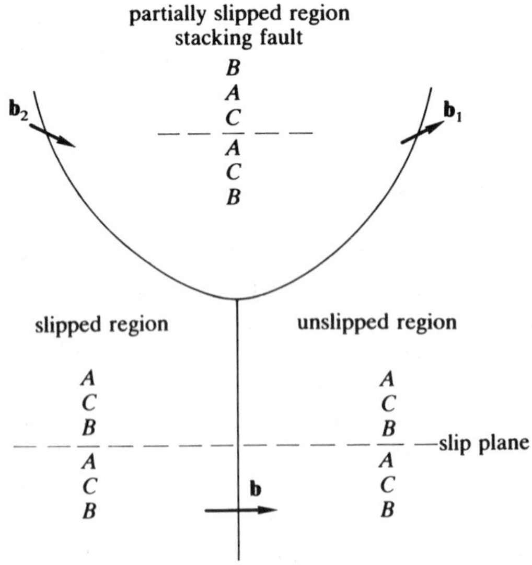


**Figure 4.21** Dissociation into partial dislocations introduces a sheet of stacking fault. The perfect and partial dislocations are shown together with the appropriate stacking sequences.

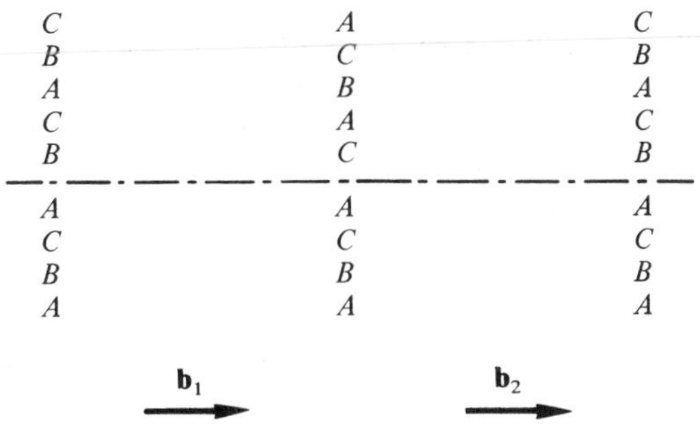

| | | |
|---|---|---|
| *C* | *A* | *C* |
| *B* | *C* | *B* |
| *A* | *B* | *A* |
| *C* | *A* | *C* |
| *B* | *C* | *B* |
| — · — · — | — · — · — | — · — · — |
| *A* | *A* | *A* |
| *C* | *C* | *C* |
| *B* | *B* | *B* |
| *A* | *A* | *A* |

$\mathbf{b}_1$ ⟶ $\mathbf{b}_2$ ⟶

**Figure 4.22** The sequence of planes in the perfect crystal and the effect of partial slip producing the stacking fault.

Suppose now that our crystal has the slip plane in an *A* layer; the passage of the partial dislocation $\mathbf{b}_1$ causes all the material above the slip plane to move one partial step in the sequence so that a *B* plane becomes a *C* plane, a *C* plane an *A* plane and so on. In the partially slipped region our sequence becomes as in Fig. 4.22. In fact where only the partial dislocation $\mathbf{b}_1$ has passed we have a small element of hcp structure sandwiched in the fcc sequence: this is known as a stacking fault. Now when the second partial dislocation $\mathbf{b}_2$ comes along it causes a further step in the material above the slip plane, so that a *C* plane becomes a *B* plane, a *B* plane an *A* plane etc., and this clearly restores the stacking to the correct sequence by removing the fault. Since the stacking fault disturbs the lattice periodicity, it has a definite surface energy and this determines how far the partial dislocations separate. If the stacking fault energy is large enough, it may prevent dissociation into partial dislocations altogether. In this case we should rewrite the energy criterion as

$$b^2 < b_1^2 + b_2^2 + \Delta,$$

where $\Delta$ represents a contribution from the stacking fault energy.

Stacking faults are important in that they hinder cross-slip, i.e. the possibility for a screw dislocation to move out of its plane and into another parallel plane; this might be necessary if there were some obstacle in the path of the dislocation.

### 4.2.3 *Work hardening*

Single crystals deform easily because they contain dislocations that can move under small shear stresses. It is clear from Fig. 4.7 that a dislocation in equilibrium has a symmetrical structure and that its movement requires that work be done to surmount the barrier that each row of atoms in its path presents. This inherent resistance to dislocation motion is called lattice friction or the Peierls-Nabarro force. Another aspect is that so far we have explicitly assumed that a dislocation is a localized line defect; since the extra half plane of atoms is accommodated within two or three atom planes, we call this a narrow dislocation. However, we know that in metals the bonding is essentially isotropic and there is no reason why the distortion produced by the extra half plane should not be relaxed over a larger number of neighbouring planes: this produces a wide dislocation (Fig. 4.23). We expect wide dislocations to move more easily than narrow ones and to be less dependent on the

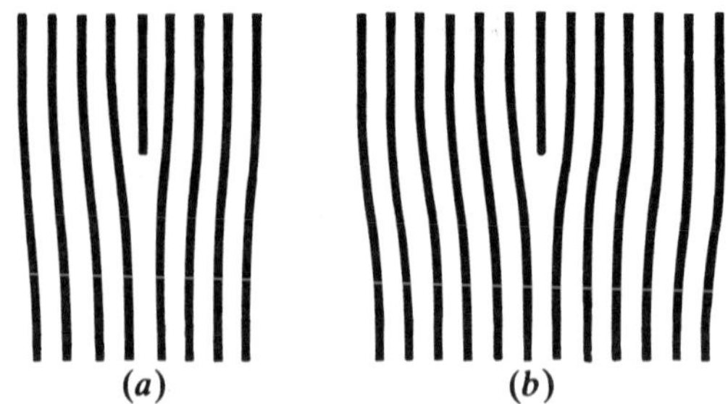


**Figure 4.23** We expect dislocations to be narrow (*a*) in ionic and covalently bonded insulators and wide (*b*) in the more isotropic metals.

temperature. As easy glide proceeds, the dislocations cross one another, leading to jogs, which slow down their motion. In addition, dislocation reactions may produce immobile dislocations causing additional anchoring. Metals with low stacking-fault energy such as Cu can cross clip only with difficulty, so deformation gradually becomes more difficult. As new sources of dislocations become operative (e.g. via the Frank-Read mechanism or increased internal stresses), the dislocation density increases, leading to an accentuated interaction; the stress fields of the dislocations overlap and the motion of individual dislocations becomes more and more difficult, leading to the rapid increase in stress necessary to maintain the rate of strain. The interaction of dislocations may cause microholes (Fig. 4.24); these acts as stress concentrators and eventually lead to rupture of the specimen.

The mechanical properties of alloys lie beyond our present scope, but it is readily appreciated that the dilation of the lattice around a dislocation may provide favourable sites for the accumulation of substitutional alloying elements (i.e. those that occupy lattice sites) and even more so for solutes that occupy interstitial positions (i.e. squeezed in between the ordinary lattice sites, as is carbon in iron). One can clearly expect alloying to affect the lattice friction and lead to an increased critical shear stress. The collection of atmospheres of solute around a dislocation may also lead to yield-point phenomena, which, although arising in many systems, are most pronounced in carbon steels (Fig. 4.25). One possible explanation of this particular

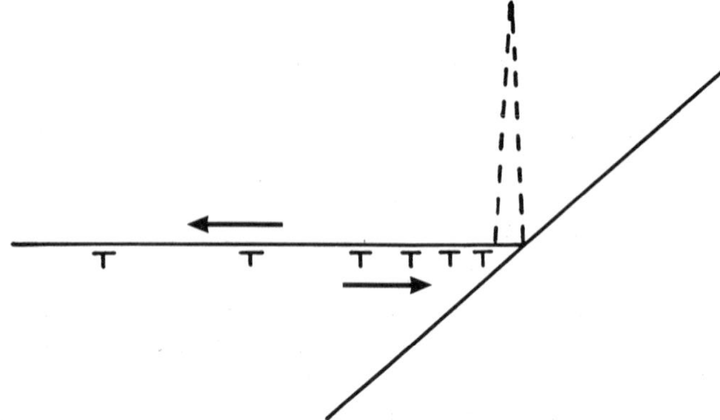

**Figure 4.24** Dislocations in the same slip plane may pile up at an obstacle, e.g. a grain boundary (see Fig. 4.10*b*), causing an incipient crack. The arrows show the shear stresses acting in the slip plane.

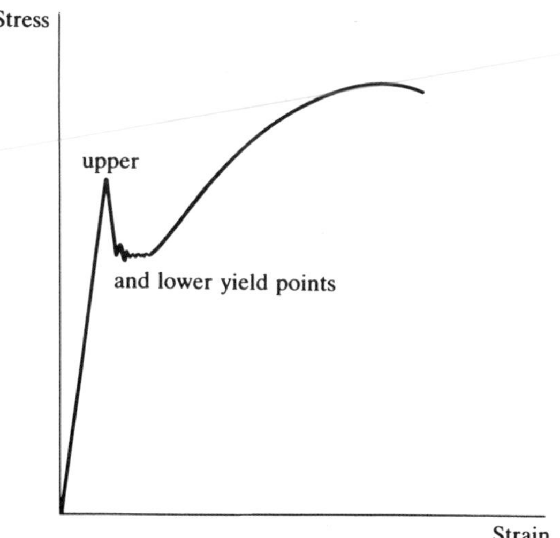


**Figure 4.25** Yield-point phenomena arise in ordinary mild carbon steels. A greater stress is needed to initiate yield than is required to produce continued deformation.

effect is that the stress needed to separate the dislocation from its atmosphere of carbon atoms is greater than that required to maintain its motion after separation, leading to the occurrence of upper and lower yield points. If the tensile test is halted after yield has occurred and the load reapplied, there is no yield drop because the carbon atoms are no longer around the dislocation. However, if we give the carbon atoms time to diffuse (i.e. we 'age' the specimen) then a renewed loading again reveals the yield point. The effect has practical disadvantages, particularly in the pressing of steel sheet as used in motor car bodies, because it leads to uneven deformation and a rough surface.

Particles in the form of separated phases affect the motion of dislocations and lead to increased yield stresses, the detailed behaviour depending sensitively on the sizes and distribution of the particles. One particularly important application of this effect is known as precipitation hardening.

## 4.3 Mechanical Behaviour of Non-Metals

Dislocations may arise in any crystal structure (and even in metallic glasses). In substances with localized and strongly directed bonds, we expect them to be narrow and experience large lattice friction. Silicon and germanium are therefore hard materials with large critical shear stresses and limited ductility, but they become plastic at high temperatures.

Dislocations also arise in ionic salts and oxides like MgO and $Al_2O_3$. Again we expect the dislocations to be narrow, and such substances are rarely plastic at room temperature, but deform readily at high temperatures. It is customary to test these materials under compression or by bending. We normally consider the ionic salts to be rather brittle materials and they cleave readily. However, LiF has been studied in great detail and when pure has been found to be quite soft with a critical shear stress of about 120 g $mm^{-2}$, not very different from a pure metal. The presence of 800 ppm Mg

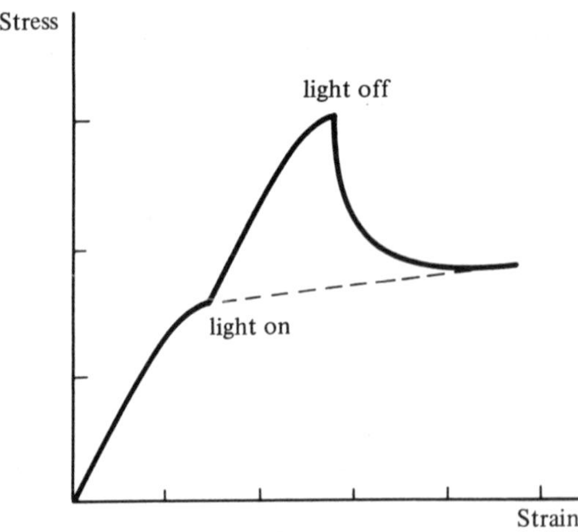


**Figure 4.26** The effect of illumination on the flow stress of a ZnO crystal. (After Carlsson and Swensson 1970.)

is, however, sufficient to raise the critical shear stress to about 1000 g mm$^{-2}$. Whereas the impure LiF cleaves readily, this is not the case for the pure salt. Experiments with LiF have fully substantiated the dislocation theory of plasticity. Glide occurs on the $\{110\}$ planes and in $\langle 110 \rangle$ directions, which, as well as satisfying criteria mentioned earlier, avoids bringing ions of the same sign together.

A particularly unusual effect is observed in ZnO, which has hexagonal structure and deforms relatively easily by basal plane slip even at room temperature. When illuminated by intense white light the flow stress (Fig. 4.26) is doubled. If the light is switched off, the flow stress returns to the lower value. The origin of this peculiar behaviour (which is observed in certain other semiconductors too) lies in the fact that it is almost impossible to prepare stoichiometric ZnO, and usually excess Zn is present as interstitial $Zn^+$ ions. On the other hand, the dislocations, which in basal slip contain atoms of the same kind along the actual line of dislocation, are charged because electrons freed from the $Zn^+$ attach themselves to the unsaturated (or dangling) bonds in the dislocation line. There are therefore Coulomb attractive forces between the excess $Zn^+$ ions and the negatively charged dislocations that augment the lattice friction. On illumination, $Zn^+$ ions in the immediate neighbourhood of the dislocation are ionized to $Zn^{2+}$, and the Coulomb forces between dislocations and Zn ions are further strengthened, leading to a higher flow stress. When the light is switched off, the $Zn^+$ ions are reformed and the 'friction' reduced.

## 4.4 Point Defects

### 4.4.1 *Lattice vacancies*

We have already seen how lattice vacancies and interstitial atoms may be produced by non-conservative motion of dislocations (Fig. 4.17). The point defects are, however, thermodynamically stable and can be produced in other ways. A crystal containing $N$ atoms that, although finite, is perfect in every other way, has zero configurational entropy. If in some manner we remove $n$ atoms from their proper sites, thereby forming $n$ vacancies, these may be distributed among the $N$ possible sites in

$$\frac{N!}{n!(N-n)!}$$

ways, leading to a configurational entropy (chemists would say entropy of mixing)

$$S_c = k_B \ln \frac{N!}{n!(N-n)!}. \tag{4.6}$$

This entropy is sufficiently large (and furthermore $dS/dn \to \infty$ as $n \to 0$) that, although it costs an energy $E_v$ to form each vacant lattice site, the free energy of the imperfect crystal is less than that of the perfect one at finite temperatures. If $\Delta F$ is the difference in free energy, we can calculate the equilibrium concentration of vacancies at a given temperature $T$ by minimizing $F$ with respect to the variable $n$. Thus we write

$$\begin{aligned} \Delta F &= \Delta U - T\,\Delta S \\ &= nE_v - Tk_B \ln \frac{N!}{n!(N-n)!}, \end{aligned} \tag{4.7}$$

and putting

$$\frac{d\,\Delta F}{dn} = 0 \tag{4.8}$$

leads to

$$\frac{n}{N-n} = e^{-E_v/k_BT}. \tag{4.9}$$

Equation (4.9) is readily obtained from (4.7) and (4.8) if Stirling's formula $\ln x! = x \ln x - x$ is used to simplify the entropy term. Since $n/N$ is at most $\approx 10^{-3}$, equation (4.9) may safely be written

$$\frac{n}{N} = e^{-E_v/k_BT}. \tag{4.10}$$

However, although vacancies may be thermodynamically stable, this does not necessarily mean that they will arise (diamond is a stable form of carbon but it is scarce and difficult to prepare by artificial means). To make a vacancy in an infinite perfect lattice demands that at the same time we create an interstitial atom (Fig. 4.27), and this is a considerable mechanical hindrance. However, real crystals are finite and,

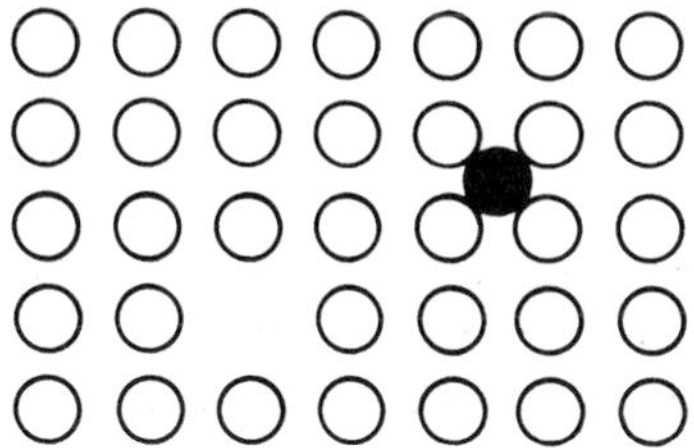

**Figure 4.27** Two important point defects are the lattice vacancy and the interstitial atom.

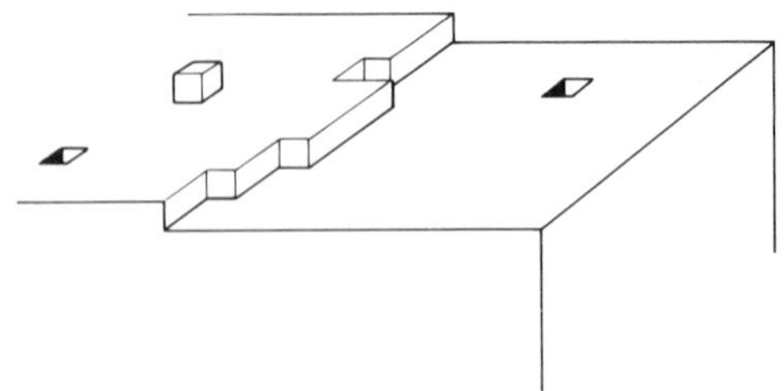

**Figure 4.28** An idealized picture of the surface of a crystal, assuming cubic atoms. We expect the surface layer to be incomplete together with the presence of surface holes and freely moving surface atoms in equilibrium with the vapour.

even if they are otherwise perfect, they possess a surface. Normally in this book we are trying to understand the properties of bulk matter and assume that the surface is so remote that it has no effect on the behaviour. The surface is, however, an important attribute, especially in chemical reactions of solids with gases and other condensed phases. The properties of solid surfaces are at present a rapidly developing research field, but one that we shall not pursue here (see Chapter 14). For our present purpose, we must consider the surface as a defect and one that allows lattice vacancies to be created in a simple manner. We may, in a stylized fashion, represent the surface as in Fig. 4.28. Even if the crystal were perfect and had plane sides, we could well imagine that fluctuations in the thermal energy of the atoms might allow an atom to jump from its surface site to become an isolated surface atom, leaving a surface vacancy behind. This surface vacancy could in turn be occupied by an atom coming from below, and in this way the vacancy might wander or diffuse into the body of the crystal. There is therefore little mechanical hindrance to the formation of vacancies when surfaces are present. It is clear, however, that their formation is easier the more thermal energy the atoms possess, and in practice they are introduced by annealing samples at temperatures approaching the melting point.

As we have already indicated, vacancies can move through the crystal, and this demands a certain energy (there is a lattice friction for the movement of a vacancy analogous to that for the dislocation). We speak of an energy of migration $E_m$. In fact usually $E_m < E_v$, but for metals both quantities are of order 1 eV atom$^{-1}$. The interstitial atom is also a point defect, but, as we have already implied, its energy of formation is so high, about 5 eV atom$^{-1}$, that we may neglect its occurrence when discussing thermally produced defects.

### 4.4.2 *Diffusion*

Although lattice vacancies occur in small equilibrium concentrations, they are not insignificant features of solids. It is intuitively obvious that their presence must facilitate diffusion in both pure metals and alloys. Nevertheless, the exact mechanisms through which mass transport takes place in solids in the presence of temperature or concentration gradients are not fully established, but if we concentrate on self-diffusion we can appreciate that one mechanism could be vacancy-controlled. Self-diffusion is the transport of matter in an elemental substance, e.g. the diffusion of Cu in Cu metal. We need the vacancy and its movement. Experi-

**Table 4.1** Some values of $E_v$, $E_m$, $E_D$ for vacancies in fcc metals[a]

| | Cu | Ag | Au | Pt | Al |
|---|---|---|---|---|---|
| $T_m$ (K) | 1356 | 1233 | 1336 | 2046 | 931 |
| $E_v$ (eV) | 1.07 | 1.09 | 1.0 | 1.3 | 0.78 |
| $E_m$ (eV) | 0.88 | 0.83 | 0.78 | 1.21 | 0.56 |
| $E_v + E_m$ (eV) | 1.95 | 1.92 | 1.78 | 2.51 | 1.34 |
| $E_D$ (eV) | 1.83 | 1.97 | 1.82 | 2.51 | 1.39 |
| ($n/N \times 10^5$ at $T_m$) | 4.5 | 3.5 | 17 | 63 | 6 |

[a] Formation and migration energies for vacancies in pure metals and certain alloys are given in Takumara (1965).

mentally we know that self-diffusion is temperature-activated, and the diffusion coefficient is written as

$$D = D_0 e^{-E_D/k_B T}. \tag{4.11}$$

Furthermore, $E_D$ is in the range 1–3 eV and it may be determined with a relative accuracy of about 0.05 eV, but experimental results may vary over a range $\pm 0.2$ eV; values of $D_0$ on the other hand may differ by more than a factor of 10. (Two tabulations of diffusion data are Lazarus 1960 and Smithells 1967.) The vacancy mechanism implies that

$$E_D = E_v + E_m. \tag{4.12}$$

The data in Table 4.1 show that for certain face-centred cubic metals (4.12) is indeed true, lending support to the hypothesis. Diffusion is an extremely important phenomenon, which controls the composition of alloys and the high temperature creep strength of metals; the vacancy is therefore a significant quantity in such phenomena. In polycrystalline materials diffusion also arises within grain boundaries, and this leads to a more rapid transport of materials than bulk diffusion, particularly at low temperatures.

If the interstitial has an energy of formation of about 5 eV then we readily calculate that their concentration is insignificant (of order $10^{-19}$) at the melting point. Thus intrinsic interstitial atoms cannot play any part in self-diffusion. However, many important alloys have interstitial components, notably carbon in steels. The interstitial atom is found to have a very high diffusion coefficient; we therefore expect that if interstitial atoms are formed in pure metals (as might occur during deformation), they will also migrate very quickly ($E_m \approx 0.1$ eV) and disappear at sinks provided by grain boundaries and dislocations.

### 4.4.3 *Thermal expansion and heat capacity*

If there is very little relaxation of the atoms surrounding a lattice vacancy then the presence of vacancies causes a reduction in density of a pure metal compared with that calculated on the basis of the lattice spacing. Conversely we may say that at high temperatures the bulk coefficient of thermal expansion must be larger than that of the lattice (Fig. 4.30). It is straightforward to show that if $L$ is the macroscopic

**Box 4.1. Diffusion**

Diffusion is the transport of atoms by thermal motion in the presence of a concentration gradient. If the atoms are all similar (except for the use of radioactive isotopes) then we speak of self-diffusion, whereas if there are two different atomic species then we have impurity or chemical diffusion. Consider an assembly of planes in an elemental substance (Fig. 4.29) and let there be a concentration gradient in the x direction with $dc/dx < 0$. Let two adjacent planes be in regions with concentrations $c_1$ and $c_2$ respectively, where the latter

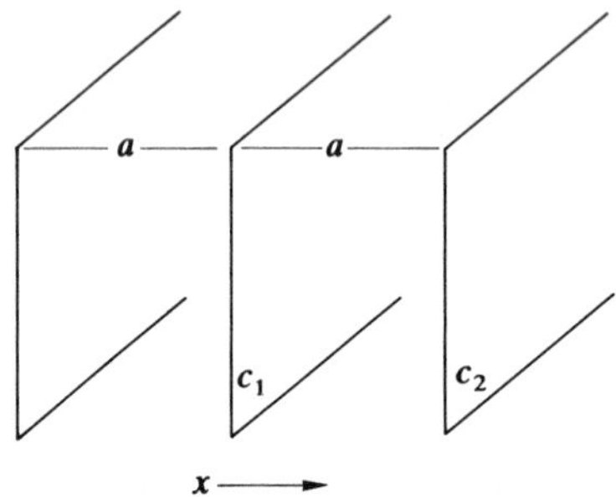


**Figure 4.29**

are only infinitesimally different. Unit areas of these planes then contain $ac_1$ and $ac_2$ atoms, $a$ being the interplanar spacing. We assume that all the atoms vibrate with the same frequency $v$. Without any knowledge of the atomic mechanisms involved, we postulate that in an element of time $\delta t$ each atom moves to the right $v\ \delta t$ times and an equal number to the left; furthermore that the probability that the atom jumps to either right or left at each encounter with a neighbouring plane is $P$. The net flux of atoms is then

$$\delta n \approx v\ \delta t\ a(c_1 - c_2)P,$$

but

$$c_2 = c_1 + a\frac{dc}{dx},$$

so

$$\frac{\delta n}{\delta t} = j \approx -va^2P\frac{dc}{dx}.$$

$j$ is the material current density, which we write as

$$j = -D\frac{dc}{dx}. \qquad (1)$$

In three dimensions

$$\mathbf{j} = -D\nabla c. \qquad (2)$$

$D$ is called the coefficient of diffusion and experimentally is found to be independent of concentration, but very dependent on temperature according to

$$D = D_0 e^{-E_D/k_BT}, \qquad (3)$$

*continued*

$E_D$ being the activation energy for diffusion and $D_0$ a constant (data for some metals are given in Table 4.2). Diffusion serves to eliminate concentration gradients and to promote homogeneity. We must therefore find the changes in concentration in a volume element to be exactly compensated by the net flux of atoms in or out of the volume element. If the crystal structure is to be maintained without the creation of holes or the accumulation of excess atoms then

$$\frac{dc}{dt} = -\frac{dj}{dx},$$

**Table 4.2**

| | $E_D$ (eV atom$^{-1}$) | $D_0$ (cm$^2$ s$^{-1}$) |
|---|---|---|
| Ag | 1.9 | 0.8 |
| Cu | 1.85 | 0.6 |
| Au | 1.86 | 0.15 |
| Mg | 1.41 | 1.0 |
| Al | 1.4 | 1.7 |
| Pb | 1.26 | 1.2 |
| $\alpha$-Fe | 2.6 | 2 |
| $\gamma$-Fe | 2.82 | 0.2 |
| Ni | 3.0 | 2.5 |
| Carbon in iron | | |
| $\alpha$-Fe | 0.87 | 0.02 |
| $\gamma$-Fe | 1.4 | 0.1 |

or in three dimensions

$$\frac{dc}{dt} = -\nabla \cdot \mathbf{j}. \tag{4}$$

Combining (2) and (4), we obtain

$$\frac{dc}{dt} = \nabla \cdot (D\nabla c),$$

which for constant $D$ becomes

$$\frac{dc}{dt} = D\nabla^2 c. \tag{5}$$

Equations (2) and (5) are known as Fick's equations.

The experimental study of diffusion usually demands the solution of (5) under appropriate boundary conditions (for further details see Shewmon 1963).

We consider every diffuse jump that an atom makes to be a random independent event. Any given atom therefore moves randomly in its diffusive motion and may move in any direction with equal probability. This behaviour is an example of a 'random walk' in statistics; it then turns out that after a time $t$ an atom will be found a distance $x$ from its starting point, where (Shewmon 1963)

$$\langle x^2 \rangle = 2Dt. \tag{6}$$

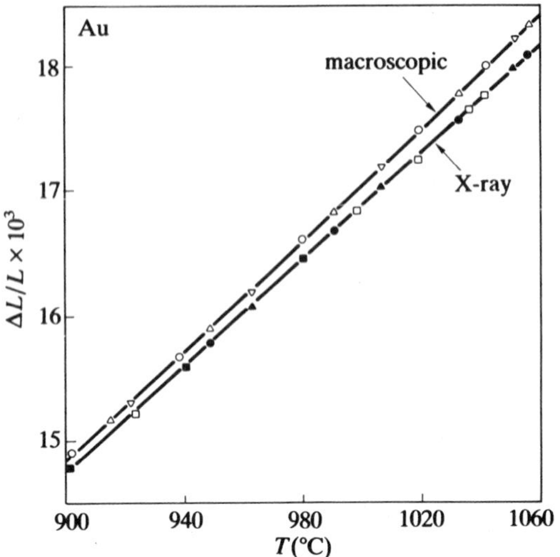


**Figure 4.30** The macroscopic and atomic (i.e. X-ray) thermal expansion coefficients are not identical. (After Simmons and Balluffi 1962.)

length and $a$ the lattice parameter then

$$\left(\frac{n}{N}\right)_T = 3\left[\left(\frac{\Delta L}{L}\right)_T - \left(\frac{\Delta a}{a}\right)_T\right], \tag{4.13}$$

which shows that we may determine the concentration of vacancies by simultaneous measurements of the changes in macroscopic length and lattice parameter. Such measurements have shown that the vacancy is the only significant thermal defect in metals.

The more vacancies that are formed, the more energy required and the more the internal energy of the metal depends upon the defect concentration. We write this contribution as

$$\Delta U_v = nE_v = E_v N \mathrm{e}^{-E_v/k_B T},$$

whence the extra contribution to the heat capacity due to creation of vacancies becomes

$$\frac{\mathrm{d}}{\mathrm{d}T}\Delta U_v = \frac{NE_v^2}{k_B T^2}\mathrm{e}^{-E_v/k_B T}. \tag{4.14}$$

This is readily observable, as Fig. 4.31 shows, and in principle allows a determination of $E_v$.

### 4.4.4 Quenched metals

Non-equilibrium concentrations of vacancies may be obtained by rapidly cooling (quenching) metal wires or foils from high temperatures. The cooling must be sufficiently rapid that no significant change in concentration occurs. When cold the vacancies cannot move. By quenching from a series of carefully controlled temperatures, different concentrations of defects may be obtained. These defects can, at

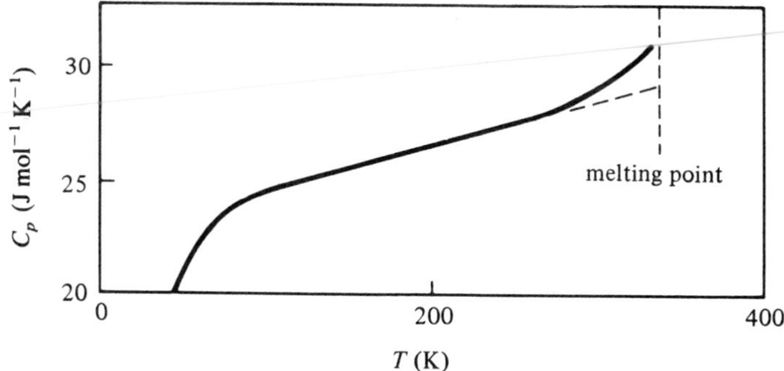


**Figure 4.31** The excess heat capacity associated with the equilibrium production of vacancies in potassium. (After MacDonald 1955.)

low temperatures (4 K), contribute markedly to the electrical resistance of a chemically pure metal and in direct proportion to their concentration. This provides a convenient method of measuring $E_v$, since we may write

$$\Delta R_T = Cn_T = CN\mathrm{e}^{-E_v/k_B T}. \tag{4.15}$$

$\Delta R_T$ is the incremental increase in resistance at 4 K after quenching from the temperature $T$, and $C$ is a constant. Clearly

$$\ln \Delta R_T = \ln CN - \frac{E_v}{k_B T}, \tag{4.16}$$

and $E_v$ may be obtained directly from the gradient of a graph of $\ln \Delta R_T$ against $1/T$ (Fig. 4.32).

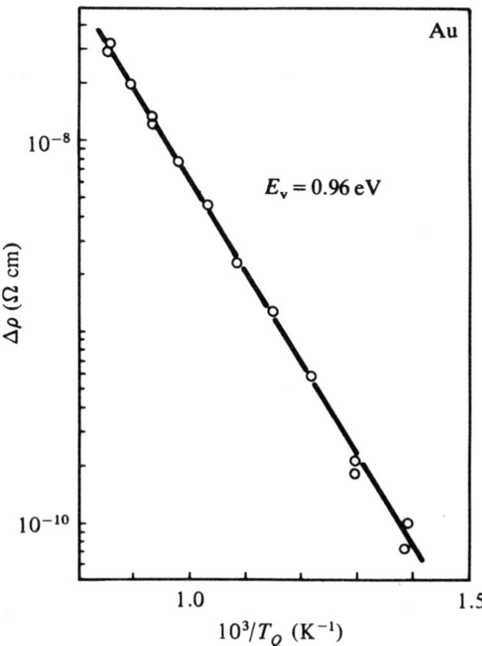


**Figure 4.32** The determination of $E_v$ from resistivity data; $\Delta\rho$ is the increment of resistivity, at 4 K, produced by quenching from $T_Q$. (After Bauerle and Koehler 1957.)

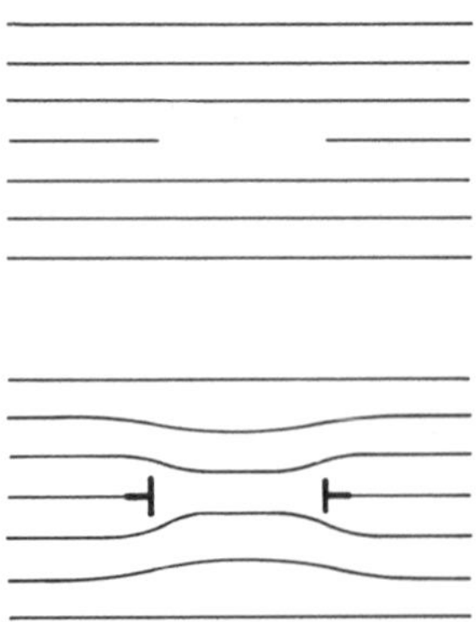

**Figure 4.33** Vacancies may collect to form planar clusters.

It is in this way that values of $E_v$ have usually been obtained. If, after quenching, the specimens are gradually warmed then the vacancies begin to move and the equilibrium concentration is established. The kinetics of this recovery process observed via the electrical resistance allows the migration energy to be determined.

The above experiment must be performed with care, and it has been found that it is best if temperatures not too near the melting point are used, otherwise vacancy complexes (e.g. divacancies and larger clusters) may form, both at the high temperature and presumably during quenching. Examination of quenched foils under the electron microscope has clearly demonstrated the presence of dislocation loops created as a result of quenching. These may have at least two possible origins. Vacancies may collect into planar clusters and the surrounding atom planes collapse to form a ring of dislocation (Fig. 4.33). A second possibility is that the large internal pressure associated with a non-equilibrium concentration of vacancies leads to the spontaneous creation of dislocation loops.

### 4.4.5 Radiation damage

Experiment has shown that to displace an atom from a lattice site to an interstitial position using particle radiation requires an energy of about 25 eV. From the scientific viewpoint, the ideal way to do this is by electron bombardment because 1 MeV electrons can produce isolated vacancy-interstitial pairs (also called Frenkel defects). These defects are well defined and suitable for accurate study. Heavier particles such as neutrons, protons, deuterons and ions cause much more disruption of the lattice, leading to complicated damage patterns and pronounced changes in the physical properties, but these are not always easy to analyse.

### 4.4.6 Point defects in non-metals

This is an extremely large field of study and there is a whole menagerie of defects in non-metals because from the simple defect many more complicated ones can be derived either by clustering or association with impurity species such as $O^{2-}$ and other ions. Non-metallic compounds can depart from stoichiometry, leading to pronounced changes in physical properties such as density, diffusion, electrical conductivity, optical and dielectric behaviour. We shall confine attention to two of the simplest and best known features of defects in ionic salts, namely the *F* centre† and ionic conductivity.

In principle we can consider the formation of both cation and anion vacancies in an ionic salt by annealing at a high temperature, but the poor thermal conductivity of the salt prevents their retainment by quenching – the specimen would be destroyed. Instead, anion vacancies are formed by introducing excess metal into the salt by annealing KCl, say, in an atmosphere of K vapour. The reaction is

$$K \rightarrow K^{+} + \boxminus$$

anion vacancy
+ electron

The metal atom becomes an ion and an anion vacancy is created, the latter having an associated positive charge; at low temperatures the electron created during the formation of the metal ion therefore becomes weakly bound to the anion vacancy; the unit is known as an *F* centre because it colours the otherwise transparent crystal. The electron bound to the vacancy has a quasi-hydrogen-like spectrum and its ionization energy corresponds to a photon in the optical range. Optical excitation of the trapped electron in the anion vacancy gives rise to a broad absorption band; at room temperature that for KCl arises around 5600 Å, corresponding to a photon energy of about 2.3 eV. Another way to form the simple vacancy defect, this time an empty cation site, is to introduce a small amount of divalent metal ions into the salt. We make a mixed salt $K_{1-x}Ca_xCl$. To maintain charge neutrality, each $Ca^{2+}$ ion must replace two $K^+$ ions. A cation vacancy therefore accompanies each divalent ion that is introduced into the lattice (Fig. 4.34).

We can well appreciate that electrical conduction in an ionic salt demands the movement of ions, and if the lattice is perfect this cannot readily arise; at ordinary temperatures such salts are insulators. They become conducting only when diffusion may take place; this leads to a conductivity exponentially dependent on temperature in the same manner as the diffusion coefficient. Clearly, empty lattice sites facilitate both diffusion and electrical conductivity, so the presence of ions of different valence can cause significant increase in the low-temperature conductivity.

In recent years much attention has been given to the construction of batteries using solid electrolytes, particularly for electrically driven vehicles. There are many problems involving suitable electrodes as well as the chemical and physical stability of the electrodes and electrolytes. But above all, the solid electrolyte must have a reasonable conductivity $>1\ \Omega^{-1}\ \text{cm}^{-1}$ at ambient temperatures, i.e. at or slightly above room temperature. The observation of a much larger than ordinary ionic conductivity in AgI above 147°C sparked off much interest; several substances are now known to show good conductivity (i.e. log $\sigma$ in the range 0–1) at high temperatures, and on account of this they have been called 'superionic conductors'. In addition to AgI they include $RbAg_4I_5$ and similar compounds, as well as those based on the $\beta$-aluminas ($Na_2O{\cdot}11Al_2O_3$). So far, all these fast ionic conductors display phase transitions, and the high conductivity is associated with a high-temperature phase in which a particular ion becomes very mobile. In AgI it is the silver ion that moves, and it appears as though the Ag ions acquire liquid-like properties and can diffuse rapidly in the rigid lattice of I ions. In the $\beta$-aluminas the

† The *F* comes from the German word for colour: 'Farbe'.

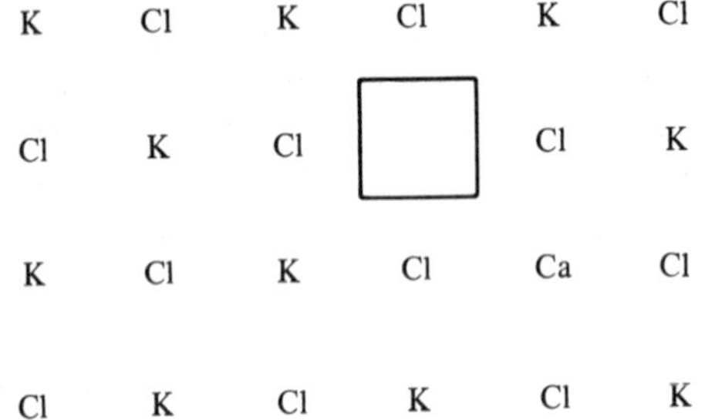


**Figure 4.34** A cation vacancy accompanies each Ca atom introduced into KCl.

Na or equivalent ion moves rapidly in a two-dimensional network of channels, again assuming liquid-like behaviour. In fact, the superionic solids have electrical conductivities more appropriate for molten salts, so the pseudo-liquid character of the fast ion is a good physical picture of the situation. This is further borne out by the fact that the transition from a poorly conducting to a highly conducting phase (e.g. at 147°C for AgI) is accompanied by a large heat of transformation which is approximately half the usual heat of melting of such a salt.

As yet, superionic behaviour has not been obtained at room temperature and the development of practical batteries presents formidable problems.

## References

BAUERLE, J. E. and KOEHLER, J. S. (1957) *Phys. Rev.* **107**, 1493.

CARLSSON, L. and SWENSSON, C. (1970) *J. Appl. Phys.* **41**, 1652.

LAZARUS, D. (1960) *Solid State Physics*, Vol. 10 (ed. F. SEITZ and D. TURNBULL ), p. 71. Academic Press, New York.

MACDONALD, D. K. C. (1955) *Defects in Crystalline Solids.* The Physical Society, London.

READ, W. T. (1953) *Dislocations in Crystals.* McGraw-Hill, New York.

SHEWMON, P. G. (1963) *Diffusion in Solids.* McGraw-Hill, New York.

SIMMONS, R. O. and BALLUFFI, R. W. (1962) *Phys. Rev.* **125**, 862.

SMITHELLS, C. J. (ed.) (1967) *Metals Reference Book, II.* Butterworths, London.

TAKUMARA, J. I. (1965) *Physical Metallurgy* (ed. R. W. CAHN ), p. 681. North-Holland, Amsterdam.

## Further Reading

CAHN, R. W. and HAASEN, P. (Editors), *Physical Metallurgy*, 3rd revised edition (1983) North-Holland Physics Publication, Amsterdam, containing: WOLLENBERGER, H. J. *Point Defects*, pp. 1139–1221; HIRTH, J. P. *Dislocations*, pp. 1223–1258; HIRSCH, P. B. (Editor), *The Physics of Metals 2, Defects.* (1975) Cambridge University Press, Cambridge.

## Problems

**4.1** Give concise definitions of (a) dislocation, (b) the Burgers vector, (c) critical shear stress, (d) partial dislocation and (e) stacking fault.

**4.2** A screw dislocation in a crystal runs parallel to the $z$-axis. Its Burgers vector is in the positive $z$-direction. During deformation, it is cut by an edge dislocation moving in the $xy$ plane. Choose a Burgers vector for the edge dislocation and then illustrate the situation before and after the dislocations pass through one another.

**4.3** Which is the more favourable direction for the following dislocation reaction:

$$\tfrac{1}{2}a[110] = \tfrac{1}{6}a[11\bar{2}] + \tfrac{1}{6}a[112] + \tfrac{1}{6}a[110]?$$

**4.4** When a dislocation, e.g. an edge dislocation, moves completely across the slip plane and out of the crystal it produces an amount of slip **b**. Assuming that the amount of slip produced while the dislocation moves within the crystal is proportional to the relative displacement of the dislocation in the slip plane, derive an expression for the force acting on an arbitrary length of dislocation when the crystal is subject to a shear stress $\tau$.

**4.5** A single crystal of zinc, in the form of a cylinder with right-circular cross-section 5 mm$^2$, is subjected to axial strain and glide bands are observed. A close inspection of the crystal showed that the normal to the glide plane made an angle 30° with the strain axis whereas the glide direction differed by 120° from this axis. The first glide band appeared when the axial load was 2.06 N. What is the critical shear stress of the zinc crystal? On which plane and in which direction (expressed in the usual indices) does glide arise in a zinc crystal?

**4.6** Using a carefully drawn diagram, demonstrate that a regular linear array of edge dislocations gives rise to a low-angle boundary. Derive the misorientation in terms of the dislocation strength and separation.

**4.7** What system of interacting dislocations produces (a) a row of interstitial atoms, (b) a row of lattice vacancies?

**4.8** A screw dislocation running vertically through a set of crystal planes contains a jog. Describe how the jog may affect the movement of the screw dislocation. Remember that the screw dislocation does not have a fixed slip plane. Draw diagrams showing the situation when the screw together with the jog move rigidly in a direction (a) parallel and (b) perpendicular to the jog. If the condition of rigid movement is relaxed, what do you think will happen?

**4.9** How would you expect dislocation climb to be affected by a lattice vacancy concentration that is (a) greater than the equilibrium value, (b) less than the equilibrium value? How might such concentrations arise?

**4.10** What determines the concentration of lattice vacancies in a crystal? For copper, the migration energy of a vacancy is 0.8 eV and furthermore the self-diffusion coefficient of copper at 700 K and 1000 K is $3.43 \times 10^{-15}$ and $1.65 \times 10^{-11}$ cm$^2$ s$^{-1}$ respectively. What are the vacancy concentrations at these two temperatures?

**4.11** A gold wire is heated to successively higher temperatures. At each temperature, the wire is rapidly quenched into water and thereafter the resistance is measured at 4 K. The extra resistivity $\Delta R$ due to this rapid cooling is tabulated below together with the quench temperature $T$:

| $T$ (°C) | $\Delta R$ ($\mu\Omega$ cm) |
|---|---|
| 597 | 0.0013 |
| 647 | 0.0022 |
| 697 | 0.0048 |
| 747 | 0.0078 |
| 797 | 0.0120 |
| 847 | 0.0200 |
| 897 | 0.0300 |

What quantitative information may be obtained from the above data?

**4.12** Sometimes alloys are made by sintering compacts of metal powders. In the present case an alloy of Cu and Ni with the atomic composition CuNi is to be prepared. The particles have uniform size with diameter 0.1 mm. Assuming that the particles are homogeneously mixed prior to pressing and that their mutual diffusion rates are equal and given by $D_0 = 0.1\ \text{cm}^2\ \text{s}^{-1}$, $E_D = 2$ eV, estimate the times required to produce an homogeneous alloy when the compacts are heated to 727°C and 927°C. Note that Cu and Ni form what is called a miscible binary system, i.e. they dissolve one another in the fcc structure completely over the whole range of compositions from pure Cu to pure Ni.

**4.13** Using the data provided in the text, calculate (a) the self-diffusion coefficient of iron at 850°C, (b) the diffusion coefficient of carbon in iron at 850°C, (c) the value of $\langle x^2 \rangle$ for the cases (a) and (b) when the sample is maintained at 850°C for $10^5$ s, i.e. somewhat over 24 h, and (d) the diffusion coefficient of carbon in iron at 950°C.

**4.14** Given that you are able to perform a diffusion experiment during a period of 30 days and that you are able to analyse a diffusion layer with thickness of order $10^{-3}$ cm, what is the smallest diffusion coefficient that could be studied? If the sample is Cu, what is the lowest temperature at which this experiment can be made? If the temperature is increased to 950°C, how does this affect the time needed to achieve the same result? Only rough estimates are required.

**4.15** Suppose a vacancy and an interstitial atom are created as a single event – such a combination is called a Frenkel defect. If $N$ is the total number of atoms, $N'$ the number of available interstitial sites, $U$ the energy of formation and $S$ the associated change in thermal entropy per Frenkel defect, calculate the equilibrium concentration of such defects at a temperature $T$.

# 5

# Lattice Vibrations

For an ideal monoatomic gas maintained at constant volume† the internal energy is determined wholly by the kinetic energy of the atoms

$$dU = dQ = C_v \, dT,$$

where $U$, $Q$ and $C_v$ have their usual thermodynamic significance. Each atom has three degrees of freedom‡ and, by Maxwell's law of the equipartition of energy, the total energy per atom is $\frac{3}{2}k_B T$, leading to a heat capacity $\frac{3}{2}R$. The solid formed from such a gas possesses six degrees of freedom per atom, because we must also include the potential energy, and we find the heat capacity to be $3R$, which corresponds to very nearly 25 J mol$^{-1}$ K$^{-1}$. This is in keeping with the law of Dulong and Petit first formulated in 1819. However, by the end of the 19th century it was well known that the heat capacity of simple solids such as the metallic elements decreased rapidly at low temperatures, apparently approaching zero near 0 K. It was also known that diamond had an exceptionally low heat capacity at room temperature (Fig. 5.1).

## 5.1 The Einstein Model

In 1907, Einstein produced a theory of heat capacity based on Planck's quantum hypothesis (1901). He assumed that each atom of the solid vibrates about its equilibrium position with an angular frequency $\omega$. Each atom has the same frequency and vibrates independently of other atoms. A mole of solid is therefore assumed to be composed of $3N$ one-dimensional oscillators. From quantum mechanics, we know that a linear harmonic oscillator has the energy spectrum

$$E = (n + \tfrac{1}{2})\hbar\omega. \qquad (5.1)$$

† Attention is confined to the heat capacity at constant volume and we consider only changes to the internal energy brought about by temperature.
‡ The number of degrees of freedom is the number of squared terms in the expression for the energy.

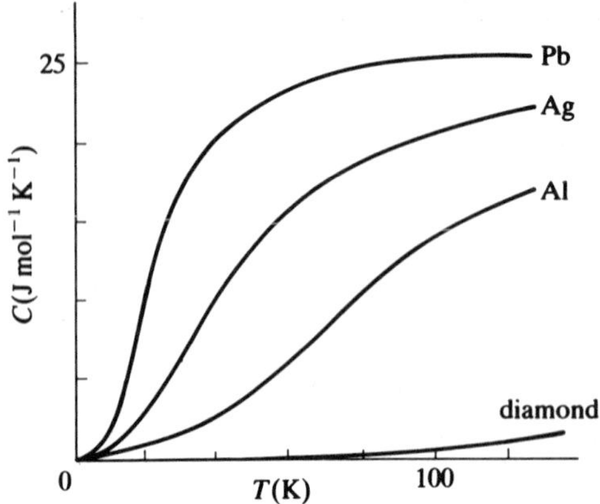


**Figure 5.1** The variation of heat capacity with temperature for certain elements.

The zero-point energy $\frac{1}{2}\hbar\omega$ is not significant for the heat capacity because we are only interested in the change in the internal energy brought about by a change in temperature; we may choose our zero of energy at $\frac{1}{2}\hbar\omega$ so that $E_n = n\hbar\omega$.

The occupancy of this level is determined by the Boltzmann factor

$$f(E_n) = \mathrm{e}^{-n\hbar\omega/k_\mathrm{B}T} \Bigg/ \sum_{n'=0}^{\infty} \mathrm{e}^{-n'\hbar\omega/k_\mathrm{B}T}, \tag{5.2}$$

and the total energy of the solid becomes

$$U = \frac{3N \sum_{n=0}^{\infty} n\hbar\omega \mathrm{e}^{-n\hbar\omega/k_\mathrm{B}T}}{\sum_{n=0}^{\infty} \mathrm{e}^{-n\hbar\omega/k_\mathrm{B}T}}. \tag{5.3}$$

The sums may be evaluated if we recognize that

$$\frac{\mathrm{d}}{\mathrm{d}x} \ln \sum_{n=0}^{\infty} \mathrm{e}^{-nx} = -\frac{\sum_{n=0}^{\infty} n\mathrm{e}^{-nx}}{\sum_{n=0}^{\infty} \mathrm{e}^{-nx}} = \frac{\mathrm{d}}{\mathrm{d}x} \ln (1 - \mathrm{e}^{-x})^{-1},$$

and, putting $x = \hbar\omega/k_\mathrm{B}T$, we may differentiate the term on the right in the standard manner to obtain

$$U = \frac{3N\hbar\omega}{\mathrm{e}^{\hbar\omega/k_\mathrm{B}T} - 1}. \tag{5.4}$$

At high temperatures $\hbar\omega \ll k_\mathrm{B} T$, and we write $\mathrm{e}^{\hbar\omega/k_\mathrm{B}T} \approx 1 + \hbar\omega/k_\mathrm{B}T$, leading to

$$U = \frac{3N\hbar\omega}{\hbar\omega/k_\mathrm{B}T} = 3Nk_\mathrm{B}T = 3RT$$

and

$$C = \frac{\mathrm{d}U}{\mathrm{d}T} = 3R,$$

in agreement with experiment.

At low temperatures $\hbar\omega \gg k_B T$ and $e^{-\hbar\omega/k_B T} \gg 1$, leading to

$$U = 3N\hbar\omega e^{-\hbar\omega/k_B T},$$

whence

$$C = 3Nk_B \frac{\hbar^2\omega^2}{k_B^2 T^2} e^{-\hbar\omega/k_B T}$$

$$= 3R\left(\frac{\theta_E}{T}\right)^2 e^{-\theta_E/T}. \tag{5.5}$$

$\theta_E$ has been written for $\hbar\omega/k_B$. It is called the *Einstein temperature* and characterizes the solid. The general expression for the Einstein heat capacity is obtained by differentiating (5.4), giving

$$C = 3R\left(\frac{\theta_E}{T}\right)^2 \frac{e^{\theta_E/T}}{(e^{\theta_E/T} - 1)^2}. \tag{5.6}$$

The Einstein heat capacity is plotted in Fig. 5.2 together with experimental data for silver, assuming $\theta_E = 160$ K. The general agreement is very good, but the model fails at low temperatures, predicting a more rapid decrease with temperature than is found in practice. Furthermore, we must look upon $\theta_E$ as an *ad hoc* parameter that cannot satisfactorily be expressed in terms of other physical constants. Nevertheless, Einstein's successful application of the quantum hypothesis to the thermal behaviour of solids was a most significant achievement; taken together with his almost simultaneous introduction of the light quantum and the interpretation of the photoelectric effect, he established the importance of the quantum hypothesis for phenomena other than that of thermal radiation.

## 5.2 The Debye Model

The Einstein approach is too simple. The atoms do not vibrate independently but form a system of coupled oscillators. We therefore expect a frequency spectrum

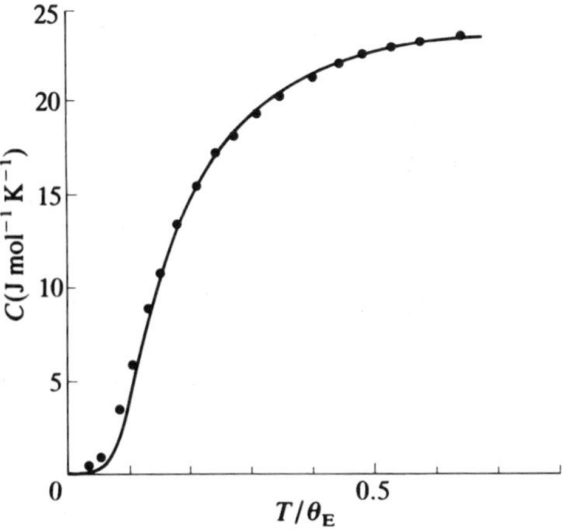


**Figure 5.2** The Einstein expression for the heat capacity fitted to data for Ag with $\theta_E = 160$ K. The fit is very good except at the lowest temperatures.

containing altogether $3N$ normal modes. Debye, in 1912, assumed that the frequency spectrum was that of an elastic continuum with a high-frequency cut off $\omega_{max}$ determined by the fixed number of vibratory modes (we shall discuss these features in greater detail later).

Each mode contributes an Einstein term, so that the heat capacity may be written

$$C_D = \int_0^{\omega_{max}} N(\omega) C_E(\omega)\, d\omega, \tag{5.7}$$

where $N(\omega)$ is the frequency spectrum and $C_E(\omega)$ is the Einstein heat capacity of an oscillator of frequency $\omega$.

The agreement with experimental data at low temperatures is now much improved (Fig. 5.3). At low temperatures we expect only the low frequencies to be excited, the atomic vibrations have large wavelength and the discrete atomic character of the solid becomes less important so the continuum approximation is appropriate. We define a *Debye temperature*

$$\theta_D = \hbar\omega_{max}/k_B \tag{5.8}$$

and for $T < 0.1\theta_D$ the following expression can be deduced from (5.7); see Problem 5.2:

$$C_D = \frac{12}{5}\pi^4 R\left(\frac{T}{\theta_D}\right)^3. \tag{5.9}$$

The continuum approach can only be an approximate description of the atomic motion, and one would prefer to calculate the frequency spectrum in terms of the crystal structure and the forces between atoms. This is a difficult task, but one that theoreticians, notably Born and von Kármán, were engaged upon at the very time Debye produced his theory. However, such calculations provide much more information than the frequency spectrum of collective oscillations; they show the relation

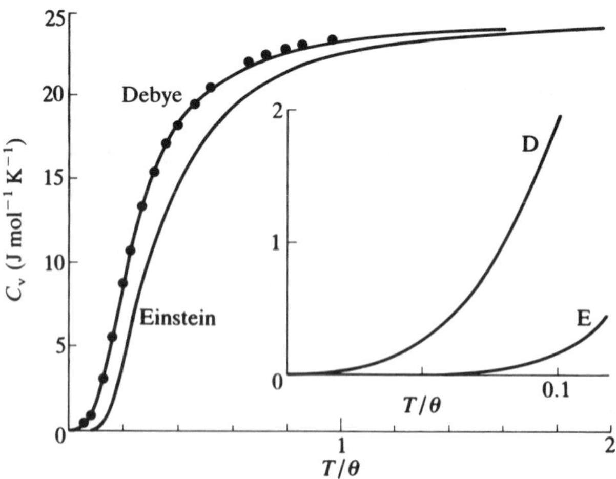


**Figure 5.3** The Debye expression for the heat capacity fitted to data for Ag with $\theta_D = 215$ K. Here we also compare the Einstein and the Debye expressions when $\theta_E = \theta_D$. In particular, the inset diagram shows the failure of the Einstein values near the absolute zero of temperature.

between frequency and wavenumber in specific directions, the so-called *dispersion relations.*

In wave motion, the phase and group velocities are defined as

$$v_p = \frac{\omega}{k}, \qquad v_g = \frac{d\omega}{dk}, \tag{5.10}$$

where $k = 2\pi/\lambda$ is the modulus of the wave vector $\mathbf{k}$.

For any homogeneous medium, we associate $v_p^{-1}$ with the absolute refractive index, and when the latter varies with $\mathbf{k}$ dispersion results.

A dispersion relation is therefore an equation relating $\omega$ and $k$. The most common example of dispersion is of course the separation of a pencil of white light into the colours of the spectrum when the light enters a glass prism. However, there was no good method of determining the dispersion relations for lattice vibrations until about 1955, when techniques based on the inelastic scattering of neutron beams from research reactors allowed the first thorough study of their properties. Thus, although theoretical attempts to calculate dispersion relations for atomic vibrations were being made in 1912, no really general advance was made in this subject until the post-war years owing to the lack of experimental data. The heat capacity, depending on the average total energy of the solid, does not allow us to deduce the details of the atomic motion. In the past 25 years, sophisticated models for describing the vibrational properties of solids have been successfully developed. However, we shall not consider the details of real behaviour but discuss an elementary problem – that of the one-dimensional chain of coupled point masses. Although this is of no use in accounting for the quantitative behaviour of any real solid, it is important because it shows us the principal form of the behaviour and allows us to appreciate experimental results. Furthermore it introduces us to new concepts of great importance in the study of wave propagation in periodic potentials. The reader should therefore make every effort to understand the new ideas introduced with this example. We begin by recalling the way in which a simple elastic rod executes longitudinal oscillations.

## 5.3 The Continuous Solid

Consider a rod of elastic material with cross-section $A$, density $\rho$ and arbitrary length. If this rod is caused to vibrate longitudinally then an element of original length $\Delta L$ at an arbitrary position $L$ becomes of length $\Delta(L + u)$ (Fig. 5.4*a*). If $E$ is Young's modulus for the material of the rod then the stress $\sigma$ and force $F$ acting at the point $L$ are related to the strain by

$$F = \sigma A = AE\left[\frac{\Delta(L+u) - \Delta L}{\Delta L}\right]_{\Delta L \to 0} = AE\,\frac{du}{dL}.$$

Under longitudinal vibration the local strain varies with position along the rod, and our element of original size $\Delta L$ experiences a net force $\Delta F$ and is accelerated according to

$$\Delta F = AE\,\frac{d^2u}{dL^2}\,\Delta L = \rho A\,\Delta L\,\frac{d^2u}{dt^2}.$$

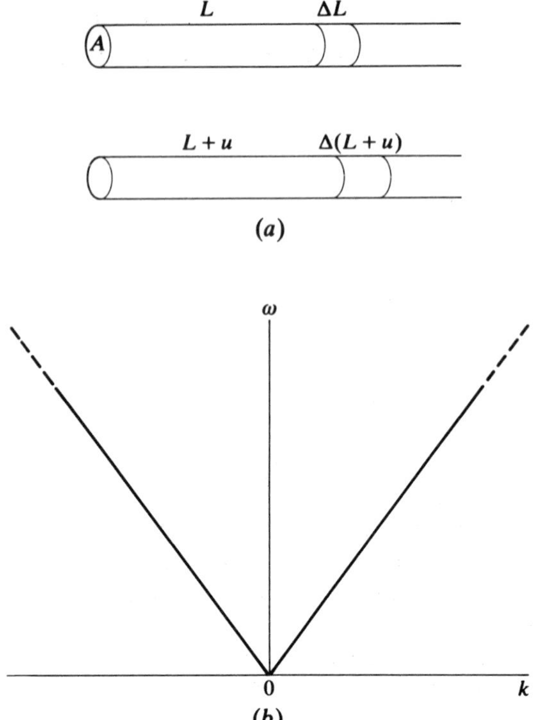


**Figure 5.4** (*a*) The extension of a continuous solid. (*b*) Dispersion-free behaviour of the elastic continuum.

The equation of motion becomes

$$\frac{d^2u}{dt^2} = \frac{E}{\rho}\frac{d^2u}{dL^2}. \tag{5.11}$$

This is a wave equation and the elastic vibrations propagate with velocity $v = (E/\rho)^{1/2}$. Clearly $v$ is independent of wavelength, and, writing $v = \omega/|\mathbf{k}|$ (where $\mathbf{k}$ is, as usual, the wave vector with modulus $2\pi/\lambda$), we find a linear relationship between $\omega$ and $\mathbf{k}$ (Fig. 5.4*b*). Note that there is no restriction on the size of either $\omega$ or $\mathbf{k}$, and in principle infinitesimally short-wavelength vibrations may be propagated. We also see that the phase and group velocities are equal. There is no dispersion. Sometimes the area of the rod is not treated explicitly (as for elastic strings), and then we invoke the mass per unit length of rod or string, $\rho'$. We then write

$$\frac{E}{\rho} = \frac{AE}{\rho'} = \frac{c}{\rho'}.$$

$c$ is called the force constant of the rod or string, and the elastic waves propagate with a velocity

$$v = \left(\frac{E}{\rho}\right)^{1/2} = \left(\frac{c}{\rho'}\right)^{1/2}.$$

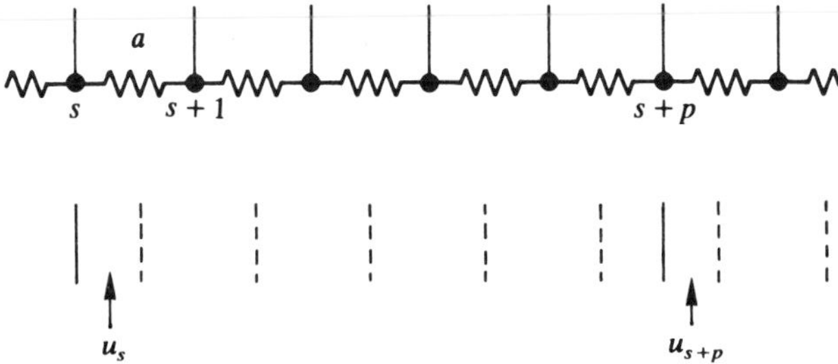


**Figure 5.5** A linear chain of identical point masses.

## 5.4 The Linear Lattice

Suppose now that we imagine a linear lattice of atoms that are at rest, each atom being at a distance $a$ from its neighbours (Fig. 5.5). If this appears artificial, we may think of sets of parallel planes in a real crystal. The longitudinal vibrations of such a set of planes reduce to those of a linear lattice. For small vibrational amplitudes we assume that the force on any one atom is proportional to its displacement relative to all the other atoms. We assume the *harmonic approximation.* In the notation of Fig. 5.5, we may then write for the resultant force on a chosen atom $s$ (the atoms of the chain are assumed numbered)

$$F_s = \sum_p c_p(u_{s+p} - u_s). \tag{5.12}$$

All the atoms in the chain interact with any given atom $s$; $p$ takes both positive and negative values and the force constant $c$ depends on $p$. We expect $c_p$ to be large for nearest and next nearest neighbours (i.e. $p = 1$ and 2 respectively) and to decrease rapidly as $p$ increases.

If each atom has mass $M$ then the equation of motion of the particular atom $s$ is

$$M\frac{\mathrm{d}^2 u_s}{\mathrm{d}t^2} = \sum_p c_p(u_{s+p} - u_s). \tag{5.13}$$

From experience, we expect harmonic vibrations, so we write

$$u_s = u\mathrm{e}^{\mathrm{i}(kx_s - \omega t)}, \qquad x_s = sa, \tag{5.14}$$

which represents a travelling wave of amplitude $u$. Clearly

$$u_{s+p} = u\mathrm{e}^{\mathrm{i}[k(s+p)a - \omega t]}. \tag{5.15}$$

The equation of motion becomes

$$Mu(\mathrm{i}\omega)^2\mathrm{e}^{\mathrm{i}(ksa - \omega t)} = \sum_p c_p u(\mathrm{e}^{\mathrm{i}k(s+p)a} - \mathrm{e}^{\mathrm{i}ksa})\mathrm{e}^{-\mathrm{i}\omega t},$$

which simplifies to

$$-M\omega^2 = \sum c_p(\mathrm{e}^{\mathrm{i}kpa} - 1). \tag{5.16}$$

Symmetry demands that $c_p = c_{-p}$, leading to

$$-M\omega^2 = \sum_{p>0} c_p(\mathrm{e}^{\mathrm{i}kpa} + \mathrm{e}^{-\mathrm{i}kpa} - 2),$$

or

$$\omega^2 = \frac{2}{M}\sum_{p>0} c_p(1 - \cos kpa) = \frac{4}{M}\sum_{p>0} c_p \sin^2 \tfrac{1}{2}kpa. \tag{5.17}$$

The qualitative significance of this exercise is maintained even when we restrict the interaction to that of nearest neighbours only, so, for convenience, we put $p = 1$ and obtain the following dispersion relation:

$$\omega^2 = \frac{4c_1}{M} \sin^2 \tfrac{1}{2}ka. \tag{5.18}$$

This is plotted in terms of $\omega$ and $k$ in Fig. 5.6: it should be compared with the behaviour of the homogeneous string (Fig. 5.4). We note

(a) there is a maximum vibrational frequency $\omega = 2(c_1/M)^{1/2}$, and

(b) the behaviour is periodic in $k$ with period $2\pi/a$.

The allowed frequencies $\omega$ vary in a periodic manner with $k$, but the repetition provides no new information over that contained in a given interval of $k$ with measure $2\pi/a$. Which interval should we choose? Clearly it is sensible to associate $\omega = 0$ with $k = 0$; $k = 0$ corresponds to a vibration of infinite wavelength and implies that the atoms are at rest or move rigidly without relative displacements. This then requires that we choose the interval $-\pi/a \leqslant k \leqslant +\pi/a$ as the significant one. What about values of $k$ that lie outside this interval, e.g. $k > \pi/a$? A little later we shall argue that it is better to avoid this question by using another interpretation. However, if we consider these larger $k$ values, they do not lead to new values of $\omega$, and the wave forms possess nodes and maxima that are not associated with the positions of the atoms. Thus with $k = 4\pi/a$ we obtain the situation depicted in Fig. 5.7. One can obtain similar figures for $\omega > 0$, but this diagram shows in a simple manner the unphysical character of the solutions when $k = |\mathbf{k}| > \pi/a$. We note that the interesting interval of $\mathbf{k}$ is $-\pi/a$ to $+\pi/a$, and this corresponds to a measure $2\pi/a$, which is none other than the base vector of the associated reciprocal lattice. Now in the real-space lattice we cannot distinguish any atom (or plane of atoms) from any other. Our choice of origin is purely arbitrary. Similarly, we expect all the atoms of the lattice to experience the same vibrational motion; *in fact the vibrational modes are a property of the lattice not of the individual atoms.* The

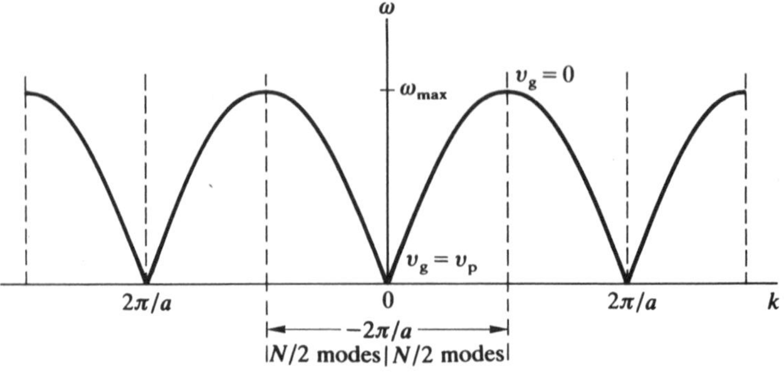


**Figure 5.6** The dispersion relationship for the linear chain of identical point masses with interaction restricted to that between nearest neighbours.

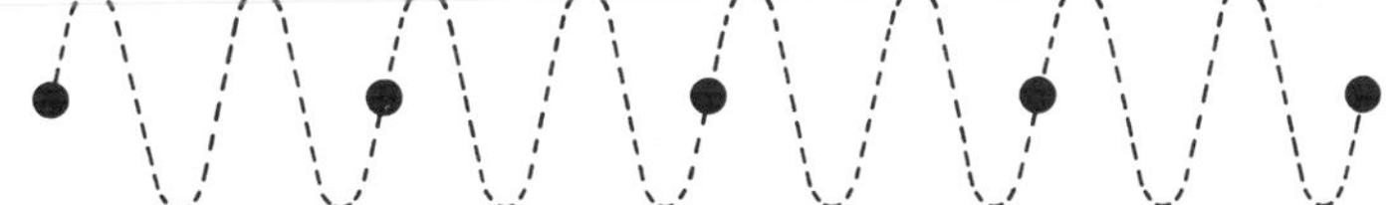

**Figure 5.7** The wave profile for $\omega = 0$, $k = 4\pi/a$. The wave form coincides with the positions of the masses only at certain nodes.

relationship between $\omega$ and $\mathbf{k}$, the dispersion relation, is a property of the lattice but is portrayed in the reciprocal lattice. We find that $\mathbf{k}$ space and reciprocal space are one and the same, not just dimensionally, but also with regard to scale. Now the propagation of waves in the real-space lattice is described in $\mathbf{k}$ space, and we must assert that, just as every cell in the real-space lattice is equivalent, so must we expect every cell in reciprocal space (i.e. $\mathbf{k}$ space) to be equivalent. If we find that the allowable frequencies are described within a unit cell of the reciprocal lattice then every such cell must contain identical information. However, whereas in the real-space lattice we normally use the ordinary primitive cell, in the reciprocal lattice we require the Wigner-Seitz cell (see Fig. 2.3).

The unphysical character of waves with $\lambda < 2a$, i.e. $|\mathbf{k}| > \pi/a$, can be circumvented if we accept the arbitrary nature of any choice of origin in the reciprocal lattice; thus we could label our periodic $\omega - k$ behaviour of the linear chain in the manner of Fig. 5.8.

This has several advantages.

(a) It shows that each cell of $\mathbf{k}$ space contains the same information.

(b) It demonstrates that physically the maximum allowable value of $|\mathbf{k}|$ is determined by the lattice spacing $|\mathbf{k}_{\max}| = \pi/a$; in other words there is a minimum allowable wavelength of $2a$.

(c) When $k \approx 0$, we find that $d\omega/dk = \omega/k =$ constant, similarly to the case for the continuous string. This is to be expected since $k \approx 0$ corresponds to vibrations

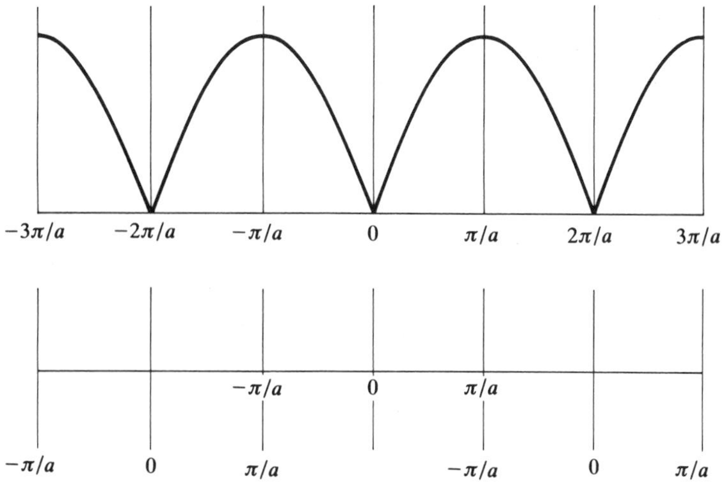


**Figure 5.8** If we use the fact that the origin of the reciprocal lattice is arbitrary, we may label the $k$ axis as in the lower figure and restrict the range of $k$ to $-\pi/a \leqslant k \leqslant \pi/a$.

with long wavelength, so long that the discrete atomic structure may be neglected. This is the case for the propagation of sound waves, and the velocity of sound controls the initial slope of the dispersion graph.

(d) It shows that when $k = +\pi/a$ and $-\pi/a$ we do not have two separate modes of vibration, but these two $k$ values represent the same mode. Furthermore we can never excite $k = \pi/a$ without introducing $k = -\pi/a$. This happens because the vibrational waves are subject to diffraction, and Bragg's law applies just as for the diffraction of X rays or neutrons. In the linear lattice, we can only have waves propagating along the lattice and the Bragg angle is always $\pi/2$, so that clearly the diffraction condition becomes

$$\lambda = 2d \sin \tfrac{1}{2}\pi = 2a. \tag{5.19}$$

In other words Bragg reflection arises when $k = \pm\pi/a$. Thus, even if we excited only the state $k = \pi/a$, we should obtain $-\pi/a$ through Bragg reflection. By a process of detailed balancing, we end up with equal amounts of $+\pi/a$ and $-\pi/a$ waves. This is just the situation that gives rise to a standing wave, and we can appreciate that the group velocity $d\omega/dk$ is zero because a standing wave does not transport energy.

In addition to longitudinal vibrations, the linear lattice supports transverse displacements leading to two independent sets (in mutually perpendicular planes) of vibrations that can propagate along the lattice. In the linear lattice, these transverse modes are degenerate.

Since the forces acting in a transverse displacement are different (they are in fact weaker) from those in a longitudinal one, they give rise to a new branch of dispersive modes lying below the longitudinal branch (Fig. 5.9). Normally any arbitrary displacement of the atoms in the chain excites many of these different modes. It is clear that one could extend this analysis to two- and three-dimensional structures

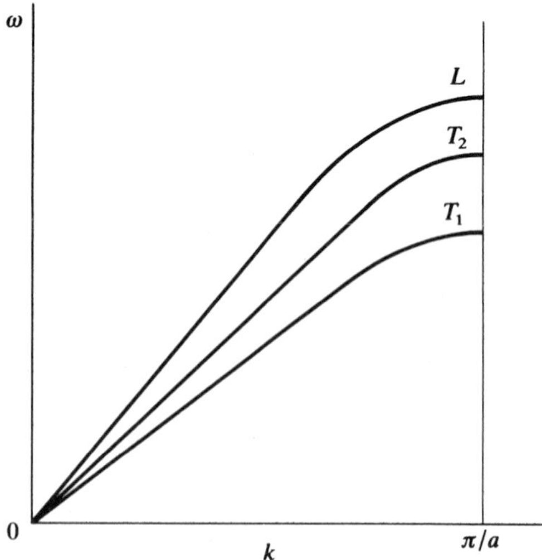


**Figure 5.9** Longitudinal and transverse modes. The transverse modes may be degenerate as in the cubic structures.

and calculate the behaviour of real solids. The snag is that we must then know the various force constants $c_p$. However, such calculations are not necessary for our present purpose. We need to appreciate the general features of lattice vibrational spectra rather than the details appropriate to particular specimens. We have already described three-dimensional lattices as collections of planes. Each family of planes can execute longitudinal and transverse vibrations, and there are particular dispersion relations connecting $\omega$ and $k$. The behaviour is describable in the three-dimensional Wigner-Seitz cell of the reciprocal lattice. Any study, either theoretical or experimental, of the dispersion relations throughout the whole reciprocal cell demands much labour and usually attention is confined to important symmetry directions. Experimental data for Al are shown in Fig. 5.10; one immediately notes the similarity with those of our very elementary model (Fig. 5.9). We can at once appreciate the important aspects in terms of the discrete atomic character of the structure producing maximum values of $k$ and $\omega$. By a combination of selected measurement and interpolation, the data may be extended to cover the whole of the reciprocal cell.

## 5.5 Counting Modes

Einstein treated each vibrating atom as a quantum oscillator and thereby estimated the heat capacity of the solid. However, if we have access to the complete dispersive behaviour throughout the reciprocal lattice cell then we should be able to calculate the vibrational frequency spectrum and thereby the lattice heat capacity. We shall now do this and first show that *in* **k** *space the modes are distributed uniformly with a constant density.*

First, however we must comment on terminology. A mode is a vibration of given wave vector **k**, frequency $\omega$ and energy $E = \hbar\omega$. Sometimes we use the term 'state' instead of mode, particularly when discussing the behaviour of electrons. We may define quantities $N(\omega)$, $N(E)$ and $N(k)$ that are the densities of modes or states, i.e. the number of modes or states at a particular neighbourhood ($\omega$, $E$, $k$) per unit interval of $\omega$, $E$ or $k$ in unit measure of the sample, e.g. per unit volume, per atom or per mole.

Suppose we ask how many modes or states are to be found within an interval between two modes ($\omega$, $E$, $k$) and ($\omega + \Delta\omega$, $E + \Delta E$, $k + \Delta k$). Let this number be $\Delta N$. The we find that

$$\Delta N = N(\omega)\,\Delta\omega = N(E)\,\Delta E = N(k)\,\Delta^3 k,$$

$N(\omega)$ and $N(E)$, which are simply related, differ markedly from $N(k)$, the latter being significant because it is independent of $k$. It is through $N(k)$ that we arrive at $N(\omega)$ or $N(E)$. $\Delta^3 k$ is the volume of **k** space enclosed between the two energy surfaces $E$ and $E + \Delta E$.

Real solids have finite size and to avoid the effects of the surfaces we assume *periodic boundary conditions.* Thus in our macroscopic specimen we select a sample element well within the specimen and of a size large compared with atomic dimensions. We then demand that opposite ends of our chosen element vibrate with the same amplitude and phase. In terms of our linear lattice, this means the boundaries of our sample are atoms $s$ and $s + N$ (Fig. 5.11). It has length $Na$, where $N$ is a large number. We now demand that atoms $s$ and $s + N$ vibrate in phase and with the

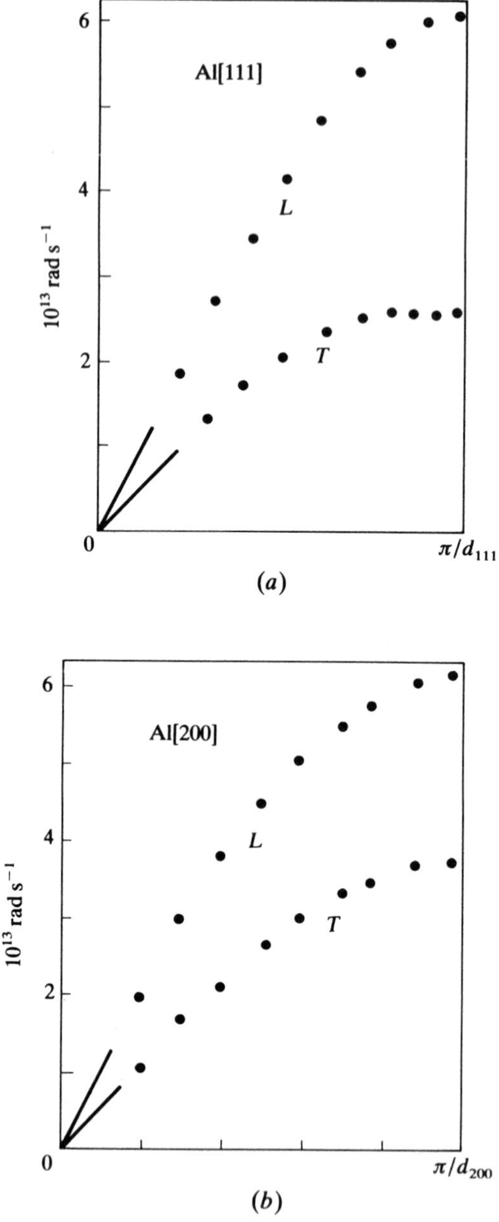


**Figure 5.10** Lattice vibrational modes along [111] and [200] for Al; note the general similarity to the behaviour of the linear atomic chain. (After Stedman and Nilsson 1966.)

same amplitude. Another way of looking at the problem would be to imagine the $N$ cells of the sample bent into a continuous ring. In either case we have no surface problem. Recalling the expression for the vibrational amplitude (5.14), we write

$$u_s = u_{s+N},$$

$$u\,e^{iksa} = u e^{ik(s+N)a},$$

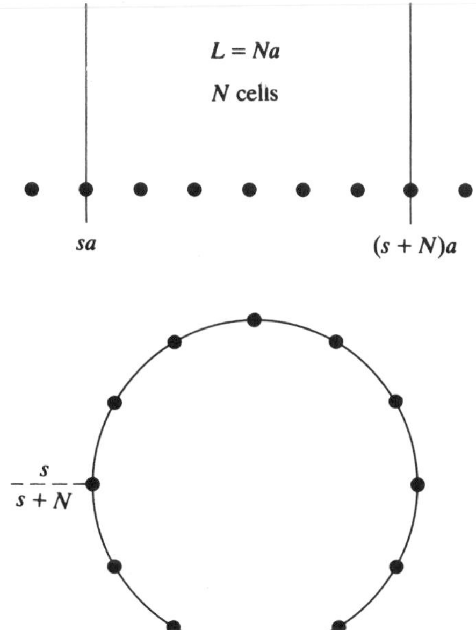


**Figure 5.11** Periodic boundary conditions applied to the linear chain. We demand that atoms $s$ and $s + N$ vibrate with the same amplitude and phase; this is most easily imagined if we form an endless circular chain.

which requires that

$$e^{ikNa} = 1,$$
$$kNa = n2\pi, \quad n \text{ integral}.$$

The separation of two adjacent modes is the smallest value of $\Delta k$ that can arise, and

$$\Delta k = \Delta n \frac{2\pi}{Na}.$$

Clearly, since $n$ is an integer, $\Delta n = 1$, and the smallest value of $\Delta k$ is

$$\Delta k_{\mathrm{m}} = \frac{2\pi}{Na}. \tag{5.20}$$

$\Delta k_{\mathrm{m}}$ is independent of $k$ and determined solely by the size of the sample. Thus for a given sample there is a uniform distribution of modes in **k** space.

The density, i.e. the number of modes per unit interval of **k** space, is $1/\Delta k_{\mathrm{m}}$ or $Na/2\pi$.

The density of modes in one-dimensional **k** space is

$$\frac{Na}{2\pi} = \frac{L}{2\pi}, \tag{5.21}$$

$L$ being the length of the sample.

For a linear chain of length say 1 cm and lattice spacing of order $10^{-8}$ cm, we find $\Delta k_{\mathrm{m}}$ to be of order 1 $\mathrm{cm}^{-1}$; but the Wigner-Seitz cell of the reciprocal lattice has a size of order $10^8$ $\mathrm{cm}^{-1}$. Thus there are some $10^8$ modes evenly distributed throughout the one-dimensional Wigner-Seitz cell. Although the distribution of modes is discrete, they are so closely spaced that they form what is called a quasicontinuous distribution, i.e. one that is discrete but for all practical purposes continuous.

We associate one longitudinal and two transverse modes with each **k** value; for the linear lattice the total number of modes in the sample is 3 × density of states in **k** space × size of Wigner-Seitz cell in the reciprocal lattice. In other words

$$3\,\frac{Na}{2\pi}\,\frac{2\pi}{a} = 3N, \tag{5.22}$$

so that altogether our linear lattice contains $3N$ modes.

We can develop this argument to two- and three-dimensional lattices, applying the periodic boundary conditions to the principal symmetry axes. Thus if our specimen has dimensions $N_1\mathbf{a}$, $N_2\,\mathbf{b}$ and $N_3\,\mathbf{c}$, where **a**, **b** and **c** are the base vectors of an orthorhombic primitive cell, then again our vibrational modes are distributed uniformly throughout the Wigner-Seitz cell of the reciprocal lattice with a density

$$\frac{1}{\Delta^3 k_{\mathrm{m}}} = \frac{N_1 N_2 N_2\, abc}{8\pi^3} = \frac{V}{8\pi^3}, \tag{5.23}$$

where $V$ is the volume of the sample. This density applies to each branch of the vibrational modes and for all Bravais lattices. We should therefore imagine the modes as points uniformly distributed throughout the primitive cell of **k** space, and each point is associated with a volume $\Delta^3 k_{\mathrm{m}}$.

The number of points depends on the volume of the sample – if this seems inconvenient, one can get around it by talking of the density of states per primitive cell of sample, which is a constant quantity independent of macroscopic size.

> The density of modes per primitive cell of a three-dimensional sample is
>
> $$\Omega/8\pi^3$$
>
> per branch of modes, where $\Omega$ is the volume of the primitive cell in the direct lattice.

Henceforth we shall often use the word ‘state’ as well as ‘mode’. We shall use the symbol $N(k)$ to denote the density of states in **k** space, and the context will dictate whether we use unit volume of sample or the volume of the primitive or unit cell.

Normally we are more interested in the quantity $N(\omega)$, the frequency spectrum, which represents the density of states per unit frequency interval per given volume of sample. This we now determine.

Each point in **k** space corresponds to a given vibrational frequency. We can connect all points with the same frequency to produce a contour of constant frequency, and for a three-dimensional sample this is a surface. For small frequencies, and therefore small **k** values, we have seen that there is no dispersion; for any given branch of the spectrum the velocity of propagation is isotropic; a surface of constant frequency is then a sphere. At larger energies the anisotropy of dispersion causes greater and greater distortions from spherical form, but the constant-frequency surfaces always possess the symmetry of the crystal. Eventually the surface must contact the boundaries of the Wigner-Seitz cell. The contours of constant frequency always make normal intersections with the cell boundaries in order that the continuity conditions between adjacent cells be fulfilled.

Figure 5.12 shows a small portion of two adjacent surfaces of constant frequency corresponding to frequencies $\omega$ and $\omega + d\omega$. Note that owing to dispersion these surfaces are not normal to the direction of the wave vector **k**. Consider a small pill box bounded by the surfaces $\omega$ and $\omega + d\omega$ centred around the point **k**. The unit vector normal to the frequency surfaces is **n**. The pill box has area $dS\,\mathbf{n}$.

*For a given vibrational branch* the pill box of volume $d^3k$ contains $N(k)\, d^3k$ states, i.e., from (5.23),

$$\frac{V}{8\pi^3}\, d^3k \text{ states.}$$

Thus between the surfaces $\omega$ and $\omega + d\omega$ we must have

$$\int_{S_\omega} \frac{V}{8\pi^3}\, d^3k \text{ states per branch,} \tag{5.24}$$

where the integral is taken over the shell of constant energy $\omega$ and thickness $d\omega$. If, however, the density of states per unit frequency range is $N(\omega)$ then $N(\omega)\, d\omega$ is the same quantity as given in (5.24). Thus

$$N(\omega)\, d\omega = \frac{V}{8\pi^3} \int_{S_\omega} d^3k. \tag{5.25}$$

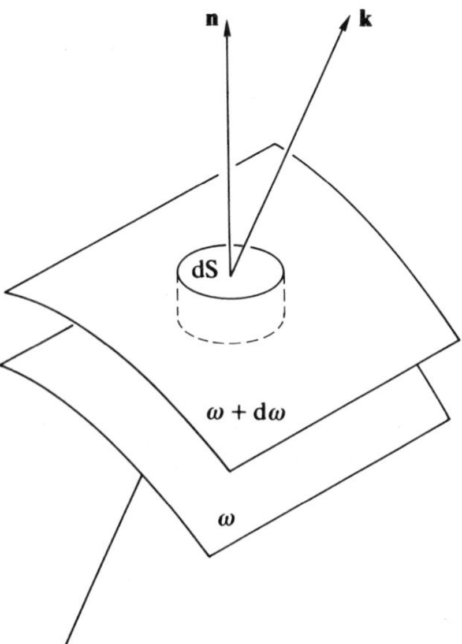


**Figure 5.12** A small element of **k** space bounded by surfaces of constant frequency $\omega$ and $\omega + d\omega$.

But the volume of the pill box is given by

$$d^3k = dS\ \mathbf{n} \cdot \mathbf{dk} \tag{5.26}$$

and

$$d\omega = \mathbf{\nabla_k}\,\omega \cdot \mathbf{dk}. \tag{5.27}$$

Since we are dealing with surfaces of constant frequency, there can be no component of $\mathbf{\nabla_k}\,\omega$ tangential to them and only the normal component of $\mathbf{\nabla_k}\,\omega$ is finite. Thus

$$d\omega = |\mathbf{\nabla_k}\,\omega|\,\mathbf{n} \cdot \mathbf{dk}, \tag{5.28}$$

and, substituting in (5.25) and (5.26), we obtain

$$N(\omega) = \frac{V}{8\pi^3} \int_{S_\omega} \frac{dS}{|\mathbf{\nabla_k}\,\omega|} \tag{5.29}$$

$$= \frac{V}{8\pi^3} \int_{S_\omega} \frac{dS}{v_g}. \tag{5.30}$$

Equation (5.29) is a result of great importance in solid state physics. We shall use it not only in the present connection, but also when we consider the properties of electrons in solids. If we write $\hbar\omega = E$ then we can also express (5.29) in terms of the energy:

$$N(E) = \frac{V}{8\pi^3} \int_{S_E} \frac{dS}{|\mathbf{\nabla_k}\,E|}. \tag{5.31}$$

We have been able to establish very general principles and formulae concerning the vibrational states of crystals on the basis simply of their periodic structure. Of course the application of (5.29) demands that we know the detailed form of the contours of constant frequency which are derived from the dispersion relations. These have now been measured (and in certain cases calculated) for many elements.

We have shown the example of Al in Fig. 5.10. Now each longitudinal or transverse branch gives rise to its own frequency spectrum, and those corresponding to the data of Fig. 5.10 are shown in Fig. 5.13. Note how the transverse modes have lower frequencies than the longitudinal ones; the latter arise as a rather sharp band just below the cut-off frequency. Near $\omega = 0$ we see the similarity to the Debye spectrum (Fig. 5.14); this is understandable since Debye assumed the dispersion law $\omega = v_g k$ to be valid and this is the case at very low frequencies, but we must expect different velocities for the longitudinal and transverse modes.

Spectra such as those of Fig. 5.13 may be used to calculate the heat capacity as a function of temperature, and good agreement with experiment has been obtained. It will be noted that the spectrum of Fig. 5.13 shows characteristic peaks and cusps indicating singular behaviour. Such features arise in all phonon spectra and originate at points in the Wigner-Seitz cell where $\mathbf{\nabla_k}\,\omega = 0$. They are known as van Hove singularities.

## 5.6 The Debye Model Revisited

As an instructive exercise, we may use the analysis of the previous section to evaluate the Debye heat capacity. We neglect the difference between longitudinal and

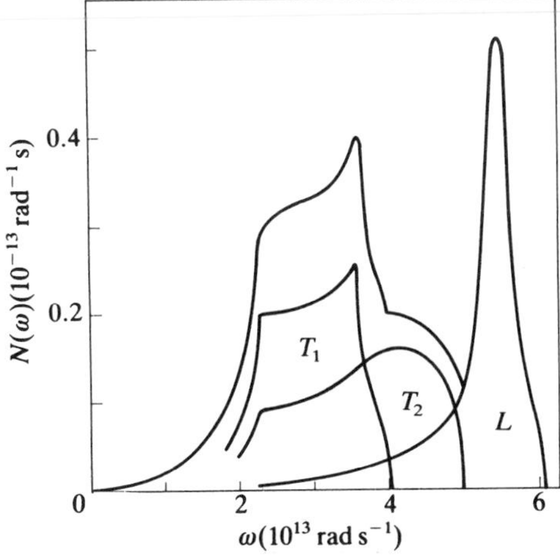


**Figure 5.13** The measured vibrational spectrum for Al. (After Stedman *et al.* 1967.)

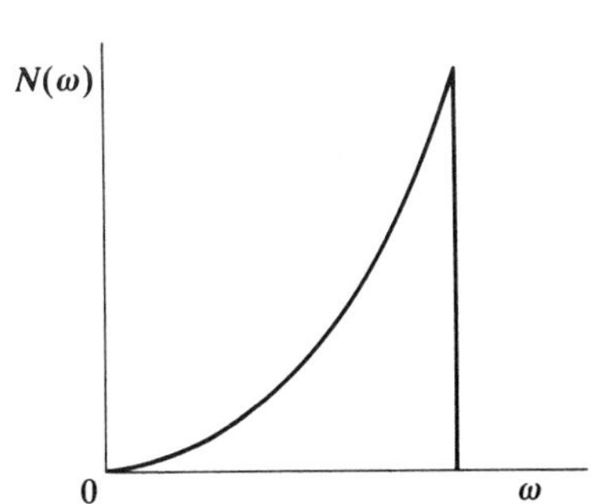


**Figure 5.14** The Debye spectrum.

transverse modes and assume a single linear isotropic dependence of $\omega$ on $k$. This produces spherical surfaces of constant frequency. We have seen that for *each branch* of modes

$$N(\omega) = \frac{V}{8\pi^3}\int \frac{\mathrm{d}S_\omega}{\nabla_{\mathbf{k}}\,\omega} = \frac{V}{8\pi^3 v_\mathrm{g}}\int \mathrm{d}S_\omega .$$

Furthermore,

$$\int \mathrm{d}S_\omega = 4\pi k^2, \qquad \omega = v_\mathrm{g}\, k,$$

so that

$$N(\omega) = \frac{V}{8\pi^3 v_\mathrm{g}}\, 4\pi k^2 = \frac{V}{2\pi^2}\,\frac{\omega^2}{v_\mathrm{g}^3}\,. \tag{5.32}$$

The total number of modes is $3N$, where $N$ is the number of atoms (primitive cells) in the sample. Thus for each of the three identical branches of modes

$$\int_0^{\omega_{\max}} N(\omega)\,\mathrm{d}\omega = N = \frac{V}{6\pi^2}\,\frac{\omega_{\max}^3}{v_\mathrm{g}^2},$$

whence

$$\omega_{\max}^3 = \frac{6N\pi^2}{V}\, v_\mathrm{g}^3, \tag{5.33}$$

where $v_\mathrm{g}$ is the velocity of sound in the sample and is independent of $k$ in the Debye model. The heat capacity according to Debye is then calculable using the above expressions for $N(\omega)$ and $\omega_{\max}$ in (5.7). We see that the Debye approximation replaces the integration over the Wigner-Seitz cell of the reciprocal lattice and the true dispersion by isotropic behaviour and a limiting spherical frequency surface. Approximations valid at low and at high temperatures may be evaluated (see Problem 5.2).

## 5.7 Acoustic and Optical Modes

Our previous discussion assumed the sample to be a pure element in which the atoms occupy a Bravais lattice; all atoms are then equivalent and must experience the same motion. For the linear lattice (except in the immediate neighbourhood of a node) we expect the particles in a given short section, large enough to contain several atoms, to be moving in the same direction at any given instant of time. This is the situation when sound propagates in a solid, and the vibrational modes described in Figs. 5.9 and 5.10 are therefore known as the *acoustic branches.*

Suppose, however, that our sample has a structure based on two different kinds of atom, e.g. CsCl, or two similar atoms occupying non-equivalent positions, as in Mg, which has hcp structure. The crystallographic unit cell now contains two different atoms. In the linear lattice this can be represented by two different masses as in Fig. 5.15. Now we must assume different displacements for the different atom types, and a similar analysis to that in Section 5.4 leads to the following equations (nearest-neighbour interaction):

$$M_1 \ddot{u}_s = c_1(v_{s+1/2} + v_{s-1/2} - 2u_s),$$

$$M_2 \ddot{v}_{s+1/2} = c_1(u_{s+1} + u_s - 2v_{s+1/2}).$$

If we assume that

$$u_s = u e^{i(ksa - \omega t)},$$

$$v_{s+1/2} = v e^{i[k(s+1/2)a - \omega t]}$$

then we obtain the following dispersion relation with nearest-neighbour interaction:

$$M_1 M_2 \omega^4 - 2c_1(M_1 + M_2)\omega^2 + 4c_1^2 \sin^2 \tfrac{1}{2}ka = 0. \tag{5.34}$$

The latter is readily solved to give

$$\omega^2 = c_1\left(\frac{M_1 + M_2}{M_1 M_2}\right) \pm \left[c_1^2\left(\frac{M_1 + M_2}{M_1 M_2}\right)^2 - \frac{4c_1^2}{M_1 M_2}\sin^2 \tfrac{1}{2}ka\right]^{1/2} \tag{5.35}$$

or

$$\omega_\pm^2 = A \pm (A^2 - B \sin^2 \tfrac{1}{2}ka)^{1/2}. \tag{5.36}$$

There are now two solutions for $\omega^2$, providing two distinctly separate groups of vibrational modes. The first group, associated with $\omega_-^2$, contains the acoustic modes found earlier. The second groups arises with $\omega_+^2$ and contains the *optical modes*; these correspond to the movement of the different atom sorts in opposite directions, it is a contra-motion whereas the acoustic behaviour is motion in unison (see Problem 5.8). The name 'optical modes' arises because in ionic crystals, like CsCl, they cause an electric polarization and can therefore be excited by light, which as a result is strongly absorbed. However, these optical modes occur in all structures

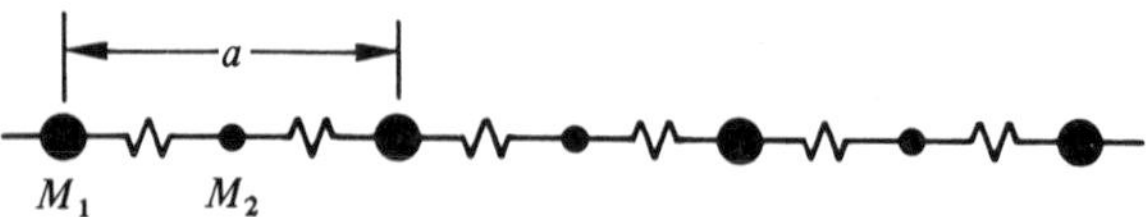


**Figure 5.15** A diatomic linear chain.

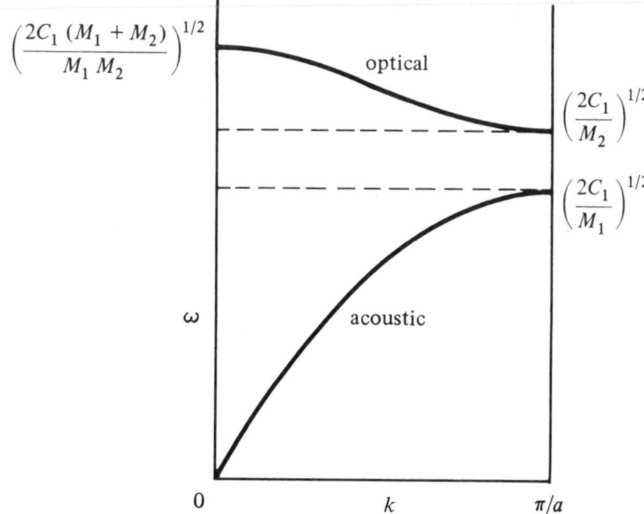


**Figure 5.16** Optical and acoustic modes. The optical modes lie at higher frequencies and show less dispersion than the acoustic modes. The limiting frequencies are obtained directly from (5.35) using the appropriate $k$ values.

with two or more different atoms (chemical or structural difference). As Fig. 5.16 shows, the optical modes lie at higher frequencies than the acoustic branches and the dispersion is much weaker. Generally in a substance characterized by $p$ atom sorts there occur one group of acoustic $(L + 2T)$ modes together with $p - 1$ groups of optical $(\mathrm{L} + 2T)$ modes. The different groups are separated by intervals of frequency that are prohibited for collective vibrations. The crystal acts as a mechanical filter with passing and stopping bands.

## 5.8 Attenuation

Returning to the simple dispersion curve of Fig. 5.6, we may ask what happens when we attempt to excite a lattice vibration with frequency greater than the maximum value arising at the cell boundary. We have seen that, (5.18),

$$\omega^2 = \frac{4c}{M}\sin^2 \tfrac{1}{2}ka$$

and

$$\omega_{\max} = 2(c/M)^{1/2},$$

corresponding to the maximum value of $\sin^2 \frac{1}{2}ka$.

If $\omega > \omega_{\max}$ then $\sin^2 \frac{1}{2}ka > 1$, and this can only be considered if we allow $k$ to become complex. For $\omega > \omega_{\max}$ we therefore write

$$k = \pm\left(\frac{\pi}{a} + \mathrm{i}\alpha\right). \tag{5.37}$$

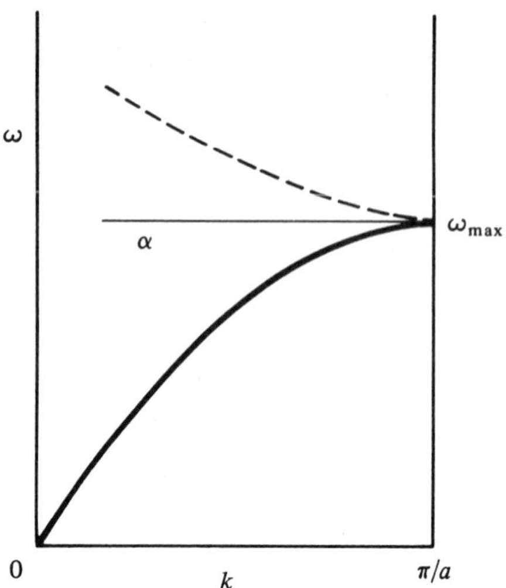


**Figure 5.17** For $\omega > \omega_{max}$ strong attenuation arises. $\alpha$ is the attenuation coefficient, which increases markedly when $\omega$ exceeds $\omega_{max}$. $\alpha$ is plotted horizontally to the left.

On insertion in (5.14), we see immediately that the amplitude of vibration is damped by a factor $e^{-\alpha x_s}$; in the harmonic approximation $\alpha$ is zero when $\omega \leqslant \omega_{max}$, but increases rapidly as $\omega$ becomes larger than $\omega_{max}$ and so prevents the occurrence of normal modes (Fig. 5.17). It is straightforward to show that (5.37) leads to the dispersion relation

$$\omega^2 = \frac{4c}{M} \cosh^2 \tfrac{1}{2}\alpha a. \tag{5.38}$$

## 5.9 Phonons and Quantization

The collective motion of a periodic arrangement of atoms has been described in terms of normal modes, each being a travelling wave of the form (5.14)

$$u_s = u e^{i(ksa - \omega t)}.$$

By analogy with the case of light waves and photons, together with a knowledge of the quantum properties of the harmonic oscillator, we associate a quantum of energy $\hbar\omega$ with a mode of frequency $\omega$ (Fig. 5.18). The energy content of a lattice mode is associated with the amplitude of vibration $u_s$ and corresponds to an integral number of quanta. The quanta of lattice vibrations are known as *phonons*.

A discrete vibrational state or mode has a well defined wave vector $\mathbf{k}$ and an application of de Broglie's principle implies a linear momentum $\mathbf{p} = \hbar\mathbf{k}$. However, the only true mechanical momentum that can arise is a rigid motion of the whole

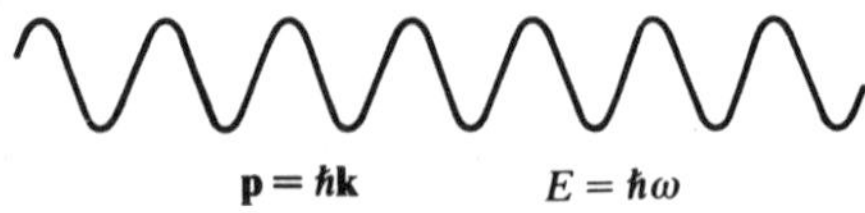


**Figure 5.18** The quantum of lattice vibration is called the phonon.

specimen. The vibrational modes transport energy, but they are completely described by the *relative motion* of the individual atoms, whose average displacement is zero; they cannot therefore contain a net momentum and the quantity $\hbar\mathbf{k}$ is not to be identified with the conventional linear momentum.

On the other hand, vibrational modes interact and this interaction demands that not only energy be conserved but also the quantity $\hbar\mathbf{k}$ – the latter is therefore often called the *crystal momentum* or pseudomomentum. Two (or more) interacting modes, or phonons as we shall now call them, obey the following conservation laws (Fig. 5.19):

$$\left.\begin{aligned}\hbar\omega_1 + \hbar\omega_2 &= \hbar\omega_3,\\ \hbar\mathbf{k}_1 + \hbar\mathbf{k}_2 &= \hbar\mathbf{k}_3 + \hbar\mathbf{G}.\end{aligned}\right\} \tag{5.39}$$

(The presence of $\hbar\mathbf{G}$ is explained in the next section.)

## 5.10 Brillouin Zones

So far, our discussion of phonons has been couched in terms of the Wigner-Seitz cell of the reciprocal lattice. When considering the propagation of waves in periodic structures, it is more customary to call this cell the first *Brillouin zone*, and it is the only zone of physical significance for lattice vibrations. In later work concerning electrons in periodic potentials, we shall consider higher Brillouin zones, and reference to the second zone is often made in the case of phonons. We must return to Fig. 5.6 and the periodic variation of $\omega$ with $k$. Suppose, contrary to what we said in Section 5.4, that we choose an arbitrary zero in the reciprocal lattice and let $k$ vary monotonically, taking higher and higher values. Our significant interval is still $2\pi/a$ and $\mathbf{k}$ space is divided into a series of regions centred around our chosen origin. Brillouin suggested that this division be made as in Fig. 5.20. Each interval of similar shading corresponds to a particular interval $2\pi/a$, but only the central region corresponds to a continuous part in the linear lattice. Successive regions are numbered 1, 2, 3 etc.; these are the 1st, 2nd, 3rd, etc., Brillouin zones. Clearly a given vibrational mode or phonon characterized by a specific frequency can now be represented by different wave vectors in the different zones. Suppose that $\mathbf{k}$ is an arbitrarily chosen vector (not necessarily in the 1st zone); then we can always express $\mathbf{k}$ as

$$\mathbf{k} = \mathbf{k}_1 + \mathbf{G}, \tag{5.40}$$

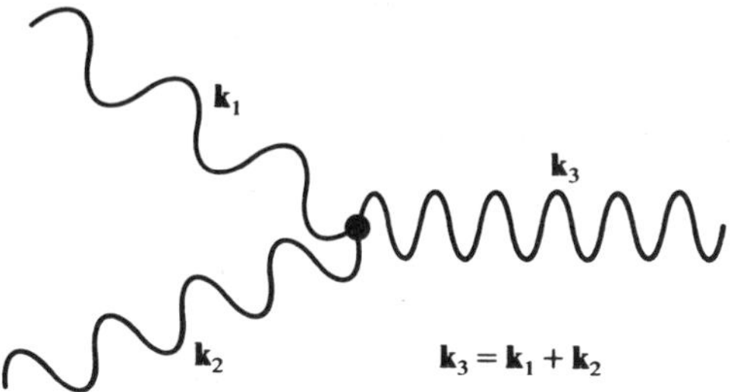


**Figure 5.19** Phonons interact, conserving energy and wave vector.

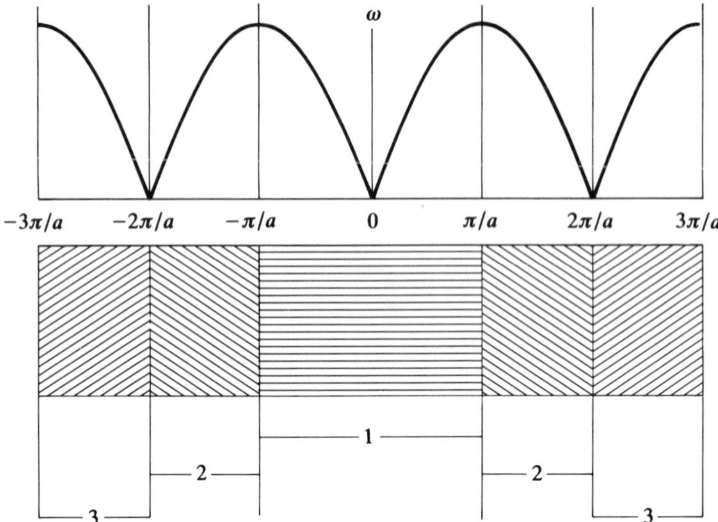


**Figure 5.20** The concept of the Brillouin zone in the linear lattice. The first three zones are shown. The second and higher zones always consist of separate parts symmetrically placed about the origin. All the zones have the same total size.

where $\mathbf{k}_1$ is the value appropriate to the particular mode in the first zone and $\mathbf{G}$ is, as usual, a vector of the reciprocal lattice. This means that the wave vector of a phonon is indeterminate with regard to any reciprocal lattice vector $\mathbf{G}$. Our earlier description may be called that of the repeated first Brillouin zone, whereas the present is that of the extended zone scheme. However, for the moment it is sufficient for us to remember that the first Brillouin zone is identical with the Wigner-Seitz cell of the reciprocal lattice.

The zone boundaries are determined by the planes bisecting the lines joining a particular point in the reciprocal lattice with its neighbours. These planes define $\mathbf{k}$ values for which Bragg reflection arises, and this allows us to express the equation for the zone faces in a simple manner. For Bragg reflection, $\Delta\mathbf{k} = \mathbf{k}' - \mathbf{k} = -\mathbf{G}$ (Fig. 3.1*b*) and $|\mathbf{k}'| = |\mathbf{k}|$, so we obtain the equation for the zone faces, assuming that the structure factor is finite,

$$2\mathbf{k} \cdot \mathbf{G} - G^2 = 0, \qquad S_{hkl} > 0, \tag{5.41}$$

or

$$\frac{\mathbf{k} \cdot \mathbf{G}}{G} = \tfrac{1}{2}G$$

(see Figs. 5.21–5.23). The boundaries of the higher zones also satisfy (5·41).

## 5.11 Inelastic Neutron Scattering

The photoelectric effect clearly demonstrates the interaction between light quanta and a mechanical particle – the electron. Similarly, phonons may interact with one another, light quanta and mechanical particles provided that the conditions (5.39)

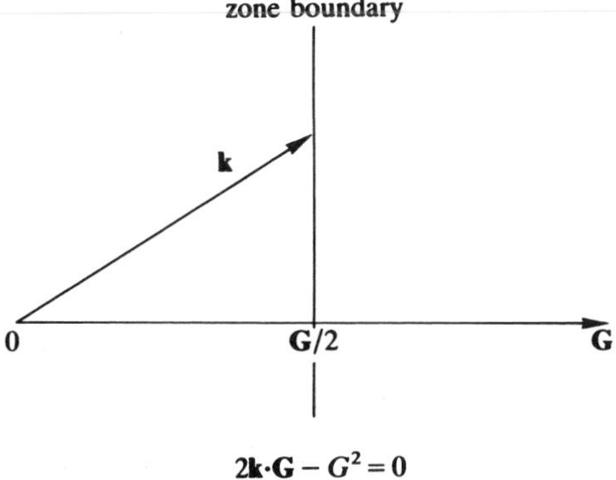


**Figure 5.21** The equation for the zone boundary. It is derived from the condition for lattice diffraction, but the structure factor must be finite.

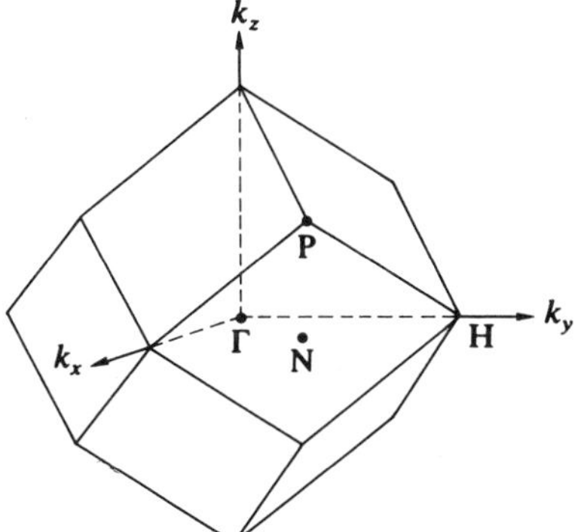


**Figure 5.22** The first Brillouin zone for the bcc lattice. Significant symmetry points are marked by the conventional symbols. Γ is always used to denote the centre of a zone.

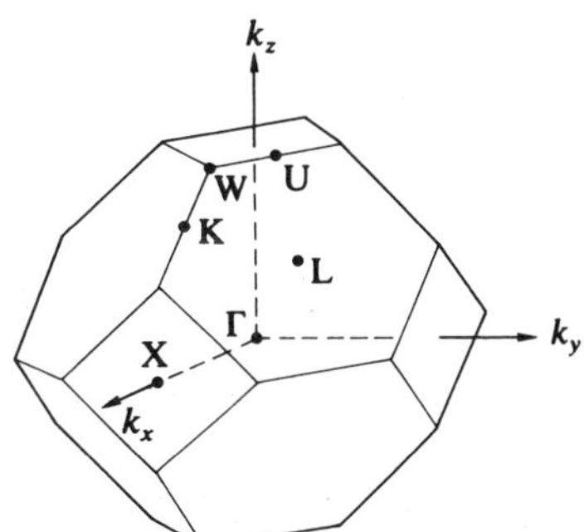


**Figure 5.23** The first Brillouin zone for the fcc lattice and the conventional symbols for important symmetry points.

are satisfied. A phonon can therefore exchange energy with electromagnetic or particle radiation, and this provides a mechanism for the experimental determination of phonon dispersion curves. The method is to illuminate the sample with a collimated probing beam of well defined energy. Certain photons or particles will excite phonons or themselves accept quanta from them. The probability that only a single phonon is exchanged is so much greater than other possibilities that we confine attention to this case. In such an interaction the inelastically scattered radiation leaves the specimen with altered energy and wave vector. We now need to distinguish between the probe and the phonons, *so henceforth we shall represent the phonon by* $(\omega\mathbf{q})$ and reserve $(\nu\mathbf{k})$ for photons or particles. Our equations (5.39) become

$$\mathbf{k} - \mathbf{k}' = \mathbf{q} + \mathbf{G}$$

$$\frac{\hbar^2 k'^2}{2m} - \frac{\hbar^2 k^2}{2m} = \pm\hbar\omega \quad \text{(particles)},$$

$$h\nu' - h\nu = \pm\hbar\omega \quad \text{(electromagnetic radiation)}.$$

**Table 5.1**

| | Energy (eV) | Frequency (rad $s^{-1}$) | Approximate wavelength (Å) | Approximate wave vector ($Å^{-1}$) |
|---|---|---|---|---|
| Phonon | 0.013 | $2 \times 10^{14}$ | 3 | 2 |
| X-ray photon | 4100 | | 3 | 2 |
| Electron | 16 | | 3 | 2 |
| Neutron | 0.009 | | 3 | 2 |

If we can measure the changes in energy and wave vector of the probing rays then we can determine related values of $\omega$ and $\mathbf{q}$. However, let us consider the quantities involved. A typical frequency in the middle of the acoustic branch has an energy roughly equivalent to half the Debye temperature, and $\theta_D$ for many metals is about 300 K. Thus

$$\hbar\omega = k_B \times 150 = 0.013 \text{ eV}$$

and

$$\omega \approx 2 \times 10^{14} \text{ rad s}^{-1}, \quad q \approx 2 \text{ Å}^{-1}, \quad \lambda \approx 3 \text{ Å}.$$

We now consider the data shown in Table 5.1. Although possible in principle, practical aspects ordinarily prevent the use of X rays or electrons in this process. The X-ray photon with energy of about 4000 eV has a width of order 1 eV, very much larger than the energy exchange that can arise with phonons. The energy of the electron may be accurately controlled, but the cross-section for the event of interest is much smaller than for other scattering processes; furthermore, the strong absorption of electrons in solids prohibits their use.† Neutrons, on the other hand, are ideally suited in terms of energy and wave vector but also on account of their lack of charge and very penetrating character. They are the only radiation that allows the complete study of phonon dispersion. The principal features of the experiment are illustrated in Fig. 5.24.

## 5.12 Thermal Properties

In real solids the forces between atoms are not truly harmonic. The potential energy varies with interatomic separation as shown schematically in Fig. 5.25. The higher the temperature, the more phonons that are excited and the more pronounced the effects of anharmonicity. This leads to changes in the equilibrium separation of the atoms. The lattice therefore expands. Thermal expansion arises as a direct result of the asymmetrical dependence of potential energy on atomic separation, i.e. the anharmonicity. It is therefore not surprising that the variation of the coefficient of thermal expansion with temperature parallels that of the heat capacity (Fig. 5.26). When discussing the thermal behaviour of solids, it is important to remember that *the energy content of the lattice resides in the phonons and not in particular atoms.*

† Electrons are, however, suitable for the study of surface vibrational modes.

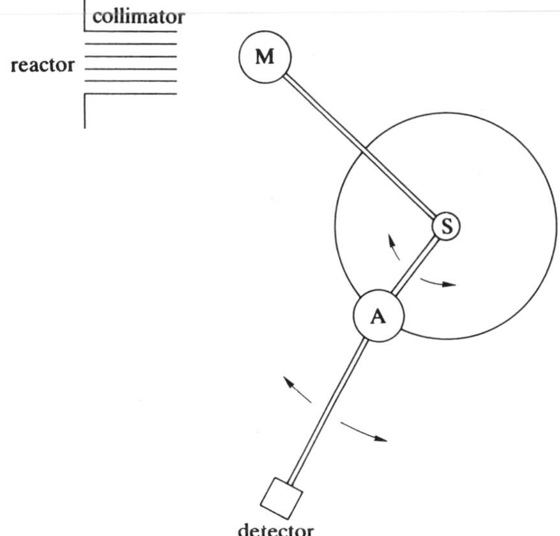


**Figure 5.24** A schematic arrangement of a 'triple-axis' or 'three-crystal' spectrometer. Thermal neutrons from the reactor are collimated and fall on a large crystal M, of say Cu. By Bragg reflection from this crystal, a monochromatic beam is obtained; to provide a choice of wavelength, the whole apparatus may be rotated about M (1st axis). The monochromatic neutrons impinge on the single-crystal sample S. The neutrons are scattered by the sample and the scattered beams are analysed in a third crystal A, using Bragg reflection to determine their wavelengths. In order to cover the scattering in the various directions, the analyser crystal and detector must be rotated around S (2nd axis); furthermore, to analyse the scattered beams, the detector must be rotatable about the analyser crystal (3rd axis).

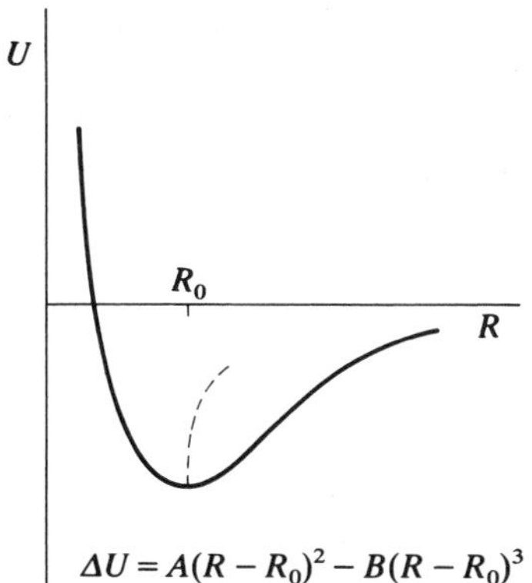


**Figure 5.25** Anharmonicity in the pair potential.

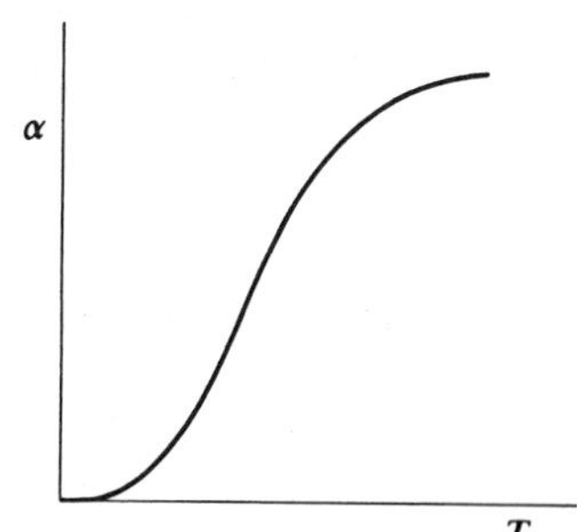


**Figure 5.26** The temperature variation of the coefficient of thermal expansion mirrors that of the heat capacity.

This may seem paradoxical, but the point is that we cannot convey thermal energy to a particular atom (we exclude the case of isolated impurity atoms). We cannot cause an atom to vibrate without exciting a collective oscillation, a phonon. In a pure crystal, and within the harmonic approximation, two or more phonons may pass through one another without interaction. At the surface of the sample they would be reflected and continue to exist in unchanged form. In real solids we find imperfections and anharmonicity, which cause phonons to interact, to be scattered and to decay. Phonons therefore have finite lifetime and we must think of *phonon wave packets with particle-like properties.* These phonons behave very much like particles in a gas: they constantly interact (collide) and are in a kinetic equilibrium at a given temperature. That this is the case is particularly evident in the thermal conductivity. In non-metals heat is conducted solely by the phonons and one might expect thermal energy to be transported at a speed corresponding roughly to that of sound. This may be the case in nearly perfect single crystals at very low temperatures. However, the mutual interactions of phonons lead to mean free paths of order 100 Å, and the phonon gas therefore conducts heat in a similar manner to an ordinary gas. The energy diffuses through the gas and this is a slow process. In fact, Debye, long before the concept of the phonon was developed, applied the results of simple kinetic gas theory to solids with remarkable qualitative success.

The elementary kinetic theory of gases gives the following expression for the thermal conductivity $K$:

$$K = \tfrac{1}{3}C\langle v\rangle\Lambda, \tag{5.42}$$

where $\langle v\rangle$ is the average particle velocity, $\Lambda$ the mean free path and $C$ the heat capacity per unit volume. Since $\langle v\rangle$ varies little, the important quantities are $C$ and $\Lambda$. At low temperatures few phonons are excited and they have low frequencies and long wavelengths; thus the gas is of low density and we find large mean free paths. On the other hand, $C = 0$ at 0 K and increases proportionally to $T^3$. We expect the latter aspect to be the dominant feature at low temperatures. At high temperatures larger than $\theta_D$, $C$ is essentially constant and the phonon gas has greater density, leading to a greater phonon interaction and correspondingly shorter $\Lambda$, the more so the higher the temperature. Thus at high temperatures we expect $K$ to decrease with temperature. Somewhere in the intermediate range of temperature, $K$ attains a maximum. This is borne out by experiment (Fig. 5.27). The maximum value of $K$ depends on the specimen size because in a pure crystal the dimensions of the sample must set an upper limit to $\Lambda$, which may vary from say 50 Å to 50 mm. At the upper range of temperature one finds

$$K \sim T^{-1}.$$

This, as indicated above, is understandable if we assume

$$\Lambda \sim \langle n\rangle^{-1},$$

where $\langle n\rangle$ is the average density of phonons controlled by the Bose-Einstein distribution law (5.4):

$$\langle n\rangle \sim \frac{1}{e^{\hbar\omega/k_BT} - 1} \xrightarrow[\hbar\omega \ll k_BT]{} \frac{k_BT}{\hbar\omega}$$

Thus

$$K \sim \Lambda \sim \langle n\rangle^{-1} \sim T^{-1},$$

in keeping with the observed behaviour.

Impurities of different mass (and this includes isotopes of the same element) cause disruptions in the periodic distribution of masses and $\Lambda$ becomes shorter, leading to poorer thermal conductivity. Similarly, polycrystalline samples are poorer thermal

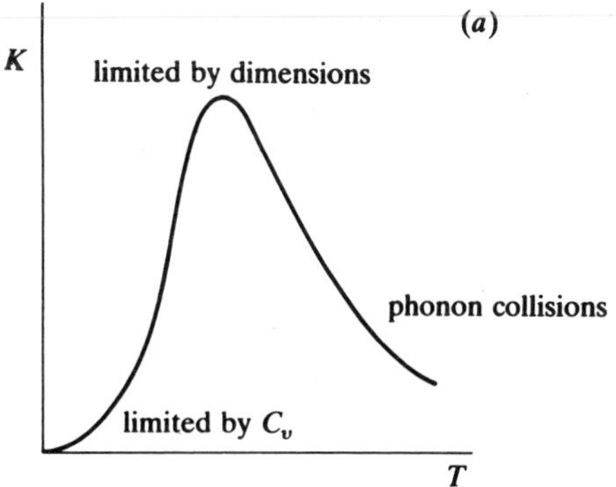


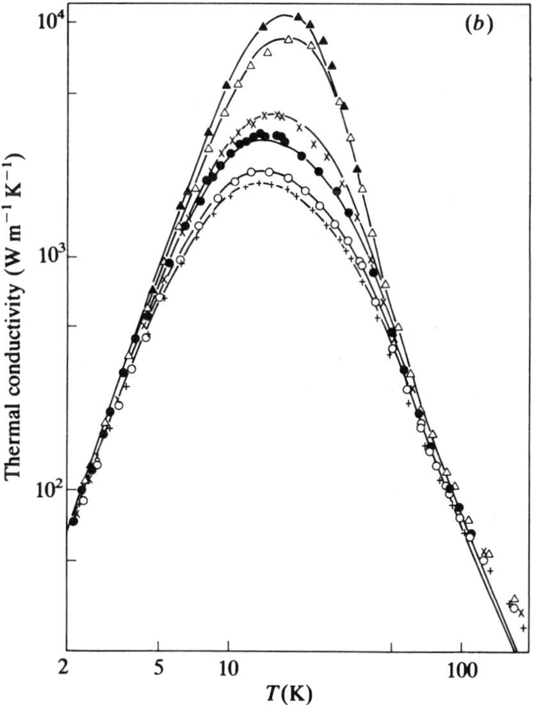


**Figure 5.27** (a) The principal form for the variation of thermal conductivity. (b) Experimental data for LiF crystals containing different amounts of the isotope $^6$Li: ▲, 0.02% $^6$Li; △, 0.01%; ×, 4.6%; ●, 9.4%; ○, 25.4%; +, 50.1%. (After Berman and Brock 1965.)

conductors than single crystals owing to scattering at the grain boundaries, and the maximum value of $\Lambda$ is determined by the crystallite size and not the external dimensions of the sample.

An unusual feature of phonon scattering is provided by the requirement that the crystal momentum be conserved. Two phonon packets of well defined wave vector may interact to produce a single resultant phonon, but both energy and phonon momentum must be conserved. Thus $\mathbf{q}_1 + \mathbf{q}_2 = \mathbf{q}_3$, but, provided $\mathbf{q}_3$ remains within the first Brillouin zone, there is nothing unusual in this reaction and it does not affect the conduction of heat. Such a process is called a normal or 'n' process. Suppose, however, that $\mathbf{q}_1$ and $\mathbf{q}_2$ are both $> \frac{1}{4}\mathbf{G}$. Then it is clearly the case that $\mathbf{q}_3$ may be $> \frac{1}{2}\mathbf{G}$; it therefore lands outside the first Brillouin zone and must be found in an adjacent zone centred around a nearest-neighbour point of the reciprocal lattice (or, in the language of the extended zone scheme, it ends up in the second zone). However we choose to describe this situation, the important point is that $\mathbf{q}_3$ becomes oppositely directed to the vectors $\mathbf{q}_1$ and $\mathbf{q}_2$; $\mathbf{q}_3$ may be represented in the original Brillouin zone by $\mathbf{q}_3 - \mathbf{G}$ (Fig. 5.28). Such a reversal of the crystal momentum is only understandable in the zone picture. The behaviour is described as an umklapp or 'u' process and is clearly significant in lowering the thermal conductivity. We may well ask where does the 'missing momentum' go. The answer is that it goes to the crystal structure as a whole.

The largest pseudomomentum that may be associated with a phonon is $\hbar q_{max}$ and $q_{max} \sim |\mathbf{G}|/2$ where $\mathbf{G}$ is a base vector of the reciprocal lattice. This means that two phonons cannot ordinarily combine to produce a single phonon with $q > q_{max}$. On the other hand a solid crystalline substance of macroscopic dimensions may have arbitrarily small velocity and momentum. Such a body may move rigidly without affecting the relative motion of its constituent atoms. The umklapp process is possible because the excess pseudomomentum in a phonon reaction may be taken up by a rigid recoil of the sample, but this is only allowable if the recoil momentum

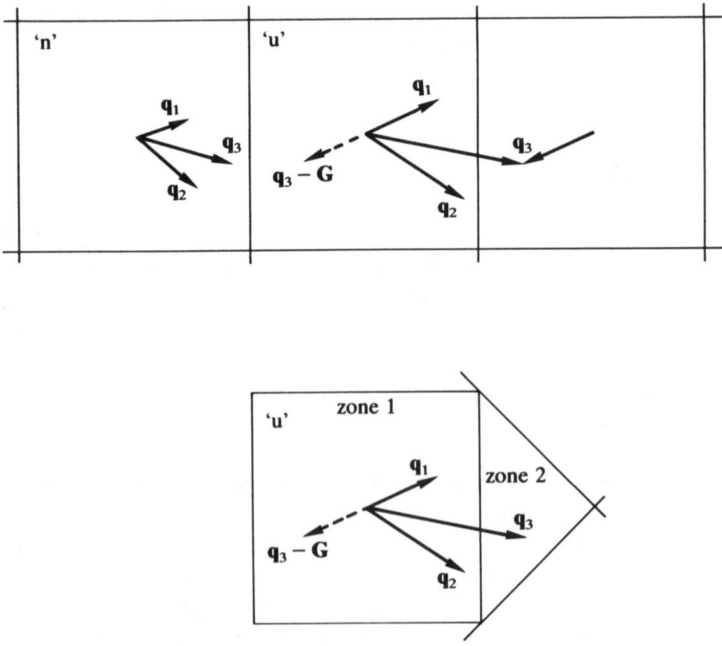


**Figure 5.28** Normal (n) and umklapp (u) processes. The latter process is shown in the repeated first zone scheme, but also using the concept of the second zone – which in this case comprises four triangular areas, of which only one is shown (the second and higher zones will be described in greater detail in Chapter 7). The 'u' process is a direct consequence of the translational symmetry of the lattice.

occurs in units of $\hbar G$ because such increments, owing to the translational symmetry of the lattice, leave the internal degrees of freedom unaffected. The crystal structure is both a source and a sink for crystal momentum in units of reciprocal vector **G**. This property appears in all processes that involve exchanges of energy between particles (including photons) and the crystal. The crystal acts as a momentum buffer, enabling energy and momentum balance to be achieved. This is also evident in Bragg's law for X rays in the form (3.8).

The umklapp process requires that the interacting phonons have a wave vector of at least $\mathbf{G}/4$ which means that their energies are $\approx k_B \theta_D/2$, $\theta_D$ is associated with the maximum value of $q$ namely $\mathbf{G}/2$. The probability that a phonon has energy $k_B \theta_D/2$ may be assumed proportional to $\exp(-\theta_D/2T)$. So where umklapp collisions are dominant, for $T \ll \theta_D$, the thermal conductivity decreases with increasing temperature according to $\exp(\theta_D/2T)$, but eventually, as described earlier, at high temperature the $1/T$ dependence prevails.

There is an extensive discussion of the thermal conductivity of solids in Berman (1976).

## References


BERMAN, R. (1976) *Thermal Conduction in Solids.* Oxford University Press, Oxford.

BERMAN, R. and BROCK, J. C. F. (1965) *Proc. R. Soc. Lond.* **A289**, 46.

STEDMAN, R. and NILSSON, G. (1966) *Phys. Rev.* **145**, 492.

STEDMAN, R., ALMQVIST, L. and NILSSON, G. (1967) *Phys. Rev.* **162**, 549.

WALKER, C. B. (1956) *Phys. Rev.* **103**, 547.


## Further Reading


DOVE, M., *Lattice Dynamics* (1993) Cambridge University Press, Cambridge.

For a detailed discussion of specific heat data for Cu, Ag and Au, see: MARTIN, D. L. (1987) *Can. J. Phys.* **65**, 1104.


## Problems

**5.1** Which are the three elements with the highest Debye temperatures and which three have the lowest values?

**5.2** Using equations (5.7), (5.32) and (5.33), obtain the expression for the Debye heat capacity. Then reduce it to a more manageable form using $z = \hbar\omega/k_B T$ and $\theta_D = \hbar\omega_{max}/k_B$. Try to find limiting values for the heat capacity at high and low temperatures. In the latter instance it is more convenient to determine the total energy and then differentiate with respect to temperature. Furthermore, assume $z_{max} \to \infty$ at low temperatures. The following standard integral is needed:

$$\int_0^\infty \frac{z^3}{e^z - 1} \, dz = \frac{\pi^4}{15}.$$

**5.3** Using the Debye approximation as described above, show that

(a) the heat capacity of a two-dimensional lattice at low temperature varies as $T^2$.

(b) the heat capacity of a linear lattice at low temperature varies as $T$.

**5.4** Show that for a simple elastic spring the force constant for transverse vibrations is much less than that for longitudinal vibrations.

**5.5** Following the example of the linear chain (Section 5.4), determine the frequency spectrum and sketch its form.

**5.6** Repeat the discussion of the linear lattice (Section 5.4), but introduce an elastic interaction between next-nearest neighbours so that $c_2 = \frac{1}{2}c_1$. Illustrate the resulting dispersion curve and sketch the frequency spectrum.

**5.7** Phonons are elastic waves. Sound waves are phonons. Heat waves are phonons. Sound propagates with a velocity of order $10^5$ cm s$^{-1}$ but 'heat waves' propagate much more slowly in a diffusive manner. Explain this difference in behaviour.

**5.8** Continue the study of longitudinal vibrations of the diatomic chain begun in Section 5.7 by (a) determining the phase velocity for the acoustic modes near $k = 0$ and (b) demonstrating that in the optical modes the different masses move in opposite directions (i.e. show that the ratio of the displacement amplitudes, $u/v$, is negative). What is the significance of the fact that for $k = \pi/a$ the limiting frequency of the acoustic mode involves only $M_1$ and that of the optical mode only $M_2$ ($M_1 > M_2$)? (Hint: how does $u/v$ behave?)

**5.9** Lattice vibrations are normally the source of changes in the internal energy of a solid with temperature, but other contributions can arise from transitions between close-lying electronic multiplet levels and at very low temperatures nuclear hyperfine structure can make significant contributions (see Section 15.4 for a short account).

**5.10** Consider the following two-dimensional close-packed lattice and its unit cell:

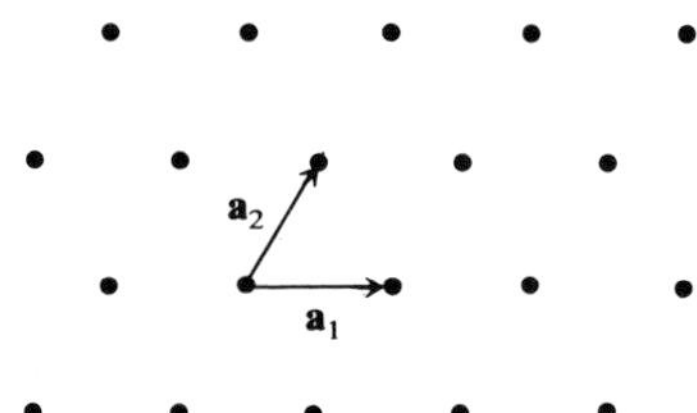


Primitive cell vectors $\mathbf{a}_1$ and $\mathbf{a}_2$ are shown. Determine the first Brillouin zone appropriate to this lattice. Arrange that the vectors $\mathbf{A}_1$ and $\mathbf{A}_2$ are correctly oriented. Then, using the nearest-neighbour approximation for a vibrating net of point masses, determine the dispersion equation (let the central mass point have coordinates $m\mathbf{a}_1$, $n\mathbf{a}_2$). Calculate the frequency at two non-equivalent symmetry points on the zone boundary. (Assume transverse vibrations.)

**5.11** At room temperature a certain class of diamond conducts heat four times better than copper and at 50 K an $Al_2O_3$ crystal has a thermal conductivity 60 W cm$^{-1}$ K$^{-1}$, which is larger than the best value for copper at any temperature. Attempt to explain why this is so.

**5.12** Glass is composed of an irregular arrangement of Si and O ions with average separations of about 10 Å. The thermal conductivity of glass is about 0.01 W cm$^{-1}$ K$^{-1}$ at room temperature. The thermal conductivity of glass increases with temperature. Explain.

**5.13** We call the Wigner-Seitz cell of the reciprocal lattice the first Brillouin zone. How does the volume of this zone compare with that of the Debye sphere, i.e. a sphere with $q_{max}$ determined by $\theta_D$?

**5.14** Using the dispersion properties of the linear lattice (which are similar to those of a three-dimensional lattice in a specific symmetry direction, e.g. [100] or [110]), demonstrate that an n or u process cannot occur if all the phonons involved have the same polarization, i.e. all $L$ or all $T$ polarization, but that a change of polarization character is unavoidable if the energy and wave vector are to be conserved. (Hint: make a copy of the $L$ and $T$ dispersion properties as given in Fig. 5.10 on tracing paper so that you may see, by laying the trace over the original figure, how two $T$ modes may combine to produce an $L$ mode.)

**5.15** The figure below shows a sector of the (001) section of the first Brillouin zone for Al. It is called the irreducible sector of the Brillouin zone. It is the least portion of the zone that, with regard to its shape and the symmetry of the reciprocal lattice, allows us to construct the complete section of the Brillouin zone. In experiment or calculation it is sufficient to determine $\omega(q)$ only over this region of $q$ space. Contours of constant frequency for the longitudinal vibrational modes are drawn. (These data were obtained in a study of the diffuse scattering of X rays from aluminium (Walker 1956).) Using the symmetry properties of the zone, this irreducible sector allows us to realize the dispersion properties throughout the complete section. The scale of $q$ in the [200] and [220] directions is linear and you must mark off your own division of $q$ (what are the values of $q$ at the zone boundaries?). The frequency is given in units of $10^{13}$ Hz.

Plot $\omega$ as a function of $q$ in the [200] and [220] directions and estimate the velocity of sound in these two directions.

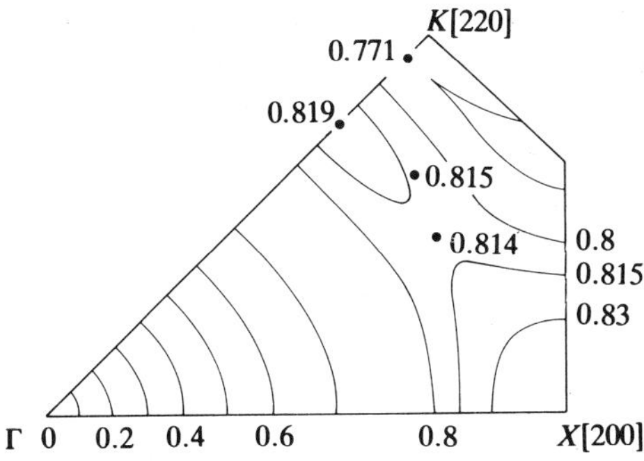

6

# Metallic Behaviour and the Free Electron Gas

Most solid pure elements are metals, which is to say that they conduct electricity – the better so the lower the temperature. Metals are opaque and reflect light excellently, they are plastic and exhibit work hardening. In this chapter, we shall attempt to understand some of these characteristic properties in terms of the simplest model – the ideal free electron gas.

Experimentally we know that a metal obeys Ohm's law accurately, no matter how weak the applied electric field. Deviations are observed only at very high field strengths. Metallic conductivity does not require an activation energy, as is the case for electrical conductivity in ionic salts or semiconductors like Si. We conclude that certain electrons in metals must be free to move through the sample and cannot be permanently bound to particular atoms. We think of these 'free electrons' as confined to the metal in a manner analogous to that whereby molecules of an ordinary gas are confined to a container such as a glass bottle. For electrons, the container is a potential barrier arising at the surface of the metal (Fig. 6.1). Within the metal, we shall find that the free electrons constitute a gas of uniform density.

It is clear that the free electrons must be the valence electrons because these are the most loosely bound to the parent atoms. In fact, a simple, but in many respects quite accurate, picture of a metallic crystal (of aluminium, say) is that of a giant molecule in which all the atoms are covalently bound to one another by the valence electron gas, but without any directional character to the bonds – a sort of 'isotropic covalency'. We can measure the lattice parameters of metals and ionic compounds and estimate the sizes of metallic atoms and the appropriate ions. Confining attention to the simpler metals, the ionic volume is found to be only 10–15% that of the atom. Each ion gives rise to a deep potential well and those of adjacent ions overlap to produce an effective potential similar to that shown in Fig. 6.2. Each valence electron moves in the combined potential of all the ions and all the other valence electrons. Each atomic cell of the pure metal crystal is of necessity an electrostatically neutral unit and any given electron moves in an effective potential that is smoother than that of Fig. 6.2. The simplest approach is to assume it to be constant. This is of course a very big assumption, but it simplifies the problem enormously and, as we shall see, still allows contact with real behaviour. We emphasize that this

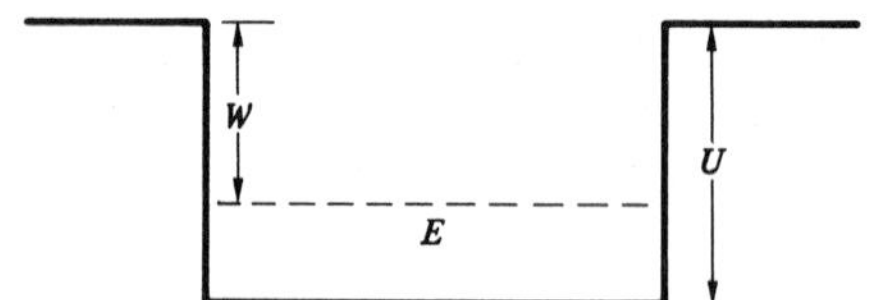


**Figure 6.1** The constant potential box in one dimension. $U$ is the potential in which all the electrons move and $E$ their kinetic energy measured from the bottom of the box.

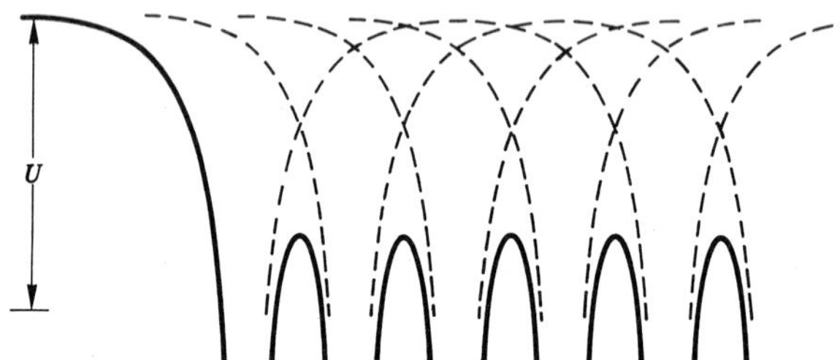


**Figure 6.2** The form of the potential through a line of atoms in a solid.

constant potential is the effective potential that a single valence electron experiences and is the resultant average potential arising from interactions with all the ions and all the other electrons including their mutual Coulomb and exchange interactions. It was first examined by Sommerfeld in 1928.

## 6.1 Fundamental Properties of the Free Electron Gas

The Schrödinger equation for an electron in the ideal free electron gas is

$$\left(\frac{-\hbar^2}{2m}\nabla^2 + U\right)\psi = \mathrm{i}\hbar\frac{\mathrm{d}\psi}{\mathrm{d}t}, \tag{6.1}$$

where $U$ is the constant potential energy possessed by every electron. The general solution to (6.1) may be written

$$\psi_+ = A\mathrm{e}^{\mathrm{i}(\mathbf{k}\cdot\mathbf{r}-\omega t)} + B\mathrm{e}^{-\mathrm{i}(\mathbf{k}\cdot\mathbf{r}+\omega t)}, \tag{6.2a}$$

$$\psi_- = A\mathrm{e}^{\mathrm{i}(\mathbf{k}\cdot\mathbf{r}-\omega t)} - B\mathrm{e}^{-\mathrm{i}(\mathbf{k}\cdot\mathbf{r}+\omega t)}, \tag{6.2b}$$

where $\omega = W/\hbar$, $W$ being the energy eigenvalue. The components of $\psi$ represent oppositely directed travelling waves. In a box of constant potential and finite dimensions, the wave function must have zero amplitude close to the boundaries of the box, and this leads to a standing wave; (6.2$b$) with $A = B$ is the appropriate wave function. However, standing waves may be avoided by the device of periodic boundary conditions (Section 5.5) and (just as for phonons) we choose to write the wave function as the travelling wave

$$\psi = A\mathrm{e}^{\mathrm{i}(\mathbf{k}\cdot\mathbf{r}-\omega t)}, \tag{6.3}$$

allowing **k** to take positive and negative signs. (As an exercise, show that periodic boundary conditions applied to an element of a linear free electron gas lead to a density of states in wave-vector space $N(k) = L/2\pi$.) Substituting (6.3) into (6.1) leads to

$$\frac{\hbar^2 k^2}{2m} = W - U = E \qquad (U < W < 0). \tag{6.4}$$

$E$ is the kinetic energy of an electron with energy eigenvalue $W$. It is convenient to measure $E$ from the bottom of the potential box. Henceforth we shall not need to take explicit account of the time dependence of the wave function.

Writing

$$|\mathbf{k}| = \left(\frac{2mE}{\hbar^2}\right)^{1/2}, \tag{6.5}$$

we find that the wave vector is the quantum number for the electrons in the gas. Since all the electrons in the gas possess the same constant potential energy, the various electron states can only differ with regard to their kinetic energies. The different states of increasing $k$ value are therefore filled progressively in proportion to the number of electrons present.

The electrons are confined to the sample of volume $V$, so normalization of the wave function leads to

$$\int_{-\infty}^{+\infty} \psi\psi^* \, \mathrm{d}V = A^2 V = 1.$$

Therefore

$$\psi = V^{-1/2} \mathrm{e}^{\mathrm{i}\mathbf{k}\cdot\mathbf{r}} \tag{6.6}$$

The probability of finding the electron in an element of volume $\mathrm{d}V$ is $\psi\psi^* \, \mathrm{d}V$, i.e. $\mathrm{d}V/V$. The free electrons thus constitute a gas of uniform density and are described by plane-wave functions with specific wave vectors **k**. By direct analogy with our treatment of phonons, we introduce the periodic boundary conditions and associate specific electron states with each allowable wave vector. Again each state occupies a volume $\Delta^3 k = 8\pi^3/V$ of reciprocal space, leading to the same density of states in **k** space that we found in Section 5.5:

$$N(k) = V/8\pi^3 \tag{6.7}$$

Although phonons and electrons are very different objects, in this respect they have the same physical basis, both being described by plane waves. Whenever physical behaviour is characterized by wave functions of the form (6.3), we find (6.7) applies: this result is valid for all periodic structures and continua, the latter being the limiting case of periodicity when the period becomes zero.

### 6.1.1 *The Fermi wave vector*

The electron has spin angular momentum appropriate to a quantum number $s = \frac{1}{2}$ and it can take two and only two orientations relative to an arbitrarily directed weak reference magnetic field. Each allowable **k** state is therefore doubly degenerate for electrons. Equation (6.4) and the isotropy of the constant potential imply that the surfaces of constant energy in **k** space are spheres. If our ideal gas contains a total of $N$ electrons then at absolute zero the lowest $\frac{1}{2}N$ levels are filled and there is a maximum value of the wave vector, corresponding to the highest occupied level. This maximum value is known as the Fermi wave vector and is denoted by $k_F$. Clearly

$$\frac{N}{2} = \frac{V}{8\pi^3}\frac{4\pi}{3}k_F^3,$$

giving

$$k_F = (3\pi^2 n)^{1/3}, \tag{6.8}$$

we have written $n$ for the electron density $N/V$. There is a corresponding maximum kinetic energy for the electrons, the Fermi energy

$$E_F = \frac{\hbar^2 k_F^2}{2m}. \tag{6.9}$$

### 6.1.2 *The free electron energy band*

Any interval between $k$ and $k + dk$ corresponds to a spherical shell bounded by surfaces of constant energy $E$ and $E + dE$. Suppose this shell contains $dN$ electron states; then it is clear from our previous discussion that

$$\frac{dN}{2} = \frac{V}{8\pi^3} 4\pi k^2 \, dk.$$

We use (6.4) and change the variable from $k$ to $E$ to obtain

$$dN = \frac{V}{\pi^2\hbar^3}(2m^2E)^{1/2} \, dE. \tag{6.10}$$

We now define a *density of states* $N(E)$, which is the number of electron levels (spin included) per unit energy range available in the sample. Then, if $dN$ is the number of electron levels in the interval of energy $E$ to $E + dE$,

$$dN = N(E) \, dE. \tag{6.11}$$

Comparison of (6.10) and (6.11) leads to

$$N(E) = \frac{V}{\pi^2\hbar^3}(2m^3E)^{1/2}. \tag{6.12}$$

We see that the density of states is proportional to $E^{1/2}$ and to the volume of the sample (Fig. 6.3). We could of course normalize to unit volume, but this has the inconvenience that $N(E)$ takes very large values. It is more appropriate to refer to the atomic volume, and we therefore define the *density of electron states per unit energy range per atom* as

$$N(E) = \frac{\Omega}{\pi^2\hbar^3}(2m^3E)^{1/2} = \text{constant} \times E^{1/2}, \tag{6.13}$$

where $\Omega$ is the atomic volume.

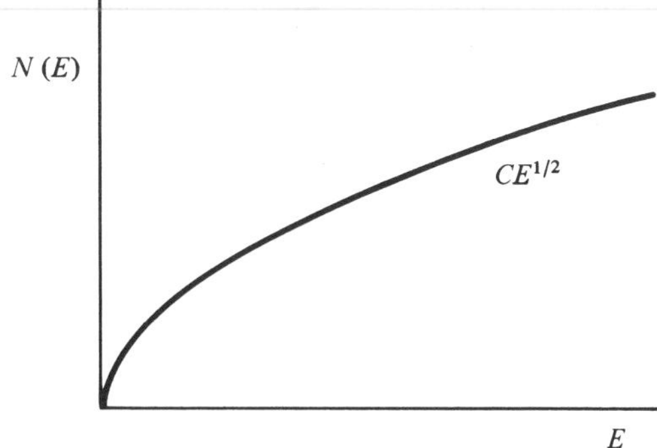


**Figure 6.3** The parabolic energy band for free electrons.

## 6.1.3 *Charge density oscillations*

The ideal free electron gas may be thought of as a homogeneous distribution of negative charge compensated electrostatically by a uniform background of positive charge. The latter is carried by the ions and contains essentially the whole of the system's mass. We think of the ions as a stationary background through which the electrons move. Now the electron gas is a dynamic system and we might therefore expect fluctuations in the average uniform density. Suppose that in a small and, for convenience, cylindrical volume the electrons undergo a rigid displacement $\Delta$ (Fig. 6.4). Since the positive background is stationary, this leads to a dipole moment of strength $\mathbf{p} = -ne\boldsymbol{\Delta}Al$ corresponding to a polarization (dipole moment per unit volume) $\mathbf{P} = -ne\boldsymbol{\Delta}$. This polarization, created in an overall neutral assembly of electrons and ions, becomes the source of an electric field $\mathbf{E}$ so that

$$\mathbf{E} = \frac{-\mathbf{P}}{\varepsilon_0} = \frac{ne\boldsymbol{\Delta}}{\varepsilon_0},$$

$\varepsilon_0$ being the vacuum permeability.

The equation of motion of any electron in the density fluctuation is then

$$m\frac{\mathrm{d}^2\boldsymbol{\Delta}}{\mathrm{d}t^2} = -e\mathbf{E} = -\frac{ne^2}{\varepsilon_0}\boldsymbol{\Delta},$$

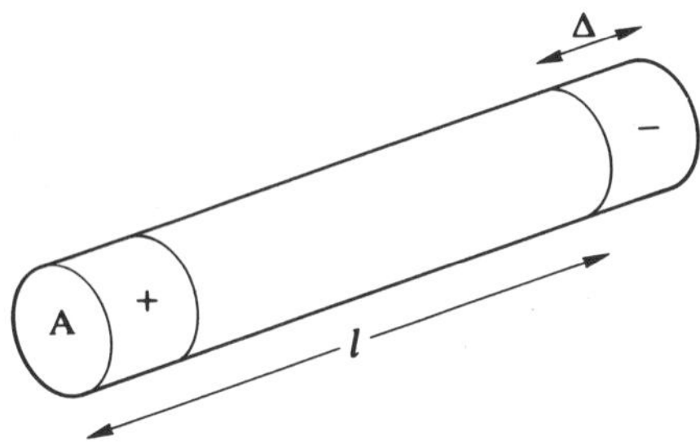


**Figure 6.4** A simple charge density fluctuation in the free electron gas. Only the electrons are mobile in the stationary background of positive charge.

which is an equation for harmonic motion with frequency

$$\omega_p^2 = \frac{ne^2}{\varepsilon_0 m}. \tag{6.14}$$

We see that the electron gas can perform longitudinal charge density oscillations, which are called plasma oscillations. The term 'plasma' was coined to describe the conventional ionized gas that exists in discharge tubes, but it is clear that it is appropriate in the case of the electron gas. The quantum of plasma oscillation, $\hbar\omega_p$, is called a *plasmon.*

We have now established the basic features of the ideal electron gas:

*Fermi wave vector* $\qquad k_F = (3\pi^2 n)^{1/3}$,

*Fermi energy* $\qquad E_F = \dfrac{\hbar^2 k_F^2}{2m}$,

*density of states per unit energy per atom*

$$N(E) = \frac{\Omega}{\pi^2}\frac{1}{\hbar^3}(2m^3 E)^{1/2},$$

*plasma energy* $\qquad \hbar\omega_p = \hbar\left(\dfrac{ne^2}{\varepsilon_0 m}\right)^{1/2}$.

## 6.2 Numerical Values of $k_F$, $E_F$, $N(E)$ and $\hbar\omega_p$

Clearly there is only one independent variable, the electron density $n$, which may also be written $n_0/\Omega$, where $n_0$ is the valence. An alternative representation is to use the radius $r_0$ of the spherical volume available to each electron. Thus

$$\tfrac{4}{3}\pi r_0^3 = n^{-1}. \tag{6.15}$$

It is conventional to express $r_0$ in terms of the Bohr radius $a_0$ (that of the electron orbital in the hydrogen atom). We define

$$r_s = r_0/a_0. \tag{6.16}$$

Table 6.1 gives some typical values of $n$ and $r_s$ for simpler metals. Note the limiting values for Be and Cs.

It might be thought a daring enterprise to apply our free electron gas formulae to real substances. First we must establish the valence. This is straightforward for the simple metals such as Na, Mg, In and Pb. We just use the conventional chemical valence. But we cannot do this for the transition metals, rare earth metals or actinides, so the following numerical application is restricted to the simple metals. The relevant quantities are listed in Table 6.1. We see that $E_F$ and $\hbar\omega_p$ are of similar magnitude but that $\hbar\omega_p$ is always somewhat larger than $E_F$.

We may also write

$$k_B T_F = E_F. \tag{6.17}$$

The most energetic electron has the Fermi energy and we can appreciate its very large value if we calculate the equivalent temperature $T_F$, called the Fermi temperature. This is in the range $10^4$–$10^5$ K. The immensity of the Fermi energy (and

**Table 6.1** Free electron data for selected metals

| | $n_0$ | $n$ ($10^{28}$ m$^{-3}$) | $r_s$ | $k_F$ ($10^{10}$ m$^{-1}$) | $E_F$ (eV) | $\hbar\omega_p$ (eV) |
|---|---|---|---|---|---|---|
| Li | 1 | 4.63 | 3.37 | 1.11 | 4.71 | 8.0 |
| Na | 1 | 2.53 | 4.0 | 0.92 | 3.15 | 5.91 |
| K | 1 | 1.32 | 4.0 | 0.75 | 2.04 | 4.27 |
| Rb | 1 | 1.07 | 5.33 | 0.70 | 1.78 | 3.85 |
| Cs | 1 | 0.87 | 5.71 | 0.64 | 1.54 | 3.46 |
| Be | 2 | 24.63 | 1.87 | 1.93 | 14.3 | 18.5 |
| Mg | 2 | 8.6 | 2.66 | 1.37 | 7.12 | 11.0 |
| Ca | 2 | 4.6 | 3.28 | 1.11 | 4.68 | 8.0 |
| Zn | 2 | 13.14 | 2.31 | 1.57 | 9.44 | 13.4 |
| Al | 3 | 18.07 | 2.08 | 1.75 | 11.66 | 15.8 |
| Pb | 4 | 13.19 | 2.31 | 1.57 | 9.46 | 13.5 |

thereby the plasmon energy) is immediately apparent. The occupied energy levels of the electron gas at 0 K extend from essentially zero kinetic energy to the Fermi energy. As we have seen, the levels are distributed according to the function $N(E)$, which describes a parabolic energy band (Fig. 6.3). The term 'band' is used because the available energy levels in all real samples are so closely spaced that they form what is usually called a 'quasicontinuous distribution', i.e. something that although discrete is for all practical purposes continuous.

We write

$$N(E) = \frac{V}{\pi^2}\frac{1}{\hbar^3}(2m^3E)^{1/2} = CE^{1/2},$$

and it is sufficient to remember the energy dependence – the form of the constant factor can readily be derived if required. Often we do not need to know its exact value, as the following calculation of the average energy of an electron at 0 K illustrates:

$$\langle E\rangle = \int_0^{E_F} N(E)E\,\mathrm{d}E \Big/ \int_0^{E_F} N(E)\,\mathrm{d}E$$
$$= \int_0^{E_F} CE^{3/2}\,\mathrm{d}E \Big/ \int_0^{E_F} CE^{1/2}\,\mathrm{d}E,$$

whence

$$\langle E\rangle = \tfrac{3}{5}E_F. \qquad (6.18)$$

We choose to normalize $N(E)$ to the atomic volume and to measure energy in electron volts. If we know the valence $n_0$ then, as the following example demonstrates, we can readily calculate $N(E)$ in terms of $n_0$ and $E_F$. Clearly

$$\int_0^{E_F} N(E)\,\mathrm{d}E = \tfrac{2}{3}CE_F^{3/2} = n_0,$$

giving

$$C = \frac{3}{2}\frac{n_0}{E_F^{3/2}},$$

whence

$$N(E) = \frac{3}{2}\frac{n_0}{E_F^{3/2}} E^{1/2}. \tag{6.19}$$

In particular

$$N(E_F) = \frac{3}{2}\frac{n_0}{E_F}. \tag{6.20}$$

This is a convenient expression for estimating $N(E_F)$. Consider aluminium, for which $E_F = 11.6$ eV and $n_0 = 3$. Then $N(E_F) = \frac{3}{2}(3/11.6) = 0.39$ eV$^{-1}$ atom$^{-1}$. This is a typical order of magnitude of $N(E_F)$ in a simple metal. A mole of aluminium (27 g) contains $6 \times 10^{23}$ atoms and the molar volume is 10 cm$^3$; thus in 10 cm$^3$ of aluminium at the Fermi level there is a density of levels corresponding to $0.39 \times 6 \times 10^{23} = 2.34 \times 10^{23}$ levels per eV per mole and their average separation is $4.3 \times 10^{-24}$ eV. We can very well appreciate their quasicontinuous character and the applicability of the term 'energy band'. It is also clear why we prefer to refer $N(E)$ to the atom.

## 6.3 Comparison with Experiment

Can we in any way compare our simple estimates of $E_F$, $N(E_F)$ and $\hbar\omega_p$ with the properties of real metals? First of all we must recognize that many measurements are made at room temperature and we must consider the effect of temperature on the electron gas. It is clear physically that the average electron energy is so large compared with the thermal energy associated with all normally obtainable laboratory temperatures that we can expect the electron gas to be insensitive to temperature changes. Electrons obey the Pauli principle, no two electrons within an atom having the same set of quantum numbers. In the case of the electron gas, this means that two electrons may not have the same components of $(k_x, k_y, k_z)$ and the same direction of spin. Because of this, the occupation of electron states is governed by the Fermi–Dirac statistical distribution, and the probability that an energy level $E$ is occupied is determined by

$$f(E) = \frac{1}{e^{(E-E')/k_B T} + 1}. \tag{6.21}$$

$E'$ is a characteristic parameter of the gas (the chemical potential), and for the free electron gas it is weakly temperature-dependent:

$$E' \approx E_F\left[1 - \frac{\pi^2}{12}\left(\frac{T}{T_F}\right)^2\right]. \tag{6.22}$$

We shall not need to take account of this weak temperature dependence, and we can therefore replace $E'$ by $E_F$ so that

$$f(E) = \frac{1}{e^{(E-E_F)/k_B T} + 1}. \tag{6.23}$$

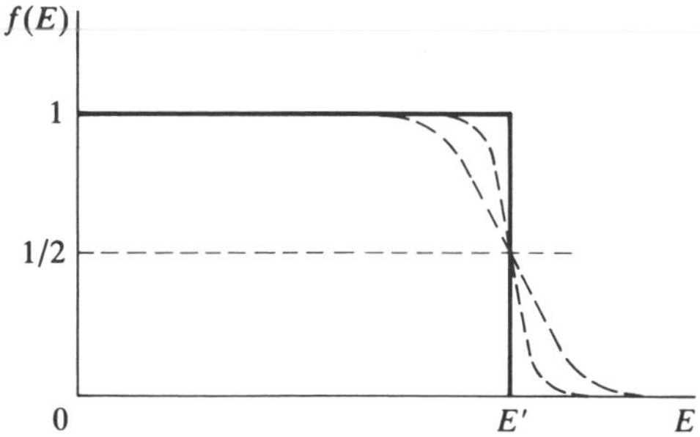


**Figure 6.5** The Fermi-Dirac distribution function. The full curve shows the step cut-off appropriate to 0 K. At finite temperatures, as indicated by the broken lines, the probability that a state of given energy near $E'$ is occupied varies in a more gradual manner, the more so the higher the temperature. However, at all normal laboratory temperatures, the change in $f(E)$ near $E'$ is still very sharp because $k_B T \ll E'$.

Curves illustrating the probability that an electron level is occupied are shown in Fig. 6.5. At room temperature there is a slight width to the transition from filled to unfilled states, but this can only be about 0.05 eV, which is very small compared with $E_F$; the Fermi cut-off therefore remains sharp. It is customary to call the sharp boundary between occupied and unoccupied states in **k** space the *Fermi surface*; as we have seen, the ideal electron gas has a spherical Fermi surface, but in real metals complicated shapes arise.

### 6.3.1 Plasmons

It is understandable that with $\hbar\omega_p \approx 10$ eV, plasma oscillations, i.e. charge density oscillations in the electron gas, cannot occur spontaneously; it costs far too much energy to create a plasmon, a quantum of such an oscillation. However, plasmons can be excited by the passage of energetic electrons through a metal. Thin films of metals (of order 1000 Å) are transparent to electrons, but it is found that on passage through the film some of the electrons lose energy in discrete amounts. Such energy losses are also found when electrons are reflected from solid metal surfaces. In aluminium, for example, the losses are about 15, 31 and 46 eV, which correspond very closely to $\hbar\omega_p$, $2\hbar\omega_p$ and $3\hbar\omega_p$. Similar good agreement is found for other such metals, and the simple model of an ideal electron gas appears to be relevant.

### 6.3.2 X-ray spectroscopy

Suppose an aluminium sample is made the target of an X-ray tube and we arrange to ionize only the highest-lying core level (Fig. 6.6); in the case of aluminium this means the LIII level. The empty electron state in the L level may be filled by a valence electron falling into it. In doing so, it emits radiation with a photon energy of about 70 eV, a so-called soft X-ray. The emitted X ray may be analysed using a vacuum grating spectrometer. Now if the valence electrons occupy a band of energy levels, the emitted soft X rays must have a distribution of wavelengths that reflects

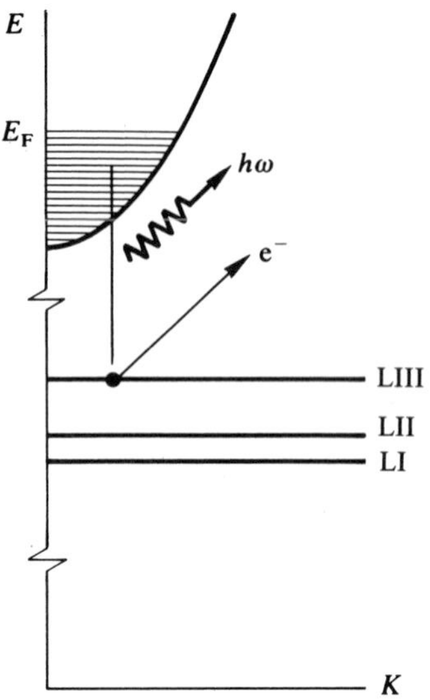


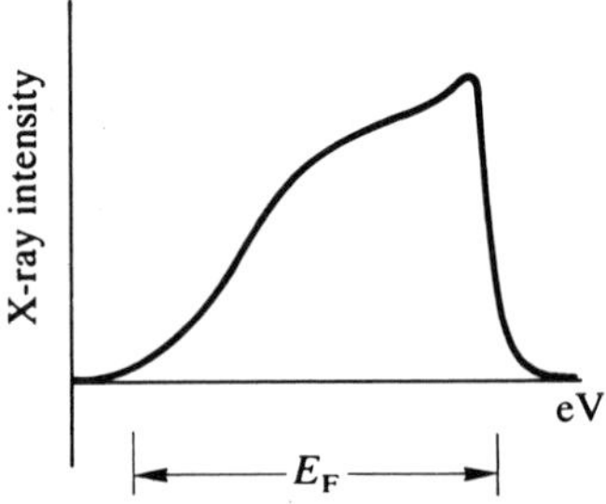


**Figure 6.6** The origin of the soft X-ray spectrum in a metal and a schematic illustration of its general form. A high-lying core level is excited and the empty state is filled by an electron from the valence band, the energy being released as a photon.

the energy spread of the valence electrons. This experiment thus allows us to (a) test whether a band of electron levels exists, and (b) measure the occupied portion of the band – the band width, which is clearly a measure of $E_F$. In principle one should also be able to derive the distribution $N(E)$, but this requires an accurate knowledge of transition probabilities. The experiments should be performed in very high vacuum, $\leqslant 10^{-10}$ torr, to prevent oxidation or other contamination of the specimen. The experimental results give ample confirmation of the existence of valence electron bands, and the band widths for the simpler metals are in surprisingly good agreement with free electron estimates of $E_F$. There is also a clear indication of a sharp Fermi cut-off (Fig. 6.7).

A complementary experimental method is to use monochromatic X rays, or ultraviolet rays, to expel photoelectrons from the sample. The X-ray photons are usually Al $K\alpha_{12}$ 1486.6 eV or Mg $K\alpha_{12}$ 1253.6 eV. These photon energies are sufficiently high to expel core electrons as well as valence electrons. Some of the photo-

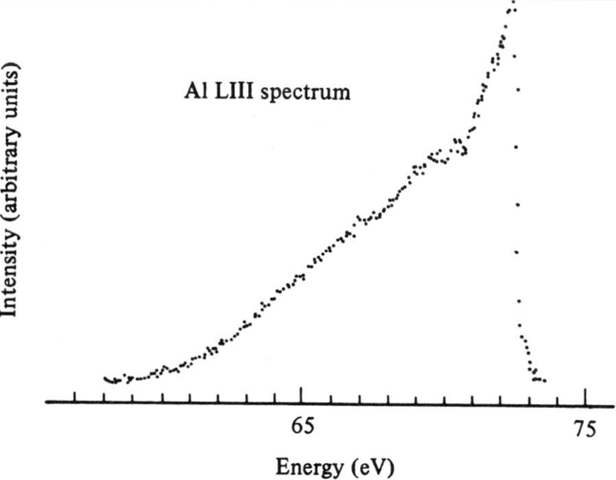


**Figure 6.7** A typical soft X-ray spectrum for a simple metal. The band width compares favourably with the free electron value for $E_F$ (see Table 6.2). (After Aita and Sagawa 1969.)

emitted electrons do not lose energy in escaping from the sample, and they leave the metal with a certain kinetic energy (Fig. 6.8).

$$\mathrm{KE} = \hbar\omega - \text{binding energy} - \Phi. \tag{6.24}$$

The binding energy is the depth below the Fermi level from which the electron originated, and $\Phi$ is the work function, i.e. the minimum energy required to remove

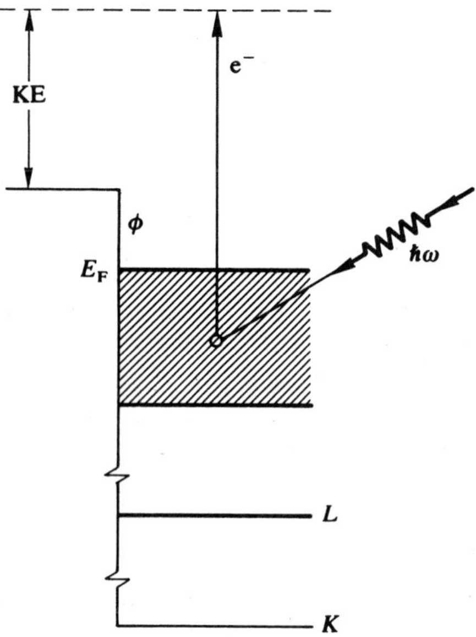


**Figure 6.8** The principle of a photoemission experiment. A photon ejects an electron, which, without loss of energy, leaves the sample with a characteristic kinetic energy.

an electron from the metal. The kinetic energy is obviously the greater the lower the binding energy of the electron, and those electrons originating near the Fermi level will possess the largest value. An X-ray photoemission spectrum appears as in Fig. 6.9. Sharp peaks arise, corresponding to photoemitted core electrons, and at lower binding energies there is a continuous distribution of electrons, corresponding to the band of occupied valence states. The kinetic energies are analysed in the simplest manner using a retarding potential, but there are more sophisticated procedures based on electron optics. Ultraviolet light ($h\nu < 50$ eV) may also be used and is very effective for the study of valence electrons, but the detailed interpretation of the spectrum is often more involved than for X-ray photoemission spectroscopy. Again ultrahigh-vacuum conditions are essential and the details of experimental technique are more complicated than the description of principles may imply. We shall have more to say about these experiments later. It is sufficient for now to note that they provide convincing evidence that the valence electrons of simple metals occupy energy bands with a width similar to the free electron values of $E_F$. It is difficult from these measurements to derive the exact shape of the valence electron bands in simple metals but sufficient evidence exists that there is a strong resemblance to the free electron parabola.

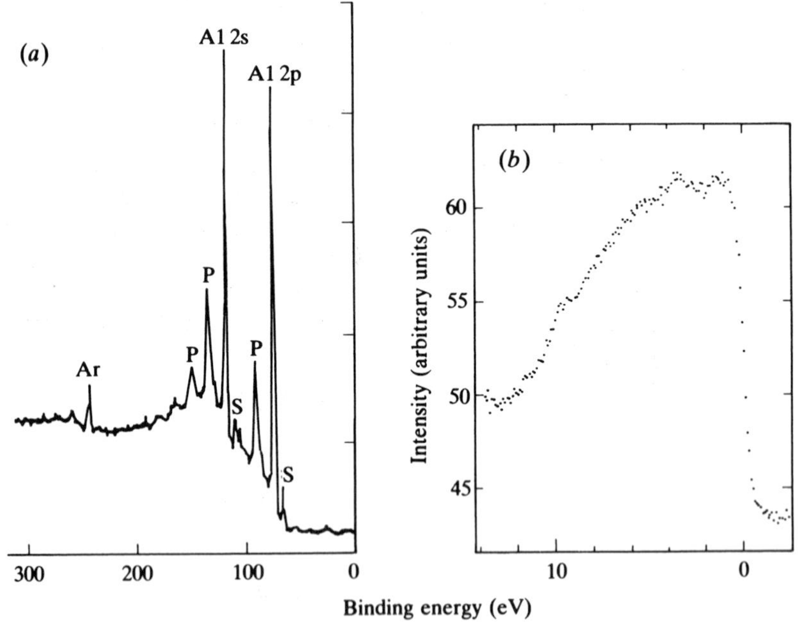


**Figure 6.9** (a) X-ray photoemission spectrum for Al taken with Mg $K\alpha$ radiation. The atomic 2s and 2p peaks are easily discerned. The extra peaks (P) are caused by certain electrons incurring energy losses (in integral units of the plasmon energy) during their passage out of the metal. Additional weak satellite peaks (S) originate from extraneous radiation in the source. The photoionization cross-section for the valence electrons is very small and their spectrum is barely detectable. The valence electrons are better studied using ultraviolet light, $\hbar\omega \leqslant 50$ eV. (b) shows the valence band for Al taken with Mg $K\alpha$ X rays, but with an exposure of about 72 h. The sharp Fermi cut-off (modified by instrumental broadening) is clearly seen. The valence band width is found to be about 13 eV. ((a) after Muillenberg 1979; (b) after Baer and Busch 1973.)

### 6.3.3 Electronic heat capacity

At the close of the last century Drude attempted an explanation of metallic behaviour in terms of a classical electron gas. In several ways it was remarkably successful, but there was one particularly severe criticism. If the electrons behave as independent gas particles then the law of equipartition of energy should apply and the electrons each ought to possess an energy $\frac{3}{2}k_B T$, leading to an electronic heat capacity of $\frac{3}{2}n_0 R$ per mole of metal ($n_0$ is the valence). In practice we find almost no difference between the heat capacities of metals and insulators. The simple quantum approach to the free electron gas offers a convincing explanation of this classical discrepancy.

We consider electrons occupying a band of energy levels with a distribution $N(E)$. At temperature $T$ the total kinetic energy is

$$\Sigma = \int_0^\infty N(E)Ef(E)\,\mathrm{d}E,$$

and the heat capacity becomes

$$C_e = \frac{\mathrm{d}\Sigma}{\mathrm{d}T}.$$

The evaluation of $\Sigma$ and $C_e$ is complicated by the presence of the Fermi function (for a complete discussion see Ashcroft and Mermin 1976). We choose merely to quote the result, which is

$$C_e = \tfrac{1}{3}\pi^2 k_B^2 N(E_F)T. \tag{6.25}$$

We can, however, approximate this result in a simple fashion. Thermal energies are so small compared with the average energy of the electrons in the gas that they only allow a change in the occupation of energy levels within a thin strip of width $2k_B T$ in energy, centred about the Fermi level. Within this narrow interval we may assume the density of states to be constant and given by $N(E_F)$. Furthermore, $f(E_F)$ is always equal to $\frac{1}{2}$ (Fig. 6.5). Thus the number of electrons receiving thermal energy equal to $\frac{3}{2}k_B T$ per electron is just

$$N(E_F) \times 2k_B T \times \tfrac{1}{2},$$

and

$$\Sigma = \tfrac{3}{2}N(E_F)k_B^2 T^2,$$

which gives a value of $C_e$ close to (6.25):

$$C_e = 3k_B^2 N(E_F)T. \tag{6.26}$$

Remember that our definition of $N(E_F)$ refers to the density of states per eV per atom. Our value of $C_e$ refers to the atom as unit. We can readily obtain $C_e$ per unit mass or per mole by using the appropriate multiplying factor. Incidentally, we note that *this result does not require $N(E)$ to be of free electron form but applies generally, whatever the band shape.*

In principle this provides us with a means to measure $N(E_F)$ in metals. We rewrite (6.25) as

$$C_e = \gamma T. \tag{6.27}$$

The electronic heat capacity depends linearly on $T$. Furthermore, one readily estimates from the free electron gas formulae that $\gamma$ is small, of order 1 mJ mol$^{-1}$ K$^{-2}$. Normally the electronic heat capacity will be completely swamped by the much larger lattice heat capacity. Only at very low temperatures (say $<10$ K, and lower with decreasing Debye temperature) will it be observable.

The total heat capacity including the Debye term may now be written as

$$C = \gamma T + \alpha T^3,$$

i.e.

$$\frac{C}{T} = \gamma + \alpha T^2. \tag{6.28}$$

Assuming that $\gamma$ and $\alpha T^2$ are of comparable size, a plot of $C/T$ against $T^2$ yields a straight line and the intercept on the ordinate gives $\gamma$ (Fig. 6.10).

The electronic heat capacities of many pure metals and alloys have now been determined. They provide a valuable indication of $N(E_F)$. However, owing to various interactions between the valence electrons, but more significantly between the valence electrons and the ions, care has to be taken in the detailed interpretation of the data.

The following equation allows a rapid assessment of the apparent $N(E_F)$ measured in eV$^{-1}$ atom$^{-1}$ from $\gamma$ values in mJ mol$^{-1}$ K$^{-2}$:

$$\gamma = 2.36N(E_F). \tag{6.29}$$

Even so, experimental data for the simple metals are often in fair agreement with the estimates of free electron gas theory. When differences arise, it is customary to reconcile theory and experiment by using the electron mass as an adjustable parameter. Thus, if $\gamma_0$ is the theoretical free electron value and $\gamma_{exp}$ the measured value, we define an effective mass $m^*$ by

$$\left(\frac{m^*}{m}\right) = \frac{\gamma_{exp}}{\gamma_0}. \tag{6.30}$$

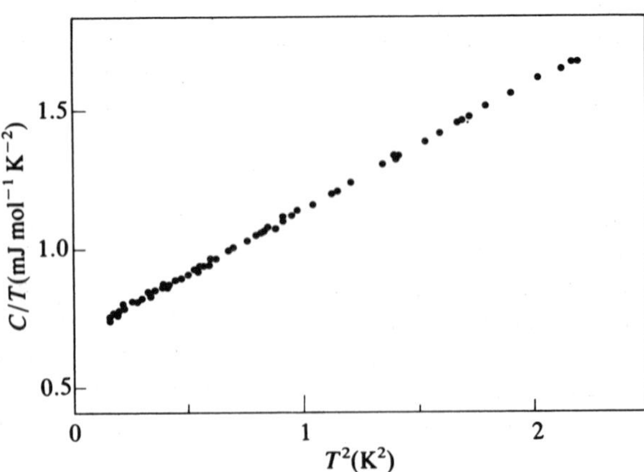


**Figure 6.10** A graph of $C/T$ against $T^2$ allows the determination of $\gamma$. The data shown are for pure Au. (After Martin 1968.)

**Table 6.2** $\gamma$ (mJ mol$^{-1}$ K$^{-2}$)

| | | | |
|---|---|---|---|
| Na | 1.38 | Ti | 3.35 |
| K | 2.08 | V | 9.26 |
| Mg | 1.3 | Cr | 1.4 |
| Al | 1.35 | Mn | 9.2 |
| Pb | 2.98 | Fe | 4.98 |
| Cu | 0.7 | Co | 4.73 |
| Ag | 0.65 | Ni | 7.02 |
| Au | 0.73 | Pt | 7.0 |

Insertion of the effective mass in the expression of $\gamma_0$ then gives agreement with experiment. Equation (6.30) implies the following relationship:

$$\left(\frac{m^*}{m}\right) = \frac{N(E_F)_{\text{true}}}{N(E_F)} = \frac{\gamma_{\text{exp}}}{\gamma_0}, \tag{6.31}$$

but again we must take care not to neglect the influence of other interactions when using such a simple analysis. The use of an effective electron mass is a common expedient in reconciling experiment with theory (or rather vice versa). We solve no problems this way, but it is convenient practice and we shall later find good reasons as to why an electron should exhibit a mass different from its ordinary rest mass. When we study experimental $\gamma$ values of different metals we notice that whereas for the simple metals $\gamma \approx 1$ mJ mol$^{-1}$ K$^{-2}$, values for the transition metals are considerably larger, in the range 5–10 mJ mol$^{-1}$ K$^{-2}$ (Table 6.2). We shall later find a straightforward explanation for this difference.

### 6.3.4 *The magnetic susceptibility*†

Although this was unknown to Drude, the electron possesses a permanent magnetic moment, the Bohr magneton $\mu_B$, and a classical electron gas should therefore exhibit a susceptibility varying rapidly with temperature ($\kappa \propto T^{-1}$); this is not observed. Again the quantum mechanical approach provides the answer. In the presence of an external field $B_0$, the free electron has potential energy $-\mu_B B_0$, $\mu_B$ being the Bohr magneton (see Appendix 11.2). $\mu_B = e\hbar/2m = 5.766 \times 10^{-5}$ eV T$^{-1}$, so if $B_0$ takes a value 10 T, a by no means insignificant strength, then

$$\mu_B B_0 \approx 6 \times 10^{-4}\,\text{eV}.$$

We see immediately that magnetic fields of ordinary strength, of order 1 T, can only cause slight changes in the energy of an electron gas. We can calculate the magnetic energy by imagining the valence band to be divided into two equal halves, one of each spin direction. An external magnetic field causes each of these two half-bands to be displaced by an amount $\mu_B B_0$, but in opposite directions. Electrons in one half-band find themselves in states lying above empty states in the other half-band, and to minimize the energy they flip spin and transfer to the other half-band. This

† See Sections 11.1–11.3 for the definitions of magnetic quantities.

continues until a balance is obtained. The external field has in this way caused a difference in occupation of the two spin bands and the gas becomes magnetized with a magnetization $M$ (dipole moment per unit volume).

Clearly, from Fig. 6.11,

$$M = \mu_B[N(\uparrow) - N(\downarrow)],$$

and this is expressed as

$$M = \frac{\mu_B}{2}\left[\int_0^{E_F+\mu_B B_0} N(E)\,\mathrm{d}E - \int_0^{E_F-\mu_B B_0} N(E)\,\mathrm{d}E\right].$$

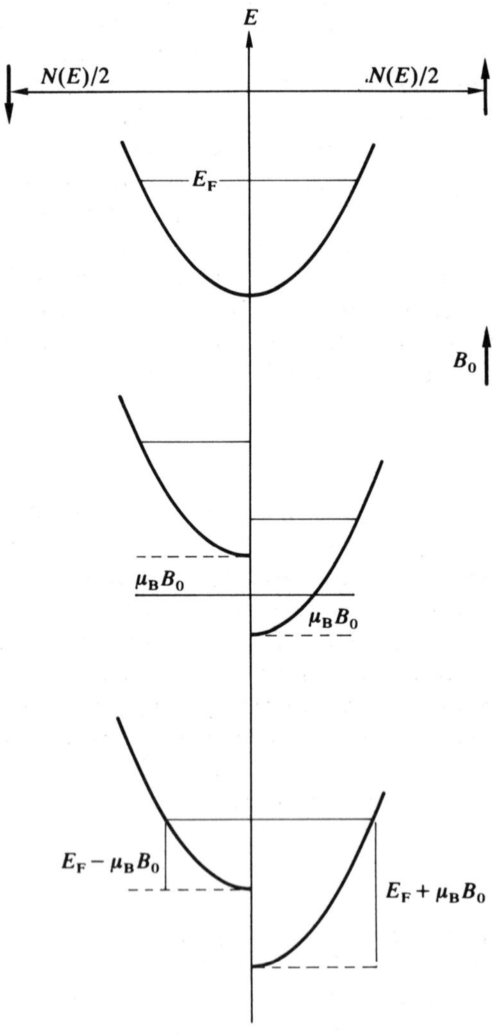


**Figure 6.11** The displacement of the bands of different spin in an applied magnetic field leads to unequal populations and a net magnetic moment. The above diagrams give a very distorted representation because $\mu_B B_0$ is only of order $10^{-4}$ $E_F$ for normal fields.

Since $\mu_B B_0 \ll E_F$, we may safely write

$$\int_0^{E_F \pm \mu_B B_0} N(E)\,dE = \int_0^{E_F} N(E)\,dE \pm \mu_B B_0 N(E_F),$$

which finally leads to

$$M = \mu_B^2 N(E_F) B_0, \tag{6.32}$$

$$\kappa = \frac{\mu_0 M}{B_0} = \mu_0 \mu_B^2 N(E_F). \tag{6.33}$$

$\kappa$ being the magnetic susceptibility per unit volume. Again this result is, in principle at least, generally applicable and is not confined to the free electron case. We find that, in contrast with the classical result, the susceptibility $\kappa$ is independent of temperature and directly proportional to $N(E_F)$.

This Pauli paramagnetism, as it is called, is wholly or partly counterbalanced by the diamagnetism that stems from the negative charge on the electron. Electrons in core orbits as well as in their free motion react to oppose the build up of the external field. The reaction produces an induced dipole moment directed against $B_0$. The susceptibility therefore has negative sign. It is also temperature independent. In all but the lightest of metals the Pauli paramagnetism is outweighed by the core diamagnetism, so it is not so important as the electronic heat capacity as a monitor of $N(E_F)$. As for the free electron gas, Landau showed that its diamagnetic contribution is numerically equal to one third that of the Pauli susceptibility. Just as for the heat capacity the transition metals are exceptional exhibiting a strong net paramagnetism which is essentially temperature independent, in qualitative agreement with (6.33). The core diamagnetism is of course a feature of all atoms in all substances.

## 6.4 Electrical Properties

The electrical conductivity is one of the most striking properties of metals and provides the impetus for introducing the ideal electron gas model. We can understand that when a metal is exposed to an applied field, an electric current should flow, but we can have little hope of appreciating how electrical resistance arises or how it varies with temperature (Fig. 6.12). We may presume that the electrons in some way collide with the ions (which are very large in comparison with the electrons), but nevertheless the free electron gas approach is completely unsuited to solving this problem and we must proceed in an *ad hoc* fashion and assume the presence of some viscous force that prevents the continued monotonic acceleration of the electron. We therefore postulate an equation of motion of the form

$$m\frac{d\mathbf{v}}{dt} + \frac{m\mathbf{v}}{\tau} = -e\mathbf{E}, \tag{6.34}$$

$\mathbf{E}$ being the electric field strength. The second term on the left is the viscous retarding component and $\tau$ is a quantity known as the 'relaxation time'. Classically $2\tau$ is the effective acceleration time of an electron or, in other words, $2\tau$ is the time between successive 'collisions'. After each collision, the drift velocity $\mathbf{v}$ produced by the field $\mathbf{E}$ is assumed to be reduced to zero and the acceleration must begin all over again. The drift velocity $\mathbf{v}$ is superimposed on the random motion of the electrons in the gas, which, for those near the Fermi level, occurs with the Fermi velocity $v_F$,

**Table 6.3** Fermi velocity and relaxation time from room-temperature conductivity and (6.37)

| | $v_F$ ($10^8$ cm s$^{-1}$) | $\tau$(RT) ($10^{-14}$ s) |
|---|---|---|
| Li | 1.29 | 0.88 |
| Na | 1.05 | 3.2 |
| K | 0.85 | 4.1 |
| Be | 2.25 | 0.51 |
| Mg | 1.58 | 1.1 |
| Al | 2.0 | 0.8 |
| Pb | 1.83 | 0.14 |
| Cu | 1.57 | 2.7 |
| Ag | 1.39 | 4.0 |
| Au | 1.39 | 3.0 |

with $\frac{1}{2}mv_F^2 = E_F$ (Table 6.3). Consider (6.34) and suppose that at a certain time $t_0$ the field is removed. Then

$$m\frac{dv}{dt} + \frac{mv}{\tau} = 0,$$

i.e.

$$\frac{dv}{v} = -\frac{dt}{\tau},$$

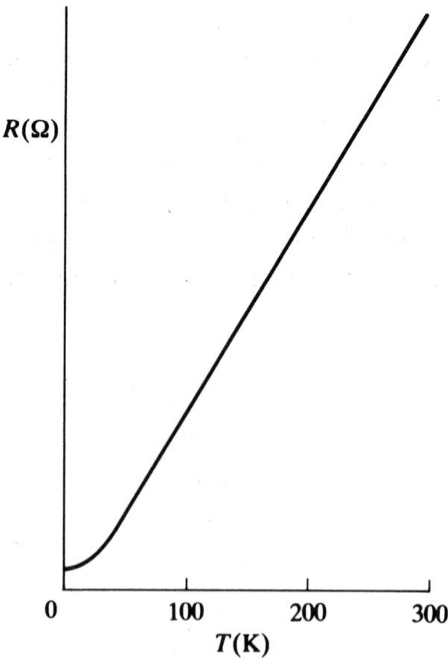


**Figure 6.12** At normal temperatures metals have an electrical resistance that varies essentially linearly with temperature, but below about 20 K it becomes constant. For good conductors like Cu or Al, the resistivity at 4 K may be only of order $10^{-4}$ times that at room temperature, but for transition metals this ratio is at best about 0.1.

whence

$$v(t) = v(t_0)e^{-(t-t_0)/\tau}. \tag{6.35}$$

The velocity decays with a time constant $\tau$. On the other hand, when a constant external field **E** is applied, experience tells us that a steady current flows in the sample; the current density must therefore be constant and this implies a constant drift velocity, the acceleration being zero. We thus arrive at the following:

$$m\mathbf{v}/\tau = -e\mathbf{E}, \quad \text{i.e. } \mathbf{v} = -e\tau\mathbf{E}/m. \tag{6.36}$$

The current density is given by

$$-ne\mathbf{v} = ne^2\tau\mathbf{E}/m,$$

so we write the conductivity

$$\sigma = ne^2\tau/m. \tag{6.37}$$

This is a frequently used expression for the conductivity and retains its significance even in more advanced treatments. The essence of the problem (and our ignorance) lies in the calculation of $\tau$ or its replacement by more fundamental quantities. For our purpose $\tau$ remains a phenomenological parameter that is to be established by experiment. A particular feature that we must explain is its rapid variation with temperature, but, as mentioned above, we cannot approach this aspect on the basis of the free electron gas.

How do we consider electrical conductivity in **k** space? We may express (6.36) in terms of **k** and write

$$\hbar\,\Delta\mathbf{k} = -e\mathbf{E}\tau, \tag{6.38}$$

corresponding to the acquirement of a drift velocity **v**. Thus all electrons in the gas are shifted in **k** space by an amount $\Delta\mathbf{k}$. Suppose **E** is directed along the positive $x$ axis. Then the whole Fermi sphere of occupied levels is shifted an amount $\Delta k$ in the direction of the negative axis, and it stays there as long as the field remains unchanged (Fig. 6.13). Now $\Delta k$, compared with $k_F$, is a small quantity, and it is immediately apparent from Fig. 6.13 that an effective redistribution of occupied electron states has occurred only in the immediate vicinity of $E_F$; we have increased the occupation of states near $-k_F$ at the expense of those near $+k_F$ and this imbalance produces the current flow. If the external field **E** is removed, the drift of charge

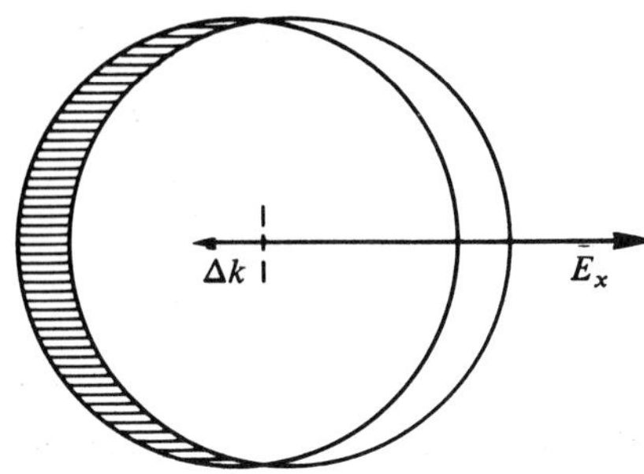


**Figure 6.13** In **k** space electrical conduction is associated with a slight displacement of the Fermi sphere about the origin. If $E_x \approx 1000$ V m$^{-1}$ and $\tau \approx 10^{-12}$ s then $\Delta k \approx 10^{-4}\,k_F$.

decays and the sphere of occupied states moves back to its symmetrical position about the origin. Suppose we apply an alternating field of the form

$$\mathbf{E} = \mathbf{E}_0 \, \mathrm{e}^{-\mathrm{i}\omega t}.$$

Then we expect the electrons to move in sympathy with the applied field and acquire a drift velocity

$$\mathbf{v} = \mathbf{v}_0 \, \mathrm{e}^{-\mathrm{i}\omega t}.$$

Equation (6.34),

$$m \frac{\mathrm{d}\mathbf{v}}{\mathrm{d}t} + \frac{m\mathbf{v}}{\tau} = -e\mathbf{E},$$

becomes

$$-\mathrm{i}\omega m\mathbf{v} + \frac{m\mathbf{v}}{\tau} = -e\mathbf{E},$$

whence

$$\mathbf{v} = \frac{-e\mathbf{E}\tau}{m(1 - \mathrm{i}\omega\tau)},$$

which leads to

$$\sigma = \frac{ne^2\tau}{m(1 - \mathrm{i}\omega\tau)} = \frac{\sigma_0}{1 - \mathrm{i}\omega\tau}, \tag{6.39}$$

$\sigma_0$ having been written for the DC conductivity. We shall make use of (6.39) in discussing the optical behaviour of the electron gas.

## 6.5 Optical Properties

Metals are ordinarily considered to be absorbers of light and very good reflectors. This is true for visible light and infrared radiation. However, in the ultraviolet region of the spectrum the reflectivity is usually very low and metals can become less absorbing than is usually thought. The optical behaviour of solids is governed by all the electrons they contain and not just those that happen to be free, so it must be understood that we can hope only to approximate the optical behaviour of the valence electrons with our electron gas model. We shall find that our results have significance primarily for the red and infrared regions of the spectrum.

We have seen in the previous section that the use of the relaxation time $\tau$ allows us to introduce static and time-dependent electrical conductivities $\sigma_0$ and $\sigma$ respectively. These conductivities describe the transport of charge through a sample under the action of an electric field produced by electrodes attached to the sample. It may be argued that electromagnetic radiation can only produce induced currents in a specimen. These currents oscillate to and fro, charge neither entering nor leaving the sample. Furthermore, these induced currents are transverse to the wave vector of the radiation.

If we assume that the wavelength of the light is very large compared with atomic dimensions, we may neglect this latter aspect and assume that the behaviour is governed by Maxwell's version of Ampère's law

$$\nabla \times \mathbf{H} = \varepsilon \frac{d\mathbf{E}}{dt} = \varepsilon_0 \frac{d\mathbf{E}}{dt} + \frac{d\mathbf{P}}{dt}. \tag{6.40}$$

Recalling our earlier discussion of the electron gas, we write

$$\mathbf{P} = -ne\mathbf{\Delta},$$

$$\frac{d\mathbf{P}}{dt} = -ne\frac{d\mathbf{\Delta}}{dt} = -ne\mathbf{v} = \sigma\mathbf{E}.$$

We think of $-ne\mathbf{v}$ as a current obtained from the field **E**, so that (6.40) becomes

$$\nabla \times \mathbf{H} = \varepsilon_0 \frac{d\mathbf{E}}{dt} + \sigma\mathbf{E},$$

$\sigma$ being given by (6.39). The latter equation may be expressed in two ways because the field may be written

$$\mathbf{E} = \mathbf{E}_0 \, e^{-i\omega t},$$

and

$$\frac{d\mathbf{E}}{dt} = -i\omega\mathbf{E}.$$

Hence

$$\nabla \times \mathbf{H} = \left(\varepsilon_0 + \frac{i\sigma}{\omega}\right)\frac{d\mathbf{E}}{dt} = \varepsilon\frac{d\mathbf{E}}{dt}, \tag{6.41a}$$

or

$$\nabla \times \mathbf{H} = (\sigma - i\omega\varepsilon_0)\mathbf{E} = \hat{\sigma}\mathbf{E}. \tag{6.41b}$$

The first description is that of a lossy dielectric, the second that of a metal with reactance, $\varepsilon$ and $\hat{\sigma}$ being complex quantities. Substituting (6.39) for $\sigma$, we obtain

$$\varepsilon = \varepsilon_0 + i\frac{ne^2\tau}{\omega m(1 - i\omega\tau)},$$

which on rationalization becomes (recalling (6.14) for $\omega_p$)

$$\frac{\varepsilon}{\varepsilon_0} = \varepsilon_r = 1 - \frac{\omega_p^2}{\omega^2 + \tau^{-2}} + i\frac{\omega_p^2\tau}{\omega(1 + \omega^2\tau^2)}, \tag{6.42a}$$

or

$$\varepsilon_r = \varepsilon_1 + i\varepsilon_2, \tag{6.42b}$$

$\varepsilon_1$ and $\varepsilon_2$ being the real and imaginary parts of the complex number $\varepsilon_r$. The corresponding expressions for $\hat{\sigma}$ are

$$\frac{\hat{\sigma}}{\varepsilon_0} = \frac{\omega_p^2\tau}{1 + \omega^2\tau^2} - i\omega\left(1 - \frac{\omega_p^2}{\omega^2 + \tau^{-2}}\right) \tag{6.43a}$$

and

$$\frac{\hat{\sigma}}{\varepsilon_0} = \sigma_1 - i\sigma_2 \tag{6.43b}$$

It can be seen that

$$\sigma_1 = \omega\varepsilon_2, \qquad \sigma_2 = \omega\varepsilon_1.$$

Quite generally, then, we expect a light wave interacting with an electron gas to produce currents that are in phase with the electromagnetic field vector and currents in quadrature with this vector. Only the in-phase components produce optical absorption via Joule heating. We may describe this energy dissipation per unit volume and unit field strength in terms of a wholly real 'optical conductivity' $\sigma(0)$, although just $\sigma$ is used when there is no risk of confusion; $\sigma(0)$ is a measure of the optical absorption. Similarly, a wholly real dielectric constant $\varepsilon(0)$ governs the in-quadrature currents. Then

$$\nabla \times \mathbf{H} = \varepsilon(0)\frac{\mathrm{d}\mathbf{E}}{\mathrm{d}t} + \sigma(0)\mathbf{E},$$

and from the previous discussion

$$\varepsilon(0) = \varepsilon_0\,\varepsilon_1, \tag{6.44a}$$

$$\sigma(0) = \varepsilon_0\,\sigma_1 = \omega\varepsilon_0\,\varepsilon_2. \tag{6.44b}$$

We see that the optical absorption may be expressed in terms of $\sigma_1$ or $\omega\varepsilon_2$. Since a refractive index $\mathcal{N}$ may be defined by

$$\mathcal{N}^2 = \varepsilon_\mathrm{r} = 1 - \frac{\omega_\mathrm{p}^2}{\omega^2 + \mathrm{i}\omega\tau^{-1}}, \tag{6.45}$$

even the imaginary part of this quantity may serve as a measure of optical absorption. We now consider the simplest situation, that of the perfectly conducting electron gas, and let $\tau \to \infty$ so that

$$\mathcal{N}^2 = \varepsilon_\mathrm{r} = 1 - \frac{\omega_\mathrm{p}^2}{\omega^2}. \tag{6.46}$$

Clearly

(a) when $\omega > \omega_\mathrm{p}$, $\varepsilon_\mathrm{r} > 0$ and $\mathcal{N}$ is wholly real;

(b) when $\omega < \omega_\mathrm{p}$, $\varepsilon_\mathrm{r} < 0$ and $\mathcal{N}$ is wholly imaginary.

What does this mean? Consider the expression for the electric field vector in the electron gas:

$$\mathbf{E} = \mathbf{E}_0\,\mathrm{e}^{\mathrm{i}(\mathbf{k}\cdot\mathbf{r} - \omega t)}.$$

Now $\mathbf{k} = \mathcal{N}\mathbf{k}_0$, where $\mathbf{k}_0$ is the wave vector in vacuum. If $\mathcal{N}$ is real then $\mathbf{E}$ is propagated without change in amplitude; there is no absorption and the metal behaves as a dielectric possessing a poor reflectivity: we note that this situation arises when $\omega > \omega_\mathrm{p}$, i.e. at high photon energies, so we can, at least in part, appreciate the poor 'metallic' properties in the far ultraviolet. On the other hand, when $\mathcal{N}$ is wholly imaginary, $\mathcal{N} = \mathrm{i}\kappa$, the wave form is

$$\begin{aligned}\mathbf{E} &= \mathbf{E}_0\,\mathrm{e}^{\mathrm{i}(\mathrm{i}\kappa\mathbf{k}_0\cdot\mathbf{r} - \omega t)}\\ &= \mathbf{E}_0\,\mathrm{e}^{-\kappa\mathbf{k}_0\cdot\mathbf{r}}\mathrm{e}^{-\mathrm{i}\omega t}.\end{aligned}$$

The amplitude decreases exponentially as the wave progresses through the gas, and after a distance $(\kappa k_0)^{-1}$ the amplitude has decreased to $\mathrm{e}^{-1}$ times its initial value.

Now in optics we learn that the intensity reflection coefficient at a boundary between two media, one of which is vacuum, is

$$|r|^2 = \left|\frac{1-\mathscr{N}}{1+\mathscr{N}}\right|^2 \quad \text{(normal incidence)}, \tag{6.47}$$

and if $\mathscr{N} = i\kappa$ then $|r| = 1$ and we obtain perfect reflectivity. For our perfectly conducting electron gas this arises when $\omega < \omega_p$.

The plasma frequency therefore marks the dividing line between optically absorbing and transmitting regions of an electron gas. Even when $\tau$ becomes finite, this result is maintained. Since $\hbar\omega_p \approx 5$–$15$ eV and the limit of visible light lies at about 3 eV, it is clear why metals ordinarily have a high reflectivity. One might ask whether the transparency at high photon energies would not be reduced by the creation of plasmons. The answer is that ordinarily the production of plasmons by light is not possible because light is a *transverse* wave motion and plasmons are *longitudinal* charge density oscillations. Ordinarily, light cannot couple to the plasmons.

However, let us return to (6.42) and express $\varepsilon_r$ in terms of its real and imaginary parts. We find that

$$\text{Re}\ \varepsilon_r = 1 - \frac{\omega_p^2}{\omega^2 + \tau^{-2}} = \varepsilon_1, \tag{6.48}$$

$$\text{Im}\ \varepsilon_r = \frac{\omega_p^2}{\omega^2 + \tau^{-2}}\frac{1}{\omega\tau} = \varepsilon_2, \tag{6.49}$$

and we have already seen that

$$\varepsilon_2 = \frac{\sigma_1}{\omega}.$$

The equivalents of these relations were first derived for the classical electron gas by Drude, but they retain their value for describing the infrared optical behaviour of metals. They are illustrated in Fig. 6.14, and typical experimental data are shown in Figs. 6.15 and 6.16. We see that at low energies there is a strong similarity between experiment and the simple theory. We can fit (6.48) and (6.49) to the data and obtain estimates of $\omega_p$ and $\tau$. Optical data for many of the simple metals in liquid form are available, and the agreement between experiment and the free electron theory is very striking. Derived values of $\omega_p$ closely match those calculated on the basis of the free electron formulae, and where differences arise we can again allow the electron mass to be adjusted and thereby introduce an *optical effective mass* for the electron.

When in liquid form the simple metals optically appear to approximate the conditions of the free electron gas better than do their crystalline phases. This is very apparent when we compare the optical spectra of liquid and solid aluminium (Fig. 6.17). The solid aluminium produces sharp characteristic absorption bands; we shall see later that these arise as a direct consequence of the periodic arrangement of the atoms in the crystal.

Other metals, notably Cu, Ag and Au, also show Drude-like behaviour at low photon energies, but at higher frequencies there are well-marked absorption edges. On the other hand, germanium, which in the pure state and at very low temperatures is an insulator, and therefore has no free electrons, is transparent in the infrared region of the spectrum, but at higher energies has an absorption spectrum

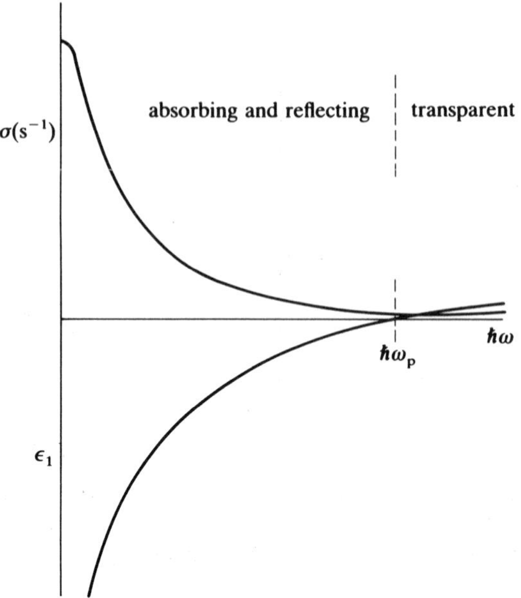


**Figure 6.14** The optical conductivity $\sigma$ and dielectric constant $\varepsilon_1$ of the electron gas according to Drude's equations.

not very different from a metal. We shall explain the main features of these spectra later, but the important point at the moment is that the simple model of the ideal electron gas gives a good, albeit phenomenological, description of the high reflectivity of metals and particularly so for the liquid state.

## 6.6 The Hall Effect

Suppose we have a conductor with an accurately formed rectangular cross-section carrying a current $I$ (see Box 6.1, pp. 164–165). A magnetic field $B_0$ is directed perpendicular to one face of the specimen, and it is then found that in the direction perpendicular to both $I$ and $B_0$ a steady potential difference arises across the specimen. This is the Hall effect. We write the *Hall coefficient* $R_H$ as

$$R_H = \frac{1}{nq}. \tag{6.50}$$

The sign of $R_H$ is determined by that of the charge carrier $q$, and its size is inversely proportional to the density of charge carriers. In metals we expect $q = -e$, and therefore $R_H$ should have a negative sign and $n$ should be the electron density. Clearly $R_H$ is readily calculable on the assumption that $n$ is the density of valence electrons. The Hall coefficient is also easily measured, even in the case of a liquid metal; some values of $R_H$ are given in Table 6.4. With one or two exceptions the liquid simple metals are found to have Hall coefficients very close to those calculated for the appropriate free valence electron gas. The solid metals on the other hand do not show such good agreement with simple theory, and it is sometimes the

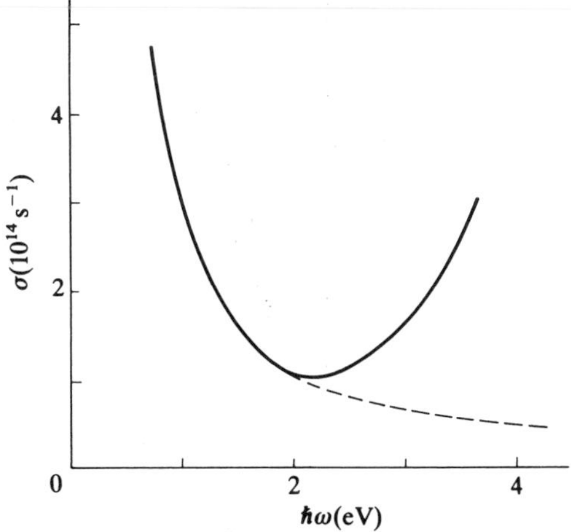


**Figure 6.15** The optical conductivity of Li metal at room temperature. The increase above 2 eV is caused by the crystal structure of Li (and will be explained in Chapter 9), but the absorption at low energies is Drude-like. (After Mathewson and Myers 1972.)

case that the measured value has a positive sign, which cannot be understood from our present standpoint. We again notice that a liquid metal follows the simple electron gas model more accurately than does a crystalline metal. Clearly the transition from the isotropic disordered distribution of ions that characterizes the liquid to the strict regular geometrical distribution typical of the crystal causes distinct changes in the detailed motion of the electrons.

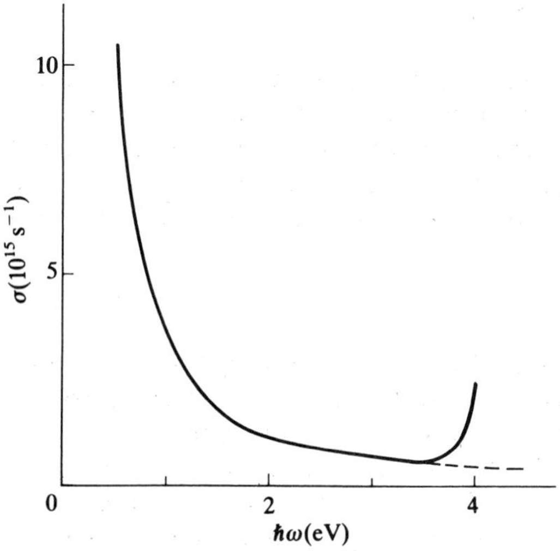


**Figure 6.16** The optical conductivity of Ag at room temperature. This metal has a wide 'Drude' range. The increase in absorption near 4 eV arises on account of the crystal structure and the presence of d electrons in this metal. (After Hunderi and Myers 1973.)

**Table 6.4** Hall coefficients

| | $R_H$ ($10^{-11}$ m$^3$ A$^{-1}$ s$^{-1}$) | | | |
|---|---|---|---|---|
| | Solid | Liquid | Free electron value | $n_0$ |
| Li | −17 | | −13.1 | 1 |
| Na | −25 | −25.5 | −25.5 | |
| Cu | −5.5 | −8.25 | −8.25 | |
| Ag | −9.0 | −12.0 | −12.0 | |
| Au | −7.2 | −11.8 | −11.8 | |
| Be | +24.4 | −2.6 | −2.53 | 2 |
| Zn | +3.3 | −5 | −5.1 | |
| Cd | +6.0 | −7 | −7.25 | |
| Al | −3.5 | −3.9 | −3.9 | 3 |
| Ga | | −3.9 | −3.95 | |
| In | | −5.6 | −5.65 | |
| Bi | $-10^4$ | −3 | −4.3 | 5 |
| Mn | −9.3 | | | |
| Fe | +2.45 | | | |
| Co | −1.33 | | | |
| Ni | −6.11 | | | |
| Mo | +12.6 | | | |
| Ta | +10.1 | | | |
| W | +11.8 | | | |

*Notes:* (a) certain simple as well as transition solid metals have positive Hall coefficients; (b) the liquid simple metals approximate the free electron values extremely well; (c) the behaviour of Bi is exceptional.
From Busch and Güntherodt (1974).

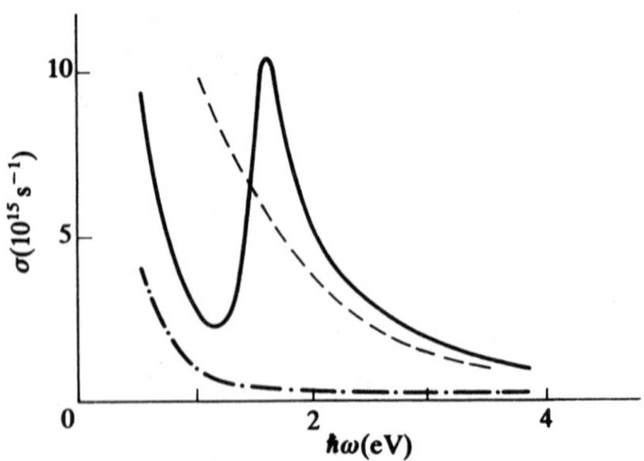


**Figure 6.17** The optical conductivity of Al shows pronounced crystal structure effects whereas the liquid phase of this metal presents a completely Drude-like behaviour. ——, total absorption at 25 K; —·—, Drude at 25 K; ‒‒‒, liquid at 900 K. (After Mathewson and Myers 1972 and Miller 1969.)

## 6.7 Thermal Conductivity

We normally associate metals with good thermal conductivity, and the simple metals are good conductors of heat. We may apply the standard result of kinetic gas theory and, just as for our discussion of the gas of phonons, write the thermal conductivity as

$$K = \tfrac{1}{3}C_e\langle v\rangle\Lambda.$$

For our electron gas, only electrons in the neighbourhood of the Fermi level can accept small thermal energies; the Pauli principle forbids energy exchanges between electrons inside the Fermi sphere because all available electron states are already occupied. We therefore write

$$C_e = \tfrac{1}{3}\pi^2 k_B^2 N(E_F)T.$$

Usually we refer $N(E_F)$ to the atom and write it as $\frac{3}{2}n_0/E_F$. In calculating $K$, we must refer to unit volume and therefore we replace $n_0$ by $n$. Thus

$$C_e = \frac{\pi^2 k_B^2}{3}\frac{3}{2}\frac{n}{E_F}T.$$

Furthermore,

$$\langle v\rangle = v_F, \quad \text{where } \tfrac{1}{2}mv_F^2 = E_F,$$

$$\Lambda = v_F\tau,$$

whence

$$K = \frac{\pi^2 k_B^2 n\tau}{3m}T. \tag{6.51}$$

Now we may replace $n\tau/m$ using our expression for the electrical conductivity

$$\sigma_0 = \frac{ne^2\tau}{m},$$

and thereby obtain the ratio

$$\frac{K}{\sigma_0} = \frac{\pi^2 k_B^2}{3e^2}T. \tag{6.52}$$

The quantity $K/\sigma_0 T$ is independent of temperature and is known as the Lorenz number (*Lorenz* was a Dane and is not to be confused with the Dutchman *Lorentz*); whereas (6.52) expresses the Wiedemann-Franz law relating the thermal and electrical conductivities and was first derived by Drude.

In pure metals, the electrons dominate the thermal conduction process, but the variation of $K$ with temperature is very similar in form to that found for an insulator, as Fig. 6.18 demonstrates. Just as in the case of the electrical conductivity, we cannot expect to understand the detailed variation of the thermal conductivity without making arbitrary decisions regarding the dependence of $\tau$ on temperature. However, the similarity between the behaviour of the thermal conductivity of crystalline insulators and metals suggests that in the crystalline metal the electrons responsible for transport of heat are also subject to both normal and umklapp scat-

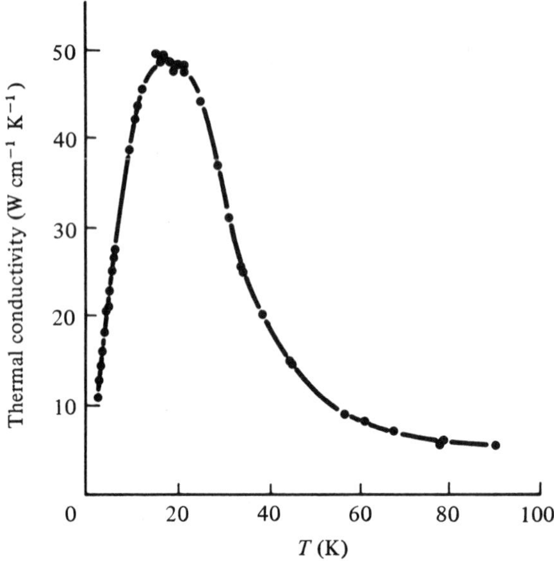


**Figure 6.18** The temperature variation of the thermal conductivity of Cu. (After Berman and MacDonald 1952.)

**Box 6.1. The Hall Effect**

The free electron gas is an isotropic system; the electrical conductivity is independent of the direction of the applied electric field. This is also the case for real metals with cubic structures, but metals with lower symmetry are anisotropic and the behaviour is governed by the conductivity tensor with nine components. Anisotropy may be introduced into the free electron gas by the application of a magnetic field **B**. Let such a field be directed along the z-axis. The equation of motion for the electrons (6.34) now becomes

$$m\frac{d\mathbf{v}}{dt} + \frac{m\mathbf{v}}{\tau} = -e(\mathbf{E} + \mathbf{v} \times \mathbf{B}).$$

In the steady state the acceleration is zero and we find

$$\mathbf{v} = -\frac{e\tau}{m}(\mathbf{E} + \mathbf{v} \times \mathbf{B}).$$

This equation may be written in terms of its x, y and z components. We readily solve for the component velocities and then write the equations for the current densities:

$$j_x = -nev_x = \frac{\sigma_0}{1+\omega_c^2\tau^2}(E_x - \omega_c \tau E_y),$$

$$j_y = -nev_y = \frac{\sigma_0}{1+\omega_c^2\tau^2}(\omega_c \tau E_x + E_y),$$

$$j_z = -nev_z = \sigma_0 E_z.$$

$\omega_c = eB/m$ is called the cyclotron frequency (see Problem 6.22).

*continued*

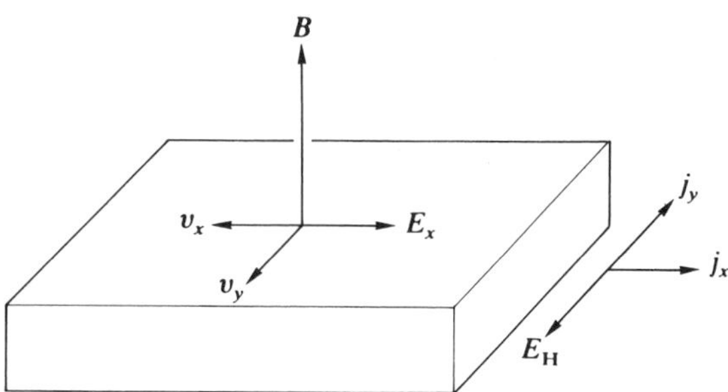


The previous equations are conveniently written in matrix form:

$$\begin{pmatrix} j_x \\ j_y \\ j_z \end{pmatrix} = \frac{\sigma_0}{1+\omega_c^2\tau^2} \begin{pmatrix} 1 & -\omega_c\tau & 0 \\ \omega_c\tau & 1 & 0 \\ 0 & 0 & 1+\omega_c^2\tau^2 \end{pmatrix} \begin{pmatrix} E_x \\ E_y \\ E_z \end{pmatrix}.$$

We see here the nine components of the conductivity tensor and note that

$$\sigma_{xx} = \sigma_{yy} = \frac{\sigma_0}{1+\omega_c^2\tau^2}, \qquad \sigma_{xy} = -\sigma_{yx} = -\frac{\sigma_0\,\omega\tau}{1+\omega_c^2\tau^2}.$$

In the Hall effect the external fields are $E_x$ and $B_z$ $(=B)$. However, the field $B$ causes a drift of electrons in the $y$ direction, but no steady current can flow in this direction because there is no circuit continuity. Instead, charge piles up on the surface of the sample, setting up an electric field $E_y$: The latter nullifies the Lorentz force that produced the original drift of charge. This 'Hall field' $E_H$ is found by setting $j_y = 0$. Thus

$$j_y = -nev_y = 0 = \frac{\sigma_0}{1+\omega_c^2\tau^2}(\omega_c\tau E_x + E_y),$$

giving

$$E_H = E_y = -\omega_c\tau E_x = -\frac{eB}{m}\tau E_x,$$

whereas

$$j_x = \frac{\sigma_0}{1+\omega_c^2\tau^2}(E_x + \omega_c^2\tau^2 E_x) = \sigma_0 E_x.$$

The Hall coefficient is defined as $R_H = E_H/j_x B$, i.e.

$$R_H = -\frac{e\tau}{\sigma_0 m} = -\frac{1}{ne}.$$

The Hall resistance is defined as

$$\mathscr{R}_H = \frac{V_y}{I_x}.$$

tering processes. Furthermore, although metals are superior to insulators in conducting heat at room temperature, at low temperatures their maximum values are comparable. Measurement of the thermal conductivity is much more tiresome and inherently less accurate than that of electrical resistivity. One might therefore, in certain practical instances, consider using (6.52) to convert electrical conductivity values into estimates of thermal conductivity. A prerequisite for such a procedure is that $\tau$ should be the same for both conduction processes.

Measured $K$ and $\sigma_0$ values for pure metals follow the Wiedemann-Franz law at and above room temperature; this suggests that the law is best obeyed in real metals under conditions of high thermal and electrical resistance. Alloying invariably increases the electrical and thermal resistance of a metal and this implies that the Wiedemann-Franz law should hold for alloys thereby allowing estimates of the thermal conductivity to be made via measurement of the electrical conductivity.

## 6.8 Final Comments

### 6.8.1 *Exchange and many body interactions*

The free electron gas model is surprisingly good in describing certain elementary features of metallic behaviour, provided we are willing to accept the limitations attendant on the use of a relaxation time in transport processes. Whereas we can estimate quantities dependent only on the energy of electrons or the density of electron states, we cannot make *a priori* estimates of resistivity or thermal conductivity, nor can we appreciate why certain elements are metals and others insulators. Nevertheless we obtain simple formulae, some of which have wide applicability. Furthermore, we know from experiment that liquids are often conductors and certain substances, e.g. Ge and Bi, conduct electricity better when in liquid form. We conclude that crystallinity is not essential for metallic behaviour and, qualitatively at least, we can perhaps understand why the model of free electrons describes liquid metals much better than solid conductors. The liquid possesses no long-range structural order and its statistical homogeneity is better approximated by the electron gas model. Experimentally we find that conduction in both solids and liquids is favoured by a large coordination number, greater than eight, and the lack of strongly directional bonds. Clearly, at this stage we cannot say more about the detailed electronic structure of solid or liquid metals other than that the *distribution of energy states for valence electrons must bear some strong resemblance to those calculated on the free electron model.* In other words, real metals must, we feel, have some quasicontinuous distribution of valence electron states, otherwise it is difficult to understand the qualitative and quantitative success of the elementary approach. One of the major problems of solid state physics has been (and still is) the determination of the exact form this quasicontinuous distribution of electron states takes in real solids and liquids. However, before pursuing this question, we make a few final remarks concerning the further development of the free electron gas model.

In the elementary model described in this chapter, the electron is assumed to move in the Coulomb field of the ions together with that arising from the Coulomb interactions with all the other free electrons; these two fields combine to produce the assumed constant potential in which every free electron moves. The result is then that the wave function of any electron is dependent only on its own coordinates and not on those of all the other electrons. Instead of solving a problem in $3N$ variables ($N$ being the number of electrons in the gas), we solve $N$ problems in three variables; for the simple free electron gas these $N$ problems are identical because all electrons experience the same potential. This procedure, whereby the problem is treated in terms of the behaviour of a single electron moving in the field of the ions and an average potential (not necessarily constant) from all the other electrons, is maintained in all present-day calculations of the energy levels for electrons in solids. The approach is known as the 'one electron approximation' or 'independent particle approximation' because the Schrödinger equation is solved for one electron at a time and not for the assembly of $N$ interacting particles that comprise the system. The method works extremely well, perhaps better than might have been expected. It is nevertheless the case that in order to solve the Schrödinger equation for a given electron in the average field of the other $N - 1$ electrons we must know their charge distributions, and this is only possible if we already know their wave functions. The approach is therefore to start by making an intelligent guess for the wave functions of these other $N - 1$ electrons and then to solve for the wave function of the chosen electron. Using this knowledge, one can then obtain new and better approximations for the other $N - 1$ wave functions. By successive iterations, better and better approximations to the $N$ wave functions are obtained, and finally one arrives at the desired situation where, for all electrons, the $\psi$ values obtained in the solutions differ insignificantly from those used as input to establish the potential. The solutions for the $N$ wave functions are consistent in that they reproduce the potential that gave these particular solutions. The wave functions and the potentials are said to be self-consistent. However, it is possible to derive energy eigenvalues without determining the wave functions, but if the latter are not calculated there is no explicit guarantee that self-consistency has been obtained. If we attempt to treat the behaviour of the $N$ valence electrons in a real metal or the ideal free electron gas as an interacting assembly of electrons then we say that we have a many ($10^{23}$) body problem. We shall return to this aspect shortly, but for the moment we continue our discussion of the free electron gas in terms of the one-electron approach. It might be thought that the model of the constant potential box is so simple and inappropriate for real metals (particularly the transition metals) that there would be little point in attempting further refinement. This is not the case. In spite of the neglect of the most characteristic feature of a crystalline substance, the periodic potential, much attention has been given to developing the model of free electrons. Since the principles underlying these developments are of general significance, we shall give a brief description (for a thorough discussion of the topics of this section see Raimes 1961).

The two important quantities that we shall attempt to describe qualitatively are the 'exchange' and 'correlation' terms.

Suppose we solve the problem of the free electron gas in the one-electron approximation. Then each electron has a wave function $\phi_i(\mathbf{q}_i)$, and we do not need to write it in explicit form. This wave function is dependent on the space and spin coordinates of the electron $i$, both of which are included in $\mathbf{q}$. We might therefore consider writing the wave function of the total assembly of $N$ electrons as a simple product of the separate one-electron wave functions:

$$\psi_N = \prod_{i=1}^{N} \phi_i(\mathbf{q}_i). \tag{6.53}$$

However, electrons are indistinguishable – we cannot label the electron occupying the state $\phi_i$ and the coordinate $\mathbf{q}_i$. We may label states and coordinates but not the electrons occupying them. This means that we may permute the $N$ electrons in their $N$ coordinates to produce $N!$ equivalent wave functions of the form (6.53). In effect we may associate any $\phi_i$ with any $\mathbf{q}_j$. Each product wave function is an equivalent description of the electron gas; even so, each of these products is independent of the order in which the electrons are assumed to occupy the various coordinates. These product wave functions therefore do not satisfy Pauli's principle, which, in addition to asserting the indistinguishability of electrons, demands that the total wave function be antisymmetric in the electron coordinates (both spatial and spin). The interchange of the coordinates of any pair of electrons must cause the wave function to change sign, and it must vanish if two electrons are given in the same coordinates. These latter requirements may be met if the total wave function is written as an $N$th-order determinant

$$\psi = \frac{1}{(N!)^{1/2}} \begin{vmatrix} \phi_1(\mathbf{q}_1) & \phi_1(\mathbf{q}_2) & \cdots & \phi_1(\mathbf{q}_N) \\ \phi_2(\mathbf{q}_1) & \phi_2(\mathbf{q}_2) & \cdots & \phi_2(\mathbf{q}_N) \\ \vdots & \vdots & \cdots & \vdots \\ \phi_N(\mathbf{q}_1) & \phi_N(\mathbf{q}_2) & \cdots & \phi_N(\mathbf{q}_N) \end{vmatrix}. \tag{6.54}$$

The prefactor is a normalizing constant for orthonormal $\phi$. This determinant tells us that a given electron may occupy any coordinate (i.e. the rows) and a given coordinate may be associated with any electron state (i.e. the columns). Evaluation of the determinant leads to $N!$ product wave functions of the earlier form, but Pauli's principle is now satisfied because the interchange of coordinates causes the interchange of two columns of the determinant and therefore a change of sign, whereas if two electrons occupy the same state, e.g. $\phi_1(\mathbf{q}_i) = \phi_2(\mathbf{q}_j)$, then two rows of the determinant are identical and it is therefore zero.

The use of the antisymmetric determinantal wave function introduces a new quantity into the physics of systems comprising two or more electrons, namely the exchange interaction.

We now turn from a system with $10^{23}$ particles to one with only two: the helium atom. Raimes (1961) treats the matter in detail, but here we describe only the principal result. For convenience, we label the states for the two electrons in the He atom as $\psi_a(\mathbf{r}_1)$ and $\psi_b(\mathbf{r}_2)$. These wave functions depend only on the space coordinate $\mathbf{r}$ because we shall not need to consider the spin explicitly. In addition to the energies of these two electrons in the field of the nucleus, there arise two energies on account of their mutual interaction. First there is a conventional Coulomb repulsive term

$$C = \iint \psi_a(\mathbf{r}_1)\psi_a^*(\mathbf{r}_1) \frac{e^2}{r_{12}} \psi_b(\mathbf{r}_2)\psi_b^*(\mathbf{r}_2)\, d\mathbf{r}_1\, d\mathbf{r}_2. \tag{6.55}$$

This energy is independent of the orientation of the electron spins. The second feature, however, is completely new and arises as a direct consequence of using the determinantal wave function and is thereby a direct result of Pauli's principle; it is called the exchange term and is denoted by $J$:

$$J = \iint \psi_a(\mathbf{r}_1)\psi_a^*(\mathbf{r}_2) \frac{e^2}{r_{12}} \psi_b(\mathbf{r}_2)\psi_b^*(\mathbf{r}_1)\, d\mathbf{r}_1\, d\mathbf{r}_2. \tag{6.56}$$

It looks almost identical with the Coulomb term except for the exchange of coordinates. Normally in the wave mechanics of electrons in atoms this exchange term arises only between pairs of electrons with parallel spins: it has a positive sign and leads to a reduction in the total energy of the atom. This is why the electronic structure of an atom is always such that, within the limitations set by Pauli's principle, the electrons produce the largest possible total spin. We can understand the reduction in the total energy by exchange as the result of electrons with parallel spins avoiding one another because of the Pauli principle – they do not approach one another as closely as do electrons with antiparallel spins, and their electrostatic interaction is thereby reduced.

If we now return to the electron gas, we might consider the introduction of the exchange interaction to be an obvious improvement. Within the one-electron approximation mentioned earlier this is fully possible, but the effect of doing so is disastrous in that the previous agreement with experiment is completely lost. Thus, the band form is no longer parabolic, and furthermore $N(E_F)$ and quantities dependent on it are completely at variance with experiment.

The reason for this disappointing result is that, although the inclusion of the exchange interaction is certainly a valid measure, it is not the only improvement that can be made to the simple model. It is also the case that electrons with antiparallel spins avoid coming too close to one another simply on account of their mutual electrostatic repulsion. Electrons avoid one another because of their similar charge and not only when their spins are parallel – the latter influence is, however, the stronger one, as Fig. 6.19 illustrates. The mutual avoidance of electrons resulting from their similar charge leads to a correlated motion, and the associated decrease in the Coulomb repulsive energy is called the correlation energy. When account is taken of the influence of both exchange and correlation then the calculated properties of the free electron gas are again in reasonable agreement with experiment. The exchange and correlation energies are essential components in any proper theoretical calculation of the energy levels for electrons in atoms and in solids, but their evaluation presents many difficulties.

Let us now return to the many-body aspects of the behaviour of electrons. An electron gas, whether ideally free or confined to a periodic lattice, is an extremely complicated physical system. The electrons in solids interact with one another and with the ions, which themselves also form a system of interacting particles. At absolute zero the ions execute zero-point motion about the lattice sites, and the electrons constitute a degenerate gas with a clearly defined chemical potential equal to the Fermi energy. Let us begin by considering the effect of temperature on the ions (in

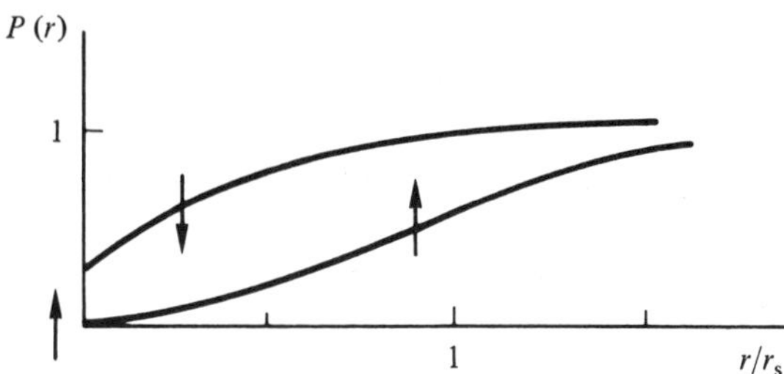


**Figure 6.19** The probability that an electron of similar spin and one of the dissimilar spin will be found in the vicinity of a given electron. (After Slater 1934.)

principle of course this is not possible because the ions interact through the electrons, so any excitation of the ions must involve the valence electrons; nevertheless we assume that we can consider the ions separately). We saw earlier in our discussion of lattice vibrations that the strong interactions between the ions demand that we describe the atomic motion not in terms of the movement of individual atoms, but by a collective motion in the form of propagating displacement waves that we call phonons. The vibratory behaviour of the lattice is also a many-body problem, although we did not mention this aspect at the time. Nevertheless, the effect of temperature is to cause collective mechanical vibrations of the lattice. In many-body terminology these are called elementary excitations. We consider the excitations of the complete system of mutually interacting particles. These elementary excitations are the phonons. We find that a phonon is characterized by a discrete energy and a well-defined wave vector that is associated with crystal momentum. Phonon wave packets may be constructed, and they interact like mechanical particles – energy and crystal momentum being conserved – we call them by the generic name 'quasiparticles' because their behaviour is 'particle-like'.

The electron gas similarly is a many-body system, a very large collection of similar particles in mutual interaction through their charges and spins. The single-particle approximation restricts attention to 'ordinary' electrons, but the many-body approach deals with the elementary excitations of the complete system of mutually interacting particles. One such excitation is the plasmon; this we also classify as a quasiparticle, but, as we have seen, plasmons are not ordinarily excited on account of their high energies. The weakest elementary excitation in the electron gas would be that created by a slight increase in temperature above 0 K. In the single-particle model certain electrons would occupy states somewhat above $E_F$ but the majority would remain unaffected. In the many-body picture, it is not possible to consider the behaviour of any one electron without at the same time asking how its behaviour affects the rest of the system; in fact the chemical potential becomes dependent on the state of excitation of the system. The problem becomes quite involved, and we shall not pursue it. Even so, and in spite of the correctness of its logical basis, many-body theory, somewhat paradoxically perhaps, has not destroyed faith in the one-electron approximation, but rather has justified its use and strengthened its standing. The reason is that many-body effects are, in general, of restricted (but quite measurable) influence and the elementary excitations may be interpreted in terms of single-particle properties modified by simple multiplicative corrections, often in the form of a contribution to the effective mass of the electron. The most important consequences of many-body interactions arise in superfluidity as observed in the isotopes of helium and the superconductivity of many metals and alloys.

Another feature that must be given brief mention is the dielectric response of freely moving electrons in solids. In electrostatics we are accustomed to consider a metal to have a dielectric constant that is infinitely large, the metal is a region of constant potential and there is no electric field within it. Nevertheless a metal is composed of essentially stationary nuclei, bound electrons and freely moving electrons. The free moving electrons can readily adjust to any charge fluctuations that may arise within the metal. Thus if we imagine an extra charge to be introduced into a metal, e.g. by implanting a bare proton into it, the electrons in the metal will move to neutralize the field of the proton and its effect will therefore be very restricted; the sample will of course carry a net positive charge, but this is uniformly distributed on its surface and produces no electric field within it. Variations in charge density and interatomic electric fields can therefore be countered by the freely moving conduction electrons. This means that the metal possesses a dielectric 'constant'. We say that charges and the fields that they give rise to are 'screened' and much less than the bare Coulomb fields that these charges would otherwise produce in a vacuum.

We use the term 'dielectric function' rather than dielectric constant because the dielectric response of the metal is dependent on both the time and spatial variation of the internal electric fields. This dielectric screening is important for the properties of metals (both solid and liquid) and not least for the occurrence of superconductivity, but it is difficult to calculate for electrons in periodic potentials so it is ordinarily used in the form for the free electron gas.

### 6.8.2 *The dielectric function*

A first approximation to the static dielectric constant of the free electron gas may be obtained in the following manner. A charge $+Q$ (e.g. a proton or foreign ion) is introduced into the initially homogeneous electron gas causing a perturbing potential $V(r)$. This leads to a change in the local electron density, $\Delta\rho$, in the immediate neighbourhood of the charge. Assuming that $eV(r)$ is small compared with $E_F$, the number of electrons involved is $eV(r) \cdot N(E_F)$. Poisson's equation may then be written

$$\begin{aligned} \nabla^2 V(r) &= -\Delta\rho/\varepsilon_0 \\ &= (e^2/\varepsilon_0) \cdot V(r)N(E_F) \\ &= \lambda_0^2\, V(r), \end{aligned} \tag{6.57}$$

where

$$\lambda_0^2 = (e^2 N(E_F)/\varepsilon_0). \tag{6.58}$$

$\lambda_0$ has the dimension reciprocal length. In the radially symmetrical perturbing potential we expect that

$$V(r) \to Q/4\pi\varepsilon_0 r \quad \text{as } r \to 0$$

and

$$V(r) \to 0 \quad \text{as } r \to \infty$$

These conditions suggest

$$V(r) = (Q/4\pi\varepsilon_0 r) \cdot \exp(-\lambda_0 r), \tag{6.59}$$

a solution that may be confirmed by substitution in (6.57). $\lambda_0$ is known as the Thomas–Fermi screening parameter and, using the numerical data of Table 6.1, we find that it is $2 \times 10^{10}\ \text{m}^{-1}$ for Al and $1.5 \times 10^{10}\ \text{m}^{-1}$ for Na. In general $\lambda_0$ is of order $10^{10}\ \text{m}^{-1}$. This means that the perturbing potential is reduced to 5% of its free space value at a distance of 0.3 nm (3Å) from the centre of disturbance; its effects are completely eliminated at a distance of two to three atomic diameters. This Thomas–Fermi approach is related to that of Debye–Hückel in the theory of electrolytes (see Feynman 1964).

It may be shown (Ziman 1972, Ashcroft and Mermin 1976) that this screening is equivalent to a spatial frequency dependent dielectric function $\varepsilon_r(q)$ of the form

$$\varepsilon_r(q) = 1 + \lambda_0^2/q^2 \tag{6.59}$$

$q$ being the wave vector of a particular Fourier component of the perturbing potential.

A more accurate one electron approximation calculation, but one that neglects electron–electron interactions, was made by Linhard, who found that at 0 K

$$\varepsilon_r(q) = 1 + \lambda^2/q^2 \tag{6.60}$$

with

$$\lambda^2 = \lambda_0^2\{1/2 + [(1 - x^2)/4x] \cdot \ln|(1 + x)/(1 - x)|\} \tag{6.61}$$

and

$$x = q/2k_F.$$

$\lambda^2$ is plotted in Fig. 6.20. $\varepsilon_r(q)$ now decreases more quickly with increasing $q$ than for the Thomas–Fermi case. The electron gas finds it more and more difficult to accommodate to rapid spatial variations in the perturbing potential. A particular feature is the infinite gradient that arises for $q = 2k_F$ and which is a direct consequence of the sharpness of the Fermi surface. Although not obvious the sharp Fermi surface also leads to oscillations in the screening charge density according to a term $\approx \cos 2k_F r/r^3$, a behaviour pattern first predicted by Friedel in a study of the scattering by a charge impurity in the electron gas; they are therefore known as Friedel oscillations, Fig. 6.21. These oscillations in screening charge density cause electric field gradients which may be felt by as many as 50–100 neighbouring atoms, and their effects are manifest in nuclear magnetic resonance experiments on dilute alloys based on copper, for example (see Section 15.3).

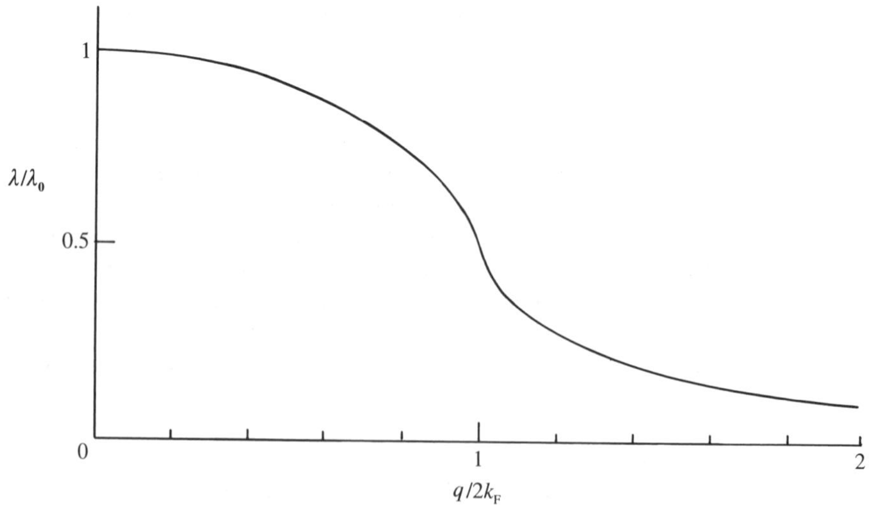


**Figure 6.20** The variation of the Linhard screening parameter $\lambda$, normalized with respect to the Thomas–Fermi $\lambda_0$, with the spatial Fourier component of potential normalized with respect to $k_F$.

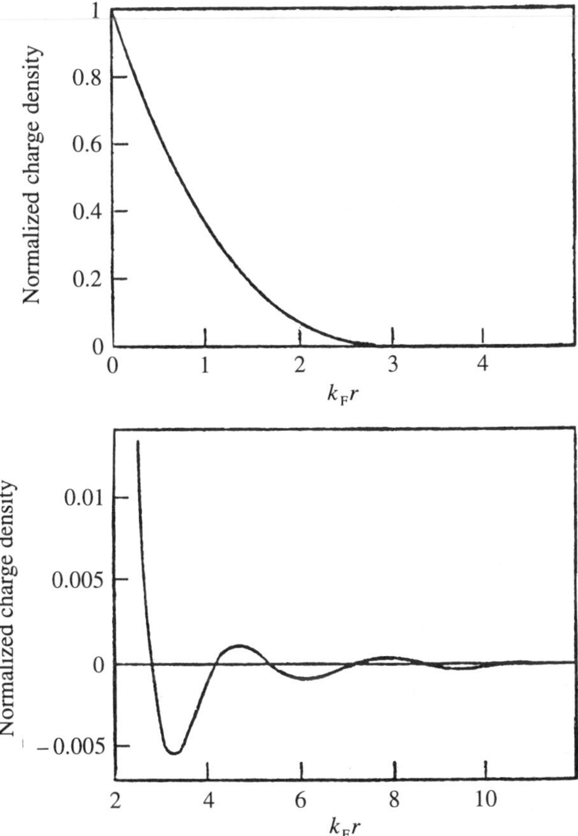


**Figure 6.21** The variation of the normalized screening charge density, arising from a perturbing charge at the origin, with distance from the origin. The Friedel oscillations are shown on an enlarged scale in the lower figure. In the presence of a magnetic moment, localized at the origin, there is an analogous variation in the polarization of the free electron spin density. Although small in amplitude the oscillations have long range, thus $k_F r = 10$ corresponds to $r \approx 1$ nm.

There is also a magnetic analogue to these charge oscillations in that an atom carrying a magnetic moment, resident usually in the d or f electrons, may polarize the spins of the sp electrons that constitute the electron gas in real systems. The polarization decays with distance from the magnetic atom, but in oscillatory fashion according to the cosine term mentioned earlier. These oscillations in magnetic polarization are associated with the names of Ruderman, Kittel, Kasuya and Yosida (RKKY). This property is particularly significant for the cooperative coupling between the localized atomic magnetic moments that arise in the rare earth metals (Section 11.5) as well as the behaviour of magnetic/non-magnetic thin film sandwich structures (Section 11.10).

We shall not discuss the dielectric function further other than to say that including the effects of time dependent disturbances allows calculation of an $\varepsilon_r(q,\omega)$ which in turn provides us with the optical properties of the electron gas as derived in simpler fashion earlier, equations (6.42). The interested reader is directed to the references mentioned above.

The bound electrons contained in the atom cores also react to internal electric fields, i.e. they are polarizable, but their polarizabilities are very small and of negligible significance compared with the effects of free or itinerant electrons.

## References

AITA, O. and SAGAWA, T. (1969) *J. Phys. Soc. Japan*, **27**, 164.

ASHCROFT, N. W. and MERMIN, N. D. (1976) *Solid State Physics*. Holt, Rinehart and Winston, New York.

BAER, Y. and BUSCH, G. (1973) *Phys. Rev. Lett.* **30**, 280.

BERMAN, R. and MACDONALD, D. K. C. (1952) *Proc. R. Soc. Lond.* **A211**, 122.

BUSCH, G. and GÜNTHERODT, H. J. (1974) *Solid State Physics*, Vol. 29 (ed. H. Ehrenreich, F. Seitz and D. Turnbull), p. 235. Academic Press, New York.

FEYNMAN, R. P., LEIGHTON, R. B. and SANDS, M. (1965) *The Feynman Lectures on Physics, vol. II* Addison-Wesley, Reading.

HUNDERI, O. and MYERS, H. P. (1973) *J. Phys.* **F3**, 683.

MARTIN, D. L. (1968) *Phys. Rev.* **170**, 650.

MATHEWSON, A. G. and MYERS, H. P. (1972) *J. Phys.* **F2**, 403.

MILLER, J. C. (1969) *Phil. Mag.* **20**, 115.

MUILLENBERG, G. E. (ed.) (1979) *Handbook of X-Ray Photoelectron Spectroscopy*. Perkin Elmer Corp., Eden Prairie, Minnesota.

RAIMES, S. (1961) *The Wave Mechanics of Electrons in Metals*. North-Holland, Amsterdam.

SLATER, J. C. (1934) *Rev. Mod. Phys.* **6**, 228.

ZIMAN, J. M. (1972) *Principles of the Theory of Solids*. Cambridge University Press, Cambridge.

## Problems

**6.1** Compare the volume of the ion to that of the atom for Na, Cs, Mg, Al, Pb, Cu and Ag (use the standard valences).

**6.2** Of the 'nearly free electron metals', Cs and Be are the extreme examples. Complete the following table:

| | $n$ | $r_s$ | $k_F$ | $E_F$ | $\hbar\omega_p$ |
|---|---|---|---|---|---|
| Cs | | | | | |
| Be | | | | | |

**6.3** At any temperature the Fermi energy is that of the highest occupied level when all the available electrons occupy the lowest available energy states. Using the formulae for the free electron gas, obtain expressions for the change in Fermi energy caused by thermal expansion or the application of hydrostatic pressure. Then calculate for both Na and Cu:

(a) the percentage change in Fermi energy per degree increase in temperature (the linear thermal expansion coefficient $\alpha$ is $7 \times 10^{-5}\ \mathrm{K}^{-1}$ for Na and $1.7 \times 10^{-5}\ \mathrm{K}^{-1}$ for Cu);

(b) the hydrostatic pressure needed to change the Fermi energy by a factor 1.000001 (the isothermal compressibilities $\kappa$ of Na and Cu are $14.7 \times 10^{-11}$ and $0.73 \times 10^{-11}\ \mathrm{Pa}^{-1}$ respectively).

**6.4** Show that for a free electron gas one may write

$$N(E) = C_d E^{(d-2)/2},$$

where $d$ is the dimensionality of the gas (i.e. 1, 2 or 3). Calculate $C_d$ for each case.

**6.5** Calculate the temperature at which the electron and lattice heat capacities are equal for (a) Al and (b) Pb.

**6.6** Name three quantities that are directly dependent on $N(E_F)$. Which do you consider best suited for an experimental determination of this quantity? For the free electron gas how are $N(E_F)$ and $E_F$ related?

**6.7** Na is a monovalent bcc metal with lattice constant 4.225 Å. What magnetic field is needed to cause a relative shift of spin bands by an amount 1% of $E_F$? What magnetic moment would arise in this case? (See Section 1.1 regarding the feasibility of magnetic fields in excess of 0.5 MG.)

**6.8** In an experiment the following values of heat capacity were determined:

| $T$ (K) | $C$ (mJ mol$^{-1}$ K$^{-1}$) |
|---|---|
| 0.5 | 1.38 |
| 1.0 | 5 |
| 1.5 | 13.1 |
| 2.0 | 28 |
| 2.5 | 52 |
| 3.0 | 87 |
| 3.5 | 136 |

What substance was under study?

**6.9** For Na the valence electrons are well described as free electrons. Calculate the ratio of the Fermi wave vector and the radius of the largest sphere that can be inscribed within the first Brillouin zone.

**6.10** Assuming free electron behaviour, what is the separation of energy levels at the Fermi level for (a) 10 cm$^3$, (b) 10 $\mu$m$^3$ and (c) 100 Å$^3$ samples of Al?

**6.11** At what wavelength will aluminium become transparent to electromagnetic radiation? What assumptions does your answer depend on?

**6.12** Si has atomic number 14 and the diamond cubic structure with lattice parameter 5.43 Å. If all the electrons in Si may be assumed to behave as free electrons under the impact of high-energy photons, estimate the refractive index of Si for X rays of wavelength 1.5 Å.

**6.13** Sodium metal displays free-electron-like behaviour. The Fermi energy is 3.2 eV, the thermal effective electron mass is 1 and the Debye temperature is 160 K. What fraction of the total heat capacity at 300 K is contributed by the electrons?

**6.14** At room temperature $\rho$ for Cu $1.78 \times 10^{-8}$ Ω m. What is the mean free path of the conducting free electrons? The lattice parameter for Cu is 3.61 Å.

**6.15** The following values of $\varepsilon_r$ were determined for Al at 300 K:

| eV | $-\varepsilon_1$ | $\varepsilon_2$ | eV | $-\varepsilon_1$ | $\varepsilon_2$ |
|---|---|---|---|---|---|
| 0.7 | 312 | 62.3 | 1.5 | 51.5 | 45.6 |
| 0.8 | 238 | 48.5 | 1.6 | 59.5 | 47.3 |
| 0.9 | 186 | 36.9 | 1.7 | 61 | 38.1 |
| 1.0 | 147 | 30.5 | 1.8 | 57.5 | 30.4 |
| 1.1 | 117 | 26.7 | 2.0 | 50.0 | 20.2 |
| 1.2 | 93 | 25.6 | 2.3 | 39.5 | 12.0 |
| 1.3 | 73 | 27.4 | 2.5 | 33.5 | 9.1 |
| 1.4 | 55.5 | 33.1 | | | |

Determine $\hbar\omega_p$ and the optical effective mass for Al. What other information may be obtained? (Hint: assume $\omega\tau \gg 1$.)

**6.16** The Hall coefficient for Hg at room temperature is $-7.6 \times 10^{-11}$ m$^3$ C$^{-1}$. At what wavelength might you expect this metal to lose its high reflectivity?

**6.17** The work function of a metal is defined as the least energy required to remove an electron from the metal into the surrounding vacuum, and is usually denoted by $\phi$. What happens when two metals with different work functions are in contact with one another?

**6.18** The Hall coefficient of liquid Al is $-3.9 \times 10^{-11}$ m$^3$ C$^{-1}$. At 77 K the electron relaxation time is $6.5 \times 10^{-14}$ s. Estimate the electrical and thermal conductivities of Al at 77 K.

**6.19** Neglecting the effects of anharmonicity and thermal expansion, give an expression for the heat capacity to be expected in the temperature interval within 50 K of the melting point of Al.

**6.20** X rays from synchrotron sources are very useful for the study of solid and liquid surfaces. For X rays with wavelength 10 Å, estimate the critical angle for total internal reflection from an Si vacuum interface, assuming that all the electrons in Si act as though they were free.

**6.21** What experimental proof do we have that the valence electrons in metals occupy energy bands? Describe the principles involved in experiments designed to study the electron energy levels in solids.

**6.22** Consider the effect of a magnetic field on a free electron gas (see Box 6.1) and show that the motion of the states in **k** space is circular with angular frequency $\omega_c = eB/m$.

**6.23** The following soft X-ray emission spectrum was obtained from a divalent free-electron-like metal:

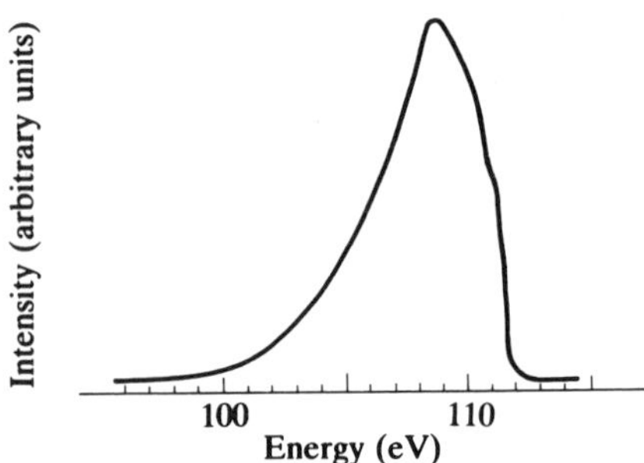


What contribution to the heat capacity would be expected from the electrons at 300 K?

# 7

# The Periodic Potential

## 7.1 Electrons in Free Atoms and in Solids

We should like to know how the electron states are distributed in real solids. To begin with, we assume the solid to be composed of similar atoms; in practice the simplest example is probably a metallic element like Na or Al. The Al atom has three valence electrons bound to a trivalent ion. The isolated ion produces a Coulomb potential (Fig. 7.1), but in an array of ions the individual potentials overlap to form a resultant similar to that of Fig. 7.2. When the valence electrons are added we obtain a neutral system and presume that the electrons occupy energy levels very similar to those calculated for the free electron gas. For the moment, however, we ask what happens to the core electrons when atoms form solids. This question is best answered by the following experiment.

We generate $K\alpha$ X rays from a vapour of Al atoms and from a lump of aluminium metal. In both cases we find a sharp $K\alpha$ line, but at slightly different energies. This shows that the $K$ and $L$ levels in both the isolated atom and the solid have similar sharp atomic character but somewhat displaced energies. We conclude that the core electrons of atoms have orbital diameters so much smaller than the interatomic separation that core electrons on adjacent atoms are unable to interact; they therefore maintain their atomic character. On the other hand, the volume available to an atom in a solid is well defined (by the primitive cell in the simplest cases) and is usually smaller than that of the free atom, which is indistinct on account of the slow decay of the radial wave function. This difference in volume, together with the overlap of the ionic potentials, produces small changes in the average charge density and thereby the average potential experienced by the core electrons, which in turn causes the difference between the X-ray photons produced in free atoms and in the solid. This difference is also apparent in the core state binding energies determined in different samples using X-ray photoemission spectroscopy. It is not difficult to appreciate that aluminium atoms in different environments, e.g. Al, $Al_2O_3$, NiAl, experience slightly different effective potentials, and given core states do not have exactly the same energies in the different substances or as in the free atom. The energy difference depends on the chemical environment; it is therefore called the

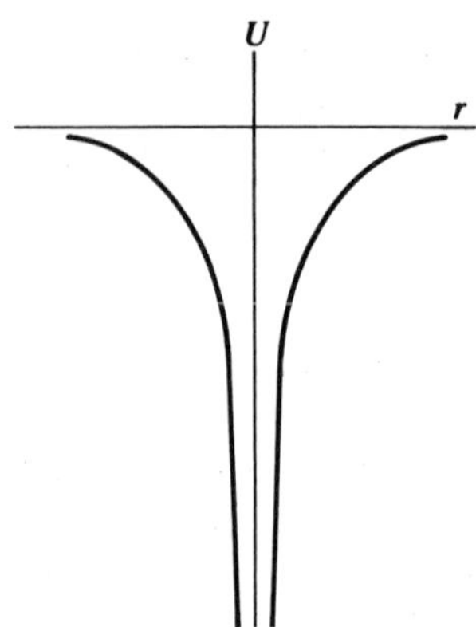


**Figure 7.1** The bare ion Coulomb potential.

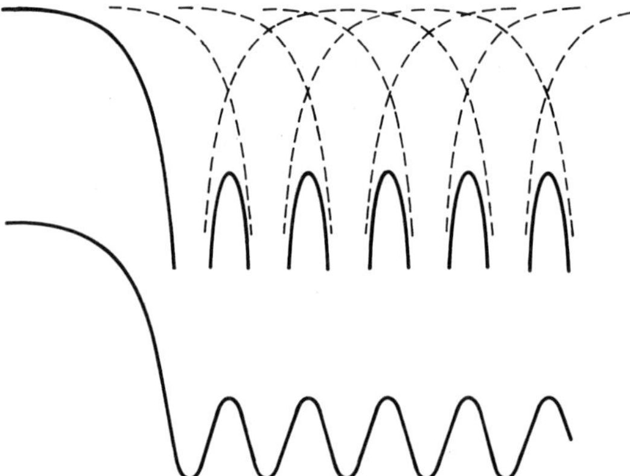

**Figure 7.2** The electrostatic potential through a line of atoms in a crystal and parallel to such a line.

'chemical shift.'† But how are the valence electrons affected by the periodic potential? Just as in our discussion of phonons, we shall not attempt to answer this question in a detailed fashion but treat a simple stylized problem that provides us with a very good physical picture of what goes on. The presentation of a detailed quantitative discussion is not necessary for our purpose.

## 7.2 The Energy Gap

In real solids, the potential experienced by a valence electron varies rapidly with position, and we should take account of exchange, correlation and, possibly, many-body effects, but what we wish to know above all is the effect of a periodic potential. The periodicity is the important aspect, and, whereas the detailed behaviour is cer-

† The shift is small, of order 1 eV, but quite measurable.

tainly dependent on the strength and Fourier content of the potential, the essential effects of periodicity should be the same for all periodic potentials. We therefore take the simplest situation and assume a weak harmonic potential of the form

$$U = U_0 + U_1 \cos \frac{2\pi x}{a}, \tag{7.1}$$

where $a$ is the lattice parameter of a linear lattice and $U_1$ is small compared with $U_0$ (Fig. 7.3).

If $U_1$ were zero then we would regain the free electron gas and the wave functions would be plane waves. This is also true for any electron with $k \ll \pi/a$ because under these conditions the wave function wavelength is much larger than the lattice spacing and the weakly periodic potential averages to the constant value $U_0$. Thus for small wave vectors we expect

$$\psi = Ae^{ikx},$$
$$E = \hbar^2 k^2/2m.$$

So for small $k$ the $E(k)$ diagram follows the free electron pattern of Fig. 7.4. As $k$ increases, we expect the effects of periodicity to become apparent. But what do we know about the propagation of waves in periodic structures? Namely that whenever $k = \pm p\pi/a$ ($p$ integral) Bragg reflection arises. Thus as $k$ approaches $\pi/a$ we expect a

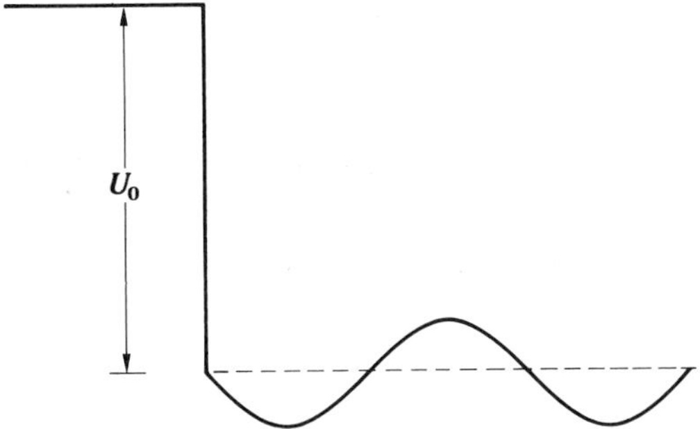


**Figure 7.3** A simple periodic potential.

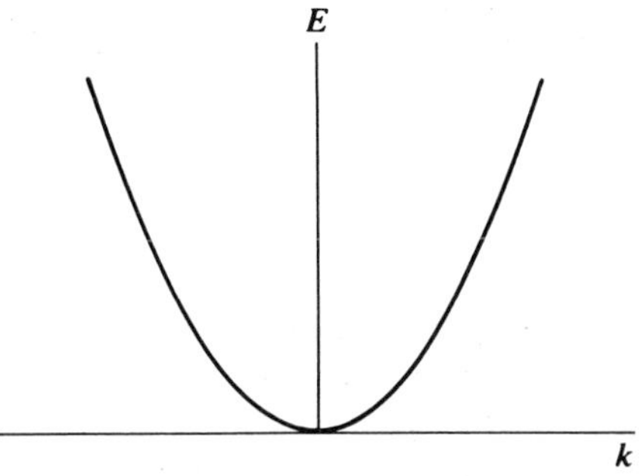


**Figure 7.4** For small values of the wave vector there is no change in the behaviour from that of the free electron gas.

back-reflected wave to appear. We then write the wave function as a linear combination of forward and backward components, and we have two choices:

$$\psi_{\pm} = Ae^{ikx} \pm Be^{-ikx}, \tag{7.2}$$

However, unless $k$ is very close to $\pi/a$, $B$ will be much smaller than $A$; but when $k = \pi/a$ we obtain successive repeated reflections, leading to equal forward and backward components,

$$\psi_{\pm} = C(e^{ikx} \pm e^{-ikx}), \tag{7.3}$$

and we find that $C = 2^{-1/2} A$. The two combinations of (7.3) lead to two charge densities proportional to

$$\psi_{+}\psi_{+}^{*} = 4C^2 \cos^2 \frac{\pi x}{a} = 2A^2 \cos^2 \frac{\pi x}{a}, \tag{7.4}$$

$$\psi_{-}\psi_{-}^{*} = 4C^2 \sin^2 \frac{\pi x}{a} = 2A^2 \sin^2 \frac{\pi x}{a}, \tag{7.5}$$

whereas in the absence of periodicity $U_1 = 0$ and

$$\psi\psi^{*} = A^2. \tag{7.6}$$

These distributions are shown in Fig. 7.5, and immediately we see that the $\psi_{+}$ combination piles the electron charge onto the ion sites whereas the $\psi_{-}$ combination places it between them. Since the ion sites carry a positive charge, $\psi_{+}$, by placing electrons in the regions of positive potential, is associated with a lower potential energy than $\psi_{-}$. Thus for a general potential, whenever the electron wave vector has a value $\pm p\pi/a$ ($p \neq 0$), Bragg reflection leads to two states described by standing wave functions formed from the two possible linear combinations. These states possess similar kinetic but different potential energies and an energy gap arises. Within this energy gap there are no eigenstates for the electrons; it is a region of energy forbidden to any electron subject only to the periodic potential.

Let the periodic part of the potential be of the form $\sum_G U(G)e^{iGx}$ with $G = \pm p2\pi/a$, $p = 1, 2, \ldots$. If all the $U(G)$ are initially zero, the electron energy is determined by (6.4) (Fig. 7.4). Suppose now that each Fourier coefficient of the potential, $U(G)$, in turn assumes a positive or negative value. Then the $E(k)$ curve

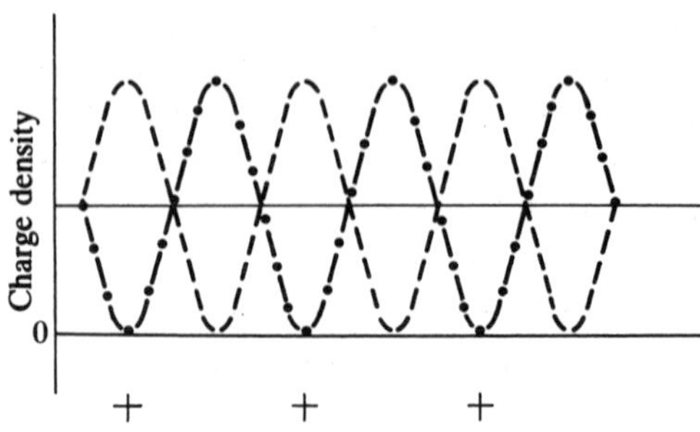


**Figure 7.5** At critical **k** values Bragg reflection of the electron arises, leading to two possible charge distributions – these are associated with different potential energies. +, ion site; ——, $\psi\psi^{*}$ free electrons; - - -, $\psi_{+}\psi_{+}^{*}$; –·–, $\psi_{-}\psi_{-}^{*}$.

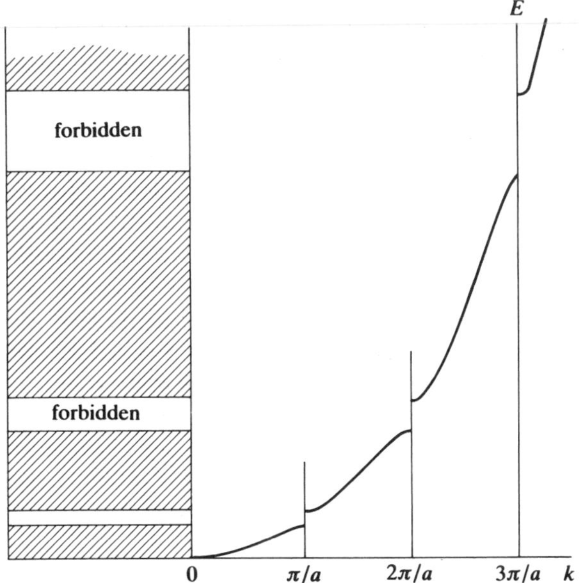


**Figure 7.6** At the critical **k** values ($|\mathbf{k}| = p\pi/a$), the $E(k)$ relation becomes discontinuous. Energy gaps associated with the two possible charge distributions, as shown in Fig. 7.5, arise. In pure materials there are no eigenstates for electrons with energies lying within these energy gaps. Note that the higher the energy of the band the larger its width in energy.

successively breaks up into segments of size $\pi/a$ that are separated by energy gaps arising whenever $|\mathbf{k}| = p\pi/a$ (Fig. 7.6). If an energy gap is to arise at $|\mathbf{k}| = p\pi/a = \frac{1}{2}G$ then the associated $U(G)$ must be finite. The energy gap then has size $2|U(G)|$ (see Appendix 7.1).

We have now learnt that a periodic potential breaks the $E(k)$ curve into discrete segments of interval $\pi/a$. This is the significant fact that allows us, in principle at least, to understand many of the characteristic features of solids. Note that there is no limit to **k** because there is no limit to the energy of the electron. In the case of phonons the wave amplitude represents a physical quantity, the amplitude of atomic vibration; we demanded that

$$|\mathbf{k}| \leqslant \pi/a$$

and limited the number of modes to $3N$. The electron wave function does not have the same real physical character as the phonon, and **k** may be as large as we please, but normally we are interested in ground-state properties and the Fermi vector then provides a natural reference point. The detailed understanding of the properties of a particular solid demands the calculation of $E$ as a function of **k** in many different directions, the determination of the energy gaps, the density of electron states and the wave functions. However, we can get a long way in the description of solids from purely qualitative arguments.

## 7.3 Brillouin Zones and Electrical Conductivity

In any real crystal there are several periodicities determined by the principal symmetry directions. Each periodicity causes Bragg reflection of electrons when these have the correct $\mathbf{k}$ values and the structure factor $S_{hkl}$ is finite (Section 3.3). Just as for phonons, we define planes of critical $\mathbf{k}$ values determined by the Bragg condition (5.41),

$$2\mathbf{k} \cdot \mathbf{G} - G^2 = 0,$$

and the volumes of $\mathbf{k}$ space bounded by the planes of critical $\mathbf{k}$ are the Brillouin zones (Section 5.10). When discussing phonons we used only the first zone, but there is no such restriction for electrons and often we are obliged to consider three or more zones. We first demonstrate particular features of the zone concept using the linear lattice because this simplifies the illustration.

The periodicity causes the segmentation of the $E(k)$ curve, and in Fig. 7.7 we have chosen an arbitrary origin and divided $\mathbf{k}$ space into a series of zones, each of measure $2\pi/a$. In principle at least, this series of zones continues indefinitely. This description, which for weak potentials bears strong resemblance to the free electron parabola, is the *extended zone scheme.* Since, as we have earlier emphasized, our choice of zero is arbitrary, we may redraw the whole diagram from every equivalent origin and thereby obtain the *repeated* or *periodic zone scheme.* Most often, we find it convenient to use the *reduced zone.* Remembering relation (5.40),

$$\mathbf{k} = \mathbf{k}_1 \pm \mathbf{G} = \mathbf{k}_1 \pm n\mathbf{G}_1,$$

we can always map the different higher segments of the extended zone scheme into the first zone and use an index to identify them. Since each segment forms an independent region of quasicontinuous energy levels, we call them bands. In this way the segment lying in the $p$th zone contains states $\mathbf{k}_i^p$. The symmetry of reciprocal space allows us to rewrite these vectors as $\mathbf{k}_{i_p}$ in the first zone:

$$\mathbf{k}_{i_p} = \mathbf{k}_i^p - \mathbf{G}, \qquad (7.7)$$

where $\mathbf{G}$ is a reciprocal lattice vector in the particular symmetry direction; $p$ then becomes a band index. It is readily seen that for the linear and two-dimensional lattices all the zones have the same size, and this is equally true for the three-dimensional case. All the basic ideas used in our counting of phonon modes via the use of periodic boundary conditions (Section 5.5) are immediately applicable, and just as before, (5.23) and (6.6), we obtain a density of wave vector states

$$N(k) = V/8\pi^3 = N\Omega/8\pi^3,$$

whereas the density of electron states is now written as

$$N(E) = 2\frac{V}{8\pi^3}\int_{S_E}\frac{\mathrm{d}S}{|\nabla_{\mathbf{k}} E|} \qquad (7.8)$$

because we must take account of spin degeneracy. Apart from the factor of 2, this is just the expression (5.31) derived earlier. We have already described the first Brillouin zone as the Wigner–Seitz cell of the reciprocal lattice, which in three dimensions has size $8\pi^3/\Omega$. The number of $\mathbf{k}$ states contained in any zone is therefore

$$N(k)8\pi^3/\Omega = N,$$

where $N$ is the number of primitive cells in the sample of volume $V$ and each cell of volume $\Omega$. *Each Brillouin zone therefore has room for one wave vector state or two*

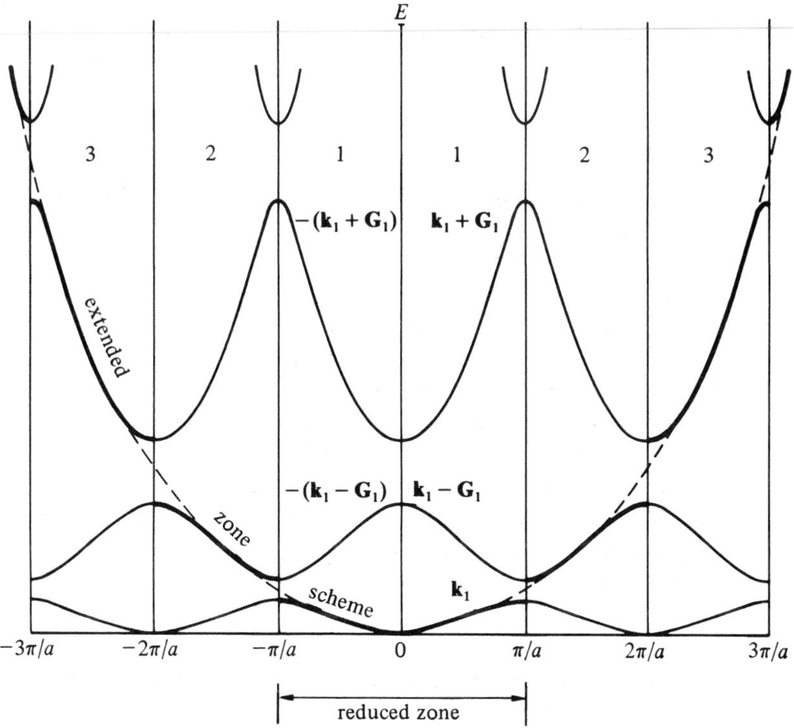


**Figure 7.7** This diagram illustrates the properties of a linear lattice, but it is also typical of behaviour along any symmetry direction in a three-dimensional crystal. If we plot $E$ as a function of monotonically increasing **k** then we obtain the thick curve, which, for a weak periodic potential, closely resembles the free electron parabola. This is the *extended zone scheme*. However, just as in our discussion of phonons, we may redraw this curve from all the possible origins of the reciprocal lattice and this leads to a series of continuous curves arranged one upon the other and extending throughout the whole of **k** space. This description is called the *periodic zone scheme*. It is sufficient if we confine attention to the thick black curve, but this extends over all the zones, of which only three are shown in the figure. In principle the electron may have any energy and an excited electron (before it leaves the sample) may be in a state within say the tenth zone. The extended zone scheme is instructive but inconvenient. We therefore use the reduced zone, i.e. all states with $|\mathbf{k}| > \pi/a$ are translated an integral number of reciprocal lattice base vectors until they land in the first zone. Thus those states that are in the interval $-3\pi/a \leqslant \mathbf{k} \leqslant -2\pi/a$ are shown as $-(\mathbf{k}_1 + \mathbf{G}_1)$, whereas those in the interval $2\pi/a \leqslant \mathbf{k} \leqslant 3\pi/a$ are written as $\mathbf{k}_1 + \mathbf{G}_1$ and $0 \leqslant |\mathbf{k}_1| \leqslant \pi/a$. All the possible **k** states may be remapped into the first zone; the description is known as the *reduced zone scheme*. The different bands may be numbered according to increasing energy. $\mathbf{G}_1$ is the base vector of the reciprocal lattice.

*electrons per primitive cell of sample.* For the linear lattice with one atom at each lattice point, the following holds:

| | first zone | second zone | third zone |
|---|---|---|---|
| monovalent element | half filled | empty | empty |
| divalent element | filled | empty | empty |
| trivalent element | filled | half filled | empty |
| quadrivalent element | filled | filled | empty |

and so on.

Now if any zone is only half filled it is possible for electrons near the Fermi energy to be accelerated by an applied electric field, and conduction can arise just as in the case of the free electron gas; we have metallic behaviour. Whenever all zones are either filled or empty the nearest available empty states are always separated from the highest occupied levels by the energy gap on the zone boundary. In a filled zone, states of $+k$ and $-k$ are equally represented and we cannot alter the balance because this demands excitation over the energy gap. If this gap is large, say 5 eV, then the substance remains unable to conduct electricity at all normal temperatures and field strengths. If the gap is finite but small then the behaviour at 0 K is that of an insulator, but at higher temperatures, such as room temperature and above, thermal excitation of electrons across the gap provides a small concentration, of order $10^{-8}$ electrons per atom, in the otherwise empty band. This low concentration of electrons produces the small specific conductivities typical of semiconductors.

*The properties of primitive linear lattices are therefore clear cut: elements with odd valence are metals, those with even valence are insulators or semiconductors.* We may ask whether a monovalent element can become an insulator; the answer is 'yes' if the linear structure is described by a basis of two atoms. The zone structure is determined by a linear Bravais lattice, but there are now two atoms to each cell, providing two electrons – just right to fill the first zone.

We conclude that the energy levels for electrons in periodic structures have the same general form in *all* solid types, but in metals electrons in the highest occupied states have immediate access to empty states whereas in insulators and semiconductors they are separated from one another by energy gaps. For metals the boundary in **k** space between contiguous occupied and unoccupied states forms a surface; in the case of the three-dimensional free electron gas this surface is spherical, but in many real metals it is usually of more complicated shape. Whatever its shape, it is called a *Fermi surface*, and during the past thirty years much experimental effort has been devoted to determining its exact form in different metals. Clearly an insulator or semiconductor has no Fermi surface.

## 7.4 Two-dimensional Lattices

Suppose we have a simple square lattice of atoms with base vector $a$; the corresponding reciprocal lattice is also square with base vector $2\pi/a$. We determine the Brillouin zone structure in the following manner:

(a) we choose a point of the reciprocal lattice as an arbitrary origin;

(b) we join this point to successive neighbouring points of the lattice and construct the perpendicular bisectors to these interpoint distances;

(c) we divide reciprocal space into areas bounded by these bisectors without their being penetrated by them.

The first Brillouin zone is readily found, and successively higher zones are formed of separate areas and the higher the zone the smaller the component parts (Fig. 7.8). The first three zones are redrawn in Fig. 7.9. Now let us ask what to expect if our atoms have 1, 2 or 3 valence electrons. First we suppose that our periodic potential defines the lattice but is so weak that there is no perturbation of the free electron

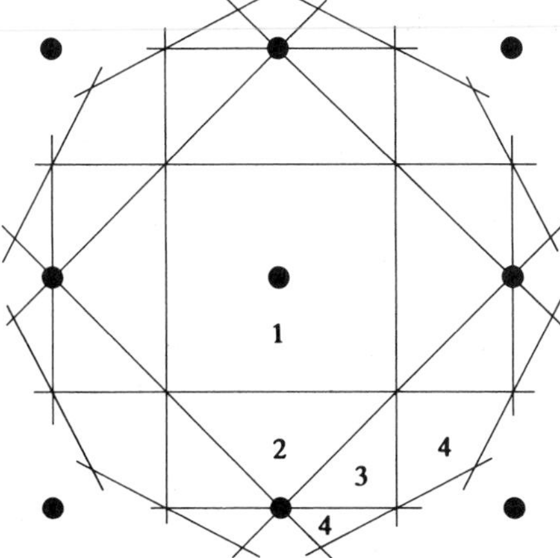


**Figure 7.8** The first four zones of the square lattice. The black dots are the reciprocal lattice points. Note that the higher the zone the smaller the component parts; nevertheless all the zones have the same size.

properties. The Fermi 'surface' then has circular form. We draw the appropriate circles corresponding to 1, 2 and 3 electrons per atom, the scale being given by our knowledge that the Brillouin zone only has room for two electrons per atom. Thus in two dimensions we have a density of electron states in **k** space of $a^2/2\pi^2$, and if $k_{F1}$, $k_{F2}$ and $k_{F3}$ are the Fermi vectors corresponding to 1, 2 and 3 electrons per atom then

$$\pi k_{F1}^2 a^2/2\pi^2 = 1, \qquad \pi k_{F2}^2 a^2/2\pi^2 = 2 \qquad \text{etc.}$$

giving

$$k_{F1} = \left(\frac{2}{\pi}\right)^{1/2} \frac{\pi}{a}, \qquad k_{F2} = \left(\frac{4}{\pi}\right)^{1/2} \frac{\pi}{a}, \qquad k_{F3} = \left(\frac{6}{\pi}\right)^{1/2} \frac{\pi}{a}.$$

In Fig. 7.9 we have constructed the appropriate Fermi circles $c_1, c_2, c_3$. We see that $c_1$ lies well within the first zone whereas $c_2$ and $c_3$ intersect zone boundaries. We

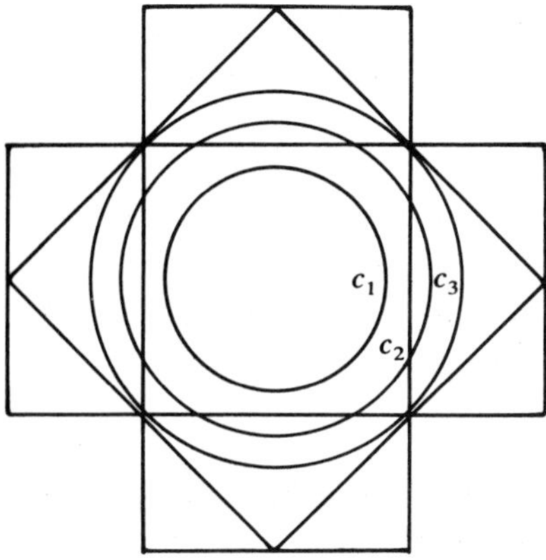


**Figure 7.9** The two-dimensional Fermi circles corresponding to 1, 2 and 3 electrons per atom drawn within the zone structure of the square lattice.

now increase the periodic potential so that energy gaps arise. Our circles $c_2$ and $c_3$ cannot now remain continuous. Any point infinitely close to, but just inside, the first zone boundary lies at a different energy from that of an adjacent point just outside that boundary and inside the second zone. The two states differ by the gap energy. A weak periodic potential therefore distorts the energy contours as shown in Fig. 7.10. *Clearly, for a two-dimensional lattice, a divalent atom does not necessarily fill the first zone.* Electrons spill over into the second zone and we maintain a Fermi surface consisting of separate portions in the two zones; the Fermi surface becomes disconnected. The system is metallic.

The simple square lattice has two principal periodicities corresponding to the [10] and [11] directions; each has its own $E(k)$ dispersion relation (Fig. 7.11). Whereas the bands never overlap when we confine attention to one symmetry direction, it is clear that the occurrence of additional different symmetry directions may

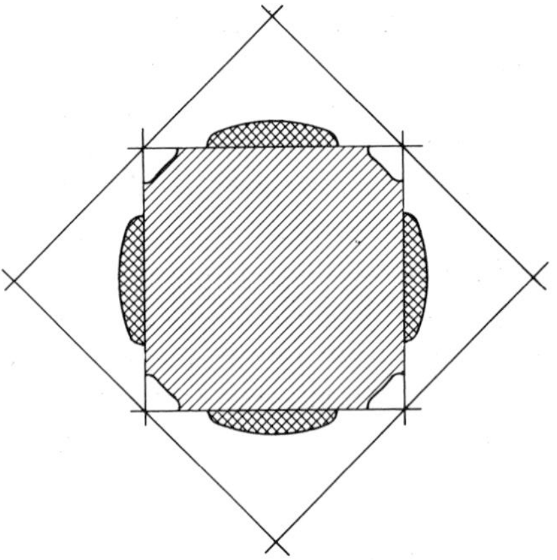

**Figure 7.10** The distortion of the Fermi circle of a two-dimensional electron gas caused by a weak periodic potential. Similar distortion is found in a plane section of a three-dimensional system.

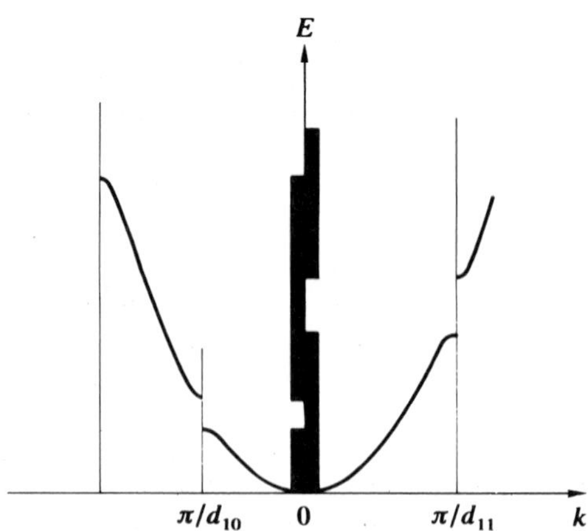


**Figure 7.11** The presence of more than one periodicity may cause overlap in the integrated density of states.

produce a continuous distribution in energy of allowed states when these are summed over all possible **k** vectors.

In Fig. 7.12 we show schematically how a weak periodic potential perturbs the energy contours as we proceed from the origin to the zone boundaries. Whereas 'necks' form on contours cutting the zone boundaries, the contours develop bumps before zone contact arises. The changed shape of the contours is a direct result of the departures of the $E(k)$ relation from the $k^2$ dependence typical for free electrons. In two dimensions our expression for the density of electron states (7.8) becomes

$$N(E) = 2\frac{A}{4\pi^2}\int_{l_E}\frac{\mathrm{d}l}{|\nabla_{\mathbf{k}}E|},$$

where $A$ is the area of the primitive cell, and if the energy contours are circles as for free electrons then it is easy to show that $N(E)$, which is proportional to the length of a contour of constant energy, is in fact a constant equal to $Am/\pi\hbar^2$. The band therefore has rectangular form (Fig. 7.13). As the potential is felt, however, the energy contours become distorted; when $k$ approaches $\pi/a$ 'bumps' arise and lengthen the contour, leading to an increase in the size of $N(E)$ over the free electron value. The effect is most acute just before zone contact. When contact is made, the contour in the first zone is broken into four pieces which become of rapidly diminishing length as the zone is filled, leading to a band shape similar to that of Fig. 7.13. We associate such a band with every Brillouin zone, but their widths become larger the higher the zone; this is most easily seen if we refer to the linear lattice (Fig. 7.6). Since each zone or band contains the same number of electrons (two), the bands must become shallower as they become broader. The detailed structure of the bands depends upon the symmetry of the crystal structure; thus if we

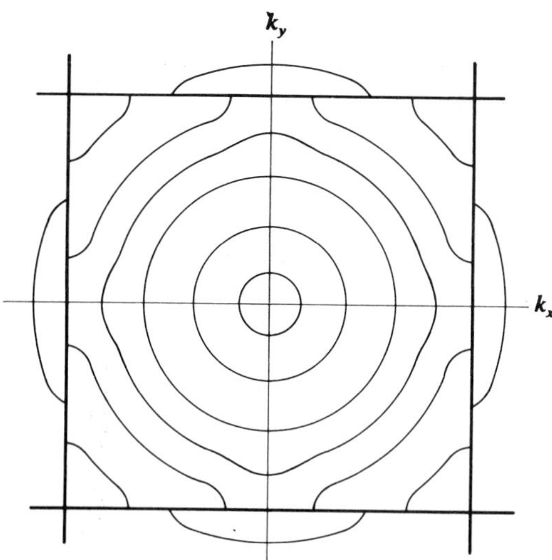


**Figure 7.12** As energy contours approach the zone boundaries, the probability of Bragg reflection increases, leading to a distortion of the free electron circles. The latter develop bumps, producing a larger circumference (a larger area in three dimensions). However, when contact with the zone boundary arises there is a decrease in the size of the contour of constant energy within a given zone.

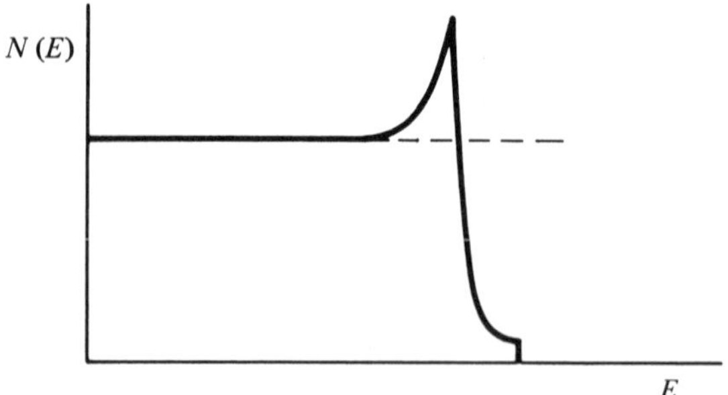


**Figure 7.13** The effect of zone boundaries on the shape of the $N(E) - E$ curve for a square lattice. The broken line is the energy band for free electrons in two dimensions.

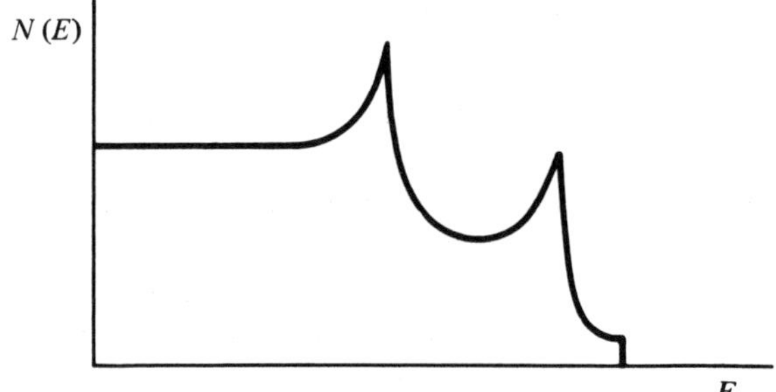


**Figure 7.14** Similar to Fig. 7.13 but for a rectangular lattice.

consider a rectangular lattice, the zones have two sets of non-equivalent boundaries leading to an additional cusp in the $N(E)$ curve (Fig. 7.14).

## 7.5 Three-dimensional Lattices

All the above features are maintained in the description of three-dimensional systems, but the illustration, even in the simplest case, becomes much more difficult. The shapes of the first zones for the bcc and fcc lattices are shown again in Fig. 7.15. Although we use the unit cell to describe the direct lattice, the Brillouin zone is a true primitive cell of the reciprocal lattice and has volume $8\pi^3/\Omega$, where $\Omega$ is the volume of the primitive cell in the direct lattice. The zone has space for two electrons per primitive cell of the direct Bravais lattice. The symbols denoting symmetry points are also shown in Fig. 7.15. Again, if the ion potential is weak, the energy contours resemble those of free electrons; they are spheres for small $k$-values. As the energy surfaces approach a zone boundary they become distorted and develop bumps that become necks when true zone contact is made. The zone for the bcc structure is bounded only by $\{110\}$ planes whereas that for the fcc has both $\{111\}$ and $\{200\}$. In the light of our previous discussion concerning the square and rectangular lattices, we might expect the band form to be somewhat more complicated for fcc substances than for those with bcc structure. Just as for the two-dimensional case, we expect the Fermi surface of polyvalent substances to extend over several zones and to be a disconnected surface. We can best illustrate this situation by showing a section of the experimentally measured Fermi surface of Pb (Fig. 7.16).

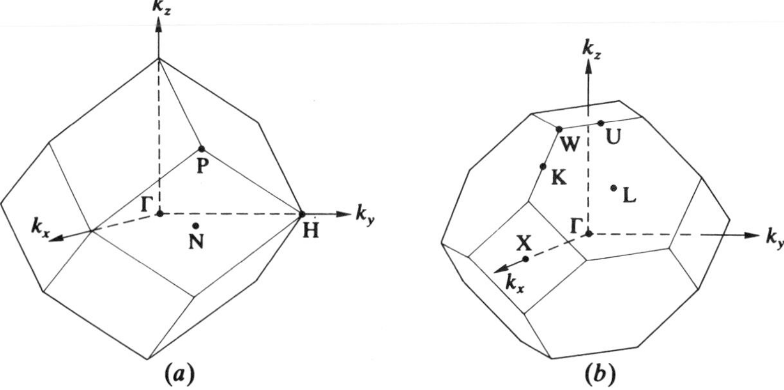


**Figure 7.15** The first Brillouin zone for the bcc (*a*) and fcc (*b*) lattices.

Surprisingly the section is quite close to circular, implying an almost spherical Fermi surface. However, we see that there is distortion in the regions of zone contact and the surface is broken up into many parts (cf. Fig. 7.10). Experiment has shown that the Fermi surfaces of the simpler metals are remarkably spherical, but their disconnected character leads to properties not encountered in the simple free electron approach. The fact that many real metals possess very nearly spherical

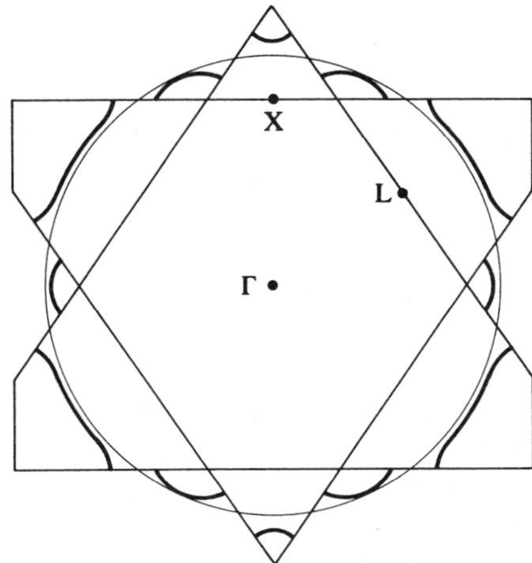


**Figure 7.16** A section, parallel to the $\{011\}$ plane, through the Fermi surface of Pb. (After Anderson and Gold 1967.)

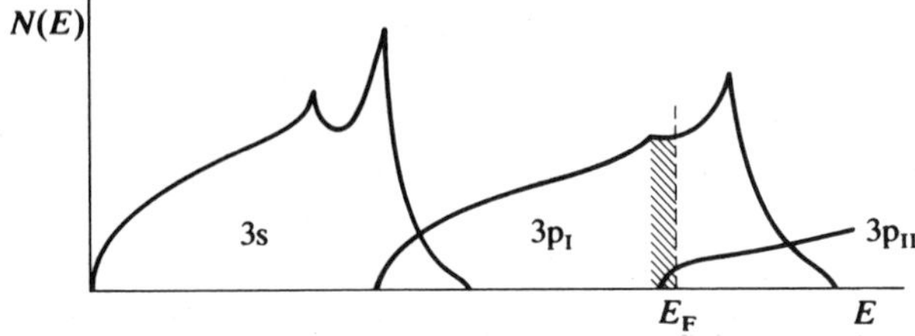


**Figure 7.17** A schematic representation of the valence electron states in Al; we expect three 3p sub-bands, of which only two are shown.

Fermi surfaces is somewhat paradoxical, because we expect the ions to present strong potentials: we shall attempt to resolve this paradox later. When constant-energy surfaces cut into two or more zones then clearly the energy bands originating in the separate zones must overlap. Thus the bands available for the valence electrons in aluminium may be described schematically as in Fig. 7.17, where the bands are labelled with the atomic states that were occupied by the electrons in the isolated atoms and from which the bands can be regarded as being derived.

## 7.6 Solution in Plane Waves

A calculation of the energy states for electrons in a real periodic potential is a formidable problem. An early approach assumed the potential to be weak and became known as the *nearly free electron approximation*; it was not expected to apply to real substances, but served as a model calculation. In recent years the approach has had unexpected practical application. We represent the periodic potential in a particular symmetry direction of a linear crystal by a Fourier series

$$U(x) = \sum_q U(q)\mathrm{e}^{\mathrm{i}qx}, \tag{7.9}$$

but the equivalence of all lattice cells demands that

$$U(x) = U(x + pa),$$

where $p$ is an integer and $a$ the lattice parameter; so

$$\sum_q U(q)\mathrm{e}^{\mathrm{i}qx} = \sum_q U(q)\mathrm{e}^{\mathrm{i}q(x+pa)}, \tag{7.10}$$

an equation valid for all $q$ and for all $p$, including $p = 1$, whence

$$\mathrm{e}^{\mathrm{i}qpa} = 1. \tag{7.11}$$

This demands that $q$ be a reciprocal lattice vector. The potential is therefore written as

$$U(x) = \sum_G U(G)\mathrm{e}^{\mathrm{i}Gx}. \tag{7.12}$$

In itself this is not surprising because physically we should expect the Fourier components to be harmonics of the basic lattice periodicity. We assume a centre of symmetry so that $U(G) = U(-G)$.

The electron wave function may also be represented by an expansion in plane waves, a Fourier series of the form

$$\psi = \sum_k c(k)\mathrm{e}^{ikx}, \tag{7.13}$$

and insertion of (7.12) and (7.13) into the time-independent Schrödinger equation leads to

$$\sum_k \lambda_k c(k)\mathrm{e}^{ikx} + \sum_G \sum_k U(G)c(k)\mathrm{e}^{\mathrm{i}(k+G)x} = E \sum_k c(k)\mathrm{e}^{ikx}. \tag{7.14}$$

$\lambda_k$ has been written for $\hbar^2k^2/2m$. If we now multiply through by $\mathrm{e}^{-ik'x}$ and integrate over the length of the crystal, we find that the orthogonality of the plane waves leaves only terms in $k'$ and $k' - G$, so that (see Appendix 7.1)

$$\lambda_{k'} c(k') + \sum_G U(G)c(k' - G) = Ec(k'). \tag{7.15}$$

$k'$ is any arbitrary wave vector and the prime becomes unnecessary. We rearrange this equation as

$$(\lambda_k - E)c(k) + \sum_G U(G)c(k - G) = 0. \tag{7.16}$$

Equation (7.16) represents an endless sequence of equations relating the required eigenvalues $E$, appropriate to the state with wave vector $\mathbf{k}$, to the Fourier coefficients of potential and those of the wave function. Knowing the $U(G)$, the equation (7.16) may be solved to provide the eigenvalues $E$, which are the roots of a determinantal equation as described in Appendix 7.1.

Returning to (7.16), the coefficient $c(k)$ may be written

$$c(k) = \sum_G \frac{U(G)c(k - G)}{E - \lambda_k}. \tag{7.17}$$

We see that $c(k)$ is small unless $E \approx \lambda_k$. We also find that the set of plane waves in the wave function contains only terms of the form $|\, k - G\rangle$ and

$$\psi_k = \sum_G c(k - G)e^{i(k-G)x}$$

$$= \left[\sum_G c(k - G)e^{-iGx}\right]e^{ikx}.$$

The wave function is a plane wave modulated by a prefactor with the periodicity of the lattice. Such a wave function is called a Bloch function and is usually written

$$\psi_k(\mathbf{r}) = U_{kG}(\mathbf{r})e^{i\mathbf{k}\cdot\mathbf{r}}. \tag{7.18}$$

For electrons moving in a periodic potential, the translational symmetry of the lattice is imposed on the wave function and the latter is always of Bloch form. Electrons in periodic potentials are often called Bloch electrons to distinguish them from the ideally free electrons.

What does such a wave function look like? Let us consider a simple metal like sodium which has one valence electron in a 3s level. In the free atom the 3s wave function has a radial distribution similar to that shown in Fig. 7.18*a*. Now the core states (1s, 2s & 2p) of the sodium atom are confined to a region which is only ~10% of the volume available to an atom in the solid sodium and the 'atomic potential' is only felt over this restricted volume outside of which the potential is approximately constant. These considerations apply to all the simpler metals. It is on this account that they are called nearly free electron like. For a given wave vector the Bloch wave function for the 3s electrons therefore takes the form depicted in Fig. 7.18*b*; we note that it has a pronounced plane wave component.

The use of equation (7.16) to determine the band structure of even the simpler metals is an impractical task because one may need between 500 and 1000 terms in the Fourier development; although correct in principle it is of no use in its present form, one needs a much better description of the potential and the wave function than the simple Fourier series.

First of all we note that the valence electron wave function has atomic character within the core of the atom and, because it is an eigenfunction of the Schrödinger equation, it must be orthogonal to the core state wave functions. The latter are the same as in the free atoms (but their energies are slightly different on account of the overlapping atomic potentials). The atomic wave functions are well known. A much better representation of the valence electron wave function for a simple metal should

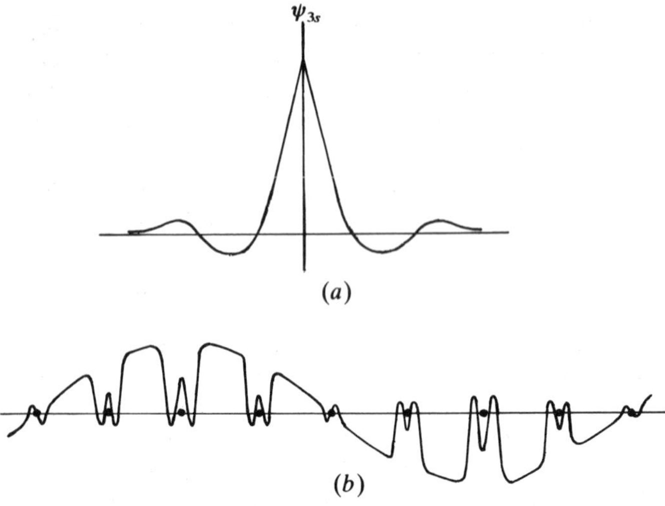


**Figure 7.18** The wave function for a 3s electron in sodium is shown in schematic fashion, (a). In sodium metal the ion core is very small compared to the atomic diameter, the Bloch functions for the 3s electrons have a marked plane wave component except for the immediate vicinity of the core region, (b).

be obtained if we therefore construct plane wave functions that are orthogonal to all the core state wave functions, a suggestion first made by Herring in 1940. Such a wave function is called an orthogonal plane wave, OPW. Relatively few OPWs are needed to describe the wave function and the computational burden is far less than that of the purely plane wave approach.

The application of this method led to an even simpler approach that allowed one to return to the use of purely plane wave 'pseudo wave functions'. Using orthogonal plane waves it was found that one could rephrase the problem in terms of an effective potential that is usually called a 'pseudopotential', (sometimes the term model potential is also used). The pseudopotential is much weaker than the true potential, partly because it is numerically small, but also because it is reduced by the dielectric screening properties of the valence electron gas. The nearly free electron approach was revitalized through the introduction of the pseudopotential.

You may well ask how a strong Coulomb potential can be replaced by a weak pseudopotential. Consider a single atomic cell in the periodic structure typical for a simple metal like Na, Al or Pb. We note that the potential is only very strong over the small core region, outside of which it is approximately constant. Now a typical valence electron has eigenenergy $\approx -10$ eV relative to the vacuum level. In the outer regions of the atomic cell this energy is composed of say $-20$ eV potential energy and $\approx +10$ eV kinetic energy. When the electron enters the core region the eigenenergy remains the same, but becomes the resultant of a large negative potential energy and a large kinetic energy; the electron velocity is large and the electron speeds through the core region. So not only is the core small but the electron traverses it at high speed and so spends comparatively little time in this region. On this account the core has only restricted influence on the valence electrons.

We can also understand, in principle at least, why the effective potential has limited size by extending the above discussion in the following way. When passing through an atomic cell the total energy of an incoming electron occupying a Bloch state must remain constant. If the wave function of the incoming electron is to change in any way its total energy must remain constant. This means that only a change in the phase of the wave function can arise. We say that the potential causes a phase shift $\eta$ when the electron passes through the atomic core. Phase changes of $p\pi$ where $p$ is an integer cause no change in the wave function because the scattering, as shown by a more detailed treatment, is dependent on sin $\eta$, so we may write for the total phase change $\eta$

$$\eta = p\pi + \Delta, \qquad 0 < \Delta < \pi.$$

As might be imagined from what has been said above, if our 'bare' core potential causes a phase shift $\eta$ such that $\Delta$ is small compared with $\pi$ then it will act as though it were a weak potential. The objective then is to express this weak effective potential in terms of its Fourier transform and find those components compatible with the lattice periodicity (i.e. the $U(G)$). These are then screened by the appropriate Fourier component of the dielectric function and together they produce the weak potential that may be used with simple mixtures of two or three plane waves as described earlier. These weak potentials may be calculated from first principles (pseudopotentials) or derived semi-empirically by finding the weakest potential necessary to reproduce a given observation like the valence spectrum of an atom (model potentials). In the past 35 years these weak potentials have found wide application to the simpler metals and even semiconductors like Si and Ge. In all these cases one can make a distinct separation between core and valence electrons. The objective is to remove the influence of the core potential and concentrate on the plane wave components of the wave function of the valence electrons in these substances. Experimentally determined Fermi surfaces, similar to that illustrated in Fig. 7.16, amply confirm this situation. (For a review see Heine 1970.)

On the other hand, if the band structure is to be determined using the full potential – and for the majority of metals this is the only approach – then the problem becomes much more complicated. It comprises primarily three parts:

(a) choice of potential;

(b) choice of basis functions;

(c) numerical computation.

If the calculation is successful then the eigenfunctions should give rise to a charge density that reproduces the original potential – the calculation should be consistent. However, it is often the case that only eigenvalues of the energy are calculated.

The fact that the simple metals have Fermi surfaces closely similar to the free electron sphere makes it of interest to compare their band structures with that of the free electron gas; to do this we need to describe the 'empty lattice'.

## 7.7 The Empty Lattice and Simple Metals

We again use our imaginary powers to control the strength of the potential. We assume a finite potential to define the lattice and then decrease it to an insignificantly low value so that the electrons become free. This is the empty lattice: it is a 'ghost' lattice, and we assume that the periodicity remains but with zero strength. What is the point of this? We want to represent the free electron energy bands in the same manner as for a true metal, taking into account the principal lattice symmetries and possible degeneracies. We have already seen that the periodicity demands that our free electrons in a lattice be described by the plane waves $\psi_{(k-G)}(\mathbf{r}) = A \exp \mathrm{i} (\mathbf{k} - \mathbf{G}) \cdot \mathbf{r}$ and that their energy be given by

$$\frac{\hbar^2}{2m} |\mathbf{k} - \mathbf{G}|^2. \tag{7.19}$$

Therefore, as we reduce the periodic potential, our real band structure must change to that of the free electron gas, the band gaps disappear and the $E(k)$ curves become parabolic. *However, we maintain the zone structure.*

We take the specific example of the bcc lattice with parameter $a$; the associated reciprocal lattice has base vectors that may be written

$$\frac{2\pi}{a}\mathbf{g}_1, \qquad \frac{2\pi}{a}\mathbf{g}_2, \qquad \frac{2\pi}{a}\mathbf{g}_3, \tag{7.20}$$

where the $\mathbf{g}_i$ are unit vectors parallel to the $x$, $y$ and $z$ axes (see Fig. 7.15*a*). We calculate the energy values $E$ in the reduced zone and write

$$\mathbf{k} = \frac{2\pi}{a}(x\mathbf{g}_1 + y\mathbf{g}_2 + z\mathbf{g}_3), \tag{7.21}$$

where

$$0 \leqslant x \leqslant 1 \quad \text{in the 100 direction,}$$
$$0 \leqslant x \leqslant \tfrac{1}{2}, \quad 0 \leqslant y \leqslant \tfrac{1}{2} \quad \text{in the 110 direction;}$$

the upper limits of $x$ and $y$ lying in the appropriate zone boundary. As Fig. 7.15 shows, the first Brillouin zone of the bcc lattice is a dodecahedron formed completely by $\{110\}$ zone boundaries. We must remember that the origin of the energy gaps is the interaction of the electron with the periodic potential and Bragg reflection. *Only* **G** *vectors that give rise to finite structure factors are significant.* For the bcc structure, the first three such **G**s are $\{110\}$, $\{200\}$ and $\{220\}$. We write **G** in its usual form

$$\mathbf{G} = \frac{2\pi}{a}(h\mathbf{g}_1 + k\mathbf{g}_2 + l\mathbf{g}_3). \tag{7.22}$$

Then (7.19) gives

$$E = \frac{\hbar^2}{2m}|\mathbf{k} - \mathbf{G}|^2 = \frac{\hbar^2}{2m}\frac{4\pi^2}{a^2}[(x-h)^2 + (y-k)^2 + (z-l)^2], \tag{7.23}$$

which is more simply written as

$$\lambda = E\frac{ma^2}{2\pi^2\hbar^2} = (x-h)^2 + (y-k)^2 + (z-l)^2. \tag{7.24}$$

The right-hand side of (7.24) is a simple expression for the energy in terms of the unit $(ma^2/2\pi^2\hbar^2)^{-1}$; this energy we write as $\lambda$.

We now demonstrate the calculation of the free electron bands in the [100] direction, taking account of a few particular **G** vectors and their multiplicities:

(a) $\mathbf{G} = 0, h = k = l = 0,$

$\lambda_1 = x^2$, which produces a single parabolic band;

(b) $\mathbf{G} = \{100\}$ is absent because the structure factor is zero;

(c) $\mathbf{G} = \{110\}$; this vector has multiplicity 12 and provides the following energy bands:

$$\lambda_2 = (x - 1)^2 + 1, \quad (110)(1\bar{1}0)\ (101)(10\bar{1});$$

$$\lambda_3 = (x + 1)^2 + 1, \quad (\bar{1}\bar{1}0)(\bar{1}10)\ (\bar{1}0\bar{1})(\bar{1}01);$$

$$\lambda_4 = x^2 + 1 + 1, \quad (011)(01\bar{1})\ (0\bar{1}\bar{1})(0\bar{1}1);$$

$\lambda_2$, $\lambda_3$ and $\lambda_4$ are all four-fold degenerate;

(d) $\mathbf{G} = \{200\}$:

$$\lambda_5 = (x - 2)^2, \quad (200);$$

$$\lambda_6 = (x + 2)^2, \quad (\bar{2}00);$$

$$\lambda_7 = x^2 + 4, \quad (020)(0\bar{2}0)\ (002)(00\bar{2});$$

$\lambda_5$ and $\lambda_6$ are non-degenerate whereas $\lambda_7$ has four-fold degeneracy.

The various bands appropriate to $\lambda_1, \ldots, \lambda_7$ are depicted in Fig. 7.19. We see immediately how the energy rises rapidly owing to the interaction with the various reciprocal lattice vectors. It is a simple matter to calculate similar curves for higher **G** values, for other symmetry directions and for different structures. In Fig. 7.20 we show the empty lattice construction in several symmetry directions for the fcc Brillouin zone together with the calculated band structure for aluminium.

We make the following remarks:

(a) Although there are significant differences, the close relationship between empty lattice and true band structures is apparent.

(b) The true potential removes some but not all of the degeneracies.

(c) In the true band structure, energy gaps appear at the zone boundaries, significant points are the X and L points at the centres of the {002} and {111} faces respectively.

(d) Certain band crossings arising in the empty lattice construction are maintained in the real band structure. This crossing or non-crossing of energy bands is governed by the symmetry of the wave functions. Crossing arises between bands of different symmetries; hybridization (mixing) and splitting between bands of similar symmetries.

(e) Except in the immediate vicinity of the zone boundaries, there is very little deviation from the free electron band form.

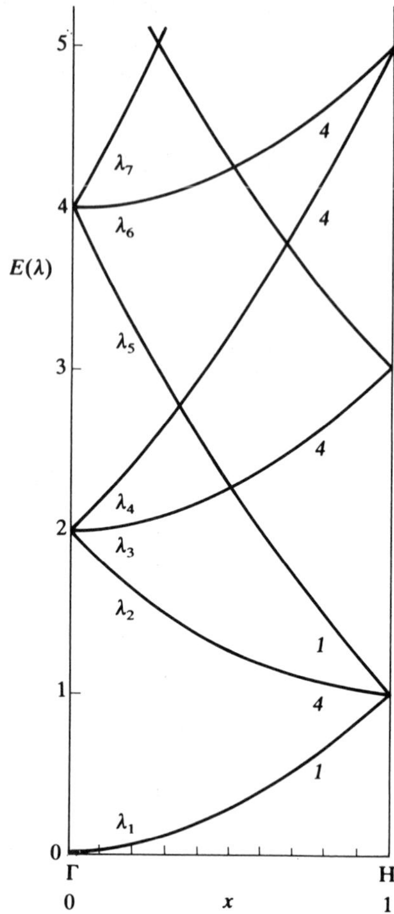


**Figure 7.19** The empty-lattice construction for free electron energy bands along [100] in the bcc structure. All the bands are of the form $E \sim (k - G)^2$. The numbers in italics indicate the degeneracy, and energy $E$ is plotted as $\lambda$ (defined in the text).

## 7.8 Core States and Band States

We have seen that in condensed solids the inner core electron levels retain their atomic character whereas the outer valence levels and higher unoccupied states form broad energy bands. But the valence electrons in, say, Be (2s shell) are core electrons in Na, whereas the Na valence electrons (3s shell), are core states in K. Thus, taking the example of the 3s shell, we expect these electron states to produce broad bands in elements with atomic number below that of $^{11}$Na, and the bands will be empty. On the other hand, for elements with atomic number greater than that of $^{19}$K, the 3s electrons are in core states with atomic character; they are of course completely filled. Thus as we proceed along the sequence of elements Na, Mg, Al, . . . , K, the 3s band must move to lower energies. (We speak of s bands and p bands, implying that the electrons have wave functions of specific symmetry. This is not the case, since s and p states becomes mixed. We should rather speak of electrons with predominantly s or p character. Provided that we remember this, we may for convenience

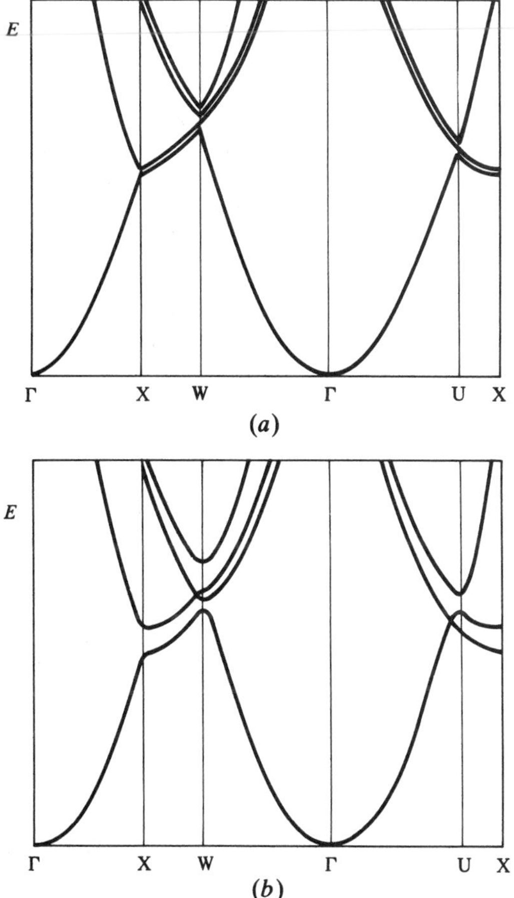


**Figure 7.20** A comparison of the empty-lattice energy bands (*a*) and detailed calculations for Al (*b*). Again the nearly free electron character of Al is confirmed. (Harrison, W., *Pseudopotentials in the Theory of Metals,* 1966, Addison-Wesley Publishing Co., Reading, Massachusetts. Figures 3.19 and 3.20. Reprinted with permission.)

use the terms 3s band or 3p band as in Fig. 7.21.) In Si the atoms are covalently bonded and the s and p states hybridize to form bonding and antibonding bands, the former being completely filled, the latter completely empty. Furthermore, these bands are separated by an energy gap of 1.17 eV, so Si is an insulator at 0 K. When we arrive at potassium the 3s and 3p levels are no longer valence states. Thus the 3s level has width of order 0.1 eV and forms part of the atomic core.

The alkali metals find little use in the pure state, although Na and K and the compound NaK have potential use as coolants in fast reactors. Their compounds with the halogens, e.g. NaCl, LiF, CsBr, in single-crystal form, find important use as optical elements such as windows and prisms in infrared spectrometers.

The principal divalent elements of groups IIA and IIB are Mg, Zn, Cd and Hg. Zn and Cd are important as alloying elements, and also as protective coatings (galvanized iron plate), as is Sn (particularly for food containers). Zn, Cd, Hg and Pb find use in batteries and other electrical components. The oxides of Zn and Pb were formerly used as paint pigments. The sulphides of zinc, cadmium and lead, for

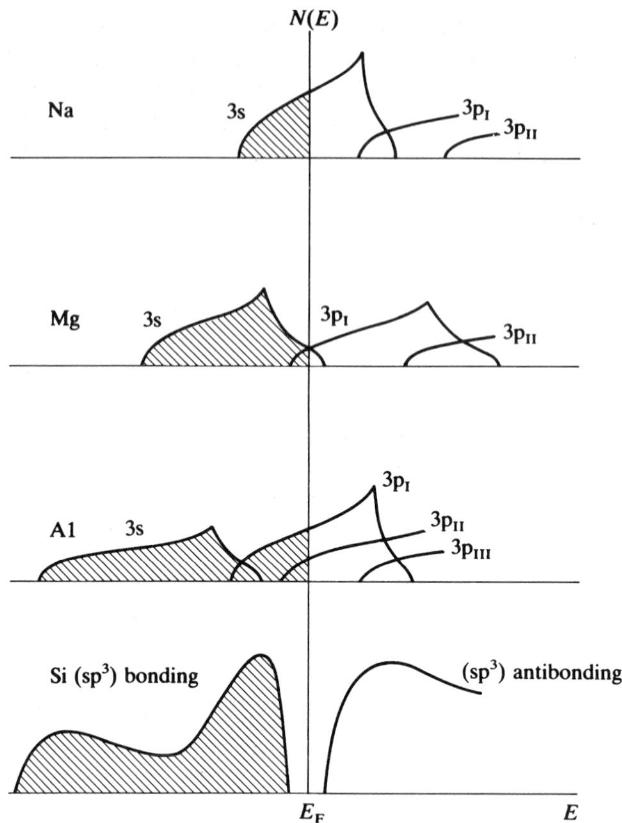


**Figure 7.21** Schematic representations of the integrated density of states for a series of neighbouring elements. It should be remembered that in Si the s and p electrons are inextricably mixed in the form of $sp^3$ hybrid orbitals, and that related hybridization will have taken place between s and p states in the other elements.

example, are semiconductors; ZnS fluoresces and PbS is an important detector for infrared radiation. In recent years, sintered compacts of ZnO particles have found important use as varistors (non-linear current-limiting devices). Aluminium when alloyed with Mg, Cu or Sn is an important constructional material, whereas its oxide $Al_2O_3$ is widely used as a ceramic insulator (both electrical and thermal) and as an abrasive (corundum), and in single-crystal form (sapphire) is the basis for the ruby laser and jewelled bearings.

## 7.9 The Transition Metals and d States†

| | | | | | | | | | | | |
|---|---|---|---|---|---|---|---|---|---|---|---|
| K | Ca | Sc | Ti | V | Cr | Mn | Fe | Co | Ni | Cu | Zn |
| Rb | Sr | Y | Zr | Nb | Mo | Tc | Ru | Rh | Pd | Ag | Cd |
| Cs | Ba | La | Hf | Ta | W | Re | Os | Ir | Pt | Au | Hg |

† Band structure calculations are reviewed by Mackintosh and Krogh Andersen (1980).

The solution of the band structure problem lies in great part in a good choice of the potential and of basis functions. Complementary to the nearly free electron approach is the tight-binding approximation. Here, although the electrons are assumed to interact and therefore occupy energy bands, the interaction is considered so weak that the wave functions are of perturbed atomic form. The electrons are considered to be tightly, but not completely, bound to their parent atoms. The tight-binding approximation finds particular application to the d electrons of transition metals.

A d electron wave function is written as a linear combination of atomic orbitals satisfying the requirements of a Bloch function

$$\psi(\mathbf{r}) = \sum_j \phi_m(\mathbf{r} - \mathbf{l}_j)e^{i\mathbf{k}\cdot\mathbf{l}_j}, \tag{7.25}$$

where $\mathbf{l}_j$ is the position vector of a given ion and $\phi_m$ an atomic orbital, $m$ being the magnetic quantum number.

The interaction between electrons is confined to those on nearest-neighbour atoms, and the evaluation of the electron energies involves overlap integrals of the form

$$\int \psi^*_{jm}(\mathbf{r}) \sum_{i\neq j} V_i \psi_{jm}(\mathbf{r})\, d\mathbf{r}, \tag{7.26}$$

which shifts the energy of the d states, and of the form

$$\int \psi^*_{jm}(\mathbf{r}) V_i\, \psi_{im'}(\mathbf{r})\, d\mathbf{r}, \tag{7.27}$$

which mixes d states on adjacent atoms. $V_i$ is the atomic potential on site $i$. The overlap integral (7.26) produces a chemical shift in the energy levels at a particular site $j$ on account of the presence of other atoms on sites $i$, whereas (7.27) says that an electron associated primarily with site $j$ spends part of its time in the vicinity of site $i$, and interacts with electrons on this site. Electrons in states $jm$ and $im'$ perturb one another and the mixing gives rise to a narrow band of quasicontinuous levels. In simple language we say that there occurs an *interatomic* $d_i - d_j$ *resonance* that spreads the initially independent degenerate d states into an energy band. The strength of the resonance depends upon the degree of overlap and therefore on the d-shell diameter and the interatomic spacing. Furthermore, since the d states comprise five orbitals, the complete d band must contain five sub-bands, each with capacity for two electrons. These d bands have rather complex structure with a detailed shape dependent on the crystal symmetry.

The previous discussion, however, neglects one of the most significant features of the d states, namely that their energies lie within those of the s valence electrons. K, Rb and Cs are monovalent metals with one valence electron in the 4s, 5s and 6s states respectively. So are Cu, Ag and Au, but the latter also contain electrons in the 3d, 4d and 5d states. In going from say K to Cu, the d states are initially empty, lying above the first 4s electron level in K, whereas in Cu they lie below this level. The s and d electrons overlap in energy. Furthermore there is no reason to believe that the s electrons are basically different in Cu than in say K; they should therefore have a large degree of plane wave character, implying an almost uniform spatial distribution. *They must therefore overlap the d orbitals.* This overlap in both real space and energy gives rise to an *intraatomic s-d resonance*, which mixes or hybridizes the s and d bands. Thus the difference between the s bands (or conduction bands, as we shall call them in future) in K and Cu is the interaction with d states in the latter metal. The effect of this resonance on the general band structure is shown in Fig. 7.22. Crudely speaking, the complete d band is narrow and acts as a wedge

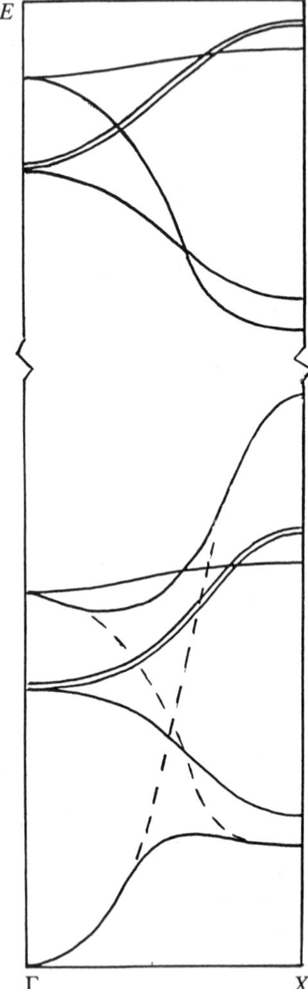


**Figure 7.22** The five d sub-bands of a transition metal atom arise on account of interatomic overlap of orbitals centred on neighbouring atoms; each of the five orbitals produces a narrow sub-band with its own characteristic dispersion (upper part of figure). It is not obvious why the different bands disperse as they do, and we take the above shapes for granted. The d electrons, however, overlap the valence electron states in both space and energy, and this leads to a hybridization or mixing of the states – this is the s-d resonance described in the text. The lower part of the figure shows the result of this mixing. Only one of the d sub-bands is strongly affected – a consequence of the symmetry dependence of the hybridization. The underlying unhybridized d sub-band and sp conduction band are shown as broken lines. This diagram shows the behaviour only in the ΓX direction in the fcc structure; Fig. 7.25 shows the complete band diagram for this crystal structure. (After Mott 1964.)

driven into the conduction band, splitting it into two parts. The resultant hybridization is related to that long recognized in chemistry as giving rise to hybrid molecular orbitals such as $d^2sp^3$.

Normally we define transition metals as only those containing incompletely filled d shells, but the empty d bands in the alkaline earths (Ca, Sr and Ba) and the filled d bands in Cu, Ag and Au lie sufficiently close to the Fermi energy to produce significant effects. Because of this, the former group may be termed incipient transition

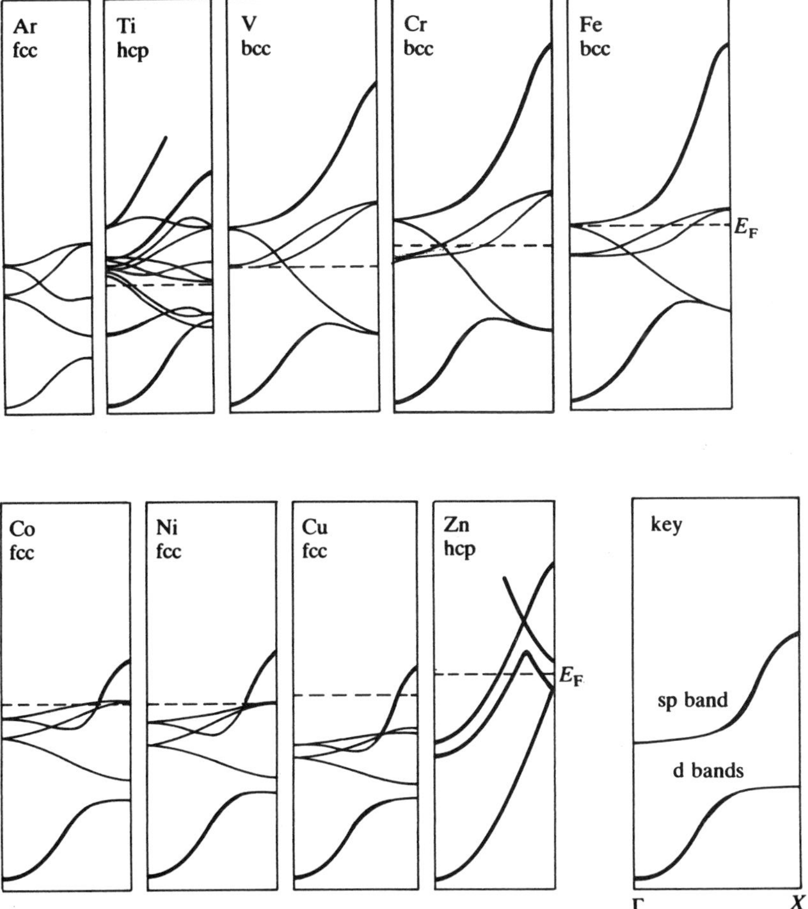


**Figure 7.23** Here the progression of the 3d band through the Fermi energy for certain elements in and near the first long series of transition metals is shown. Note that the elements have different crystal structures, but this is a minor detail (except for Ti and Zn, which have the hcp structure). The d band in the earlier elements of the series is relatively wide, but narrows and moves to lower energy as one passes from Ti to Zn. In the latter element, the d levels are completely filled and have become core states – they no longer affect the metallic behaviour. In Ar, the 3d band has not started to cross the 4s levels and there is no s-d mixing; both 3d and 4s levels are empty in this element. The filled electron states in Ar lie far below the 4s band, so this element is a transparent insulator in its crystalline form. The complex band structure of Ti and Zn arises because there are two atoms in the structure basis, causing a doubling of the number of bands. The bands of primarily sp character are denoted by the heavier lines. The energy scale may be assessed from the occupied portion of the conduction band of Cu which is $\sim$10 eV. (After Mattheiss 1964.)

metals, whereas the latter are immediate post-transition metals. On the other hand, when we come to Zn, Cd and Hg, the respective d bands have sunk sufficiently below $E_F$ that their influence is negligible. These elements are therefore simple metals with properties determined by the sp electrons. Figure 7.23 illustrates the progression of the d band through the first long transition metal series. We expect a similar pattern in the second and third series, and this similarity is amply confirmed in the observed physical properties of these metals. The bands for different metals of the same crystal structure differ primarily with regard to their width and the position of the Fermi level; they are broader for the earlier members of a series, becoming narrower as the atomic number increases. This may be appreciated in terms of the decreasing d-orbital overlap as the atomic number increases through a given series.

The d electrons have intermediate character, i.e. they are neither like free electrons nor like core electrons. Their behaviour is paradoxical, combining itinerant properties (i.e. capable of moving through the lattice) with, at the same time, a high degree of localization. It is often difficult to reconcile these two features, and occasionally we accentuate one property at the expense of the other. Since the complete d band contains ten electrons and has relatively small total width of about $5 \pm 2$ eV, the *average* density of d electron states is large, about 2 $eV^{-1}$ $atom^{-1}$, some 5 to 10 times larger than in the simple metals. However, the density of d states shows a pronounced variation through the band; this is to be expected because it is composed of five overlapping narrow bands. This detailed variation is shown in Fig. 7.24. We can immediately appreciate that the density of the d state at the Fermi level should show rapid variations as the band is progressively filled, as when proceeding from Ti to Cu.

The transition metals are very reactive chemically, often showing several oxidation states. They combine readily with non-metals such as O, C and S. Consequently, they are often difficult to obtain in a pure state. Furthermore many exhibit two or more crystal structures, each structure being stable over a specific temperature range. These features preclude the easy preparation of pure strain-free crystals and

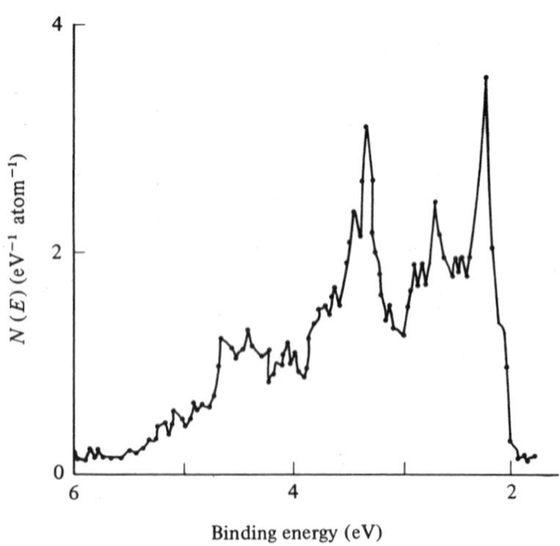


**Figure 7.24** The calculated 3d band of Cu (in integrated form), showing fine detail. (After Mueller 1967.)

complicate the study of their Fermi surfaces (which in certain cases are expected to be very complicated). Otherwise, the transition metals have often been favourite subjects for experiments because the large densities of d states give rise to readily measured electronic heat capacities and Pauli susceptibilities, and favour the occurrence of superconductivity. A few of them (but a very important class) exhibit magnetic behaviour, in particular Fe, Co and Ni, but also Cr and Mn. Even so, the d transition metals in general are not magnetic metals (in the sense that we can associate a permanent magnetic moment with each atom); this quality is much more correctly associated with the rare earth metals. The fact that the transition metals alloy both with one another and with the coinage metals is an added incentive to experiment. They are also, of course, of great importance in technology, both as metals and as compounds particularly with oxygen and carbon.

## 7.10 The Coinage Metals Cu, Ag, Au

We might very well consider these metals to be the simplest transition metals. Although the d states are filled, the d bands lie within the occupied region of the conduction band (sp band) and perturb its otherwise nearly free electron character. The Fermi surfaces of these metals have a distorted spherical form and, in contrast with the alkali metals, necks occur on the nearest zone boundary faces, i.e. at the L point (Fig. 7.15; see also Fig. 9.1). The band gaps between the first and second conduction bands that arise at the L and X points are also much larger than those for the alkali metals. In Fig. 7.25 we reproduce a complete band structure diagram for Cu together with a stylized representation of the density of states, which we shall use later when discussing physical properties. The coinage metals, particularly with regard to properties determined by $N(E_F)$, are very similar to the alkali metals. There is, however, an important exception: whereas the ion cores of the alkali

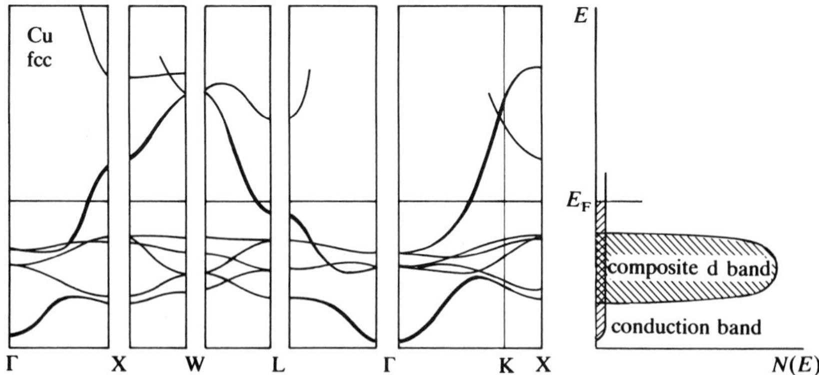


**Figure 7.25** The complete band structure diagram for Cu along the major symmetry directions (compare with the Brillouin zone of Fig. 7.15*b*). The diagram to the right is a simple schematic representation of the integrated density of levels and is convenient as a basis for the discussion of many physical properties of the later 'd' metals. We shall soon find that the sp valence electrons are responsible for the electrical conductivity of these metals; one therefore speaks of the 'conduction band'. (After Segall 1962.)

metals are small, those of the coinage metals are large owing to the presence of the complete d shells, and the ion acts as a hard sphere on account of the restrictions placed on the overlapping of filled electron shells. These metals are therefore much less compressible than the alkali metals. Cu, Ag and Au have long held an important place in fundamental studies of metals. They can be obtained in a pure state and they are very good solvents for other metals, both simple polyvalent and transition metals. In technology, Cu is by far the most important of the three. It has a special place on account of its much lower cost and is used not only where its high electrical and thermal conductivities are important, but also as a constructional material: brass when alloyed with Zn or the various bronzes when it is alloyed with small amounts of Sn, Be, Al etc. The compounds formed with O, S, N or other non-metals have been of interest primarily to solid state chemists. However, the discovery of high-temperature superconductivity in $La_{1-x}Ba_xCuO_4$ has radically changed this situation. Ag may replace Cu in certain applications where the highest electrical conductivity is required, otherwise its most important application is in the photographic process as AgBr. Au finds increasing use in the production of reliable electrical contacts in modern electronic equipment. Thus the electronics industry in the USA spends several hundred million dollars annually on Au for use in circuitry.

## 7.11 The Rare Earth Metals

These metals are characterized by their incomplete 4f shell; they are preceded by La, $5d^16s^2$ and Lu, $4f^{14}5d^16s^2$, ends the series. With the exceptions of Eu and Yb, which are usually divalent and Ce, which is sometimes quadrivalent, these elements are trivalent, and do not exhibit multiple oxidation states as do the ‘d’ metals. As Table 7.1 shows, the ionic (3 + ) radius decreases continuously as the 4f shell is filled; this is the so-called lanthanide contraction. The radius of the 4f shell is, for every element, approximately one half the ionic radius. The 4f electrons therefore belong to the ion core and experience strong intraatomic Coulomb and exchange forces. The electronic structure of the 4f shell is therefore little different in the solid from that of the free atom or ion. The lack of any interatomic 4f interaction may be appreciated from Fig. 7.26.

In contrast with the free atoms, the solid rare earth elements usually possess one 5d electron and two 6s electrons; the latter hybridize to form the valence electrons. The complete valence band must be rather complex in its detailed structure because it is built up from the six sub-bands (five from the d states, one from the s states); in practice, the situation is even more complicated because the rare earth metals favour the hcp structure, which demands a basis of two atoms to the associated primitive lattice cell. This means that the valence band involves at least ten 5d bands and two 6s bands, with the possibility of 6p bands at the higher energies. Just as in the free atoms, the 5s and 5p states remain filled. The electronic structures of the rare earth metals are given in Table 7.2 and a stylized representation of the band structure for Gd is shown in Fig. 7.27.

Note that Eu and Yb are exceptional in that they possess no 5d electron – in the former element the 4f shell is half filled and in the latter it is completely filled; thus for both elements the 4f electrons have zero orbital angular momentum. The exchange and correlation energies are particularly significant for these two elements

**Table 7.1**

| | Atomic configuration (outside the Xe core) | | Radius (Å) | |
|---|---|---|---|---|
| | X atom | $X^{3+}$ core | 4f shell | $3^+$ ion |
| La | $5d^16s^2$ | $5s^25p^6$ | 0.610 | 1.061 |
| Ce | $4f^26s^2$ | $4f^15s^25p^6$ | 0.578 | 1.034 |
| Pr | $4f^36s^2$ | $4f^25s^25p^6$ | 0.550 | 1.013 |
| Nd | $4f^46s^2$ | $4f^35s^25p^6$ | 0.528 | 1.005 |
| Pm | $4f^56s^2$ | $4f^45s^25p^6$ | 0.511 | 0.979 |
| Sm | $4f^66s^2$ | $4f^55s^25p^6$ | 0.496 | 0.964 |
| Eu | $4f^76s^2$ | $4f^65s^25p^6$ | 0.480 | 0.950 |
| Gd | $4f^75d^16s^2$ | $4f^75s^25p^6$ | 0.468 | 0.938 |
| Tb | $4f^96s^2$ | $4f^85s^25p^6$ | 0.458 | 0.923 |
| Dy | $4f^{10}6s^2$ | $4f^95s^25p^6$ | 0.450 | 0.908 |
| Ho | $4f^{11}6s^2$ | $4f^{10}5s^25p^6$ | 0.440 | 0.894 |
| Er | $4f^{12}6s^2$ | $4f^{11}5^25p^6$ | 0.431 | 0.881 |
| Tm | $4f^{13}6s^2$ | $4f^{12}5s^25p^6$ | 0.421 | 0.969 |
| Yb | $4f^{14}6s^2$ | $4f^{13}5s^25p^6$ | 0.413 | 0.858 |
| Lu | $4f^{14}5d^16s^2$ | $4f^{14}5s^25p^6$ | 0.405 | 0.848 |

and lead to their divalent metallic character whereas the other rare earth metals are trivalent. We also see that for these rare earth metals it makes sense to speak of metallic valence because we can recognize and count the valence electrons. The transitional character of the rare earths resides in the 4f electrons, and these have

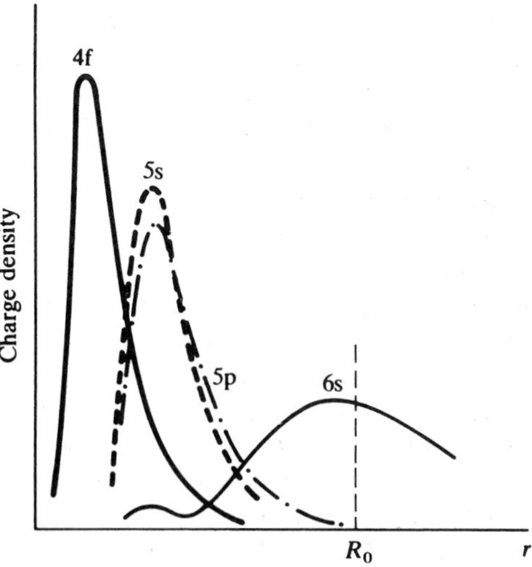


**Figure 7.26** The distribution of electron charge density in a rare earth metal atom. The 4f electrons lie within the core of the atom and are shielded by the filled 5s and 5p shells – there is only slight overlap with the valence electrons. $R_0$ is the atomic radius in the metal. On the other hand, the energies of the 4f electrons are much larger than those of the 5s and 5p electrons and of the same order as those of the valence electrons.

**Table 7.2** Electronic configuration of the solid rare earth metals

| Z | Metal | | Electronic configuration | | | Crystal structure |
|---|---|---|---|---|---|---|
| | | | 4f | 5d | 6s | |
| 57 | Lanthanum | La | 0 | 1 | 2 | dhcp |
| 58 | Cerium[a] | Ce | 1 | 1 | 2 | fcc |
| 59 | Praseodymium | Pr | 2 | 1 | 2 | dhcp |
| 60 | Neodymium | Nd | 3 | 1 | 2 | dhcp |
| 61 | Promethium | Pm | — | — | — | |
| 62 | Samarium | Sm | 5 | 1 | 2 | rh |
| 63 | Europium | Eu | 7 | 0 | 2 | bcc |
| 64 | Gadolinium | Gd | 7 | 1 | 2 | hcp |
| 65 | Terbium | Tb | 8 | 1 | 2 | hcp |
| 66 | Dysprosium | Dy | 9 | 1 | 2 | hcp |
| 67 | Holmium | Ho | 10 | 1 | 2 | hcp |
| 68 | Erbium | Er | 11 | 1 | 2 | hcp |
| 69 | Thulium | Tm | 12 | 1 | 2 | hcp |
| 70 | Ytterbium | Yb | 14 | 0 | 2 | fcc |
| 71 | Lutetium | Lu | 14 | 1 | 2 | hcp |

dhcp = double hcp, stacking sequence ... *ABAC ABAC* ..., rh = rhombohedral.

[a] Some of the properties of Ce, especially under high pressure, seem to indicate mixed-valence character in this element.

only weak interactions with the outer 5d and 6s electrons. This is in direct contrast with the conventional transition metals, where the 3d, 4d or 5d electrons are inextricably mixed with the sp electrons, causing the transitional character to appear in the valence band and thereby making the latter (with regard to transport properties at least) an ill-defined quantity.

We consider the 4f electrons to have atomic-like character; there are filled states lying below the Fermi energy and empty ones above this energy. Because of the atomic character of the 4f shell, the rare earth metals are characteristically magnetic, with atomic moments very similar to those found on the ions or free atoms. The pronounced magnetic character of the rare earth metals strongly influences many of the other physical properties, e.g. transport properties, hyperfine interactions and heat capacities.

The rare earth metals are extremely reactive and oxidize readily. They form compounds with elements such as P, B, N and S, and these compounds are of considerable current interest owing to their transport (often semiconducting), optical and magnetic properties. Their intermetallic compounds with Fe and Co are important magnetic materials.

We have implied that the valence of a rare earth metal atom is a well-defined quantity even when the atom forms part of a solid metal, compound or alloy. From a knowledge of the physical properties – and here atomic size is a very good indication – we can determine the valence. A matter of much current interest is that of

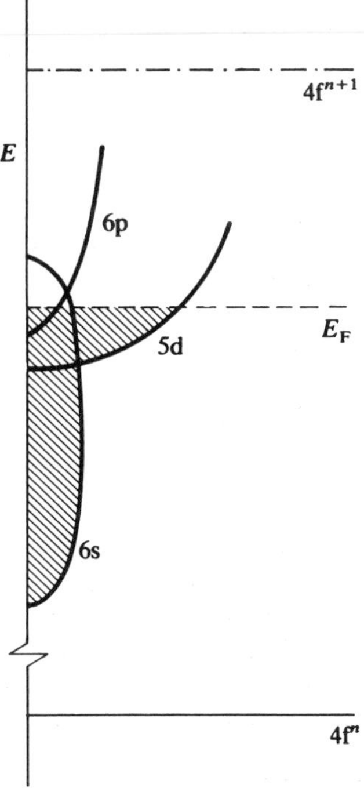


**Figure 7.27** A schematic energy diagram for a rare earth metal. The valence electrons are usually the two 6s electrons and the single 5d electron, which together form a hybridized valence band. The 4f levels have essentially atomic character; the different rare earth elements have similar valence bands associated with different atomic 4f configurations. Care must be taken in attempting to superimpose excitations between different 4f occupations on single-particle band structure diagrams.

'mixed' or 'fluctuating' valence. Thus in, for example, $SmB_6$ it appears that the Sm atoms have a non-integral valence and it is thought that each atom fluctuates, spending part of its time in tripositive and part in dipositive valence states, thereby providing the appropriate non-integral average value.

In many instances, experimental work with rare earth compounds and alloys, particularly those with semiconducting properties, is hampered by the difficulties that arise in preparing stoichiometric single crystals. Even the pure metals are difficult to obtain in a highly pure condition, and this has restricted the study of their Fermi surfaces by the more conventional techniques. From the theoretical point of view there are also significant problems. The 4f levels, being localized, lie outside the scope of the usual one-electron band structure calculation, with its emphasis on the Bloch function. So, although attempts to calculate the band character of the 6s5d states have been made, these calculations provide little information regarding their position in energy relative to the 4f levels. There has therefore been strong interest in recent years in attempts to resolve this problem experimentally by photoelectron spectroscopy.

At present the annual world production of rare earth elements is about 4000 tons, but much of this is in the form of misch-metal, which is an alloy containing 50% Ce and 30% La, the rest being various rare earth elements reflecting the composition of the original ore. Misch-metal has metallurgical uses: it promotes machinability in cast iron, and when added to steel it binds the sulphur and phosphorus into refractory rare earth compounds, thereby preventing the occurrence of manganese sulphide. The hot-rolled steel plate has better isotropy and improved mechanical strength. Some 3500 tons of misch-metal are produced annually. Another more mundane use, this time of an alloy of 75% misch-meal and 25% Fe known as ferrocerium, is in the production of flints for cigarette lighters. More scientific uses of rare earths are as catalysts in petroleum crackers and as additives in high quality optical glasses. In recent years, much attention has been paid to the $RT_5$ compounds where R is a rare earth element and T a transition metal. Thus $SmCo_5$ was formerly the world's best permanent magnet material, but its applications were limited by the availability and cost of both Sm and Co (the latter is in fact a scarce material, world production is 35 000 tons/year, of which 70% comes from Zaire). The search for alternative alloys led to the discovery of the peculiar properties of $LaNi_5$ : this alloy dissolves hydrogen to an amazing extent, and when saturated the alloy contains more hydrogen than the same volume of liquid hydrogen. More important, however, is the fact that by controlling the temperature of the alloy the hydrogen may be quickly released or quickly readsorbed. These features make it suitable as a storage medium in a hydrogen-based energy system. Although La is abundant and cheap, Ni is costly, and attempts are therefore being made to find alternative compositions. A recent result of rare earth metallurgical research was the discovery of a permanent magnet material of exceptional quality in $Nd_2Fe_{14}B$. Without doubt, many new uses will be found for these elements, and although expensive they are no longer rare.

## 7.12 The Actinide Metals

Ac, Th, Pa, U, Np, Pu, Am, Cm, Bk

Of the above metals, Th, U and Pu have been most studied. They are all radioactive, several dangerously so, although it is possible to obtain isotopes such as $^{238}U$ and $^{242}Pu$ where these dangers are minimized. Beyond U, the metals must be prepared via nuclear reactions because they do not occur naturally. Weighable amounts of actinides are available only up to einsteinium, Es. Furthermore, since single crystals are essential for much scientific work, it is not surprising that most attention has been paid to Th, U and Pu. Pu in particular is a very complicated metal and occurs in no fewer than six different crystalline phases below its relatively low melting point of 640°C. Regarding the pure metals, there is little direct information on their band structure, other than for Th for which Fermi surface data exist. At the present time, it is considered that the heavy actinides, Am and beyond, have their 5f levels strongly localized, and these elements resemble the rare earths. In the lighter actinides these 5f electrons are thought to be itinerant, and together with the 6d and 7s states form a strongly hybridized conduction band. Apart from their importance in the nuclear energy industry, these metals have had little practical application; even Th finds only limited use in the doping of tungsten filaments and as a laboratory ceramic in the form of its oxide. On the other hand, these elements are important subjects for investigation from a fundamental point of view.

## Appendix 7.1 The Nearly Free Electron Approximation

Our starting point for the derivation of (7.15) is

$$\sum_k \lambda_k c(k)\mathrm{e}^{\mathrm{i}kx} + \sum_G \sum_k U(G)c(k)\mathrm{e}^{\mathrm{i}(k+G)x} = E \sum_k c(k)\mathrm{e}^{\mathrm{i}kx}. \tag{7.28}$$

We multiply throughout by $\mathrm{e}^{-\mathrm{i}k'x}$ and integrate over $x$ from 0 to $L$, the dimension of the crystal:

$$\sum_k \lambda_k c(k) \int_0^L \mathrm{e}^{\mathrm{i}(k-k')x}\,\mathrm{d}x + \sum_G \sum_k U(G)c(k) \int_0^L \mathrm{e}^{\mathrm{i}(k-k'+G)x}\,\mathrm{d}x = E \sum_k c(k) \int_0^L \mathrm{e}^{\mathrm{i}(k-k')x}\,\mathrm{d}x. \tag{7.29}$$

Consider

$$I_{kk'} = \int_0^L \mathrm{e}^{\mathrm{i}(k-k')x}\,\mathrm{d}x = \frac{1}{\mathrm{i}(k-k')}\left[\mathrm{e}^{\mathrm{i}(k-k')x}\right]_0^L \tag{7.30}$$

The allowed values of $k$ and $k'$ are determined by the periodic boundary conditions applied to our linear lattice; these require that

$$k = p\frac{2\pi}{L} \quad \text{and} \quad k' = p'\frac{2\pi}{L},$$

where both $p$ and $p'$ are integers, so $\mathrm{e}^{\mathrm{i}(k-k')L} = \mathrm{e}^{\mathrm{i}2\pi(p-p')} = 1$. Thus when $k = k'$

$$I_{kk'} = L,$$

and when $k \neq k'$

$$I_{kk'} = 0.$$

Equation (7.29) now becomes

$$\lambda_{k'} c(k') + \sum_G U(G)c(k' - G) = Ec(k'). \tag{7.31}$$

The prime on $k$ is no longer needed and we arrive at (7.16):

$$(\lambda_k - E)c(k) + \sum_G U(G)c(k - G) = 0. \tag{7.32}$$

We show how (7.32) may be used to determine the behaviour of an electron moving in a periodic potential characterized by a single Fourier coefficient $U(G_1)$, a situation similar to that given in (7.1) but with $U_0 = 0$. Equation (7.32) introduces a sequence of equations involving $k, k \pm G_1, k \pm 2G_1, k \pm 3G_1, \ldots, k \pm nG_1, \ldots$. We shall cut off the sequence at $k \pm 3G_1$. We first write (7.32) based on $k$, which automatically introduces wave vectors $k - G$ and $k + G$ (we write $G$ for $G_1$):

$$(\lambda_k - E)c(k) + Uc(k - G) + Uc(k + G) = 0.$$

When we write the equations for $k - G$ and $k + G$ these lead to new wave vectors $k - 2G$ and $k + 2G$, and so on. The equations required are as follows, arranged symmetrically about the starting equation:

$$\begin{aligned}
(\lambda_{k-3G} - E)c(k - 3G) &+ Uc(k - 4G) + Uc(k - 2G) = 0,\\
(\lambda_{k-2G} - E)c(k - 2G) &+ Uc(k - 3G) + Uc(k - G) = 0,\\
(\lambda_{k-G} - E)c(k - G) &+ Uc(k - 2G) + Uc(k) = 0,\\
(\lambda_{k} - E)c(k) &+ Uc(k - G) + Uc(k + G) = 0,\\
(\lambda_{k+G} - E)c(k + G) &+ Uc(k) + Uc(k + 2G) = 0,\\
(\lambda_{k+2G} - E)c(k + 2G) &+ Uc(k + G) + Uc(k + 3G) = 0,\\
(\lambda_{k+3G} - E)c(k + 3G) &+ Uc(k + 2G) + Uc(k + 4G) = 0.
\end{aligned}$$

We can continue indefinitely, and no difficulty arises if we take account of higher coefficients $U$, the equations just become larger. We can arrange the above equations according to coefficient $c$; they form a system of linear equations for the $c$s. These equations have non-trivial roots for the $c$s only if the appropriate determinant is zero; this determinant is easily found and has the form

$$\begin{vmatrix}
\lambda_{k-3G} - E & U & \cdot & \cdot & \cdot & \cdot & \cdot\\
U & \lambda_{k-2G} - E & U & \cdot & \cdot & \cdot & \cdot\\
\cdot & U & \lambda_{k-G} - E & U & \cdot & \cdot & \cdot\\
\cdot & \cdot & U & \lambda_k - E & U & \cdot & \cdot\\
\cdot & \cdot & \cdot & U & \lambda_{k+G} - E & U & \cdot\\
\cdot & \cdot & \cdot & \cdot & U & \lambda_{k+2G} - E & U\\
\cdot & \cdot & \cdot & \cdot & \cdot & U & \lambda_{k+3G} - E
\end{vmatrix} = 0. \tag{7.33}$$

$k$ ranges over $-\pi/a \leqslant k_i \leqslant \pi/a$, and for each particular $k_i$ there is an associated $\lambda_i$ and a determinant as in (7.33). The solution of (7.33) for given $k_i$ leads to seven eigenvalues $E_{ip}$ $(p = 1, \ldots, 7)$. When plotted as functions of $k_i$, these $E_{ip}$ describe the first seven bands in the reduced zone scheme. The solution of (7.33), even with only one coefficient of potential, is cumbersome. If we are only interested in the electrons in the first two bands, i.e. those with wave vectors $k$ and $k - G$, then a consideration of the sizes of the various $\lambda$ compared with the values of $E$ in question shows that the coefficients $c$ associated with $k \pm 3G$, $k \pm 2G$ are going to be very small. Our single Fourier component does not merit including these terms. We may restrict attention to the two by two matrix containing only $k$ and $k - G$, for, as shown below, even $c(k + G)$ is negligible.

We therefore determine the energy values at the zone boundary where $k_i = G/2 = \pi/a$ and neglect terms for which $c(k_i)$ is small.

The coefficients $c(k)$ are given by (7.32):

$$c(k) = \frac{\sum_G U(G)c(k - G)}{\lambda_k - E};$$

only those coefficients for which the $\lambda_k - E$ are small are significant. Now on the zone boundary for which $k > 0$, $k = \frac{1}{2}G$, so our coefficients are

$$\begin{array}{lll} c(K), & k = \frac{1}{2}G, & \lambda_k \approx E, \\ c(k - G), & k = \frac{1}{2}G - G = -\frac{1}{2}G, & \lambda_k \approx E, \\ c(k + G), & k = \frac{1}{2}G + G = \frac{3}{2}G, & \lambda_{k+G} > E. \end{array}$$

Continuation shows that only the coefficients $c(k)$ and $c(k - G)$ need be considered, and our determinantal equation can be truncated to

$$\begin{vmatrix} \lambda_{k-G} - E & U \\ U & \lambda_k - E \end{vmatrix} = 0,$$

which leads to the following equation for $E$:

$$(\lambda_{k-G} - E)(\lambda_k - E) - U^2 = 0.$$

On the zone boundary

$$\lambda_k = \lambda_{G/2} = \lambda_{k-G} = \lambda;$$

so the equation becomes

$$(\lambda - E)^2 - U^2 = 0,$$

giving $E = \lambda \pm |U|$, i.e. the energy gap at the zone boundary is $2|U|$.

## References

ANDERSON, J. R. and GOLD, A. V. (1967) *Phys. Rev.* **139**, A1459.

BAER, Y. (1989) Private communication.

HARRISON, W. (1966) *Pseudopotentials in the Theory of Metals.* Benjamin, New York.

HEINE, V. (1970) *Solid State Physics*, Vol. 24 (ed. H. Ehrenreich, F. Seitz and D. Turnbull). Academic Press, New York.

MACKINTOSH, A. and KROGH ANDERSEN, O. (1980) *Electrons at the Fermi Surface* (ed. M. Springford). Cambridge University Press, Cambridge.

MATTHEIS, L. F. (1964) *Phys. Rev.* **134**, A970.

MOTT, N. F. (1964) *Adv. Phys.* **13**, 325.

MUELLER, F. M. (1967) *Phys. Rev.* **153**, 659.

SEGALL, B. (1962) *Phys. Rev.* **125**, 1099.

## Further Reading

ASHCROFT, N. W. *Lessons from Six Decades of Selectively Successful Band Theory*, Proc. Int. Conf. Teaching Modern Physics, Condensed Matter (1989) pp. 133. World Scientific Publishing Co., Singapore.

## Problems

**7.1** Using Fig. 7.8, show that the several pieces of the fourth Brillouin zone may be translated to completely occupy the first zone. Use coloured crayon or shading so that the initial and final positions of each part are readily recognizable.

**7.2** Calculate the electron concentrations (electrons per atom) required for the Fermi sphere to contact the zone faces in the bcc and fcc structures.

**7.3** In a free electron approximation, what are the minimum energy gaps required on the zone boundaries of the Brillouin zone for a divalent fcc substance in order that the first zone be completely filled. Give numerical values for Ca, $a = 5.58$ Å.

**7.4** Repeat the calculation of Section 7.7 for the empty lattice but for the fcc case and the [111] direction.

**7.5** A linear lattice of atoms is characterized by the following dispersion relation for the valence electrons:

$$E = A - B \cos ka.$$

$E$, $k$ and $a$ have their usual meanings; $A$ and $B$ are constants. Determine $E$ versus $k$ and $N(E)$ versus $E$ for the valence electrons. Illustrate with diagrams.

**7.6** An alloy with components $X$ and $Y$ has composition $XY$ and occurs as a linear structure. The alloy may have a disordered structure wherein each site is occupied by an $X$ or $Y$ atom with equal probability or in an ordered form where every $X$ atom has $Y$ atoms for neighbours. If the disordered structure has a band form similar to that of Problem 7.5, what is the band form for the ordered alloy? Illustrate with a diagram.

**7.7** Describe what is meant physically by (a) intraatomic s-d resonance and (b) interatomic d–d resonance.

**7.8** In the light of the discussion in the present chapter, draw diagrams showing the qualitative form of the integrated energy band structures (i.e. $N(E)$ versus $E$ curves) for the following elements: Ti, Mo, Pt, Cd and Tl. What kind of experimental data would you choose to support your opinions?

**7.9** Cr and Gd are both transition metals. Describe the similarities and differences in their electronic structures when in the form of free atoms and as condensed metals.

**7.10** How does band theory allow us to appreciate the occurrence of the principal forms of condensed matter, i.e. insulators, semiconductors and metals?

**7.11** Construct the first four Brillouin zones for a rectangular space lattice with primitive cell dimensions a and b (=4a).

# 8

# The Cohesion of Pure Metals

Now that we have a reasonably good physical picture of the energy band structure of pure metals, we are in a better position to attempt a simple appreciation of their cohesive energies, and in so doing we return to the discussion of Chapter 1.

## 8.1 The Simpler Metals

We begin with the alkali metals. These, as we have seen, have energy bands very similar to those of free electrons; we do not need to consider the structure in **k** space but follow how the band of valence electron levels broadens, owing to greater orbital overlap, as the average interatomic spacing decreases (Fig. 8.1). There are two important effects:

(a) the broadening of the initially sharp atomic states on account of the mutual interactions of orbitals located on different atoms, which we associate with bonding and antibonding resonances; and

(b) the shift in the centre of gravity of the levels owing partly to the overlap of the atomic potentials, but also to the character of the broadening.

It is clear from the simple situation described in Fig. 8.1 that for any element with a single valence electron (or more generally a half-filled sub-band) the condensed state is stable with regard to the vapour of independent atoms because only states in the lower half of the band are occupied. This remains true even in the absence of any shift of the centre of gravity of the levels to lower energy than found in the free atom. We can understand this cohesion in terms of the electrostatic attraction between the uniformly distributed electron gas and positive ions, an attraction that is partly offset by the kinetic energy of the electrons. A proper calculation of the cohesive energy demands a careful consideration of the exchange and correlation contributions, but this can now be done with good accuracy for the alkali metals. The cohesive energy may be expressed as a function of $r_s$ (Section 6.2), and it increases with decreasing $r_s$, i.e. increasing density of the electron gas. On passing from the monovalent alkalis to the divalent alkaline earth metals, we might expect the valence band to become filled and thereby annul the binding effect described previously. There are, however, several additional features to be considered. There is a possibility of a chemical shift of the centre of gravity of the levels to a lower energy than in the free atom, but it is also the case that, before the s states are completely filled, the lower states of the first p sub-band become occupied, leading to a strong bonding influence. Furthermore, the electron gas is considerably denser in these metals than in the alkalis. The overall effect is an increase in the cohesive energy compared with that of the preceding alkali metal (Fig. 8.2), an increase that may be augmented by the presence of empty d bands close to the Fermi level (as in Ca for example); we return to this matter later.

The trivalent elements Al, Ga, In and Tl may be understood in a similar manner to the alkali metals because, although the s states are filled, the lowest p band is only half filled, so only bonding orbitals are occupied. Again the $r_s$ values are relatively small and the trivalent elements have larger cohesive energies than the divalent metals.

The quadrivalent elements Si and Ge may be understood in terms of the covalent bond (discussed in Chapter 1), which is a result of a clear division of electron states

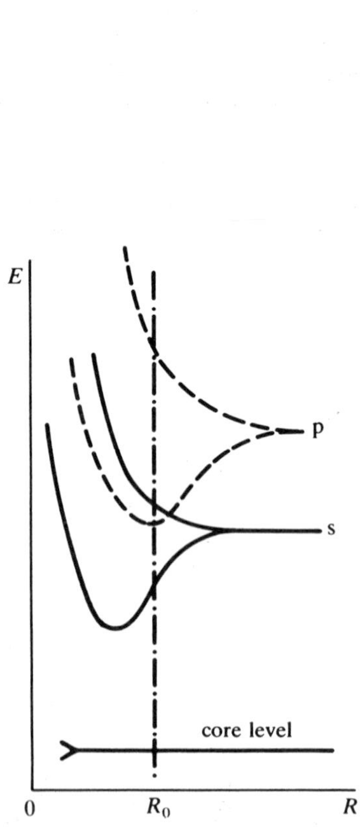


**Figure 8.1** The dependence of electron energy levels on interatomic spacing. $R_0$ is the equilibrium atomic separation.

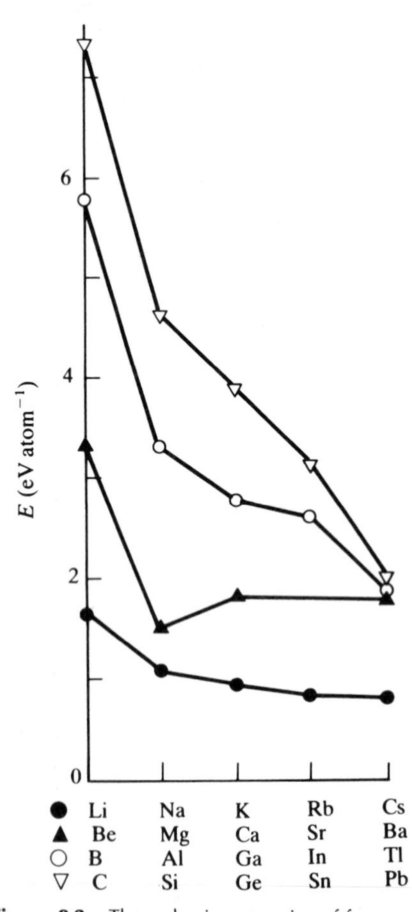


**Figure 8.2** The cohesive energies of four series of 'sp' elements. (From Kittel 1986.)

into those of bonding and antibonding character. One might ask why the elements Pb and Sn occur as true metals rather than as covalently bonded semiconductors (although Sn in fact does this in the grey modification), and we conclude that for Pb it must cost too much energy to promote an s electron into a third p state in order to form the $sp^3$ hybrid orbitals. The metallic electron structure with an approximately uniform gas of electrons described by plane wave functions provides the state of least energy. On account of the even number of electrons, the binding is relatively weak, but again $r_s$ is considerably smaller than for the alkali metals, so the cohesive energy is larger for Pb than for these metals.

## 8.2 The Transition Metals

It is immediately clear from Fig. 8.3 that not only are the binding energies of these metals much larger than those of the simpler sp metals, but there is also a similar characteristic variation throughout each series of elements. This must be attributed to the presence of the d electrons. We have seen that these electrons occupy a relatively narrow band of levels formed by summing the states in the five sub-bands. Again, we maintain that the interatomic resonances between d states localized on adjacent atoms lead to the banding of the levels, and we associate the lower half of the band with bonding states and the upper portion with antibonding states. The exact distribution of levels depends upon the crystal symmetry, and the separation into those of bonding and antibonding character is perhaps most pronounced in the bcc structure (Fig. 8.4). The width of the d band becomes narrower as we proceed from left to right along any series, but in our very qualitative discussion this aspect is neglected. It is clear that as the d band is progressively filled, electrons first occupy bonding levels, i.e. levels lying below the energy of the degenerate atomic d level. The increase in cohesive energy in the progressions from Sc to V, Y to Mo and La to W can at once be understood. The effect should be most pronounced when

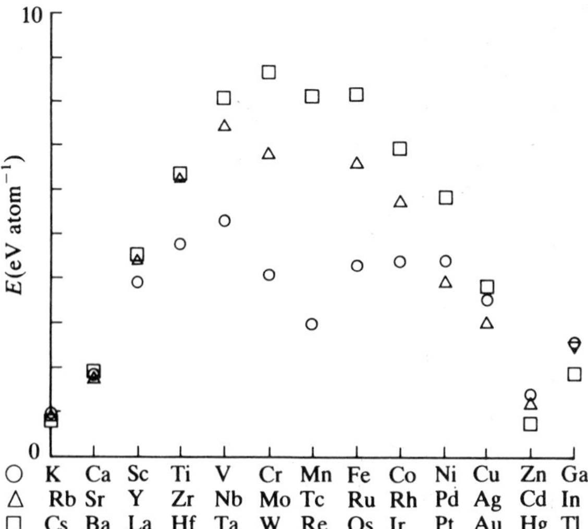


**Figure 8.3** The cohesive energies of the three series of 'd' transition metals together with those for certain neighbouring elements. (From Kittel 1986.)

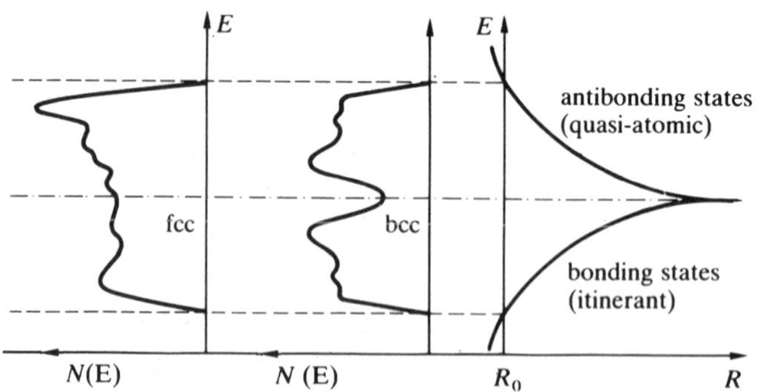


**Figure 8.4** Schematic 'd' bands for the fcc and bcc structures, showing the effect of interatomic separation $R$ and the division into bonding and antibonding portions.

the appropriate d band is half filled and this is approximately so for Mo and W, whereas Cr and Mn deviate markedly from the expected pattern. Note also that the cohesive energy for any trio of isoelectronic elements increases in the order 3d < 4d < 5d, which may be understood in terms of the increased interatomic overlap associated with an increase in orbital diameter with atomic number.

After the d bands are half filled, further addition of electrons causes occupancy of the antibonding levels, and this necessarily produces a decrease in the cohesivity. When we arrive at Cu, Ag and Au, with their complete d bands, we might conclude that they no longer make any marked contribution to the binding energy, but we discuss this aspect in greater detail below.

The conduction electrons of course also contribute to the cohesive energy of the transition metals, but their effect is small compared with that of the d electrons.

## 8.3 Post-Transition Metals

For the present purpose these are primarily Cu, Ag and Au together with Zn, Cd and Hg. Consider first Cu – the d band is complete and, as we hinted above, we might think it no longer significant in bonding mechanisms. On the other hand, we recognize that the conduction electrons in Cu are not so very different from those of K, say, and yet the cohesive energy of Cu is so much larger than that of K (by a factor of about 4). Admittedly, $r_s$ for Cu is much smaller than that for K, but it is not as small as that for Al, so it is unlikely that we can explain the better cohesive energy of the coinage metals solely in terms of the density of the conduction electron gas. Can the d electrons have any effect even though the d band is fully occupied? There has in fact been much speculation as to whether they might contribute via a van der Waals interaction, but this is a weak effect and hardly sufficient to provide a satisfactory explanation. Instead we must look to the repulsion of energy levels as a possible mechanism.

Unlike K, the conduction band in Cu is cut into two parts by the d band, which we may consider to act as a wedge separating upper and lower portions of the conduction band. The presence of a narrow dense band of d levels repels the conduction states, pushing those that lie above it to higher energies and depressing those that lie below it (Fig. 8.5). We presume the net effect to be that empty states

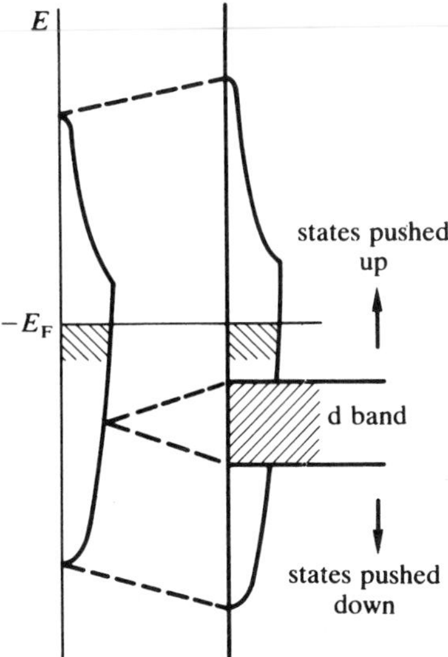


**Figure 8.5** The mechanism by which a filled d band in Cu favours binding in this element.

are repelled to higher energies, whereas occupied levels are depressed to lower energies, leading to a significant contribution to the bonding. A similar mechanism operates for Ag and Au.

We can now appreciate why on passing from a coinage metal to its neighbouring divalent element there is such a pronounced decrease in binding energy. In Zn the d electrons lie 14.5 eV below $E_F$ and are just below the bottom of the conduction band. Apart from any very weak van der Waals attraction, they can only influence the cohesion in a negative way by pushing the whole of the conduction band to higher energy.

The binding is therefore in no way aided by the filled d band. The situation is aggravated by the divalency of Zn. For these divalent elements, we conclude that the d band is detrimental to bonding.

On passing from Zn to Ga, Cd to In and Hg to Tl, there is an increase in cohesive energy; the d electrons have now become part of the ion core and are too far removed in energy to be troublesome, and we again adopt our free electron gas considerations and expect the trivalent metal to be more stable than the divalent one.

Now we can return to our discussion of K and Ca; apart from the decrease in $r_s$, it is difficult to appreciate why Ca is more stable than K, but here again we encounter d electrons. In Ca the d band is empty, but it lies with its lowest levels just *above* $E_F$. If we invoke a repulsion of levels, this can only cause a more stable binding by depression of the occupied 4sp levels to lower energies.

## 8.4 The Rare Earth Metals

These elements are usually divided into the light (Ce-Eu) and the heavy (Gd-Lu) rare earths. The former are the more difficult to understand because, although the 4f

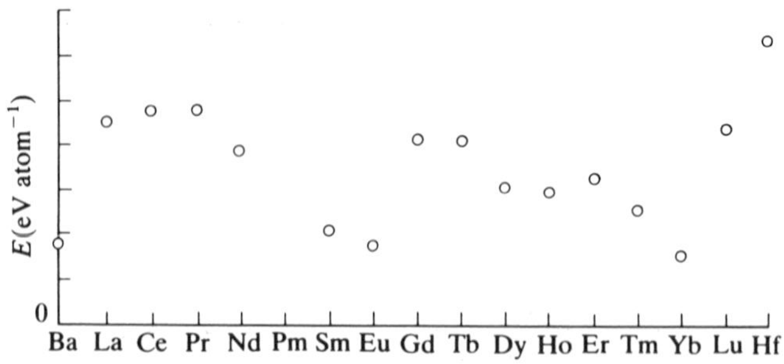


**Figure 8.6** The cohesive energies of the rare earth metals. (From Kittel 1986.)

electrons are considered strongly bound to the ions, the orbitals have larger diameter than in the heavier elements of the series and are therefore very susceptible to the local crystal fields. There is still considerable discussion regarding the possibility of quasidelocalized f electrons in the lighter rare earths. The simplest approach is to assume that only the 6s and 5d states form energy bands, and the number of conduction electrons is then determined by the occupancy of the 4f shell. In the condensed metallic state, we expect these elements to have three valence electrons per atom, and the energy band structure should be approximately the same for all the rare earths. Because of this, we might expect them all to possess the same cohesive energy, but, as Fig. 8.6 shows, there is in fact a noticeable 'sawtooth' pattern of variation. At first sight this may seem surprising, but we must remember that the cohesive energy is based upon a comparison of the energies of a given mass of metal and the same mass of vapour of free atoms. It so happens that the neutral rare earth atoms La, Gd and Lu are trivalent whereas the remainder are divalent (see Table 7.1). Thus in comparing the observed cohesive energies we contrast similar metallic situations (trivalent) with different configurations (divalent or trivalent) of the free atoms. This is the origin of the variation in Fig. 8.6. If, instead, we compare the energies of the metallic phases with the free atoms in the same configuration (i.e. all trivalent) then we find that the binding energies are indeed essentially equal, as we originally expected.

## Reference

KITTEL, C. (1986) *Introduction to Solid State Physics*, 6th edn. Wiley, New York.

9

# Some Physical Properties of Metals

## 9.1 The Fermi Surface†

The first Fermi surface to be determined was that of Cu (Fig. 9.1). The data were published in 1958, and since that time many experimental procedures have been developed for Fermi surface studies. Perhaps the most important method utilizes the de Haas–van Alphen effect, which is the oscillatory variation of the diamagnetic susceptibility as a function of magnetic field strength.‡ The method provides details of the extremal areas of a Fermi surface. All the principal methods of Fermi surface determination demand very pure strain-free single crystals, low temperatures ($<2$ K) and access to intense magnetic fields which nowadays are usually produced by super-conducting magnets with $\mathbf{B} \approx 1$–$10$ T (the lower field may be compensated for by working at lower temperature). A knowledge of the Fermi surface tells us how the different Brillouin zones are occupied and provides a test of band structure calculations only at a fixed energy $E_F$. A particular disadvantage is that these experiments are not possible with disordered alloys because the principles of the most accurate methods demand large electron mean free paths, which are not possible when alloying elements are present, even at concentrations as low as 0.1%.

We shall direct attention to the de Haas–van Alphen effect, and describe how it would arise in a 'spinless' free electron gas with infinitely large relaxation time and at very low temperature of about 1 K (see Box 9.1, pp. 221–224). The electrons in a free electron gas are characterized by their quantum number $\mathbf{k}$, with components $k_x$, $k_y$ and $k_z$. In $\mathbf{k}$ space contours of constant energy are spheres and for a given $\mathbf{k}$ an electron has velocity given by

$$\hbar \mathbf{v} = \nabla_{\mathbf{k}} E.$$

Let a magnetic field $\mathbf{B}$ be directed along the negative $z$ axis. Electron motion in this direction is unaffected by this field, but in the $(x, y)$ plane the Lorentz force induces

† See Cracknell and Wong (1973) and Springford (1980).

‡ See Chapter 11 for a brief description of experimental determination of magnetic susceptibility.

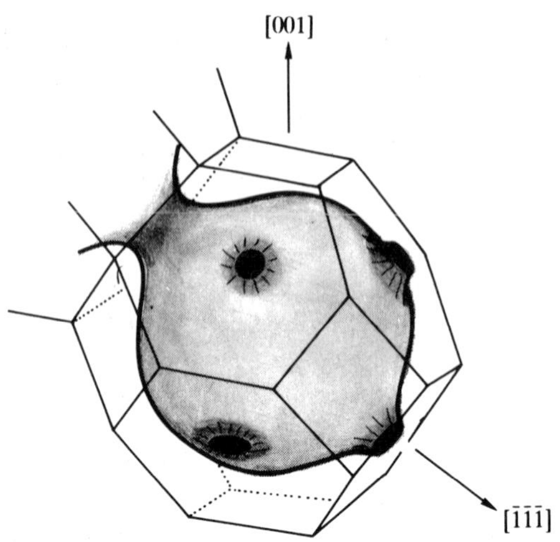


**Figure 9.1** The Fermi surface of Cu; it is mainly spherical but contact with the zone boundaries occurs at the $L$ points where 'necks' are formed.

a circular motion of the electrons. It is convenient to describe the corresponding motion in **k** space. The Lorentz force causes a representative point in **k** space to rotate in the $(k_x, k_y)$ plane with frequency $\omega_c = eB/m$. This frequency, which is known as the cyclotron frequency, is independent of **k**, so the whole system of representative points rotates about an axis through the origin of **k** space and parallel to **B**.

This regular periodic motion introduces a new quantization of the energy levels in the $(k_x, k_y)$ plane, corresponding to those of a harmonic oscillator with frequency $\omega_c$ and energy

$$E_{xy} = \hbar\omega_c(L + \tfrac{1}{2}), \tag{9.1}$$

whereas

$$E_z = \hbar^2 k_z^2/2m.$$

The quantum number $L$ takes integer values 0, 1, 2, 3, . . . . All the levels $E(L)$ have the same degeneracy $p$, and each level has acquired states, originally in the $(k_x, k_y)$ continuum, from both higher and lower energies. The total energy is almost, but not quite, unchanged. It is customary to say that the original uniform distribution of states has condensed onto the new degenerate discrete levels. We shall assume that the latter, the so-called Landau levels, are infinitely sharp. We also ensure that $\hbar\omega_c \gg k_B T$ by making **B** large and $T$ small. The net effect of the magnetic field is to change the original 'onion-like' distribution of constant energy contours that existed before the field was applied into a 'leek-like' structure. The separation of Landau levels and therefore their degeneracy increase as the magnetic field increases. We find the representative points in **k** space distributed on cylinders concentric with the field, each cylinder containing the same number of states. The occupied states, however, lie within a spherical envelope determined by the initial Fermi sphere and the Fermi wave vector of the electron gas (recall that the total energy is essentially unchanged because, although half the states have moved to higher energies, the other half have moved to lower energies and the compensation is almost perfect).

As the field is increased, $\omega_c$ and $p$ increase, and the cylinders expand. When the radius of a Landau cylinder expands, the number of occupied states that it contains decreases because its intersection with the spherical Fermi envelope shrinks. Eventually, when the radius becomes larger than $k_F$ it empties completely. However, as a level empties, it first causes the total energy to decrease very slightly, when the level lies less than $\frac{1}{2}\hbar\omega_c$ below the Fermi level, and then, when it passes through the former Fermi level, it causes the total energy to increase somewhat. Thus, every time a Landau cylinder passes through the spherical envelope, a regular fluctuation in the total energy of the electron gas occurs. The effect is largest just as the cylinder finally leaves the envelope because in this situation the emptying cylinder is tangential to the Fermi sphere and the number of states simultaneously involved is a maximum.

Thus in a monotonically increasing magnetic field, the total energy of the electron gas pulsates (but with very small amplitude), the period becoming larger as the field **B** grows larger. For a fuller description of Landau levels see Box 9.1 (pp. 221–224).

The magnetic moment of a system with free energy $F$ is defined as $-\mathrm{d}F/\mathrm{d}B$. In the present case, it is clear that the cyclotron motion produces a magnetic moment that is directed to oppose the applied field – this is just an expression of Lenz's law. The magnetic susceptibility $\mu_0 M/B$ is therefore negative and the system is said to be diamagnetic (see Chapter 11). Thus, as the field increases, the energy fluctuates, leading to oscillations in the diamagnetic susceptibility (Fig. 9.6). Such behaviour was first observed in bismuth (in quite ordinary field strengths of order 1 T) by de Haas and van Alphen in 1930. Why is the effect so important for the Fermi surface? Keeping to the spinless free electron gas, we see that if the equatorial area of the spherical envelope is $A_0$ (in the present case $\pi k_F^2$) then, since fluctuations in the diamagnetic susceptibility arise whenever a Landau cylinder passes through the Fermi sphere, every such fluctuation is associated with an integral number of partially occupied Landau levels; in other words

$$A_0 = (L + 1)\Delta A_L,$$

where $L + 1$ is the number of partially occupied levels and $\Delta A_L$ the area in the $(k_x, k_y)$ plane between any two adjacent Landau cylinders – the latter is independent of $L$ because the cyclotron frequency is the same for all electrons and $N(k)$ is constant

**Box 9.1. Landau Levels in the Free Electron Gas**

A magnetic field **B** is directed along the negative $z$ axis. The Lorentz force acts on every electron, and in **k** space all representative points move with a velocity $\mathrm{d}\mathbf{k}/\mathrm{d}t$ tangential to the surfaces of constant energy (Fig. 9.2):

$$\mathbf{F} = \hbar\frac{\mathrm{d}\mathbf{k}}{\mathrm{d}t} = -e\mathbf{v} \times \mathbf{B}.$$

The circular motion has period

$$T = \frac{\oint \mathrm{d}k}{\mathrm{d}k/\mathrm{d}t} = \frac{2\pi\hbar k}{evB} = \frac{2\pi m}{eB}.$$

*continued*

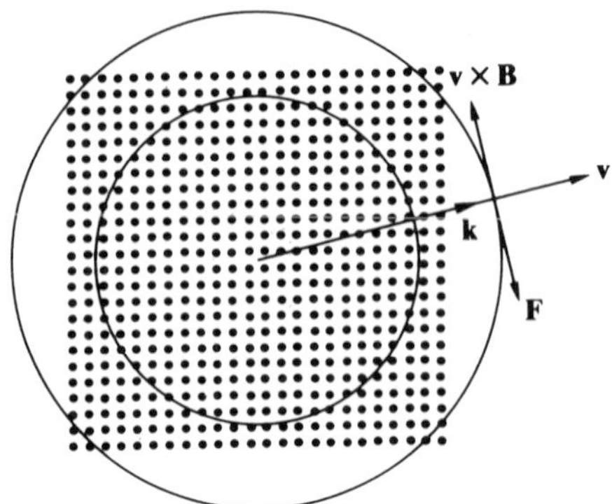


**Figure 9.2** A plane section through **k** space. The allowed electron states arise as an infinite uniform distribution of representative points, of which only a few are shown. In the presence of a magnetic field **B** directed normally into the plane section, each **k** state is subject to a Lorentz force and all the **k** states rotate with the cyclotron frequency about an axis through the origin and parallel to the field direction.

The angular frequency is

$$\omega_c = \frac{2\pi}{T} = \frac{eB}{m}.$$

$\omega_c$ is called the cyclotron frequency.

Prior to the application of the field **B**, electron states were uniformly distributed in **k** space. In the $(k_x, k_y)$ plane they were represented by an infinite square net of points, some of which are shown in Fig. 9.2. Within this plane, the field **B** causes a redistribution of the **k** states so that they lie on rings according to the new quantization introduced by the cyclotron motion (Fig. 9.3). The rings correspond to energies

$$E_L = (L + \tfrac{1}{2})\hbar\omega_c = \frac{\hbar^2 k_L^2}{2m}.$$

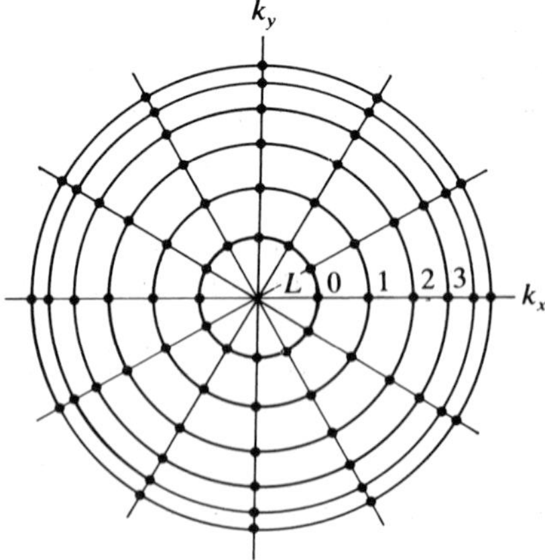


**Figure 9.3** The cyclotron motion introduces quantized levels, the Landau levels, according to $E_{x,y} = (L + \frac{1}{2})\hbar\omega_c$. These Landau levels have concentric circular form in the $(k_x, k_y)$ plane and are cylindrical surfaces in three-dimensional **k** space. Each Landau level has the same degeneracy.

*continued*

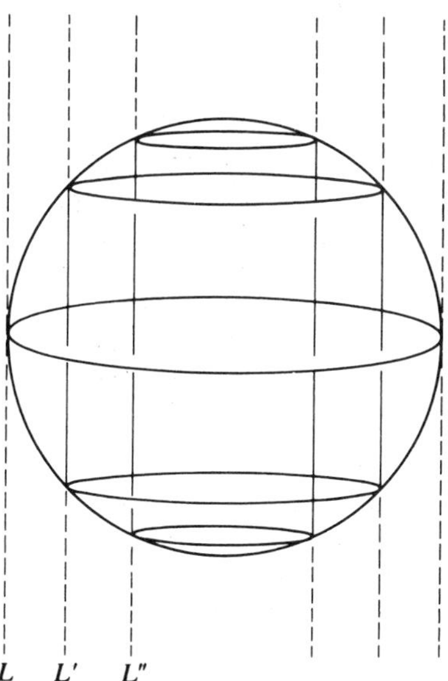


**Figure 9.4** The free electron gas has a spherical Fermi surface, but in the presence of a strong magnetic field we find the system of concentric cylindrical Landau levels. The occupied portions of these levels are determined by the original Fermi sphere. An increasing magnetic field causes the Landau cylinders to expand, and eventually they must become empty as they pass out of the Fermi sphere.

The field **B** has no influence on the $z$ motion of the electron, and the distribution of $k_z$ states has the usual quasicontinuous form. Thus in **k**-space the representative points lie on Landau cylinders, whose cross-sections are the Landau rings of Fig. 9.3. Compared with the Fermi energy, $\hbar\omega_c$ is a very small quantity and the total energy of the gas is essentially unaffected by the field. The occupied **k** states are still found within the Fermi sphere and the Landau cylinders are either partially occupied or wholly empty (Fig. 9.4).

Each Landau ring is associated with an area of **k** space. It is convenient to say that this area for the level $L$ is defined by the circles with

$$k_{L-1/2} \leqslant k \leqslant k_{L+1/2}.$$

The magnetic field causes states from both higher and lower energies to condense onto a given Landau level (Fig. 9.5). The area between two adjacent rings is

$$\pi\Delta(k_L^2) = \frac{\pi 2m\ \Delta E}{\hbar^2} = \frac{2\pi m\hbar\omega_c}{\hbar^2} = \frac{2\pi eB}{\hbar}.$$

Remembering that $N(k) = \mathscr{A}/2\pi^2$, where $\mathscr{A}$† is the area of the sample perpendicular to the field, we find the degeneracy of each Landau level per unit area of sample to be

$$p = \frac{eB}{\pi\hbar}.$$

† We use $\mathscr{A}$ here instead of $A$ to avoid confusion with the areas in **k** space.

*continued*

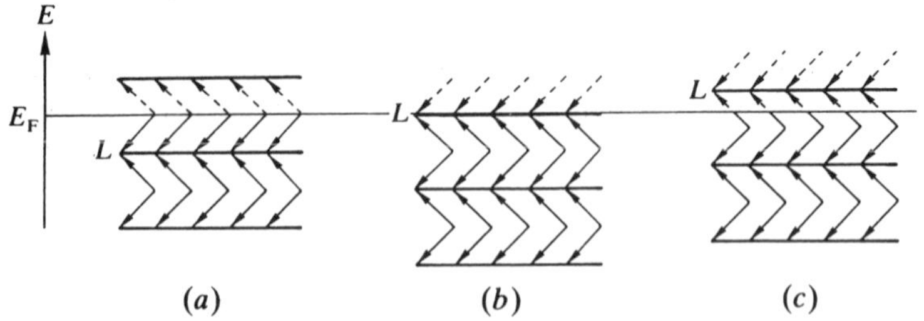

**Figure 9.5** Expressed in terms of an energy-level diagram, we find the discrete Landau levels *L*. The arrows indicate that each Landau level has received states of both higher and lower energy from the initial continuum of free electron states. The broken lines signify empty states. The passage of the Landau levels through the Fermi level causes a ripple in the total energy of the system. This is observable as a periodic change in the diamagnetic susceptibility.

When **B** increases, the rings expand and eventually they leave the Fermi sphere; in so doing they cause the total energy to fluctuate in a regular manner. Figure 9.5 shows a particular level *L* as it approaches and passes through the Fermi level. The energy is least when $E_F$ lies exactly between two Landau levels (*a*). When the level *L* is at the Fermi level, (*b*), the total energy is higher and it continues to increase even when *L* passes through $E_F$, (*c*); then the level *L* empties and there is a sudden return to a state of minimum total energy. This cycle is repeated for every level that passes through the Fermi level.

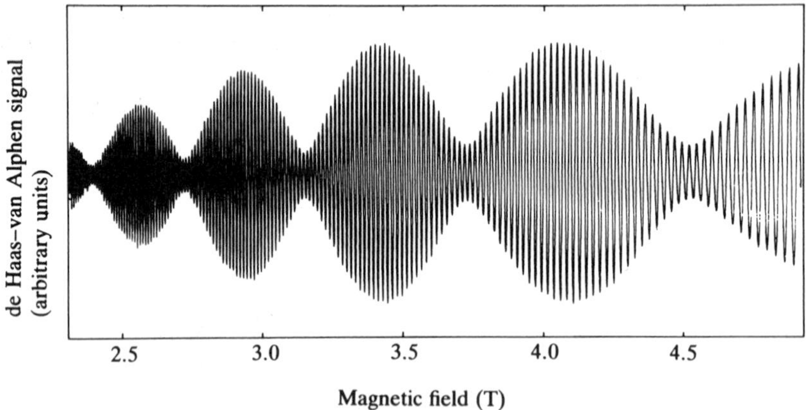


**Figure 9.6** An example of a de Haas–van Alphen signal from an iridium crystal. Among other contributions to the complete Fermi surface, there is an approximately ellipsoidal sheet of d character centred about the X point of the fcc Brillouin zone. When the external magnetic field makes an angle of 1° to the (110) plane this particular sheet (called the $X_1$ sheet) displays two closely similar extremal areas producing two oscillatory signals that give rise to beats. The two extremal areas in question differ by about 3% (there are about 35 oscillations in each beat modulation). The $X_1$ sheet (and a second similar, but larger, sheet denoted $X_2$) is associated with certain d states, and the orbits are hole orbits (i.e. the appropriate bands are almost filled); see Sections 9.6 and 9.7. In addition to these hole surfaces, Ir has two large conventional electron Fermi surface components, one s-like, the other d-like; both are centred around the Γ point. (Courtesy of O. Beckman and S. Hörnfelt.)

throughout **k** space. For free electrons (see Box 9.1)

$$\Delta A_{\mathrm{L}} = 2\pi eB/\hbar. \tag{9.2}$$

We shall derive this expression for the general case below. The period of the oscillations arises whenever $L$ changes by unity; in other words, whenever $\Delta s = 1$ in

$$A_0/\Delta A_L = s, \tag{9.3}$$

or whenever

$$\Delta\left(\frac{1}{B}\right) = \frac{2\pi e}{\hbar A_0}. \tag{9.4}$$

These oscillations in diamagnetic susceptibility and their period in terms of $\Delta(1/B)$ provide absolute knowledge of $A_0$, which is called an extremal area of the Fermi surface. In real metals, a Fermi surface may possess two or more coaxial extremal areas (as in the case of Cu along the $\langle 111 \rangle$ direction, where we find the large equatorial area, but in addition a smaller area around the neck; Fig. 9.1), each contributing its own frequency component to the diamagnetic susceptibility (see Fig. 9.6).

In real metals extremal areas are not necessarily circular, and definitely not so in the transition metals. Furthermore, the relaxation time is not infinitely large, but, provided an electron can complete several cyclotron orbits before being scattered, the cyclotron frequency is well defined as

$$\frac{2\pi}{\omega_{\mathrm{c}}} = \frac{\oint \mathrm{d}k}{\mathrm{d}k/\mathrm{d}t},$$

where $\oint \mathrm{d}k$ is the orbit, not necessarily circular, on the Fermi surface. Just as in the free electron case, there arise Landau levels; they are not circular cylinders, but cylinders adapted to the shape of the actual constant energy contours of the metal in the plane perpendicular to the magnetic field. Even so, the sensitivity to the extremal areas is maintained. Since

$$\frac{\mathrm{d}k}{\mathrm{d}t} = \frac{eBv}{\hbar}, \tag{9.5}$$

we may write

$$\omega_{\mathrm{c}} = \frac{2\pi eBv}{\hbar} \bigg/ \oint \mathrm{d}k. \tag{9.6}$$

Furthermore

$$v = \frac{1}{\hbar}\frac{\mathrm{d}E}{\delta k}. \tag{9.6}$$

We have written $\delta k$ here because we need to distinguish $\delta k$, which is normal to the energy contour, from $\mathrm{d}k$, which is parallel to the contour. Substituting for $v$ leads to

$$\omega_{\mathrm{c}} = \frac{2\pi eB}{\hbar^2}\frac{\mathrm{d}E}{\delta k} \bigg/ \oint \mathrm{d}k. \tag{9.7}$$

But $\delta k \oint \mathrm{d}k$ is $\mathrm{d}A$, an element of annular area in the $(k_x, k_y)$ plane, so

$$\omega_{\mathrm{c}} = \frac{2\pi eB}{\hbar^2}\frac{\mathrm{d}E}{\mathrm{d}A}, \tag{9.8}$$

and, because $E_{xy}$ is quantized, owing to the cyclotron motion, $\mathrm{d}A$ must also be quantized, i.e. the Landau levels are, for given **B**, separated by fixed areas. We may replace $\mathrm{d}E$ by $\hbar\omega_c$ to obtain the quantum

$$\Delta A_L = \frac{2\pi eB}{\hbar},$$

the same value that we found for the free electron gas. Again we see that the period of oscillation of the diamagnetic susceptibility is

$$\Delta\left(\frac{1}{B}\right) = \frac{2\pi e}{\hbar A_0}. \tag{9.9}$$

The complete expression for the diamagnetic magnetization of an electron gas is very complicated. The important aspect is that the magnetization is oscillatory with a period given by (9.9). It is not necessary to determine the absolute value of the diamagnetic magnetization, only its period of oscillation. The single crystal specimen, often quite minute, is placed in a carefully wound compensated pick up coil designed to react only to the specimen's induction. The external field is changed in controlled fashion. The changing diamagnetic induction produces an oscillatory voltage in the pick up coil and is processed into the output signal. In simple terms the procedure is to count the number of periods arising between two well defined values of the external magnetic field, the latter being determined by an auxiliary in situ nuclear magnetic resonance monitor often based on $Al^{27}$ (see Section 15.3). The oscillatory behaviour may be simple or complicated depending on whether one extremal area or several coaxial extremal areas contribute to the output signal. Mechanical vibrations and field inhomogeneities are sources of unwanted noise.

In order to obtain a large relaxation time and a well-defined cyclotron frequency, real samples must be as pure as possible, strain-free and in single-crystal form; single crystals are also necessary of course to isolate particular extremal areas. Scattering by phonons is minimized by working near to or below 1 K. Furthermore, if one determines how the amplitude of the de Haas–van Alphen oscillations varies with temperature, an effective mass for the electron in the particular orbit may be deduced. This effective mass is an average dynamical mass for the particular cyclotron orbit and may depart radically from the ordinary free electron mass. This is not particularly surprising, because the electron in its cyclotron orbit is not moving in free space but in the complicated periodic potential of the lattice, a feature that has not been taken account of explicitly in the analysis. It will also be appreciated that electrons in different orbits, but within the same sample, may exhibit different effective masses. This effective mass is dependent upon how $E$ varies with **k**, and we shall take up the matter again later in this chapter.

It is clear from (9.9) that the larger an extremal area of the Fermi surface, the higher the frequency of oscillation of the susceptibility and the larger the magnetic field needed to resolve it; the alkali metals demand the most stringent experimental conditions on this account, whereas the semimetal Bi (which has a very small number of free electrons, of order $10^{-4}$ atom$^{-1}$) has relatively minute extremal areas and is therefore one of the easiest substances in which to observe the de Haas–van Alphen effect, hence the initial discovery in this material. By measuring extremal areas for different orientations of the sample in the field **B**, the complete Fermi surface may be deduced.

The de Haas–van Alphen effect has a very high relative accuracy, about 1 in $10^4$, but the absolute accuracy is limited by the need to know the absolute value of the magnetic field. Since the introduction of the method in 1960, it has been applied to a wide variety of metals, particularly the 'sp' metals, as well as Cu, Ag and Au. The experimental results showed, in a most convincing manner, the nearly free electron character of the valence electrons in the simple metals, thereby justifying the 'one-electron approximation' and speeding the development and widespread use of pseudopotential theory for these metals.

The data for the coinage metals were also very significant because they finally answered the question as to whether or not the Fermi surfaces of these metals contact the zone boundaries – a question of relevance to theories of alloy formation by these metals.† Today it is difficult to appreciate how important this question appeared at that time. The past forty years has seen an enormous growth in our knowledge of the band structure of pure metals, a consequence of the parallel development of experiment and theory.

The transition metals pose greater problems. They are difficult to obtain in high purity (especially the rare earth metals) and, as we shall see later, they have inherently short relaxation times, and furthermore their Fermi surfaces are often extremely complicated. Extensive data exist for the *d* metals, but only very limited results are available for the rare earths; the reader is referred to the references already cited for further detail.

The de Haas–van Alphen effect is certainly the most important tool for studying the Fermi surface, but not the only one. The fact that both the total energy and $N(E_F)$ fluctuate in the presence of a changing magnetic field influences other properties that provide direct or indirect information about the Fermi surface. Similarly, by the observation of resonant absorption of microwave radiation in the presence of a magnetic field, the cyclotron frequency may be determined. Although of limited application to metals (because of the radio-frequency skin effect), the method is very useful for semiconductors and permits the measurement of the effective masses of the charge carriers in these substances.

Although the de Haas–van Alphen effect has been observed in certain alloys ($\ll 1\%$ of alloying element), the method, being so dependent on a long relaxation time, is totally unsuited to conventional concentrated alloys. However, in highly stoichiometric and ordered intermetallic compounds the relaxation time becomes large again and the effect observable; much attention is being devoted to these materials at the present time.

The need to obtain information about chemically disordered alloys is, to a certain extent at least, met using the technique of positron annihilation. This method, which we shall not describe in detail, is not dependent on a large relaxation time for the valence electrons. Positrons from an external source, e.g. $^{22}$Na, $^{58}$Co or $^{64}$Cu, are introduced into a single-crystal sample. Their initial energy is of order 100 keV and they penetrate several millimetres into the sample where they quickly (in about $10^{-12}$ s) attain thermal energy (i.e. an energy appropriate to the sample temperature, usually room temperature). The positron lifetime in a metal is of order $10^{-10}$ s and within this time certain positrons combine with electrons in mutual

† The credit for this discovery goes to Pippard, who first established the Fermi surface of a metal (Cu) using the anomalous skin effect.

annihilation, producing two $\gamma$-ray photons in the process. Energy and momentum are conserved. The positron has essentially zero momentum compared with that of the electron, so the momentum balance carried away by the two photons is associated wholly with the electron. Owing to momentum conservation, the two photon paths differ by a small angle (about 1 mrad) and the objective is to measure the angular correlation of the $\gamma$-ray photons. This is difficult to do in an accurate manner because a compromise must be made between signal strength and angular resolution. The data obtained reflect the momentum density of annihilating electrons in planar slices (or, in certain cases, lines) of the complete distribution of occupied states within the Fermi surface. These experimental momentum density distributions may then be compared with those derived from calculated electron band structures. It is, however, possible to identify directly certain prominent features such as the occurrence of the necks in Cu as well as to assess the sharpness of the Fermi surface.

Thus it is found that when Zn is alloyed with Cu, the Fermi surface is still clearly identifiable, being only about 10% broader than in the pure metal. The addition of Zn increases the valence electron concentration, producing a simple expansion of the Fermi surface. Another useful feature of positron annihilation is its sensitivity to the presence of point defects, particularly lattice vacancies, which produce characteristic features in the angular correlation distribution.

## 9.2 Thermal Properties

Measurements of the lattice heat capacity provide little direct information about the electronic structure of matter. Even a detailed knowledge of the phonon dispersion relations is only of indirect help. Whereas one may test a theory of electronic structure by calculation of the phonon spectrum and subsequent comparison with experimental data, it is not possible to work backwards from the experimental results and make specific conclusions about the electronic structure, except in a few rare cases that lead to information about the Fermi surface. The electronic heat capacity, however, is an important source of information regarding the density of states at the Fermi level, $N\ (E_F)$, but its usefulness is somewhat marred by the influence of one or more of the following sources of enhancement: electron-phonon, electron-electron and magnetic interactions. The measured electronic heat capacity is larger than it would be in the absence of these interactions, and unless their effect can be calculated a straightforward use of (6.25) provides us with a value of $N(E_F)$ that may be as much as twice the true value. In spite of this disadvantage, the electronic heat capacity constant $\gamma$ is a good indicator of $N(E_F)$ and how this quantity varies when a metal is alloyed.

We can now understand the large $\gamma$ values measured for the transition metals, Section 6.3.3; they arise owing to the large partial density of d states at the Fermi level, due to the narrowness of the d band and its large electron content. We illustrate this situation in Fig. 9.7, where, in a schematic manner, we compare the d bands in Pd and Ag. Pd is in fact strongly paramagnetic and on the brink of being a ferromagnetic material. The strong magnetic interactions produce an enhanced electronic heat capacity and, although the measured $\gamma$ of 9.4 mJ mol$^{-1}$ K$^{-2}$ implies that $N(E_F)$ is 4 eV$^{-1}$ atom$^{-1}$, it is considered that the true value is nearer to 2 eV$^{-1}$ atom$^{-1}$; but this is still much larger than that for pure Ag (about 0.25 eV$^{-1}$

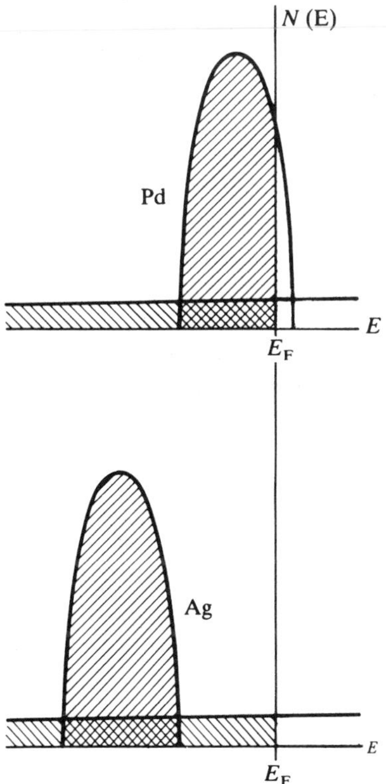


**Figure 9.7** Stylized band forms for Ag and Pd, showing the difference between d and sp bands. In Pd the d band is not quite filled and $N(E_F)$ is very large, about 2 $eV^{-1}$ $atom^{-1}$, whereas in Ag this quantity is only about 0.2 $eV^{-1}$ $atom^{-1}$.

$atom^{-1}$). It is instructive to plot $\gamma$ for the three transition metal series as is done in Fig. 9.8. We note the striking similarity in the patterns of the three series, which can only mean that the three d bands have many features in common. This similarity is also borne out in the variation of the magnetic susceptibility, the latent heat of vaporization and the cohesive energies of these elements. One can also study the electronic heat capacities of alloys of neighbouring elements and in this way obtain an indication of the variations in the d band form as a function of electron content. This is illustrated in Fig. 9.9, which also shows in a simple manner how $\gamma$ goes through a maximum when Pd is alloyed with Rh. This means that in pure Pd a maximum in $N(E)$ occurs just under the Fermi level, in agreement with calculation.

One might have expected the rare earth metals to present rather similar $\gamma$ values because the band structure is not expected to show much variation from metal to metal. However, their magnetic character together with their nuclear spins leads to large specific heats on account of hyperfine interactions; although these provide important information, they prevent accurate measurement of the electronic heat capacity. The observed values of $\gamma$ are of order 10 mJ $mol^{-1}$ $K^{-2}$. The above discussion concerned transition metals, for which the $\gamma$ values are large and readily measurable. The experimental data confirm our belief that the density of d states at the Fermi level is large, but may vary markedly from element to element. We conclude that, on the whole, they behave as expected and according to our ideas about their band structure as described in Chapter 7.

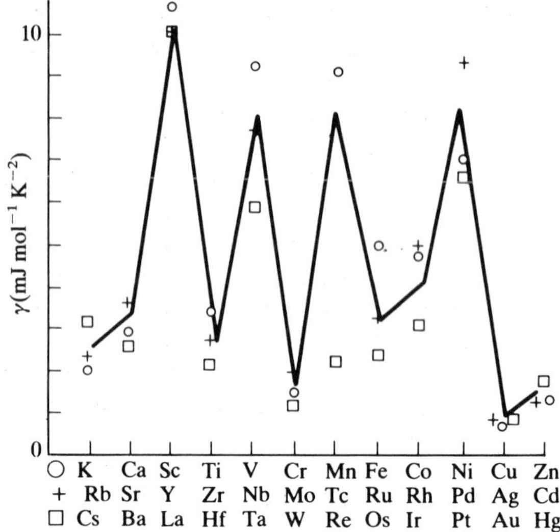


**Figure 9.8** The variation of the coefficient of the electronic heat capacity $\gamma$ through the three series of 'd' transition metals. Note the markedly similar patterns of variation, implying that the d sub-bands have the same general form in the three series of elements.

Earlier we saw that the simple metals have $\gamma$ values in reasonable agreement with nearly free electron theory. What about alloys of the simple metals or those formed by adding a polyvalent metal like Zn or Ga to Cu? There are no surprises here. Because the $\gamma$ values of the simple metals are small, of order 1 mJ mol$^{-1}$ K$^{-2}$, the changes on alloying are also very small, difficult to determine accurately and of limited significance from the band structure point view.

At this point we therefore digress a little to describe how certain alloys containing d or f electron solute atoms can produce very large contributions to the electronic heat capacity. We first consider an **AlMn** alloy and ask what happens to the 3 d electrons when an Mn atom arises as a solute atom in Al. We begin by assuming

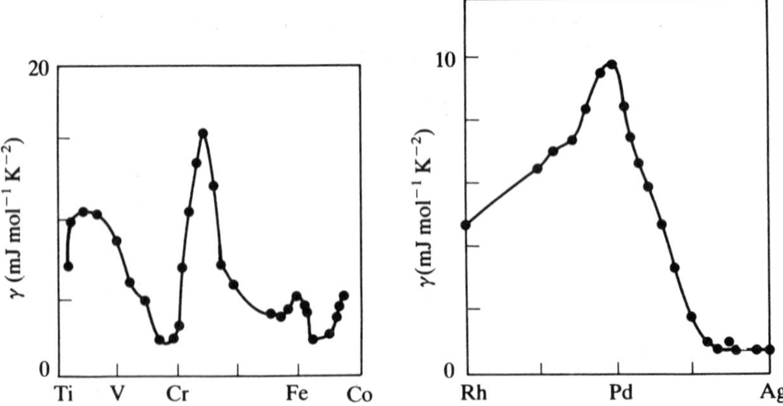


**Figure 9.9** By measuring the electronic heat capacities of alloys of neighbouring elements, one obtains an impression of how $N(E_F)$ varies with electron content and thereby an indication of the band form. (After Cheng *et al.* 1960, 1962 and Hoare and Yates 1957.)

that there are so few Mn atoms present that each may be considered to be wholly surrounded by Al atoms – this is the single impurity situation. In pure Mn metal, as in other transition metals, we have seen that the 3d band arises as a result of a strong $d_i$-$d_j$ resonance modified by a weaker but significant s-d resonance. The Al atom has no occupied d states, so the Mn d state can only resonate with the valence electron gas that permeates the whole alloy. This leads to a very narrow s-d resonance – in other words a broadened atomic d state. Since there are only 5–6 electrons in the Mn d state, this resonance must straddle the Fermi level, producing an energy level structure similar to that of Fig. 9.10. This Mn d-state resonance has a width of order 1 eV, much smaller than that of the d band in Mn metal, leading to a very large density of d state on every Mn atom. A similar situation arises with a transition metal other than Mn, and only the occupancy of the level changes to correspond to the different number of d electrons on the impurity atom. Thus if Cu were the element added to Al we should find the resonant 3d state well below the Fermi level so that it might contain all ten d electrons (or very nearly so).

Experiment has provided ample evidence for the overall correctness of the above picture. In these alloys the high density of resonant d state at the Fermi level gives rise to $\gamma$ values in the range 20–50 mJ $K^{-2}$ *per mole of transition metal.* The point we wish to make is that there is a much larger density of d state on the transition metal atom as a result of the localization of the d states to the atom and because there are no d-d interactions. If we were to increase the Mn concentration of our AlMn alloy then we should soon find Mn-Mn nearest neighbours, the d-d resonance would become operative in addition to the s-d resonance and the d levels would soon broaden into a conventional d band (but one that is still confined to the Mn atoms and avoids the Al atoms). When the d band broadens, $N(E_F)$ decreases, corresponding to a lower $\gamma$ value.

We might therefore expect $\gamma \approx 50$ mJ $mol^{-1}$ $K^{-2}$ to be close to the absolute maximum value for this quantity. It was therefore surprising when values in the range 0.1–1 J $mol^{-1}$ $K^{-2}$ were found recently in certain alloys that are now called 'heavy-fermion alloys' or 'heavy-electron alloys'. The adjective 'heavy' is used because the high $\gamma$ value implies a large thermal effective electron mass of 100–1000 times the ordinary free electron mass. That the dynamical mass is also really so large has been confirmed by de Haas–van Alphen measurements. Some alloys with these extraordinary properties are $UBe_{13}$, $UPt_3$, $CeAl_2$, $CeAl_3$ and $UAl_2$. We shall return to them in Section 11.4.3.

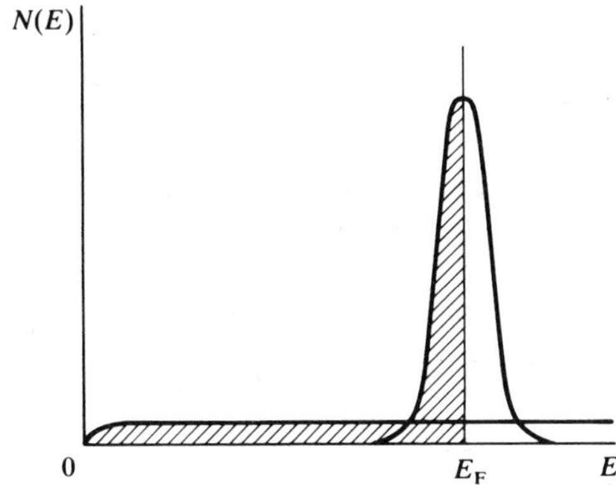


**Figure 9.10** The density of electron states associated with a localized 'd' electron resonance.

## 9.3 Magnetic Susceptibility

With the exception of the lightest, the simple metals are diamagnetic because the diamagnetism associated with the filled shells of the ion cores outweighs any Pauli paramagnetism of the conduction electrons. The transition metals, on the other hand, usually show a strong Pauli paramagnetism, which, as for the electron heats, originates in the large densities of d states at the Fermi level. Again we see the similarity of behaviour in the three series of d elements (Fig. 9.11).

The paramagnetic susceptibility usually shows a weak temperature dependence, which is associated with the large values of $|dN_d(E_F)/dE|$ that arise in the band structures of these elements.

The occurrence of ferromagnetism in Fe, Co and Ni may be regarded as arising from the strong atomic character of the uppermost d bands. As the composite d band of the first long series fills, it also narrows and the uppermost d band has very little dispersion. The more atomic character a d band has, the more likely it is that a magnetic moment will arise. The conditions appear to be best met in Fe, Co and Ni. In the second and third series the d orbitals have larger diameter and overlap more, so the occurrence of ferromagnetism is not favoured. These properties are discussed in more detail in Chapter 11.

## 9.4 Spectroscopic Studies

We have already seen that a free electron gas produces a characteristic optical absorption governed by the Drude equations (6.48) and (6.49). The valence electrons occupying unfilled bands in real metals also exhibit 'free electron' absorption, but this is usually called 'intraband' absorption because the excited electrons remain within the conduction band. However, just as in free atoms or ions, electrons in metals may absorb photons and experience quantum jumps whereby they leave the

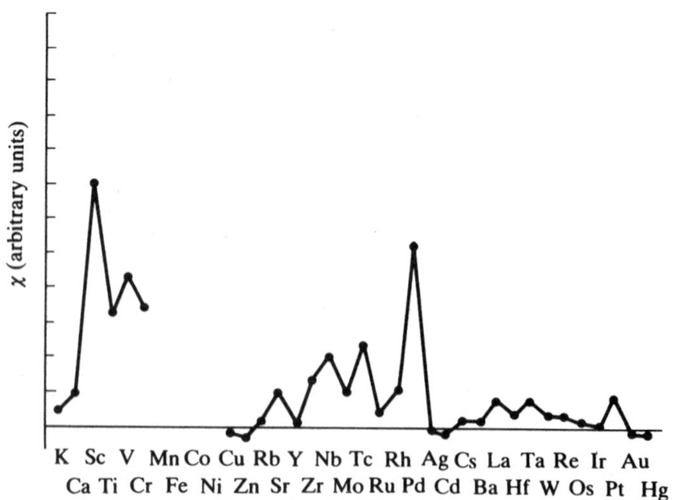


**Figure 9.11** The magnetic susceptibility of the 'd' transition metals also shows a characteristic variation through the three series of elements, but not in such a marked manner as for the electronic heat capacities.

initially occupied state $\psi_i(\mathbf{r})$ with wave vector $\mathbf{k}_i$ in one band to occupy a final state $\psi_f(\mathbf{r})$ in a higher lying empty or incompletely filled band, this is the interband absorption transition.

As is always the case, energy and wave vector must be conserved in both intraband and interband transitions; in particular

$$\mathbf{k}_f = \mathbf{k}_i + \Delta\mathbf{k}.$$

It is not difficult to see that $\Delta\mathbf{k}$, the photon wave vector, is, on the scale of the Brillouin zone, a very small quantity approaching zero. For a visible photon with energy $\sim 2$ eV ($\lambda \sim 600$ nm) we find $|\Delta\mathbf{k}| = 2\pi/\lambda \sim 10^7\ \mathrm{m}^{-1}$, whereas the Brillouin zone boundaries lie at $10^{10}\ \mathrm{m}^{-1}$; $|\Delta\mathbf{k}|$ is therefore insignificant and may safely be considered equal to zero. How then is it possible to conserve both $E$ and $\mathbf{k}$ in the optical transitions?

In the intraband transition $k_f - k_i \ll \pi/a$. Phonons have $0 \leqslant q \leqslant \pi/a$, furthermore they have very small energies ($\sim 10^{-2}$ eV). They represent a cheap source of crystal momentum; so the phonons, together with lattice defects (which have their own vibratory modes), maintain the wave vector balance in the intraband transition.

The interband transition is very different because the transition probability is finite only when

$$\mathbf{k}_f = \mathbf{k}_i \pm \mathbf{G},$$

a situation that may be illustrated in either the extended or the reduced zone scheme (Fig. 9.12). In the reduced zone scheme the transition becomes vertical and it is customary to call interband transitions vertical transitions. Such a transition demands a change in crystal momentum in units of $\hbar\mathbf{G}$ which is provided by the lattice. Recalling our earlier discussion of 'n' and 'u' processes in Section 5.12 we may describe the interband transition as an umklapp process. Phonons are not involved in the interband transition although they can produce secondary effects that are significant for certain semiconductors (see Section 10.2). The interband

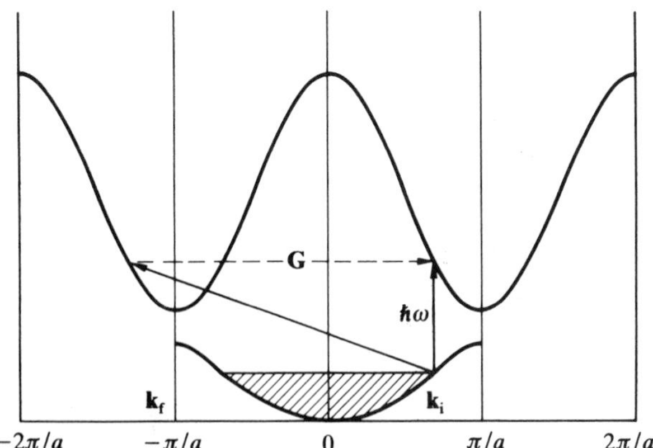


**Figure 9.12** Interband optical absorption demands a change in **k** vector for the excited electron, but photons have essentially zero wave vector compared with electrons in metals. In perfectly periodic structures, the only source of crystal momentum is the lattice itself, which acts as a momentum buffer in units of $\hbar\mathbf{G}$. The only interband optical transitions that are allowed are therefore of 'umklapp' character. In the reduced zone scheme this means that only vertical transitions are allowed.

optical conductivity may be expressed as

$$\sigma = \frac{c}{\omega^2} M^2(\text{JDS}), \tag{9.10}$$

where $c$ is a constant, $M$ is a matrix element for the electronic transition and JDS is a joint density of states. The absorption depends on the number of pairs of states per unit energy interval that are separated by the photon energy $\hbar\omega$; we call this the *joint or interband density of states.* We shall see that in certain situations, this quantity may have singular behaviour leading to characteristic absorption bands. The alkali metals (Fig. 9.13) have a simple interband spectrum with a weakly marked edge determined by the excitation of electrons from the Fermi level to the nearest empty state vertically above it. The energy gap at the zone boundary has little influence on this edge, and its position may be readily estimated from the empty lattice construction. The Drude or intraband contribution is easily recognized and allows an estimate of $\hbar\omega_p$ (Table 9.1).

These may be compared with the estimates given in Table 6.1. Furthermore, experiments show that the alkali metals become transparent to light when $\omega > \omega_p$, in keeping with our discussion in Section 6.5. Clearly, it is also possible to excite electrons in the conduction band (as well as occupied core states) to all the higher

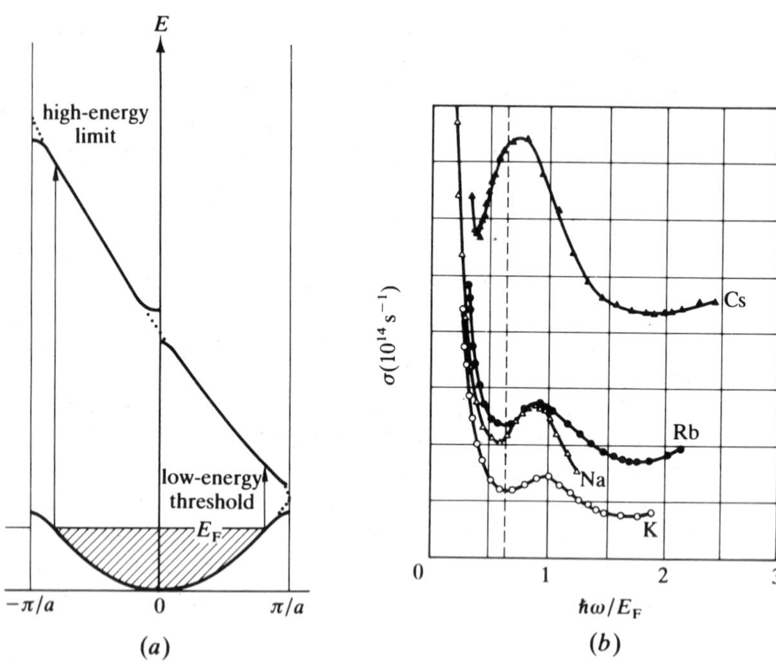


**Figure 9.13** Alkali metal spectra. In (a) three bands are shown for a nearly free electron metal. The band gaps have very limited effect on the $E(k)$ relation. In an alkali metal only the first band contains electrons, and the threshold for vertical transitions is shown as an excitation from the Fermi level. All the electrons contribute to the absorption spectrum and for the energy bands shown there would be a high-energy absorption limit: this again is a transition from the Fermi level. In (b) the observed spectra for four alkali metals are shown. The photon energy is normalized with respect to the Fermi energy. We notice the characteristic Drude region at low photon energies and a marked interband edge close to 0.64 $E_F$ (see Problem 9.6). (After Smith 1970.)

**Table 9.1** Plasmon energies (eV) for the alkali metals

| | Li | Na | K | Rb | Cs |
|---|---|---|---|---|---|
| $\hbar\omega_p$ | 6.9 | 5.7 | 3.8 | 3.4 | 3.0 |

lying empty bands, but the excitation energies are so large that they take us well outside the conventional 'optical' range, so we only consider here the excitations within say $\pm 5$ eV of $E_F$. The polyvalent simple metals have a more pronounced spectrum than the alkali metals because their Fermi surfaces cut the zone boundaries. To appreciate the significance of this event, we must consider the energy contours lying in the zone boundary plane. These contours are those formed by the intersections of the necks with the zone plane, and in the simple metals they are circular. Similarly the segments of the contours lying in the second zone also make circular intersections in the zone face. Thus, on the zone boundary plane, we have two sets of circular energy contours, one set lying in the first zone and the other in the second zone.

For a given value of **k** lying in the zone plane, these contours are separated by the energy gap specific to the zone boundary. The circular character implies that $E$ varies parabolically with the component of **k** lying in the zone boundary: this we call '**k**-perpendicular', $k_\perp$ (Fig. 9.14). The two bands lying in the zone face are parallel and separated by the energy gap. This means that as soon as the Fermi level lies within or just above the band gap, we can obtain many vertical transitions between filled and empty states that are all separated by $E_g$.† It can in fact be shown that the behaviour becomes singular, leading to an infinitely sharp absorption edge. In practice the edge is blunted and appears as a narrow band as shown for Mg in Fig. 9.15. Absorption of this kind occurs in all the polyvalent metals, but not always in such a marked way as for Mg. We immediately see that this provides us with a measure of the band gap, and the parallel band absorption, as it is usually called, is a valuable if somewhat restricted monitor of changes brought about by alloying or by temperature for example. Thus it can be demonstrated that the band gap decreases in size as the temperature is increased.

In Fig. 9.16 we again show data for Al at three temperatures, one of which is above the melting point. We notice how in the solid phase an increase in temperature broadens the band and moves it to slightly lower energy. In the liquid phase it has disappeared – as expected because there is no zone structure, only the free electron-like Drude absorption remains. It is seldom the case that the optical spectrum of a metal can receive such a simple and direct interpretation. Usually we need access to an accurate band structure diagram if a spectrum is to be interpreted satisfactorily.

Much attention has been directed to the coinage metals, and their optical spectra are shown in Fig. 9.17. They are characterized by distinct regions of intra- and interband absorption, the latter being noteworthy for the distinctive edges that arise at 2, 4 and 2.2 eV in Cu, Ag and Au respectively. The edges cause the characteristic colours of these metals and they result from transitions from the top of the d band (which, as we have already noted, is very flat) to the Fermi level.

† Consult the band diagram for Al (Fig. 7.20) and note the parallel bands along the line XW for example.

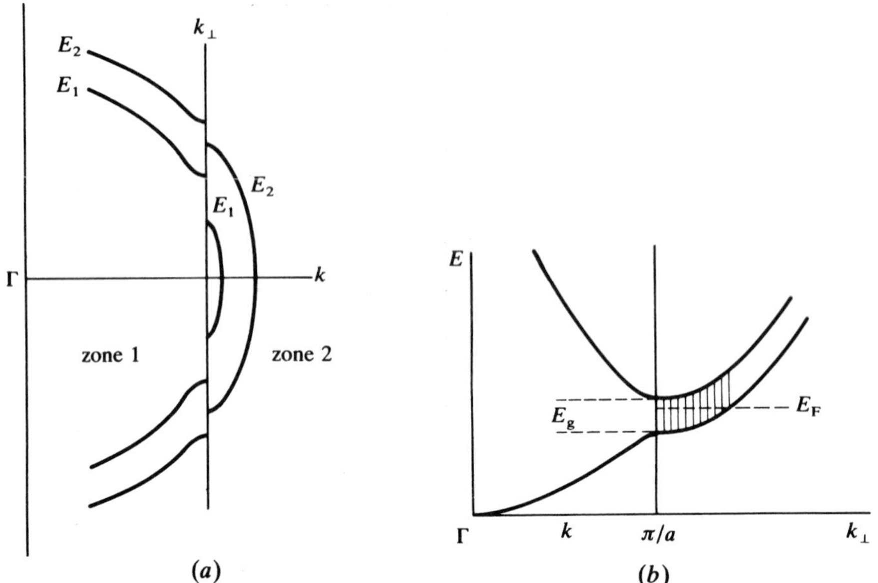


**Figure 9.14** For a polyvalent nearly free electron metal, the Fermi surface cuts the zone boundaries. Then, in a particular direction, the spherical energy contours become discontinuous at the boundary (a). We consider the variation of $E$ with $k$ in the zone boundary plane, i.e. with respect to $k_\perp$. There are two values of energy for any given $k_\perp$, one in the first and one in the second zone, the latter being at a higher energy by an amount equal to the energy gap. There are two parallel parabolic energy bands separated by $E_g$, (b). When the Fermi level enters the energy gap, or lies immediately above it, many interband transitions at the same photon energy can arise. This causes a strong interband absorption peak at the gap energy. All the polyvalent simple metals show absorption of this kind in more or less marked form.

We mention further detail only for Cu; there are also transitions from the bottom of the d band to the Fermi level (5.2 eV), as well as across the gap in the zone face near $L$ (4.8 eV). These latter two transitions combine to produce the large peak at about 5 eV. Note that we hereby obtain an estimate of the d-band width, 3.2 eV (Fig. 9.18). The intraband or Drude component of the dielectric constant can be analysed to provide an optical effective mass. We find this mass to be 1.45, 0.85 and 1.12 for Cu, Ag and Au respectively; how do these compare with the thermal masses?

An interesting feature of the spectrum of the coinage metals in the liquid state is that the d-band edge is retained, demonstrating that, even in the liquid, d electrons occupy a narrow band separate from the sp conduction states. From this we obtain an illustration of the tight binding character of the d states. Optical studies of alloys based on Cu with Ag or Au have provided important information concerning the electronic structures of alloys.

The d transition metals have rather featureless spectra that so far have provided limited information. The rare earth metals, on the other hand, have dominant features, but these have not received detailed interpretation. Optical data for the actinides Th and U have recently been obtained and support the contention that the 5f electrons are itinerant in these metals.

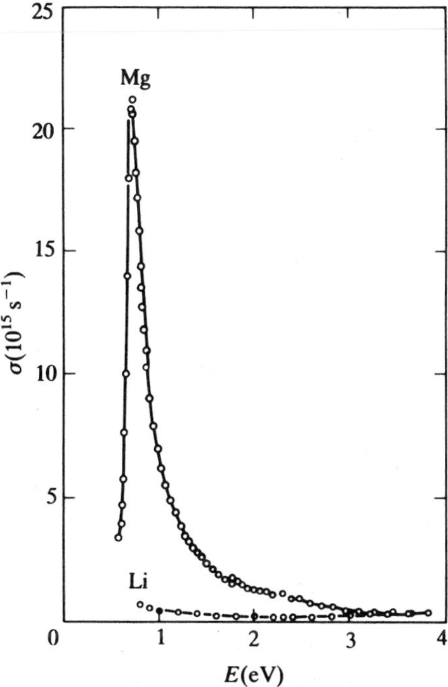


**Figure 9.15** Parallel band absorption in the $(1\bar{1}01)$ zone boundary plane of hcp Mg. Other bands arise at lower photon energies. Li has one fewer valence electron than Mg, and the lack of zone contact means that Li has a typical alkali metal spectrum. As shown here, the Li spectrum appears featureless, but this is on account of the scale used – compare with Fig. 6.14. (After Mathewson and Myers 1973.)

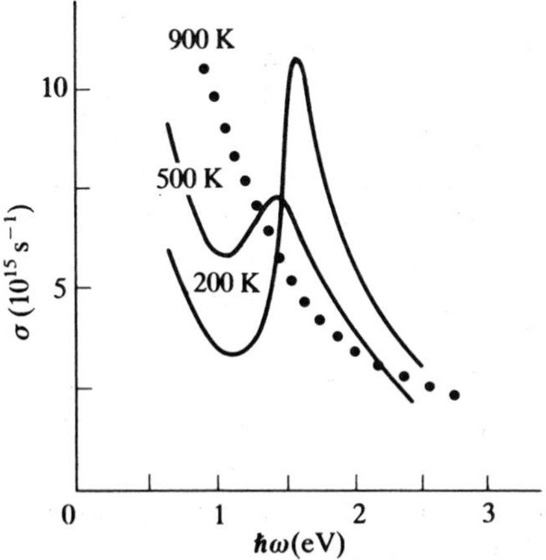


**Figure 9.16** In Al there are two parallel band absorption peaks owing to Fermi surface contact with the (200) and (111) zone boundaries. The former is at 1.5 eV while the latter, which is rather weak and not discernible in the figure, is at 0.4 eV. An increase in temperature broadens the absorption peaks and shifts them to lower energies (the band gaps become smaller). In liquid Al the bands are absent since there is no zone structure. (After Mathewson and Myers 1972 and Miller 1969.)

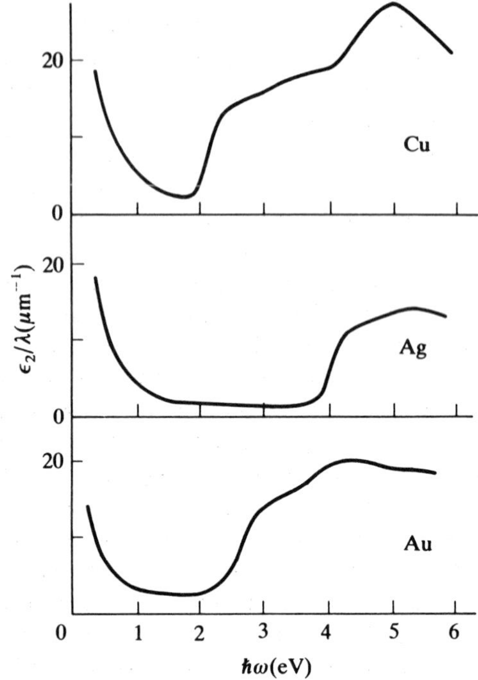


**Figure 9.17** In the coinage metals the d band is filled and therefore lies below the Fermi level. The optical spectra are dominated by the Drude absorption at low energies. The interband absorption begins abruptly at 'edges' that arise primarily in transitions from the flat top of the filled d band to the empty states just above the Fermi energy. At still higher photon energies, spectral structure develops on account of several vertical transitions between filled and empty states; see Fig. 9.18 for more detail.

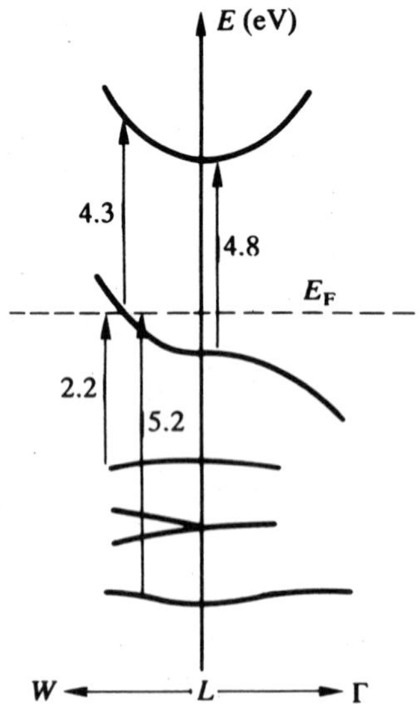


**Figure 9.18** Optical transitions arise at all **k** vectors throughout the reduced Brillouin zone, but they are often intense near symmetry points. In Cu, Ag and Au the *L* point is significant, and this figure shows the transitions that arise in Cu between the occupied states at the top and bottom of the *d* band to the Fermi energy as well as between states in the first (occupied) and second (unoccupied) conduction bands.

If, for a particular metal, we know the band structure throughout the zones (and not just along the major symmetry directions) then we can calculate the density of states as a function of energy. The latter is an integrated quantity and its experimental determination cannot lead us back to the dispersion relations. For many alloys, compounds, pure metals such as the transition metals and especially disordered or amorphous substances the integrated band form is of primary significance, and therefore much effort has been devoted to its study by soft X-ray emission and X-ray photoemission spectroscopy. We concentrate on the latter; a diagram of apparatus suitable for the purpose is shown in Fig. 9.19. Monochromatized Al $K\alpha$ X rays eject electrons from a sample and their kinetic energy is determined using an electrostatic analyser. If we assume that a significant number of the emitted electrons lose no energy while escaping from the sample (as is the case) then we expect the electron current at a particular kinetic energy $E_{KE}$ to provide us with a measure of the density of electron states at an initial state lying at an energy of $\hbar\omega - \phi - E_{KE}$ below the Fermi energy (see the earlier discussion in Section 6.3.2).

Our belief that the photocurrent provides information about the initial state (before the photon ejects the electron) resides in the fact that the X ray has a photon energy of about 1500 eV and the outer electrons lie $<20$ eV below $E_F$. When ejected they are therefore excited to very high energies $>1400$ eV and occupy states (within the metal) that comprise part of an essentially uniform continuum of band states. Any rapid variation in the photocurrent is therefore associated with rapid variations in the density of initial states. These measurements yield the position and general shape of an occupied band as well as its width (Fig. 9.20).

The X-ray photon penetrates the bulk of the sample, usually an evaporated film or thin foil in the case of elemental metals, and the energetic photoelectrons have an

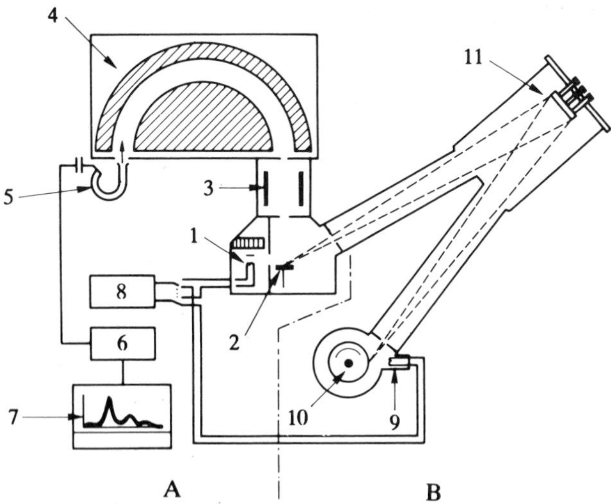


**Figure 9.19** The principal components of an X-ray photoemission spectrometer: (1) conventional water-cooled X-ray source for Al $K\alpha$ X rays; (2) specimen; (3) retarding lens for photoemitted electrons; (4) hemispherical electrostatic analyser for photoemitted electrons; (5) channeltron detector; (6) computer; (7) plotter; (8) electron gun power supply (coupled to 1 or 9); (9) electron gun for rotating water-cooled Al anode X-ray source; (10) rotating water-cooled cylindrical Al anode; (11) quartz-crystal monochromator for Al $K\alpha_1$ X rays. Part A comprises a conventional instrument and Part B is a monochromatic X-ray source that enables higher resolution to be obtained. (After Baer *et al.* 1975.)

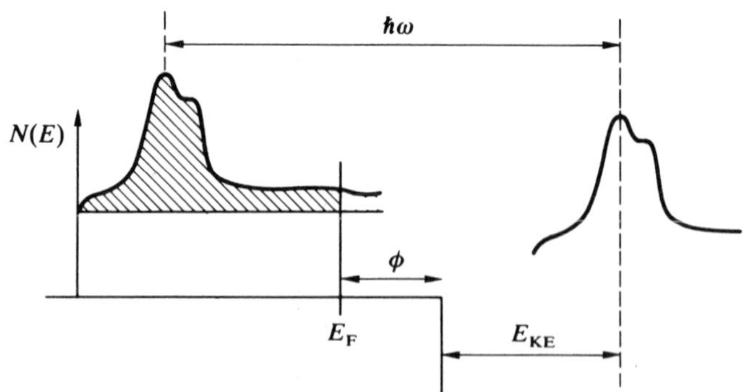


**Figure 9.20** Principles of X-ray photoemission. The kinetic energy of the photoemitted electron is measured, and the photocurrent as a function of the kinetic energy mirrors the energy distribution of the states of origin (the initial states) in the sample.

escape length of about 20 Å; it is therefore thought that they reflect the properties of the bulk material rather than those of the surface layers.

An accurate result for a Cu sample together with a calculated band form is shown in Fig. 9.21, the agreement is very striking. We particularly notice the shallowness of the sp part of the band compared with the d component. The calculated d-band envelope contains much sharper structure than is observed in practice, but it must be remembered that the resolution of the experiment is limited to about 0.35 eV (which is very good); furthermore band structure calculations are usually for 0 K and the effects of thermal vibrations and other smearing processes are not taken into account. Again we see that for Cu the total band width is about 3.5 eV, in agreement with our earlier estimate. Data for Cu, Ag and Au are shown together in Fig. 9.22. The 5d band for Au is much wider than the 3d band in Cu. Au is a very heavy metal and we know that spin-orbit coupling is significant in elements with large atomic number; thus the splitting into the $j = \frac{3}{2}$ and $j = \frac{5}{2}$ states is an important factor for the d states in Au, leading to a much broader band. It would be

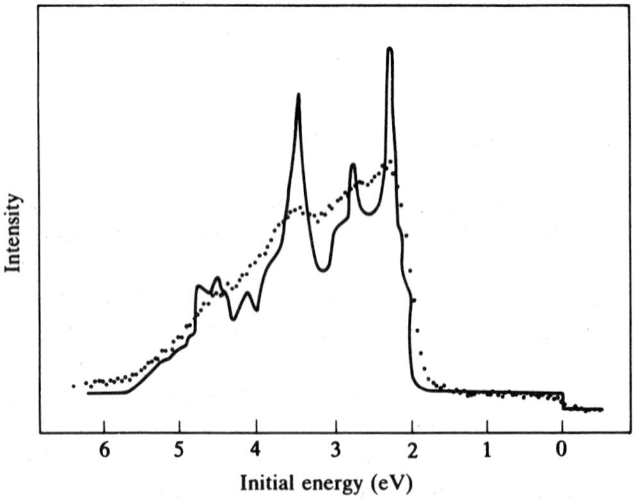


**Figure 9.21** Experimental data (· · ·) for the d band of Cu together with the calculated band form (——). (Experimental data courtesy of Y. Baer; calculated band form after Mueller 1967.)

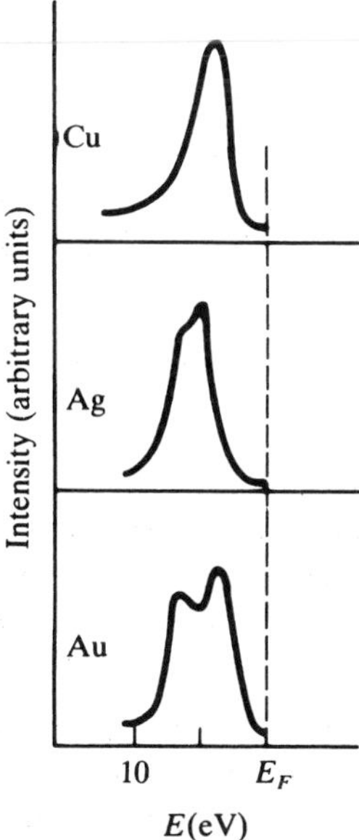


**Figure 9.22** The d bands of the coinage metals as determined by conventional X-ray photoemission spectroscopy. The zero of energy is at the Fermi energy.

wrong, however, to associate the marked double peak directly with this splitting because the separation of the double peak is much larger than the spin-orbit splitting of the free atom.

Most of the transition metals have been studied by this technique, and good agreement with regard to band form and width is found between experiment and theory. Also there are data for the rare earth metals providing information on the shape of the conduction band and the position of the 4f levels; these data are, however, complicated by the fact that after ionization the 4f shell may take one of several configurations, leading to a complex spectrum which is nevertheless very well understood.

## 9.5 Electronic Structure of Certain Alloys

Cottrell (1967) provides the following definition: 'An alloy is a metallic solid or liquid formed from an intimate combination of two or more elements. Any chemical element may be used for alloying but the only ones used in high concentrations are metals.' The various phases (crystal structures) adopted by alloys are dependent on composition, temperature and pressure. Normally behaviour at atmospheric pressure is of interest. A binary alloy phase diagram indicates the equilibrium structures that arise in a two component system for different compositions and temperatures. Such a diagram may be simple or very complicated. If an alloy system contains three or more components complication is increased. The thermodynamic background to the phase diagram is described by Cottrell (1967).

Of necessity, present interest is directed to primary substitutional solid solutions of one metal B in another metal A, the alloy having the same crystal structure as that of metal A. It is assumed that the components are randomly distributed on the lattice sites. Such a random distribution implies that there is no preference for the formation of AA, BB or AB pairs, which is very unlikely. There will usually be some tendency to short range order (AB pairing) or clustering (formation of AA and BB pairs); these tendencies become less likely the higher the temperature and they can be avoided by suitable heat treatment. Sometimes, when A and B have the same crystal structure, it happens that solubility exists over the complete composition range from 100% A to 100% B; such a system is said to be miscible. Certain alloys may, at simple compositional ratios, e.g. AB, $A_3B$, display long range order. The basic lattice structure is not affected but each of the atoms A and B occupies its own substructure (see Fig. 9.39). These superlattice structures are usually destroyed at moderate temperature ($>500°C$). It is more often the case that primary solid solutions have limited compositional stability and intermediate phases with specific crystal structures arise. Sometimes intermediate phases are stable only over a very restricted range of composition associated with simple concentration ratios like CuZn or $Ni_3Al$.

Furthermore, the different atoms are not randomly distributed but occupy preferred lattice sites in an ordered fashion and many are stable at high temperatures. They are similar to a chemical compound and are therefore called intermetallic compounds. Many such intermetallic compounds are discussed by Westbrook (1967); see also the examples in Fig. 2.13.

No overriding conclusions can be drawn from the enormous amount of data contained in the available alloy phase diagrams, although the behaviour of related alloy families can be systematized. Similarly the crystal architecture of certain intermetallic compounds, e.g. the Laves phases (see Fig. 2.13) as well as many others may be understood in terms of atomic size. Perhaps the best known empirical relationships are those associated with Hume-Rothery who, on the basis of extensive studies of alloys based on the coinage metals with the sp polyvalent metals, concluded that important factors are:

*Atomic size:* If the atomic diameters of the constituents differ by more than 15% then solubility is not favoured.

The atomic diameter may be estimated in different ways. A suitable one is to take the diameter of the mean spherical atomic volume found in the pure component: this avoids consideration of coordination numbers.

*Electrochemical factor:* The greater the difference in electrochemical character, the greater the tendency for compound formation and the more restricted the primary solid solubility.

*Relative valence factor:* Metals of low valence tend to dissolve metals of higher valence.

Transition metals are excluded from the above considerations involving valence because it is difficult to decide the valence of a transition metal in the metallic state.

An alloy in the form of a primary solid solution or intermetallic compound represents a more complicated system than the pure metal, and the question arises of the extent to which the ideas described in Chapter 7 can be successfully applied to it. Thus one can ask whether there is a well defined electron energy band structure, together with Brillouin zones and Fermi surface and so on. Later, experimental evidence will be provided to support the present contention. For the moment we maintain that the theory already described for a pure metal is, qualitatively at least, applicable to the alloy, but with one important exception. There is, in the chemically disordered alloy, no proper translational symmetry. There is an underlying lattice structure and Bragg reflection of X rays is readily observed (otherwise there would be no knowing the structure of intermediate phases), and therefore the reciprocal lattice and the Brillouin zone can be deduced. Nevertheless, chemical disorder implies lattice potential disorder and the usual argument is that this disorder causes increased scattering of the valence electrons and an uncertainty in the wave number of the electron state, implying the absence of the Bloch function. On the other hand, if Bragg reflection of X rays occurs, why should it not arise for the valence electrons in the alloy? One might expect some blurring of the Fermi surface but experimental evidence, for certain sp metal alloys at least, is that the effect of alloying is not so damaging to the principal concepts of band theory as might at first be thought. Lattice chemical disorder is a much weaker form of disorder than that typical of the amorphous or liquid state where the concepts of lattice, Brillouin zone and energy band gaps are completely absent.

### 9.5.1 sp alloys

The primary objective here is to obtain a qualitative but good physical basis for understanding the electronic structure of a simple alloy formed of two sp metals, a pseudoatom alloy, e.g. KRb, MgCd, LiMg or PbTl. It might be thought that the alkali metals would be the most suitable to start with. There are, however, only three miscible systems, namely KRb, KCs and RbCs. The alloying behaviour of the alkali metals can be attributed primarily to atomic size effects. Furthermore they are not such good systems as might at first be thought. Apart from their being very reactive, they are also of similar valence and form rather dilute electron gases. The variations in physical properties dependent on $E_F$ or $N(E_F)$ brought about by alloying would be small, difficult to measure and perhaps difficult to analyse in an unambiguous manner. It is preferable to study an alloy of two elements with different valence, because then more distinct changes in behaviour may be expected. Aluminium should be a good solvent in this case, but in practice this is not so; aluminium is actually a very poor solvent for other metals. In fact there are very few pseudoatom binary alloys with an extensive range of solid solubility. One of these is the LiMg system, which will suit our purpose.

Lithium dissolves up to 70% magnesium while maintaining the bcc structure. Consider first a very dilute solution of Mg in Li, so dilute that there are no Mg–Mg neighbours and where the bulk of the Li host is unaffected by the very small content of Mg. The situation may be idealized to that of a single Mg impurity atom. Both the solvent Li atoms and the solute Mg atom may be described by their pseudopotentials which, for convenience, are drawn as potential boxes of different depth, Fig. 9.23.

The Mg atom causes a local disturbance in the Li matrix. Given the excellent screening properties of the electron gas, Section 6.8, the host Fermi level and work function cannot be affected. The Fermi level must be the same throughout the whole alloy, including the Mg atom. But Mg is divalent and an extra electron has been added to an otherwise pure monovalent Li. The Mg valence electrons must be

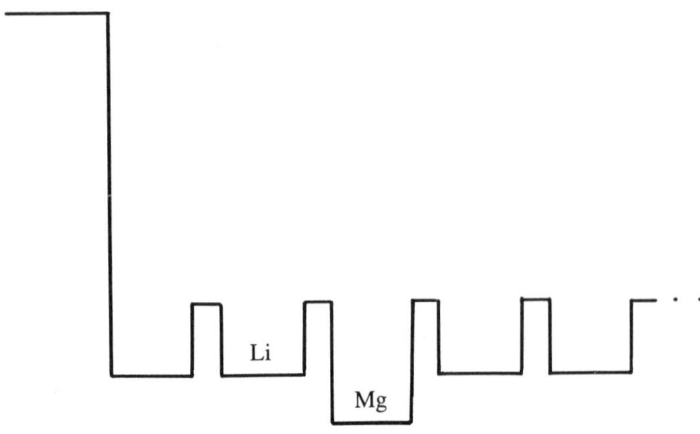


**Figure 9.23** A schematic representation of the potential in a very dilute **Li**Mg alloy.

found in the immediate vicinity of the Mg ion core; they cannot be considered as merely augmenting the uniform valence electron gas of the Li host because this implies their separation from the Mg cell, which involves a significant increase in potential energy. It would also leave the Mg cell with a positive charge, which is also incompatible with the excellent screening properties of the electron gas. The best first approximation is to assume that the Mg impurity atom in Li exists as a neutral entity just as every Li atomic cell is electrically neutral.

Perhaps the local potential well around the Mg atom is sufficiently strong to capture the two valence electrons in a bound state, but the Li electron gas would to some extent leak into the Mg cell and act as a dielectric medium to reduce the binding energy and free the two electrons, making this an unlikely possibility. How about one bound state and one free electron? This produces a strongly magnetic situation on account of the uncompensated spin on the bound electron, the arrangement requiring strong Coulomb interactions, for which there is no experimental justification. The most likely description of the Mg impurity is to say that it must remain a neutral entity but adapt to the conditions dictated by the Li host, i.e. the two valence electrons of the Mg atom must be squashed into a conduction band of the same width as that of Li, but not the same shape. The extra electron density is bunched towards the bottom of the Li conduction band, Fig. 9.24. In other words there is a local conduction band associated with the Mg atom. A similar situation arises for an aluminium impurity atom in Li: the three valence electrons must adapt to the conditions offered by the Li host. It is a Procrustean situation.

As more Mg is added to the Li the situation remains similar to that already described until the concentration is such that electrons can tunnel from one Mg atom to another. At about 20% Mg the electrons are able to build continuous paths between solute atoms, leading to itinerant states that lie below the Li conduction band but that remain confined to the Mg atoms. The solute atoms cause the alloy conduction band to deepen. Near 50% Mg the solvent and solute atoms are present in comparable amounts and it is appropriate to consider a compositionally averaged potential in which all the valence electrons from both Li and Mg move. This average potential is written $V_{vc}$.

$$V_{VC} = cV_{Mg} + (1 - c)V_{Li} \tag{9.11}$$

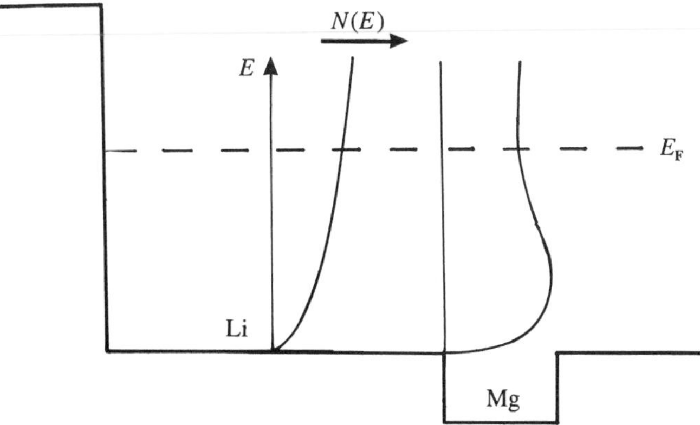


**Figure 9.24** The local conduction band associated with an Mg atom dissolved in Li.

The chemically disordered alloy is represented as an elemental metal with the same potential at each site. This is called the virtual crystal approximation and is probably quite good for estimating $E_F$ and $N(E_F)$. Even so, a very inhomogeneous system is approximated by a homogeneous one, so the next improvement is to correct the previous result by consideration of the departures from the average potential that actually arise at each lattice site, i.e. $V_{Li} - V_{vc}$ and $V_{Mg} - V_{vc}$. These inhomogeneities are essential for the treatment of scattering processes. (Later we shall see that the electrical resistance of a miscible alloy system attains a maximum value at or close to the 50% composition.)

As the Mg becomes the majority component, the virtual crystal approximation gradually becomes less and less applicable and eventually Li becomes the impurity. The situation is complementary to that of Figs. 9.23 and 9.24. The Li cell now represents a repulsive perturbation in an Mg host, Fig. 9.25.

Now it is the Li valence electron that must adapt to the Mg band structure (which for small Li contents is hcp), and in this case the Li atomic cell is surrounded by itinerant electron states to energies well below that of the Li potential box. The itinerant Mg valence states mix with those of the Li atom, causing the latter to leak into the lower portion of the Mg band. Again, in the first approximation, both Li

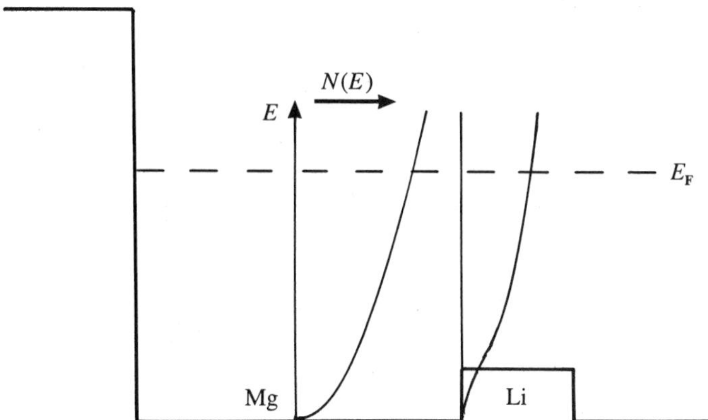


**Figure 9.25** The local conduction band associated with an Li atom dissolved in Mg.

and Mg cells remain electrically neutral, they have the same Fermi energy and the same band width, although the lower portion of the band in the Li cell is very thin in state density.

On the above qualitative but realistic picture, relatively small initial changes in $E_F$ and $N(E_F)$ are expected in dilute alloys of Li and Mg. Nevertheless, the Fermi surface should reflect the changing electron concentration, but no data are available. On the other hand, optical spectra of the bcc LiMg alloys allow a measurement of the energy band gap that arises in the (110) Brillouin zone face and how it varies with composition. This energy gap in pure Li is associated with Bragg reflection, and its presence is clearly evident in the parallel band absorption that arises in the alloys, Fig. 9.26. In fact if one calculates the absorption expected from pure Mg with bcc, structure it is found to be identical with that of hcp Mg, Fig. 9.14. This happens because the interplanar spacings of bcc (110) and hcp ($1\bar{1}01$) are the same if we assume that the atomic volume is the same in both structures. Although the absorp-

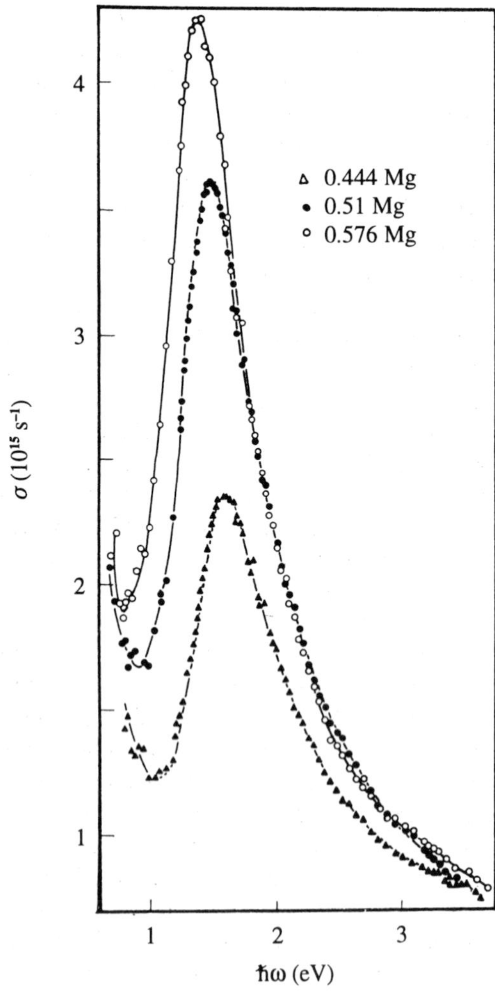


**Figure 9.26** The optical absorption of LiMg alloys in the mid-composition range showing marked parallel band absorption. Compare with Fig. 9.15. (After Mathewson and Myers 1973.)

tion line in the alloy is considerably broader than that for pure Mg, its origin is clearly identifiable. There appears to be good reason to assume that the concept of the Bloch function is still valid in this kind of alloy. The general conclusion is that chemical disorder in pseudoatom alloys is a weak disturbance, but this is not to say that it is insignificant, particularly in connection with the concept of local conduction bands associated with the different chemical species. The degree of chemical or potential disorder must increase with increasing difference in valence between solvent and solute atoms, leading to further departures from the pure metal conditions. These qualitative but general ideas may be used to attempt an understanding of other solid solutions of one sp metal in another, or sp electrons in, say, alloys of the coinage metals.

### 9.5.2 Alloys based on Cu, Ag and Au

Although attention will be focused on Cu alloys, the conclusions drawn are valid for those based on Ag or Au. The coinage metals are monovalent but possess filled narrow d bands which are overlapped by the conduction bands containing the valence electrons, see Figs. 7.25 and 9.7. The filled d states cause the ion cores to be much larger than those encountered in the sp metals.

Alloys of the form **Cu**Zn, **Cu**In, **Cu**Sn, **Cu**As and their like were studied in great detail by Hume-Rothery and coworkers in the years between the First and Second World Wars. The primary solubility limits of the above solutes in Cu were found to be smaller the larger the valence of the solute atom. This effect could be normalized with regard to the valence electron to atom ratio, $e/A$, for the limiting composition. Irrespective of the system studied the limiting composition corresponds to $e/A$ close to 1.36. Furthermore, the first solvent rich intermediate phase, the so-called $\beta$ phase with bcc structure, in the above and similar systems is stable only over a very restricted range of composition centred about $e/A$ close to 1.5. Other intermediate phases of specific structure were correlated to other $e/A$ ratios, but they will not be considered here. These remarkable results merit calling these phases 'electron alloys'. It might be thought that such general and pronounced effects should have an unambiguous and obvious cause, but as yet this appears not to be the case.

It was noticed by Mott and Jones in 1936 that $e/A$ ratios of 1.36 and 1.48 are just the values that correspond to the free electron sphere first touching the nearest Brillouin zone boundary plane for fcc and bcc structures respectively. This is the principal tangible factor. But since 1958 it has been known that the Fermi surfaces of pure Cu, Ag and Au already make contact with the nearest zone boundary, so the behaviour is not a Fermi surface–Brillouin zone effect. A possible cause that has been suggested lies in the singular behaviour of the dielectric function when $q = 2k_F$ (see Section 6.8), but this is usually considered to be a weak influence.

Whatever the true cause of the Hume–Rothery $e/A$ phases, there is no reason why we should not apply the considerations of the previous section to the valence electrons outside the d band: in other words, we expect to find neutral solvent and solute atoms and local conduction bands associated with them. How about the d band? Consider first the intermediate $\beta$ phase of the CuZn system, $\beta$ brass with the bcc structure and equiatomic composition. It may be obtained in an ordered or disordered state, but this is not important in the present connection. In pure Cu we have seen that the 3d band is centred at an energy 4.5 eV below the Fermi level and it has width of about 5 eV. In pure Zn the band is much narrower and centred at about 9 eV below the Fermi level. In all alloys between Cu and Zn the two elements have their d states at such different energies that there is no direct mixing of the states and we expect, and in fact do find in $\beta$ brass, two d bands. One is associated only with Cu atoms, the other only with Zn atoms, Fig. 9.27. This is sometimes called the split band situation. There is spatial inhomogeneity at both the conduction band and the d band level.

The d bands in $\beta$ brass are narrower than in the pure metals because there are fewer Cu–Cu and Zn–Zn neighbours in the alloy and Cu–Zn neighbours have no significant d–d interaction. Note that whereas the position of the Zn d band is little affected, the top of the Cu band is at a lower energy than in pure Cu. This is the cause of the yellow colour of $\beta$ brass, the threshold for optical transitions from the top of the d band to the vicinity of the Fermi level now lying closer to the green portion of the visible spectrum.

In Cu rich alloys a similar split d band situation holds, but the Cu d band is wider the greater the Cu concentration whereas the Zn d band becomes narrower and eventually, when Zn–Zn neighbours are too few to provide continuous hopping paths in the alloy, the Zn d band ceases to exist. It is replaced by narrow distribution of d states associated with each Zn atom.

What would happen if the two d bands of the alloying elements overlapped as in Ag–Au alloys? These alloys form a miscible system. Pure Au has a d band of width 6 eV centred about 5 eV below the Fermi level, whereas for Ag the corresponding figures are 3.5 eV and 5.5 eV. The conduction bands of these metals differ slightly, extending 7.5 eV and 9 eV below the Fermi level for Ag and Au respectively; this difference may be neglected. Photoemission experiments show that the d electrons of the constituents retain much of their individual character. A qualitative explanation of these features follows; it is based on the results of the coherent potential approximation (CPA), and leads to the following simplified picture. We neglect s–d resonance for the moment and assume that there are two coinage metals A and B with the same band width, $W$, and in the pure metals the bands are centred at energies $E_A$ and $E_B$; these latter energies are not those of the d levels in the free atoms but adjusted values after taking into account the changes in atomic volume and s elec-

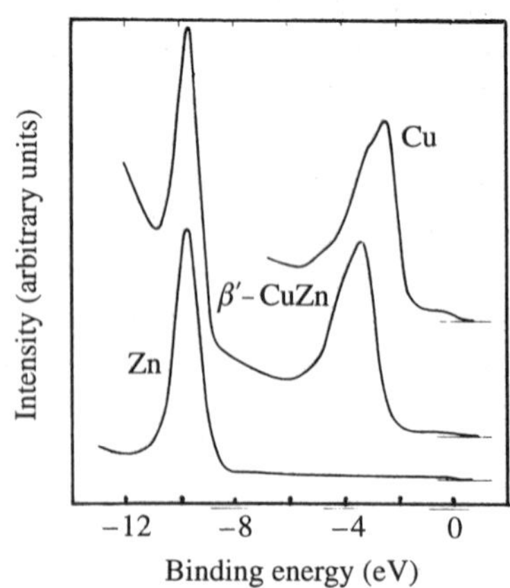


**Figure 9.27** Photoemission spectra for $\beta'$CuZn, Cu and Zn, showing the split 3d band in the alloy. (After Nilsson and Lindau 1971.)

tron density when the atoms form part of a solid. An important parameter is the ratio $R = \Delta/W$, where $\Delta = E_A - E_B$. If $R < 0.25$ then the virtual crystal approximation is applicable to the d bands, whereas when $R > 0.5$ split d bands result, but the ever present but weaker s–d interaction can demand considerable sharpening of this condition. In the interval $0.25 < R < 0.5$ there may be separate d distributions in dilute alloys, but for more concentrated alloys there will be one combined d band. In the latter instance the d electrons are to be found in the union of the two d bands appropriate to the components, but they are distributed in a non-uniform manner in both real space and energy, Fig. 9.28.

We may summarize by saying that in the pure components the tight binding overlaps produce $d_A$–$d_A$ and $d_B$–$d_B$ resonances which cause the band width $W$. In the alloy the overlapping d bands provide additional $d_A$–$d_B$ resonances which cross couple all the electrons in both solute and solvent d bands. Thus both solute and solvent d bands have a width equal to the union of the two bands and electrons from both components contribute to the state density at all energies within the composite band; however, the upper and lower portions of this composite band contain a greater proportion of the d electrons from the element which in the pure state has its d band in the higher or lower position respectively. The ever-present but weaker s–d resonance further aids the d state mixing. This result is seen even better in alloys of the coinage metals with the later 3d transition metals, in particular the AgPd and CuNi alloys, both of which form miscible systems.

Pd metal possesses a 4d band of width close to 5.5 eV and Fermi surface measurements show that it contains approximately 9.7 electrons. There are therefore 0.3 electrons per atom in the conduction band. We have already noted in Fig. 9.7 that the electronic heat coefficient $\gamma$ is very large, indicating an adjusted value for $N(E_F)$ close to 2 $eV^{-1}$ $atom^{-1}$. The Pd 4d band lies much higher in energy than the 4d band in Ag. The free Pd atom has configuration $4d^{10}$. When a Pd atom is dissolved in Ag its d states lie well above the top of the Ag d band, and the major interaction is with the conduction electron gas provided by the Ag which has $N(E_F)$ close to 0.2 $eV^{-1}$ $atom^{-1}$, so the s–d broadening mechanism is weak. Under these conditions it has been shown that the Pd d states, in a first approximation, take up a Lorentzian form and they are said to be in a resonant bound state. The large angular momentum associated with d character weakens the attractive potential of the ion core and allows the d electrons to leak, from what would otherwise be a truly bound state, into the electron gas of the matrix. The wave function is such that the d electrons spend most, but not all, of their time in the vicinity of the parent atom. This resonant bound state, in principle at least, extends over the complete conduction band; it therefore contains somewhat less than 10 electrons because the conduction band contains empty states above the Fermi level, Fig. 9.29 (recall the earlier description of **Al**Mn). Just as in pure Pd, the missing electrons are to be found in the conduction band associated with the solute Pd atom.

Photoemission experiments on dilute solutions of Pd in Ag, Fig. 9.30, combined with electron-specific heat data for AgPd (Fig. 9.9) and AgCd alloys allow one to deduce that there are about 0.35 vacant states in the Pd resonant d state, almost the same as in pure Pd metal, but the band form and $N(E_F)$ are very different for the two situations.

The ideas of the CPA may be applied to this system and, remembering the notation, $E_A$ is associated with Pd and $E_B$ with Ag. The conditions are intermediate between the split band limit and that described earlier (Fig. 9.26). As more Pd is

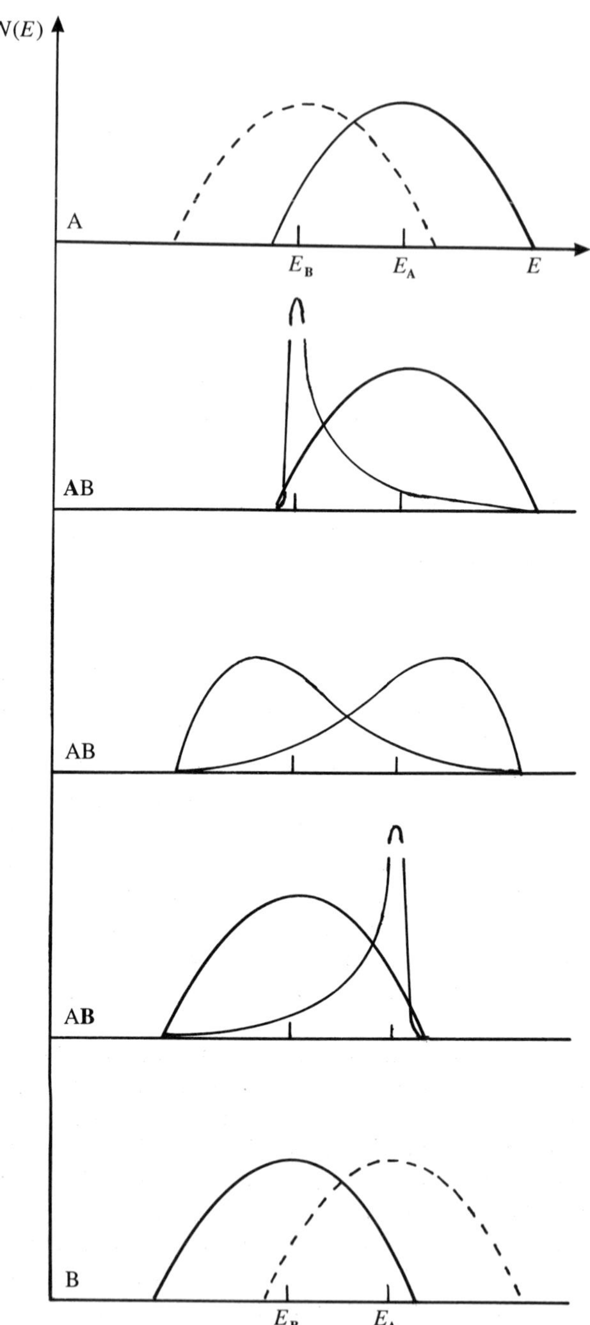


**Figure 9.28** Schematic presentation of the d-band structure in an alloy of two metals that have similar but overlapping d bands. In the alloy electrons in both components experience $d_A$–$d_B$ as well as $d_A$–$d_A$ interactions; the latter give rise to the band width $W$ in each band, whereas the former produce cross coupling of the two bands. The sketches show the density of d state per component atom; the density of states per atom of alloy is then a compositional average. (s–d hybridization is neglected.)

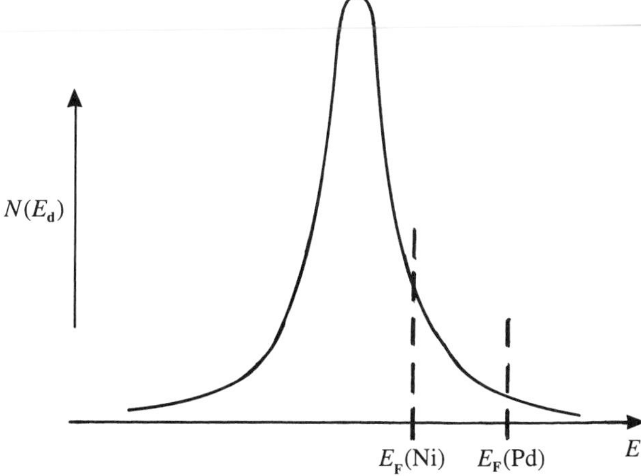


**Figure 9.29** The resonant bound density of d states as arises in a dilute **Cu**Ni or **Ag**Pd alloy. The density of d states at the Fermi level differs markedly in the two alloys because the d electron content is much smaller for Ni than for Pd.

added to the alloy, Pd–Pd interactions broaden the resonant bound state and its shape changes; eventually at about 40% Pd a conventional d band overlaps the Ag d band, and they mix with one another to produce a union as described earlier. The electronic heat data clearly indicate that as the concentration of Pd increases from 40% the state density at the Fermi level increases; this is probably due to changes in both the shape of the Pd sub-band and d electron content (but a complicating feature is that exchange interactions as well as electron–phonon interactions, both of which are strong in Pd, are also changing with composition). As the concentration of Pd increases the Ag d band becomes narrower. At very small Ag concentrations it becomes a narrow state centred at about 5.5 eV below the Fermi level and lying within the lower limit of the Pd d band. Again atomic inhomogeneity is the rule.

A similar situation arises in the miscible CuNi system. The free atom configuration of Ni, however, is $3d^8 4s^2$ and when small amounts of Ni are dissolved in Cu the associated resonant bound d state contains about 8.5 electrons, leaving 1.5 over to go into conduction states, Fig. 9.29. The lower electron content in the resonant d state leads to a larger density of d states at the Fermi level than for the corresponding Pd alloys and the electronic specific heat increases markedly with Ni content. Furthermore, the alloys become Pauli paramagnetic for Ni contents $>5\%$. These features are readily explained on the basis of the resonant bound state which always has finite, and in the case of Ni appreciable, $N(E_F)$. At about 40% Ni the alloys become ferromagnetic and the moment on the Ni atom (which is the sole magnetic carrier) increases linearly with Ni content. For pure Ni the atomic moment at 0 K corresponds to 0.6 $\mu_B$ and, as will be explained in Chapter 11, because this moment derives wholly from the uncompensated spin of the d electrons, it indicates that there are 9.4 electrons in the Ni d band and 0.6 electrons in its conduction band. Again the conclusion is that band form as well as band filling must be considered when attempts are made to understand the properties of this kind of alloy.

In Chapter 11 we shall describe how, in certain dilute alloys, e.g. **Cu**Mn, the

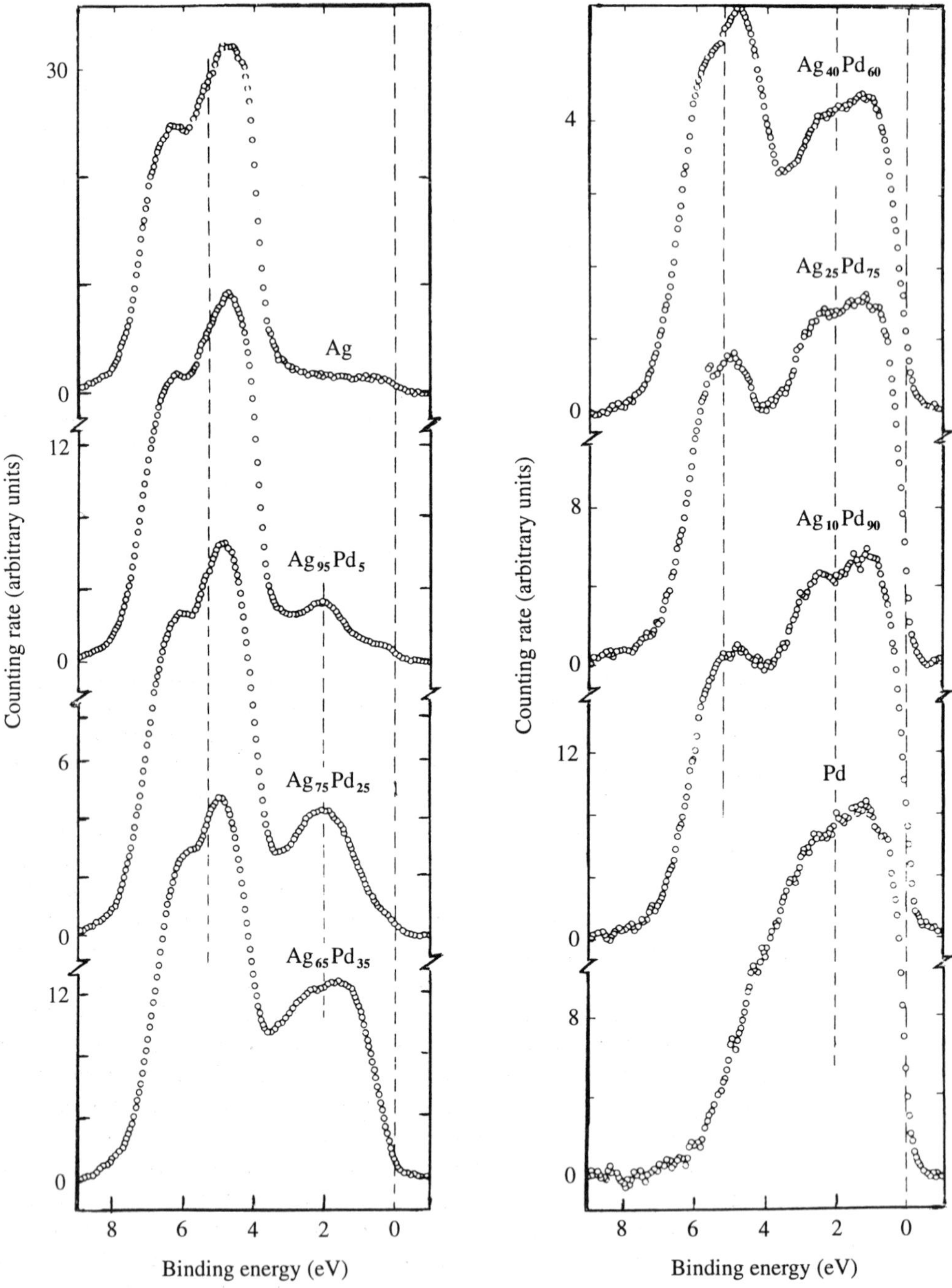


**Figure 9.30** X-ray photoemission data for AgPd alloys clearly demonstrate the individuality of the component electron structure in alloys. At all compositions the major features of the band structure may be attributed either to Ag or to Pd, as indicated by the broken lines. The separation of the d states in these alloys is insufficient to produce the split d-band situation; it is similar to that sketched in Fig. 9.28. The Fermi energy lies at zero binding energy. (After Hüfner *et al.* 1973.)

resonant bound state gives rise to strong magnetic moments on several transition metal atoms.

Alloys between two transition metals will not be discussed, but it is clear that such an alloy, particularly when the two components are close neighbours in the same series (e.g. FeCr alloys), should be suited to the approach described above. In the alloy both components have partially filled d bands and a common Fermi level. There is always significant $d_A$–$d_B$ overlap to mix the component d states, leading to the description of Fig. 9.28.

## 9.6 Electrical Resistance

The discussion of the properties of electrons in perfectly periodic lattices introduced the Bloch function; in the plane wave formulation this takes the form

$$\begin{aligned}\psi_k &= \sum_G c(\mathbf{k} - \mathbf{G})e^{i(\mathbf{k}-\mathbf{G})\cdot\mathbf{r}} \\ &= \sum_G c(\mathbf{k} - \mathbf{G})e^{-i\mathbf{G}\cdot\mathbf{r}}e^{i\mathbf{k}\cdot\mathbf{r}} \\ &= U_{\mathbf{kG}}(\mathbf{r})e^{i\mathbf{k}\cdot\mathbf{r}}. \qquad (9.12)\end{aligned}$$

There is no damping term and the wave function does not decay. Consider a nearly free electron energy band containing one electron. An electric field applied in, say, the $-x$ direction causes the electron's $\mathbf{k}$ vector to increase and, provided the lattice is perfect, the associated electric current increases. There is no electrical resistance. Eventually the electron's $\mathbf{k}$ vector reaches the Brillouin zone boundary, ($\mathbf{k} = \pi/a$), where, as a result of Bragg reflection, it is 'umklapped' back to the opposite zone boundary at $\mathbf{k} = -\pi/a$. If initially the electron had $\mathbf{k} = 0$, on application of the electric field the electron's speed ($\hbar^{-1}$ d$E$/d$\mathbf{k}$) increases with increasing $\mathbf{k}$, but as $\mathbf{k}$ approaches the zone boundary Bragg reflection becomes more and more probable, causing retardation which eventually, at the zone boundary, leads to reflection to the opposite zone boundary. At the zone boundaries the electron's real space speed is zero. Furthermore, after umklapp the field acts to reduce the electron's now negative $\mathbf{k}$ vector (i.e. make it less negative). The real space velocity is in the negative $x$ direction. As the negative $\mathbf{k}$ vector is further reduced it becomes zero again at $\mathbf{k} = 0$ and the cycle is completed. Thereafter the cycle is repeated indefinitely as long as the external field persists. In real space the electron would move to and fro in resistanceless fashion about its mean position on the $x$ axis. On the other hand, if the field is applied for a short interval of time, so that the electron just receives a jolt, then it produces a constant current. Provided the energy band is incompletely filled there will always be a net electric current associated with such a jolt. In a filled band, however, no net current arises because the mean $\mathbf{k}$ value for the complete band is zero: the substance is an insulator.

The above picture is an idealization: in practice crystals are never perfect and departures from perfect lattice periodicity distort the potential and limit the growth of the $\mathbf{k}$ vector under the influence of the applied electric field. The electron is scattered by these imperfections, but its energy is hardly affected and the change in the modulus of the wave vector is very small compared with $k_F$, so only a narrow Fermi surface skin of electron states is involved in electrical conduction, just as in our

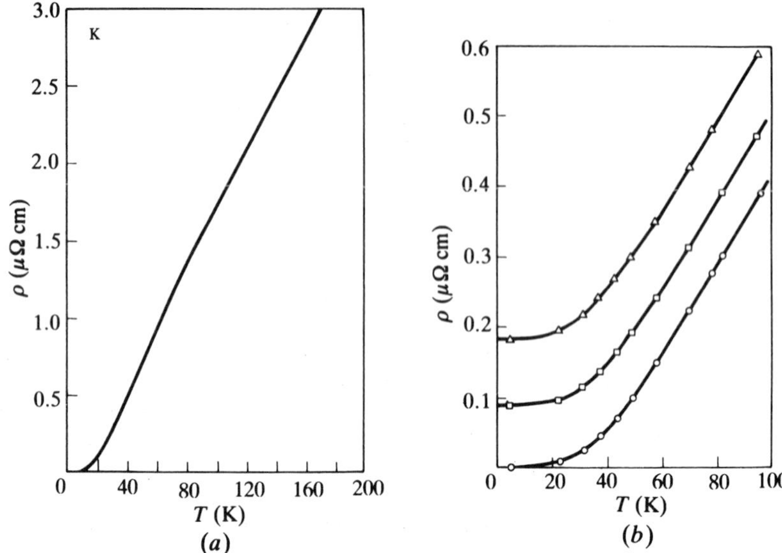


**Figure 9.31** (a) The variation of the electrical resistance of pure potassium as a function of temperature. The behaviour is almost linear over a wide range of temperature, but below about 20 K the resistance becomes constant. At high temperatures, deviations from linearity may arise, particularly for the transition metals. (b) The effect of small amounts of an alloying element (impurity) on the residual resistance of Ag: ○, 'pure Ag'; □, Ag + 0.02 at.% Sn; △, Ag + 0.5 at.% Au. (After Dugdale 1977.)

earlier discussion of the free electron gas, Fig. 6.13. What does change significantly is the direction of the electron's **k** vector. The objective therefore is to describe the origins of electrical resistance and try to explain why certain metals, notably the transition metals, are such poor conductors.

A metal such as Ag or K does have a very low though still finite resistance at 0 K. The typical variation with temperature is shown in Fig. 9.31. The dependence is essentially linear, but at very low temperatures the resistance becomes constant. This stationary value, however, depends on the purity and mechanical condition of the sample. The purer and more perfectly annealed the sample, the lower the final resistance. It is therefore believed that if a perfectly pure strain-free single crystal of a metal could be obtained then it would present zero resistance at 0 K. This is ideal resistive behaviour. Real metals retain a finite resistance at 0 K owing to residual impurities, strains and dislocations. The constant resistance at the lowest temperatures is therefore called the residual resistance. Increasing the temperature of the sample produces increased atomic vibration and a blurring of the periodicity, and the electron wave functions or wave packets formed from them are therefore scattered and the transport of charge is hindered. In simple terms we can say that the electrons collide with the phonons. The phonons have small energies compared with electrons at or near the Fermi energy, but they have large wave vectors comparable in magnitude to $k_F$. Thus the phonons allow electrons to be scattered from one point on the Fermi surface to another. Only very small energy exchanges with the phonons are involved, but these produce Joule heating. On the other hand, large changes in crystal momentum can arise (Fig. 9.32).

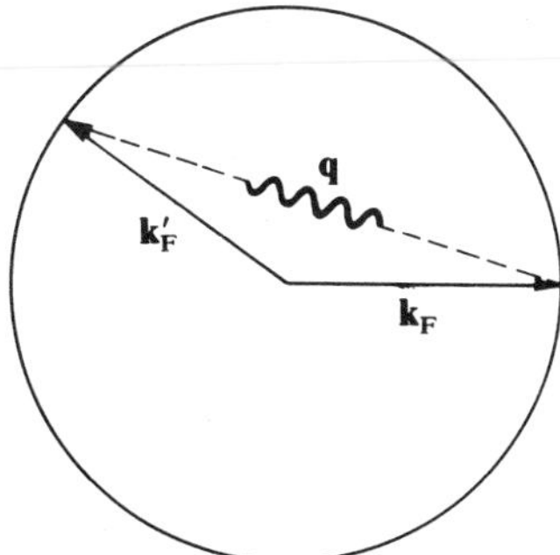


**Figure 9.32** Electrical resistance arises because conduction electrons are scattered by interaction with phonons. Energy and wave vector are conserved. Although phonons have very small energy, they possess crystal momentum of order $\hbar k_F$ ; they can therefore cause pronounced alteration in the wave vector of a conduction electron without having a significant effect on its energy.

Suppose we consider a high-temperature situation and describe the lattice vibrations in terms of the Einstein model; then we have $\hbar\omega = k_B \theta_E$ and this quantity controls the energy exchanged with the electrons. We assume the resistance is proportional to $\langle s^2 \rangle$, where $s$ is the displacement of the ions. $\langle s^2 \rangle$ is a measure of the collision cross-section that the ions present to the electrons.

Considering the $x$ component of the motion, we write

$$M\ddot{x} = -cx, \quad \text{so} \quad \omega^2 = c/M.$$

But in the Einstein model $\hbar\omega = k_B \theta_E$, so

$$c = Mk_B^2 \theta_E^2/\hbar^2.$$

Furthermore, because of the equipartition of energy,

$$\tfrac{1}{2}c\langle x^2 \rangle = \tfrac{1}{2}k_B T,$$

i.e.

$$\langle x^2 \rangle = \frac{k_B T}{c} = \frac{\hbar^2}{k_B} \frac{T}{M\theta_E^2}$$

Thus if the electrical resistance is proportional to $\langle x^2 \rangle$ then we find that

$$R \propto \frac{T}{M\theta_E^2}. \tag{9.13}$$

We expect the resistance to depend linearly on temperature above $\theta_E$ ($\approx \theta_D$).

The specific resistance $\rho$ is conventionally measured at 0°C and defined for a unit cube (i.e. unit volume) of sample. It is clear that, however convenient in practice, the specific resistance is of little fundamental interest in the comparison of different metals.

Thus Pb, with a Debye temperature of about 100 K, contains a much more dense phonon gas at 0°C than does Al, for which $\theta_D \approx 400$ K, and the different atomic masses must, according to (9.13), be allowed for. We should therefore normalize the resistivity data to take account of Debye temperature, ionic mass and atomic volume.

If this is done, we obtain the variation depicted in Fig. 9.33. We immediately find that

(a) the monovalent metals have the lowest normalized resistivity;

(b) the resistivity always increases markedly as we pass from a monovalent to its neighbouring divalent metal;

(c) the transition metals and the rare earth metals have large values of normalized resistivity.

In the light of our previous knowledge, can we understand this pattern of behaviour? We saw in our discussion of the free electron gas that the conductivity is given by (6.37),

$$\sigma = ne^2\tau/m,$$

which may be inverted to given the resistivity

$$\rho = \frac{m}{ne^2\tau}. \tag{9.14}$$

We interpret $2\tau$ as the interval of time between two successive scattering events or collisions with phonons. Only electrons near the Fermi surface (within an interval $\approx \pm k_B\theta_D$) can undergo scattering and the 'collisions' are not equivalent events because the phonons may have different wave vectors. Again we require that energy and wave vector be conserved, and the latter condition, in the periodic lattice, implies the presence of both 'u' and 'n' processes. We picture an electron in state $\mathbf{k}_F$ being scattered through an angle $\phi$ to another state $\mathbf{k}'_F$, by a collision with a phonon of wave vector $\mathbf{q}$, and we demand that

$$\mathbf{k}'_F - \mathbf{k}_F = \mathbf{q}.$$

The state $\mathbf{k}'_F$ is contained in an element of Fermi surface $dS$ and there is a definite transition probability $P(\mathbf{kk}')\,dS$ for this scattering event. The calculation of the electrical resistance demands knowledge of the electron system, the phonon system and

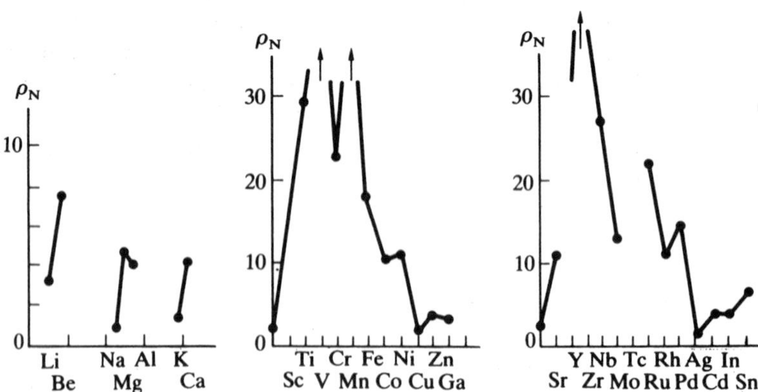


**Figure 9.33** The variation of the normalized resistivity, as defined in the text, for selected metals. Significant features are the marked increase arising when passing from monovalent to the neighbouring divalent metal and the large normalized resistivities of the transition metals. (From Ziman 1960.)

their mutual interaction in the presence of electric and magnetic fields. It is a very complex problem, and attempts at the detailed calculation of resistivity for the simplest cases do not give agreement with experiment to better than a factor of 2.

From our point of view we can obtain a reasonable physical picture and use our classical formula under the following conditions:

(a) since only electrons near the Fermi energy can contribute to the current, we assume that there is an effective number of free electrons;

(b) we define an average relaxation time for the effective free electrons according to

$$\langle \tau^{-1} \rangle \sim \left\langle \int_{\mathrm{FS}} P(\mathbf{k}_\mathrm{F}\, \mathbf{k}'_\mathrm{F})\, \mathrm{d}S \right\rangle;$$

(c) the electrons contributing to the current may have an effective mass different from that of a truly free electron; we speak of a band mass.

Clearly assumption (a) implies that $N(E_\mathrm{F})$ is an important factor in governing the conductivity. We shall not discuss (b) in any detail but merely point out that we can link the classical concept of collision time to the quantum mechanical electron-phonon scattering probability. We shall soon find good cause for the third assumption (c).

It must be emphasized that we have replaced a complex quantum mechanical interaction by three concepts that are not so separable as in the classical case. However, our assumption (a) provides us with an appreciation of the rapid increase in normalized resistivity as we proceed from a monovalent metal to the adjacent divalent metal. The monovalent metals have spherical or almost spherical Fermi surfaces with little loss of state density by zone contact. $N(E_\mathrm{F})$ is large compared with the corresponding quantity for divalent metals. The latter metals owe their conducting properties to the electrons occupying states in both first and second zones. We associate this type of Fermi surface with a reduced number of effective free electrons on account of the smaller $N(E_\mathrm{F})$. However, this argument is not readily extended to the case of trivalent and quadrivalent polyvalent metals.

We have already pointed out that the transition metals (including the rare earth metals) have large normalized resistivities, but these metals are known to have large $N(E_\mathrm{F})$ values from the contribution of the d electrons. Surely the effective number of free electrons should be large, leading to a low resistivity. How do we reconcile these contradictory viewpoints? It is true that $N(E_\mathrm{F})$ is large in these metals, but the d electrons that are responsible have large 'band masses' which neutralize their contribution to the conductivity.

### 9.6.1 *Effective band mass*

In Chapter 6 different effective electron masses, e.g. the thermal and optical masses, were introduced. The electron mass in a formula obtained by the application of a simple model was allowed to vary and thereby produce agreement with experiment; the concept is *ad hoc* and arbitrary. The effective band mass of the electron which we now introduce is a much more fundamental quantity. A linear situation is assumed and the electron is represented by a wave packet so the electron velocity is the group or energy velocity

$$v_g = \frac{1}{\hbar}\frac{dE}{dk}. \tag{9.15}$$

In the presence of an applied electric field **E**, the electron experiences a force $\mathbf{F} = -e\mathbf{E}$ and its energy increases by

$$\delta E = F\,\delta x = F v_g\,\delta t$$
$$= F\frac{1}{\hbar}\frac{dE}{dk}\,\delta t. \tag{9.16}$$

But we can also write

$$\delta E = \frac{dE}{dk}\,\delta k, \tag{9.17}$$

and, combining these last two equations, we see that

$$F = \hbar\frac{dk}{dt}. \tag{9.18}$$

Irrespective of whether or not the electron is free, (9.18) holds. We use (9.15) to obtain the acceleration

$$\frac{dv_g}{dt} = \frac{1}{\hbar}\frac{d^2E}{dk\,dt} = \frac{1}{\hbar}\frac{d^2E}{dk^2}\frac{dk}{dt}, \tag{9.19}$$

which can be written

$$\frac{dv_g}{dt} = \frac{1}{\hbar^2}\frac{d^2E}{dk^2}\left(\hbar\frac{dk}{dt}\right).$$

But, as (9.18) shows, this is equivalent to

$$F = \hbar\frac{dk}{dt} = \frac{\hbar^2}{d^2E/dk^2}\frac{dv_g}{dt}. \tag{9.20}$$

Comparison with Newton's law requires that we associate the quantity $\hbar^2/(d^2E/dk^2)$ with the mass of the electron; we call it the 'effective band mass' because it is determined by the dispersion of $E$ with **k**. The effective band mass arises because we assume the electron is acted upon only by the external field **E**. But in actual fact the electron moves under the influence of both **E** and the periodic field of the lattice: if we were to take explicit account of both fields in discussing the dynamics of the electrons then they would exhibit their ordinary mass. It is simply the case that the periodic lattice field is not treated explicitly, but represented by the effective mass of the electron. For free electrons the band mass, as one might expect, is the ordinary electron mass. In real solids, the band mass is dependent on the electron's **k** value as illustrated in Figs. 9.34 and 9.35. We note that in narrow bands, as for the 3d, 4d and 5d electrons, the effective mass is large. Near the top of a band the effective mass becomes negative, a direct result of the increasing probability of Bragg reflection when **k** approaches a zone boundary. In three dimensions the effective band mass of the electron becomes a tensor quantity.

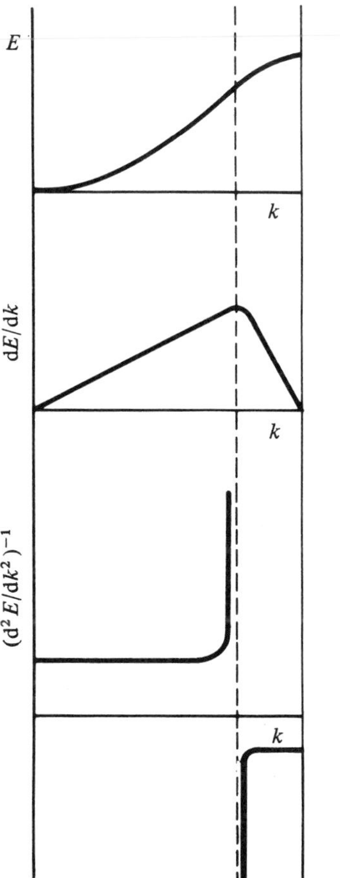


Figure 9.34 The variation of $m^* \sim (d^2E/dk^2)^{-1}$ throughout an energy band.

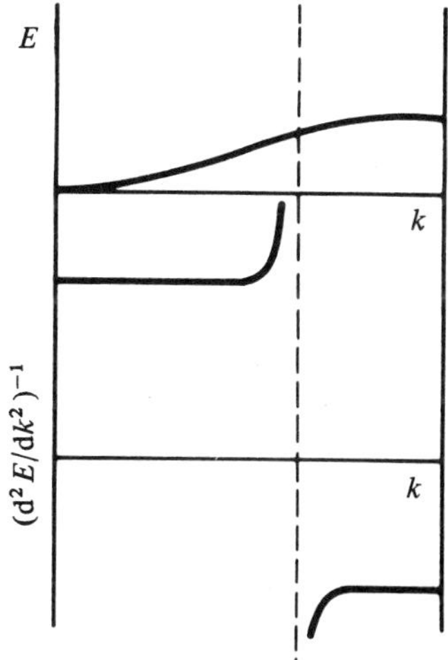


Figure 9.35 Narrow energy bands lead to large electron band masses.

Thus, despite the large density of d states near the Fermi level of a transition metal, they contribute little to the electrical conductivity because they are associated with large effective band masses. However, a closer inspection shows that this cannot be the whole story. Consider the simplified band forms for Cu and Ni shown in Fig. 9.36. The d bands are assumed superposed on the flat sp conduction bands. In both Cu and Ni we should expect to find essentially equal densities of sp electrons at $E_F$. (The ferromagnetism of Ni complicates the issue somewhat, but we neglect that aspect here.) Thus, even if the d electrons of Ni or the other transition metals made no contribution to the conductivity at all, i.e. even if they had an infinite band mass, we should still expect to find equivalent sp conductivities. Why then are the transition metals so much poorer conductors than the monovalent metals? We must look to $\langle \tau^{-1} \rangle$ for the answer.

In say potassium or copper we have a Fermi surface composed of sp states and with an almost isotropic distribution. An electron from state $\mathbf{k}_F$ is scattered to a *similar* state $\mathbf{k}'_F$. All the states at the Fermi surface have similar form. We speak of s-s scattering or potential scattering because the mechanism depends on the small changes in potential that are produced by the lattice vibrations. In Ni, or any other

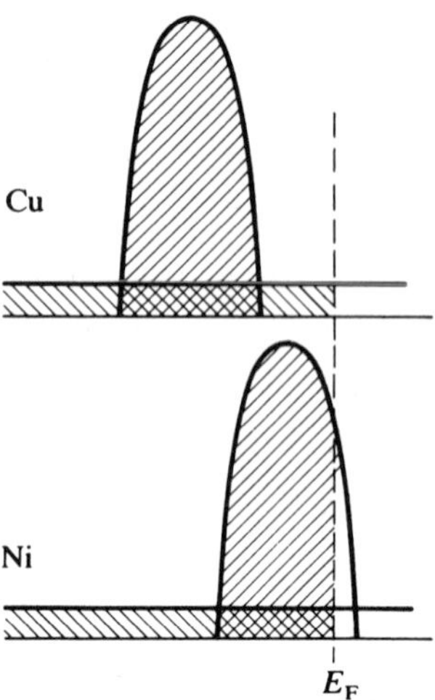


**Figure 9.36** Stylized integrated energy bands for Cu and Ni; compare with Fig. 9.7.

transition metal, the Fermi surface is more complex; in addition to the almost uniform distribution of conducting sp states, there must be large patches of d character, which, while not taking part in conduction themselves, open a new channel for the scattering of sp electrons. Thus our electron $\mathbf{k}_F$, a conducting electron, can first scatter into an empty d state, of which there are many, and then into another conducting state $\mathbf{k}'_F$ (Fig. 9.37).

Owing to the large density of d states near the Fermi energy, this is a process with a high probability and leads to a greatly reduced average relaxation time because we must now write

$$\tau^{-1} = \tau_{ss}^{-1} + \tau_{sd}^{-1}. \tag{9.21}$$

The presence of partially filled narrow d bands leads to s–d scattering and a significant reduction in the conductivity. It is for this reason that, as first pointed out by Mott, transition metals have high resistivities.

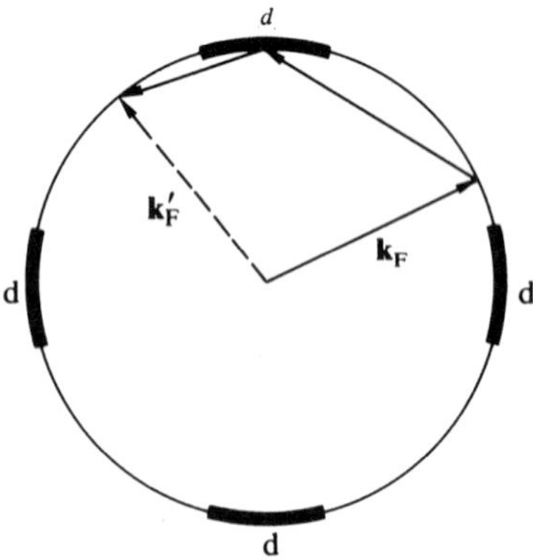


**Figure 9.37** The presence of areas of high d-state density on the Fermi surface of a transition metal introduces new processes for the scattering of conduction electrons leading to increased resistance.

Perfect conductivity is associated with perfect structural order. Any departure from crystalline perfection introduces scattering and thereby causes electrical resistance. We summarize the more important contributions to electrical resistance as follows (for a detailed discussion of these aspects see Dugdale 1977):

(a) $T \geqslant \theta_D$, thermal scattering by phonons $\rho \propto T$;
(b) $T \ll \theta_D$, small angle scattering by phonons $(k'_F - k_F \approx 0)$ $\rho \propto T^5$;
(c) electron–electron scattering $\rho \propto T^2$;
(d) impurity and defect scattering $\Delta\rho \approx$ constant;
(e) magnetic scattering processes complex behaviour.

### 9.6.2 *Electrical resistance of certain alloys*

Alloying, either in the liquid or the solid state, invariably increases the electrical resistance, and in binary miscible systems the resistance varies in a characteristic parabolic fashion (Fig. 9.38). In certain binary alloys, when the relative concentrations of the two components form simple ratios (e.g. 1 : 1, 1 : 2, 1 : 3), the atoms, which are normally randomly distributed on the lattice sites, may arrange themselves in a highly ordered superlattice (Fig. 9.39), with an attendant marked decrease in resistance (Fig. 9.40). This is a vivid indication of the significance of structural order for electrical conductivity.

When a metal is alloyed with small amounts, about 1%, of other metals, the primary effect is an increase in the residual resistance and the ideal behaviour of the solvent metal remains almost unaffected; the temperature coefficient of the solvent is therefore largely unaffected by small alloying additions; this is known as Matthiessen's rule (1864). It is, however, only approximate.

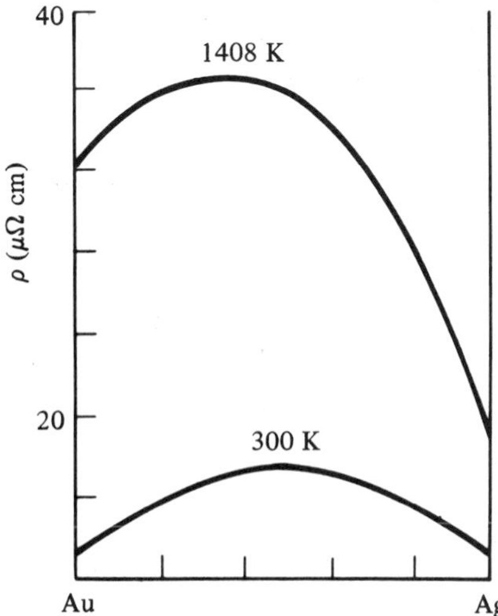


**Figure 9.38** The addition of one metal to another invariably causes an increase in the electrical resistance. This diagram shows the characteristic parabolic variation that arises in miscible Au and Ag alloys in both the liquid and solid forms, $T = 1408$ K and 300 K respectively.

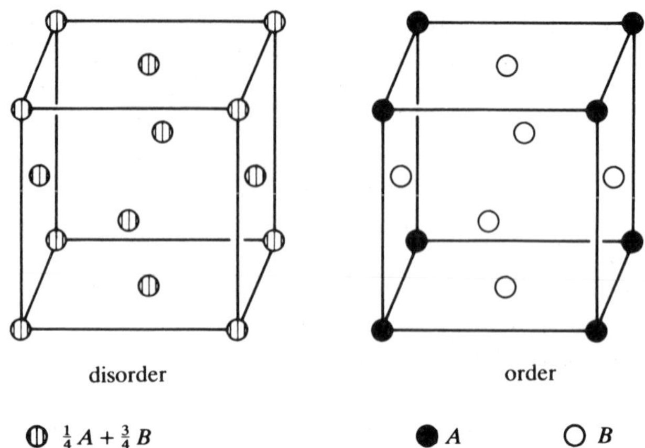


**Figure 9.39** A superlattice arises in for example a binary alloy $A_xB_y$, where $x$ and $y$ are integers, usually 1, 2 or 3. The atoms $A$ and $B$ occupy specific sites as in the ordered structure shown above. At temperatures above about 500°C, the atoms become randomly distributed on the lattice sites and the alloy is said to be disordered. The transition from ordered to disordered state is a second-order phase transition and is characterized by a large increase in the heat capacity (see Appendix 11.4).

The increase in residual resistance caused by 1% of a polyvalent metal in a coinage metal like Cu shows a simple dependence on the valence of the solute; it is found to be proportional to the square of the excess valence of the solute atom over that of the solvent, a dependence known as Linde's rule, (see Dugdale 1977). Certain transition metal solutes, however, present a very different behaviour pattern. Many years ago in the decade 1930–1940 measurements at low temperatures on what were

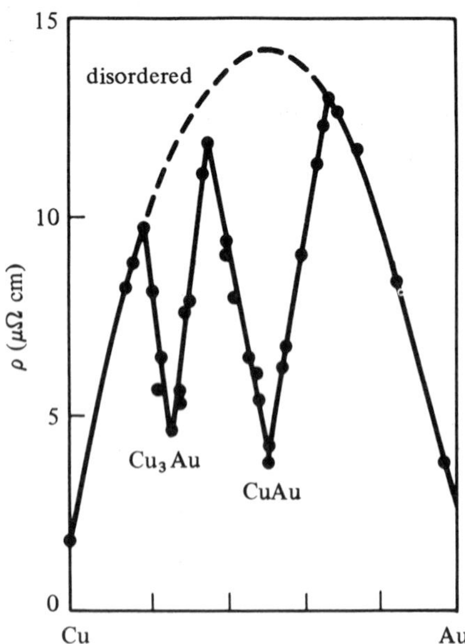


**Figure 9.40** The effect of superlattice order on the electrical resistance of Cu-Au alloys. (After Johansson and Linde 1936.)

purportedly pure metals like Au or Mg were found to exhibit a shallow minimum in the resistance at temperatures below 30 K, Fig. 9.41. After the Second World War the effect received renewed attention. Very detailed measurements showed that it was not due to the presence of grain boundaries or other purely structural defects. Eventually it became clear that the minimum arises only when transition metal impurities carrying a magnetic moment are present. In 1964 Kondo provided an explanation for this apparently innocuous behaviour.

Assuming that there is no interaction between the magnetic atoms (e.g. Fe or Mn), corresponding to extremely small concentrations, Kondo showed that the scattering of an electron by an ion carrying a magnetic moment can be spin dependent. Experiment shows that ordinarily an electron maintains its spin direction when it travels through the crystal and interacts with crystal defects or phonons. An electron of wave vector **k**, being one electron in a many electron configuration, is scattered directly to another **k**′ state of similar energy by interaction with a disturbance $V$ in the potential and with a probability amplitude dependent on a transition matrix $V_{\mathbf{kk}'}$. Direct transitions of this kind on the Fermi surface are those usually considered. An alternative less direct route to the final state **k**′ is via an intermediate state **q** for the electron. Kondo considered this second order transition and asked what would happen if the spin of the electron were to change direction when it entered the transitory state **q** and change back to its original direction on leaving it. The possibility for spin flip exists because the spin of the electron interacts with the resultant d spin on the magnetic impurity, but the latter also serves as a spin buffer and allows conservation of the total spin in the system. There are two possibilities for the transitory state **q** because it may be an unoccupied or an occupied state and it is not limited to the Fermi surface. The first possibility is obvious. The second appears unusual, but **k** and **k**′ are true initial and final states. We imagine that first an electron in an occupied state **q** is excited to the final state **k**′ and then the now empty **q** state is quickly filled by the electron in the state **k**, so the end result is that the state **k** is now empty and state **k**′ occupied. The decisive factor is that these two possibilities (state **q** empty or filled) are not equivalent; this, coupled with the fact that the wave functions for the two final situations have opposite signs, means that it is their difference that enters the calculation of the transition probability. Evaluation of the transition probability over a sharp Fermi sphere of conduction states

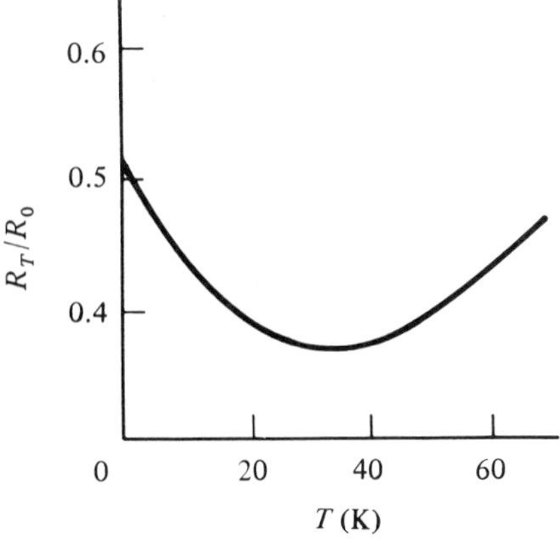


**Figure 9.41** The resistance minimum as it appears in a dilute CuFe alloy (0.1% Fe in Cu).

leads to a 'spin flip' resistance

$$\rho_{spin} \sim J^3 \ln T$$

where J is an exchange energy characterizing the interaction (the so-called s–d exchange) between the conduction electron spin and the d spin of the magnetic impurity atom. If J is negative then $\rho_{spin}$ decreases with increasing temperature. Such a decrease together with the normal low temperature resistive behaviour causes the observed shallow resistance minimum.

A negative J means that the coupling between the conducting s states and the magnetic ion causes the s electrons to set their spins so as to oppose the net spin of the d electrons on the magnetic impurity atom. At high temperatures this coupling is overcome by thermal agitation, but below a certain temperature, the Kondo temperature, which is 30 K for **Cu**Fe but only 10 mK for **Cu**Mn, the magnetic moment on the solute atom begins to be reduced by an atmosphere of solvent conduction electrons with opposed spins. The change occurs smoothly and complete compensation is attained at the absolute zero of temperature.

Kondo's explanation of the resistance minimum answered one pressing question, but the theory was incomplete and came to pose many others. The initial attempts to improve on Kondo's treatment were many and varied, leading to different predictions regarding the variation of resistivity and magnetic susceptibility near the absolute zero of temperature. In turn this led to many experimental studies of dilute alloys, but experimenters, often unwittingly, were studying the effects of magnetic impurities modified by quite strong interactions. It became clear that if the single impurity limit was to be dominant the concentration must be extremely small, preferably $<50$ ppm. Another aspect was that certain interesting systems like **Cu**Fe presented considerable metallurgical problems. The first complete theoretical solution of the Kondo problem was presented by Wilson using the computational techniques that he introduced into the study of critical phenomena.

The electrical resistance is a convenient quantity for experimental study. It can be measured accurately under a wide variety of conditions of temperature, pressure, magnetic field, chemical composition, etc. It is often used as a monitor of other phenomena, phase changes or magnetic behaviour for example. In recent years much attention has been directed to disordered systems as exemplified by the metallic glasses, the interested reader is referred to a recent introduction to this subject by Dugdale (1995).

## 9.7 The Positive Hole

The Hall effect (Section 6.6) allows us to determine the sign of the charge carrier. We expect metals to have a negative Hall constant because the current is carried by electrons. The behaviour of many solid metals is in agreement with our expectations, but there are exceptions. In particular the divalent metals Zn and Cd, as well as several transition metals, exhibit positive Hall coefficients. To appreciate the 'opposite sign', we must consider the behaviour of electrons in a nearly filled band. Consider the band diagram in Fig. 9.42, where the lower band is filled and the upper one empty. No current can flow. Suppose now that we excite one electron from the filled to the unfilled band; we may imagine this to occur by the absorption of a photon and the excited electron filling a state immediately above that which it

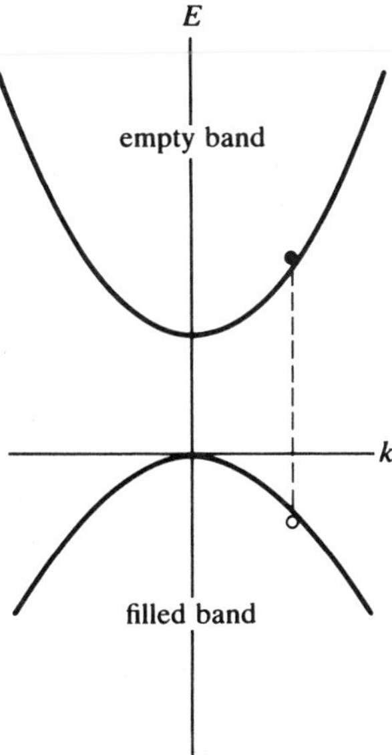


**Figure 9.42** Formation of a positive hole in a filled band of electrons. The electron is missing from the state $k = +j$ and this is the site of the hole state, but its motion is governed by its $k$ value (which is associated with the electron state at $k = -j$) and its velocity (that which an electron would have at the hole site).

occupied in the first band. In the reduced zone **k** is conserved. The single electron in the second band can conduct electricity in the manner already described. What we need to know now is the way in which the electrons in the first band react to an applied field. Consider the filled first band; it contains $N$ electrons distributed equally in states of $\pm k$, which we write $k_{\pm j}$: the band is assumed symmetric about the origin of $k$ so that $k_{-j} = -k_{+j}$. For the filled band we have

$$\sum_{i=\pm 1}^{\pm N/2} \mathbf{k}_i = 0, \tag{9.22}$$

which may be rewritten

$$\sum_{i \neq +j}^{\pm N/2} \mathbf{k}_i + \mathbf{k}_j = 0,$$

i.e.

$$\sum_{i \neq +j}^{\pm N/2} \mathbf{k}_i = -\mathbf{k}_{+j} = \mathbf{k}_{-j}. \tag{9.23}$$

*When state $+j$ is empty, the incomplete band has an effective* **k** *vector* $\mathbf{k}_{-j}$.

We can analyse the current flow in the incomplete band in a similar way; under an applied field **E** each electron acquires a velocity $\mathbf{v}_i$ and we write

$$\sum_{i=\pm 1}^{\pm N/2} (-e\mathbf{v}_i) = \sum_{i \neq +j}^{\pm N/2} (-e\mathbf{v}_i) - e\mathbf{v}_j = 0, \tag{9.24}$$

whence

$$\sum_{i \neq +j}^{\pm N/2} (-e\mathbf{v}_i) = +e\mathbf{v}_j. \tag{9.25}$$

*The complete band with state $+j$ empty may be represented by a positive charge moving with the same velocity an electron would have in state $+j$.*

The properties of the electrons in the incomplete band are equivalent to those of the vacant state $j$ if we attribute to this *vacant state*

(a) a **k** vector $\mathbf{k}_{-j}$;
(b) a velocity $\mathbf{v}_{+j}$;
(c) a charge $+ e$.

We call this vacant state a *positive hole*. Consider the acceleration of this hole state in an applied field **E**: we have

$$m_h \frac{d\mathbf{v}_h}{dt} = +e\mathbf{E}, \tag{9.26}$$

but

$$\mathbf{v}_h = \mathbf{v}_e$$

and

$$\frac{d\mathbf{v}_h}{dt} = \frac{d\mathbf{v}_e}{dt} = \frac{-e\mathbf{E}}{m_e}. \tag{9.27}$$

Using (9.27) in (9.26), we obtain

$$-e\mathbf{E}m_h/m_e = +e\mathbf{E},$$

which can only hold if $m_h = -m_e$. *We therefore attribute to the positive hole a negative electron mass.* However, we have already seen that near the top of a band the electron has a negative band mass, so our positive hole therefore has a positive band mass. Finally we can express the total force on the hole state as

$$\mathbf{F} = \hbar \frac{d\mathbf{k}_h}{dt} = e(\mathbf{E} + \mathbf{v}_h \times \mathbf{B}). \tag{9.28}$$

We may choose to describe the incomplete band either in terms of the filled electron states or in terms of the vacant states. If we choose the latter description, we find that the vacant states are equivalent to positive charges with the above properties. It is important to realize that we have an 'either-or' situation. Under equilibrium conditions we never need to consider the behaviour of electrons and holes in the *same* energy band; only when the band or sub-band is nearly filled do we need to invoke the hole. The holes then are the only charge carriers in the band and the properties of the nearly filled band are described completely by the holes. On the other hand, if the presence of holes in a band is a result of the Fermi surface extending over two zones as in a divalent metal then the second zone contains a number of electrons equal to the number of holes in the first band; these electrons and holes are independent quantities and both contribute to the transport of charge under the influence of a field **E**.

## 9.8 Electrons and Holes

It is usually the case that the Fermi surface of a metal comprises both electron and hole parts. It is important to understand exactly what we mean by this statement, so we shall first describe the behaviour of *adjacent* electron and hole states as they move in **k** space. The important properties of the electron and the hole are listed in Table 9.2. Note that the properties are given assuming that either the electron or the hole is in the state '$j$' in the same zone or band. Our purpose is to continue the previous discussion and then treat the two-dimensional case.

Consider now the band diagram of Fig. 9.43, where certain electron (•) and hole (○) states are shown; the symbol ○ indicates the position of the hole in **k** space, but the behaviour of the hole is governed by its **k** vector and this point is marked ⊙. The movement of the point ⊙ in an applied electric field must not be confused with that of the electron state • that is also at that point (because the $d\mathbf{k}/dt$ for the two particles are of opposite signs owing to their opposite charges). The movement of the electron and hole states under the action of an applied field **E** is then as shown by the arrows. For the hole, $d\mathbf{k}/dt$ is always positive, the points ⊙ always move in the same direction as the applied field, but the *sites* of the holes, points ○, move in the opposite direction. We find that the hole states and the electron states in the same band or zone move rigidly like beads on a string. If the electric field is directed along the positive $k$ axis then the movement is always in the direction of the negative $k$ axis. What happens when an electron or hole state arrives at the zone boundary? We may describe the behaviour in terms of either the reduced or the repeated zone scheme. In the reduced zone, on arriving at the zone boundary at $k = -\pi/a$, the electron or the hole state is immediately 'umklapped' back to $k = +\pi/a$ and the movement in **k** space is begun again, and so the process goes on until the field is removed. In the repeated zone scheme the picture is perhaps simpler: we must imagine an endless sequence of similar zones, the $E(\mathbf{k})$ curves are then continuous (the same electron and hole states existing in every repeated zone); the electron and hole states again move rigidly in an endless train through the whole of **k** space under the action of the electric field.

At this point it is convenient to return to the concept of the Fermi surface and ask how holes arise there. We restrict attention primarily to a two-dimensional example – the extension to three dimensions is straight-forward. Fig. 9.44 illustrates a simple square Brillouin zone, it is almost filled and for the moment we show no occupied states in the second zone. We use the repeated zone scheme, and the figure shows how hole states arise as circular areas at the corners of the individual zones (in three dimensions they would be 'hollow' spheres called 'pockets' of holes). In the presence of a magnetic field we shall again find cyclotron orbits. We analyse the

**Table 9.2**

| | Electron | Hole |
|---|---|---|
| Charge | $-e$ | $+e$ |
| Wave vector | $k_j$ | $-k_j$ |
| Band mass | $m_e = \hbar^2/(d^2E/dk^2)$ | $m_h = -m_e$ |
| Effect of field **E** | $\mathbf{F}e = \hbar(d\mathbf{k}/dt)_j = -e\mathbf{E}$ | $\mathbf{F}_h = \hbar(d\mathbf{k}/dt)_{-j} = +e\mathbf{E}$ |
| Velocity in real space | $\hbar v_e = (dE/dk)_j$ | $v_h = v_e$ |

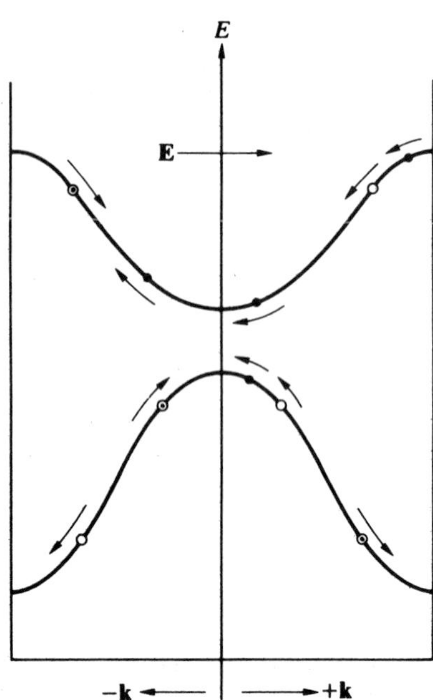


**Figure 9.43** Irrespective of where the hole state is sited, the electron and hole states, under the action of an electric field **E**, always move rigidly together in **k** space without altering their relative positions.

movement of the hole states in a manner analogous to that used earlier in one dimension and making explicit use of the properties of holes in **k** space. We ask how an electron and an *adjacent* hole state react to an applied magnetic field. The **k** vectors of the two states are shown in Fig. 9.45, as are the Lorentz forces acting on the particles (remember in particular that $\mathbf{v}_h = \mathbf{v}_e$). The movement of the electron state presents no problem, but the motion of the hole is dependent on how $\mathbf{k}_h$ changes. We put $(\mathbf{k}_h)_j = -(\mathbf{k}_e)_j$ and apply the force $\mathbf{F}_h$ to the hole vector to determine its motion. Having done this, we must see what consequences it has for the

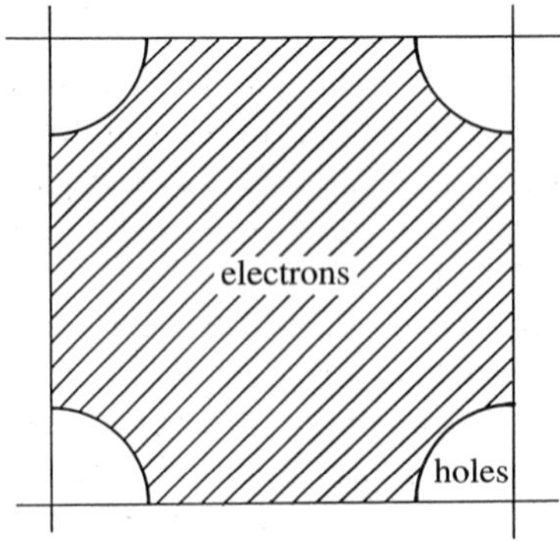


**Figure 9.44** A plane section perpendicular to (001) in a cubic zone structure with occupied states only in the first zone. The zone is almost filled, but empty states remain at the zone corners. In three-dimensional **k** space and in the repeated zone scheme these empty states form spheres of hole states. The physical behaviour is preferably described in terms of the hole states.

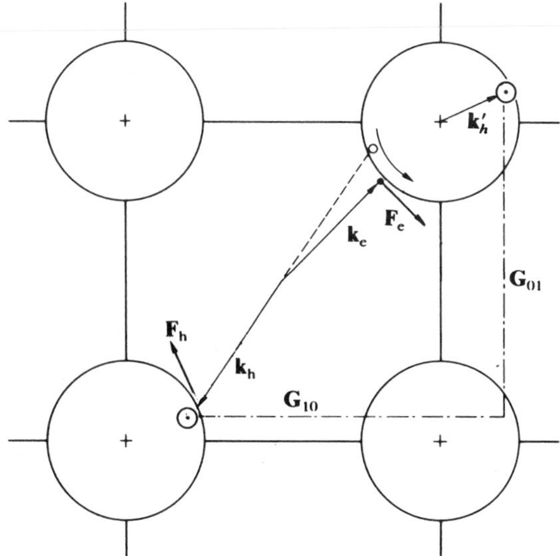


**Figure 9.45** A section through the repeated zone scheme of Fig. 9.44, showing the circular areas of hole states at the zone corners. A magnetic field is directed normally into the plane of the zone. We now ask how the hole ○ and the electron • states are affected by this field. The electron state presents no problem, $\mathbf{v}_e$ is parallel to $\mathbf{k}_e$ and $\mathbf{F}_e$ is as shown. How the hole state moves depends on $\mathbf{k}_h$ and the point ⊙. The important aspect is that the hole velocity is directed antiparallel to $\mathbf{k}_h$, leading to the force $\mathbf{F}_h$, which then tells us how ⊙ moves; to obtain the movement of the hole site ○, we must ask how $-\mathbf{k}_h$ moves. We find that electron and hole states move together. On account of the translational symmetry of $\mathbf{k}$ space, we may map $\mathbf{k}_h$ into $\mathbf{k}'_h$ using the appropriate vector $\mathbf{G}_{10} + \mathbf{G}_{01}$.

movement of the actual hole site. As the diagram shows, adjacent electron and hole states move in the same direction, anticlockwise around their common boundary in the present case. Now for such a distribution of states as shown in the figure we may concentrate attention wholly on the holes and, using the symmetry of **k**-space, we may translate $\mathbf{k}_h$ a whole reciprocal lattice vector

$$\mathbf{G} = \mathbf{G}_{10} + \mathbf{G}_{01}$$

to find the equivalent hole vector $\mathbf{k}'_h$. This allows us to consider the motion of the hole to be centred around $A$, which is a zone boundary corner and the centre of our pocket of holes. It is not difficult to imagine the equivalent three-dimensional situation and a spherical hole Fermi surface embedded in a zone that is almost filled with electrons. We have discussed the movement of adjacent electron and hole states within the same zone, but earlier we said that we always have an 'either-or' situation. In practice when considering the properties of a nearly filled energy band, we describe the behaviour solely in terms of the holes. A de Haas–van Alphen study of the three-dimensional equivalent of Fig. 9.45 would provide us with the extremal area of the hole surface and the hole mass.

It is often said that electrons and holes rotate in opposite directions in their cyclotron motion in **k** space, but this is only true when the electron and hole states are in separate bands or zones; this is always the case in practice because, as we have emphasized earlier, the transport properties of a band are interpreted in terms of either electrons or holes (even so, this does not exclude the possibility of finding both electron and hole trajectories in the same region of the Fermi surface, but in perpendicular directions).

Suppose that we have a divalent specimen and that the occurrence of holes in the first zone is a consequence of electrons spilling over into the second zone. We know how to describe the hole states, so we now concentrate attention on the electrons in the second zone. Using the repeated zone scheme, we find in two dimensions approximately elliptical regions (ellipsoidal in three dimensions) of occupied states (Fig. 9.46). The electrons in these portions of the Fermi surface have their own characteristic cyclotron frequencies, effective masses and extremal areas. These electron states are independent of the hole states in the first zone. The electron and hole surfaces arise in different zones and the representative points in **k** space rotate in opposite directions under the influence of a magnetic field.

The transition metals with their several and incomplete d sub-bands provide examples of Fermi surfaces with a complex structure of electron and hole components, whereas the divalent metals like Mg are simpler and somewhat resemble the situation of Fig. 9.46. We can now appreciate the behaviour of a divalent metal that has its electrons divided between two bands (zones) so that one is nearly filled and the other almost empty. The nearly filled band simulates positive carrier behaviour and compensates either partially or wholly the Hall effect in the conventional electrons in the almost empty band. A similar situation is readily envisaged for a transition metal where one of the sub-bands is almost filled. The detailed analysis is complicated, being dependent on the particular structure of the Fermi surface for each metal; we shall let the matter rest here. The positive hole, however, has a major role to play in the description of semiconducting behaviour, even though the latter does not possess a Fermi surface.

The hole has been described in terms of its behaviour in **k** space. What about the hole in real space? Formally we may think of a hole as a physical object with the properties listed in Table 9.2. In the presence of an electric field **E**, electrons and

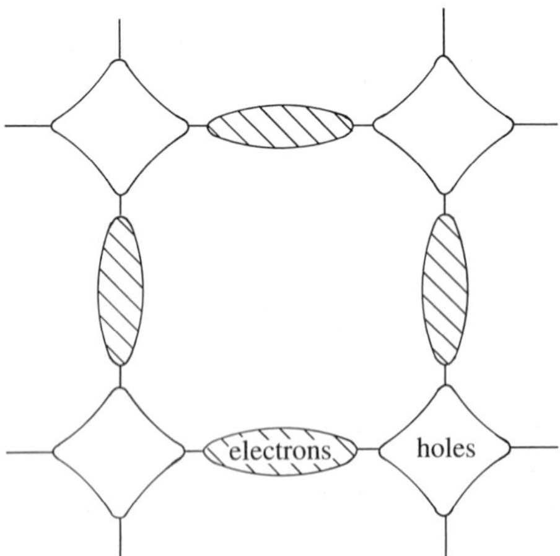


**Figure 9.46** In general the Fermi surface of a divalent metal has hole portions in the first zone and electron portions in the second zone. Using the repeated zone scheme, we may illustrate both on the same diagram. The major part of the first zone is of course filled but these states are replaced by the hole states (compare with Fig. 9.44).

holes in different bands will move in real space in opposite directions on account of their opposite charges; both contribute to the electric current. Their masses are both positive but different, as are their relaxation times (because the bands have different shapes). Formally there is no difficulty in accepting the hole concept if we remember that the behaviour of electrons in a nearly filled band may be considered as equivalent to that of positive particles that we call holes. In other words, we should not ascribe the hole a concrete existence in a metal.

In a semiconductor, however, we find a different situation. Pure Si at 0 K is an insulator because the valence electrons occupy filled energy bands (zones) and there is an energy gap separating them from the nearest empty levels. All the energy bands are either filled or empty. We may, just as before, imagine the excitation of an electron from the highest occupied level to a state in an empty band. We thereby produce a freely moving electron and a mobile hole. Si has the covalently bonded diamond structure, and to excite an electron we must break a bond, which leaves a positive charge on the Si atom with the broken bond. The positive hole is associated with this broken bond and in this case it has a concrete existence in real space as well as in **k** space. In the next chapter, we shall find that insulators and semiconductors can only conduct electricity in the presence of excitations. The charge carriers, electrons and holes (broken bonds), are both localized charges, a Coulomb field exists between them and they may in fact form a metastable combined entity, an electron-positive-hole 'molecule', which is called an exciton. However, if they are given sufficient energy, they become free and move independently. Thus when positive holes are formed by excitations they are concrete physical objects, but in metals the concept of a hole is invoked under equilibrium conditions, there is no localization of positive charge and we should not attempt to give holes a material existence.

On the other hand, just as for semiconductors, we may create an 'electron-hole pair' excitation in a metal, although, owing to the conducting properties of the metal, this excitation has only a transient existence. Thus when we create $K\alpha$ X rays, in say Al, we ionize the $K$ level. An electron is removed from the $K$ level, creating a 'core hole' in this shell, but this is rapidly filled again, leading to the production of the characteristic X radiation. Even so, the lifetimes of these transient holes in metals are often sufficiently long to perturb the electronic structure and affect spectroscopic measurements.

## References


BAER, Y., BUSCH, G. and COHN, P. (1975) *Rev. Sci. Instrum.* **46**, 466.

CHENG, C. H., WEI, C. T. and BECK, P. A. (1960) *Phys. Rev.* **120**, 426.

CHENG, C. H., GUPTA, K. P. VAN REUTH, E. C. and BECK, P. A. (1962) *Phys. Rev.* **126**, 2030.

COTTRELL, A. (1967) *An Introduction to Metallurgy.* Edward Arnold, London.

CRACKNELL, A. P. and WONG, C. (1973) *The Fermi Surface.* Oxford University Press, Oxford.

DUGDALE, J. S. (1977) *The Electrical Properties of Metals and Alloys.* Edward Arnold, London.

HOARE, F. E. and YATES, B. (1957) *Proc. R. Soc. Lond.* **240**, 42.

HÖRNFELT, S (1970) *Solid State Commun.* **8**, 673.

HÜFNER, S., WERTHEIM, G. K. and WERNICK, J. H. (1973) *Phys. Rev.* **B8**, 4511.

JOHANSSON, C. H. and LINDE, J. O. (1936) *Ann. Physik* **25**, 1.

MATHEWSON, A. G. and MYERS, H. P. (1972) *J. Phys.* **F2**, 403.

MATHEWSON, A. G. and MYERS, H. P. (1973) *J. Phys.* **F3**, 623.
MILLER, J. C. (1969) *Phil. Mag.* **20**, 115.
MUELLER, F. (1967) *Phys. Rev.* **153**, 659.
NILSSON, P. O. and LINDAU, I. (1971) *J. Phys. F: Met. Phys.* **1**, 854.
SMITH, N. V. (1970) *Phys. Rev.* **B2**, 2840.
SPRINGFORD, M. (ed.) (1980) *Electrons at the Fermi Surface.* Cambridge University Press, Cambridge.
WESTBROOK, J. (1967) *Intermetallic Compounds.* Wiley, New York.
ZIMAN, J. M. (1960) *Electrons and Phonons.* Oxford University Press, Oxford.

## Further Reading

DUGDALE, J. S. *The Electrical Properties of Disordered Metals* (1995) Cambridge University Press, Cambridge.
SHOENBERG, D. *Magnetic Oscillations in Metals* (1984) Cambridge University Press, Cambridge.
ZIMAN, J. M. (Editor), *The Physics of Metals, Electrons* (1969) Cambridge University Press, Cambridge.

## Problems

**9.1** Using the data presented in Fig. 9.8, what, approximately, are the highest and the lowest apparent values of $N(E_F)$ found for the 'd' metals? What additional aspects might demand a revision of these estimates?

**9.2** It is difficult to make significant conclusions from electronic heat data for Ag-Cd alloys but very rewarding in the case of Ag-Pd alloys. Can you give a good reason for this?

**9.3** Describe the behaviour of a hole state in a two-dimensional system. Assume an electron is missing from the state $(k_x, k_y)$ in the zone. Show that the hole state moves rigidly with the electron states under the influence of an applied field **E**.

**9.4** It is often stated that under the influence of a magnetic field electron and hole states rotate in opposite directions. Describe the conditions under which this statement is correct (a) in real space, (b) in **k** space.

**9.5** The optical spectra of Cu, Ag and Au are shown in Fig. 9.15. The spectra control the colours of these metals; describe how this is so.

**9.6** The alkali metals are known to have very small energy gaps separating the first and second conduction bands – so small in fact that they have an insignificant effect on the optical absorption of these metals. Under such conditions, determine the interband absorption threshold of these metals in terms of the Fermi energy (see Fig. 9.11). Can you think of any reason why Cs fails in this connection?

**9.7** Li metal can dissolve up to 70 atomic per cent Mg; describe qualitatively how the optical spectrum of Li must change as Mg is added. What is the significant feature?

**9.8** In the table of Hall coefficients in Chapter 6 we find that this quantity for Bi is about 1000 times larger than the free electron value. In the light of the results of the present chapter, can you suggest a possible reason for this pronounced discrepancy? Bi has a distorted simple-cubic structure with two atoms to the primitive cell. Consider only principles, not details.

**9.9** Fig. 9.40 shows the electrical resistance of disordered CuAu alloys. Cu is also completely miscible with Pd. How do you think the resistivity of disordered CuPd alloys varies with composition? Indicate on a diagram.

**9.10** Consider the energy band diagram shown below – certain electron and hole states are marked. An electric field acts as shown. Determine the movement of these states in the energy diagram and the sign of their effective masses.

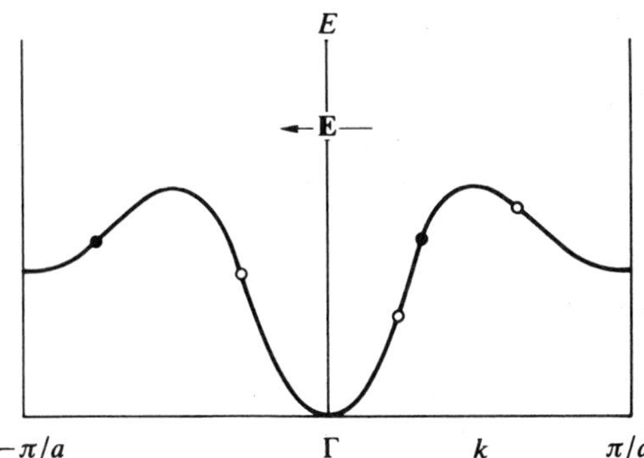


**9.11** Fig. 7.16 shows a (011) section of the Fermi surface of Pb, containing both electron and hole parts. Using the symmetry of the reciprocal lattice construct the corresponding closed sections of the closed electron and hole surfaces.

**9.12** The discussion of the Landau levels in Section 9.1 presumes the absence of the electron spin. How does the energy spectrum for free electrons in a strong magnetic field change when spin is taken into account? Are the conclusions of Section 9.1 still valid?

# 10

# Semiconductors

The importance of the electronic theory of solids as embodied in band theory is that it provides us with clear means of understanding how solids may be insulators, semiconductors or metals. This depends upon whether or not there is a Fermi surface. The existence of a Fermi surface produces metallic behaviour, whereas at 0 K, if the filled electron levels are separated from vacant ones, we have insulating properties. If the separation is large, $\geqslant 5$ eV, the substance remains an insulator at temperatures above 0 K, whereas semiconducting properties arise if the filled and empty levels lie within say 0–2 eV of one another.

In semiconductor physics, we use a distinct terminology: the filled energy band is called the *valence band*, it is after all occupied by the valence electrons; the empty band immediately above it is called the *conduction band*. The zero of energy is usually taken at the top of the valence band (Fig. 10.1). We denote the energy gap separating the two bands by $E_g$. Any process that excites an electron from the valence to the conduction band produces two charge carriers for electrical conduction: namely the electron, and the positive hole left in the valence band. Thermal and optical excitation processes are two important ways of inducing semiconducting behaviour, which will normally involve both electrons (negative 'n' particles) and holes (positive 'p' particles). Semiconductors may be pure elements like Ge or Si, but may also be compounds, for example SiC, $Cu_2O$ and GaAs. Their properties are strongly affected by the presence of impurities, defects or departures from exact stoichiometry; this made their initial study confusing, but it is on this account that many technologically important devices can be made. The properties of a perfect crystal of a pure element or perfectly stoichiometric compound are called *intrinsic* properties, whereas the influences of added impurities or defects give rise to *extrinsic* properties.

## 10.1 Intrinsic Behaviour

Ideally the intrinsic conductivity is zero at 0 K and increases with temperature owing to the thermal excitation of charge carriers (electrons, holes). We write

$$\sigma = \frac{ne^2\tau_e}{m_e} + \frac{pe^2\tau_h}{m_h}, \tag{10.1}$$

**Table 10.1** Resistivity at 20°C (Ω m)

| | |
|---|---|
| Ag | $1.6 \times 10^{-8}$ |
| Cu | $1.7 \times 10^{-8}$ |
| Fe | $10.0 \times 10^{-8}$ |
| Nichrome | $100.0 \times 10^{-8}$ |
| Ge (pure) | 0.5 |
| Si (pure) | $3.0 \times 10^{3}$ |
| diamond | $1.0 \times 10^{14}$ |

the symbols having obvious meanings. We find that the conductivity is dominated by the concentrations $n$ and $p$, which increase rapidly with temperature. The relaxation times are also temperature-dependent but much more weakly so than the carrier concentrations. Thus in semiconductors we can, to a very good first approximation, attribute the temperature variation of conductivity to the changing charge carrier concentrations. This is in direct contrast with the metallic case, where the carrier concentration is large and constant and the temperature dependence arises wholly from the variation of the relaxation time (Table 10.1).

Experimentally the conductivity of a semiconductor varies with temperature as shown in Fig. 10.2, at high temperatures $\ln \sigma \propto T^{-1}$, and this behaviour is associated with the intrinsic behaviour. Unless extreme care is taken in the preparation of a sample, sufficient impurities may be present to influence the low-temperature properties. We shall see later that impurities of valence different from that of the sample may provide a cheap (energywise) source of carriers; these are therefore readily made available at low temperatures and dominate the behaviour. As the temperature is increased, the impurity contribution saturates because of the limited concentration, and the conductivity may then decrease with temperature until the

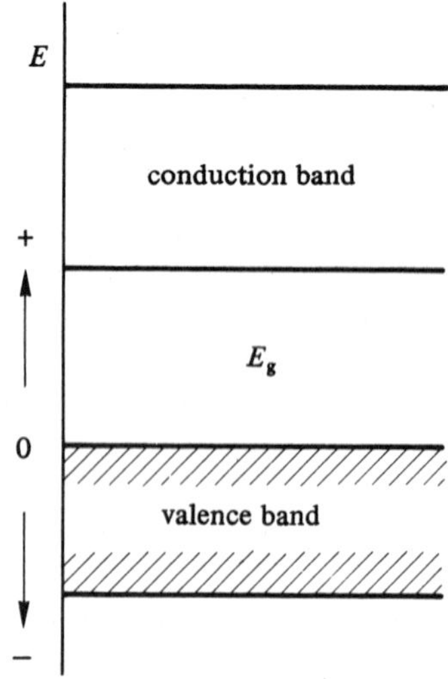


**Figure 10.1** The conventional energy band diagram for a semiconductor. The zero of energy is put at the top of the valence band.

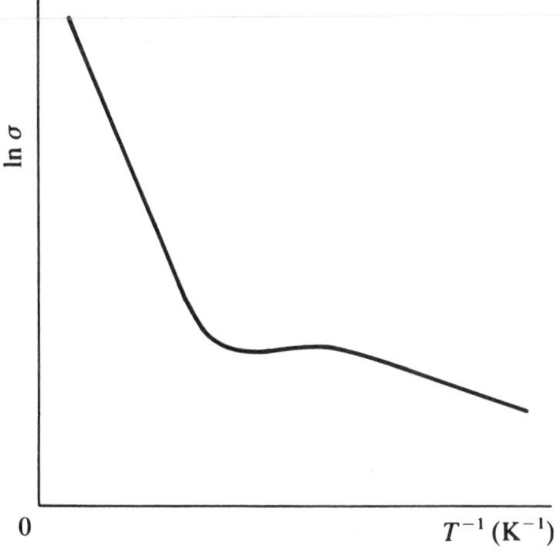


**Figure 10.2** The typical form of the variation of electrical conductivity of a semiconductor with temperature. The behaviour at high temperature implies ln $\sigma \propto T^{-1}$. At low temperatures impurities have a profound effect.

intrinsic contribution takes over. At any given temperature there are equilibrium concentrations of electrons, $n$, and holes, $p$, determined by the Fermi–Dirac distribution. But where do we place the Fermi level in a semiconductor? We must let the analysis decide this question, but we may argue that in metals the Fermi energy lies at the boundary between filled and empty states and we might expect a similar situation to exist in a semiconductor except that filled and empty states are now separated by an energy gap $E_g$. We therefore assume that the Fermi level lies at an energy $+\mu$ measured from the top of the valence band, i.e. somewhere between occupied and empty states.

For electrons, the Fermi factor is

$$f(n) = (e^{(E-\mu)/k_B T} + 1)^{-1}, \tag{10.2}$$

and when $E - \mu \gg k_B T$ it becomes

$$f(n) = e^{-(E-\mu)/k_B T}. \tag{10.3}$$

Our assumption that $E - \mu \gg k_B T$ will be justified later, but for the present we may consider the condition satisfied by making $T$ small. Equation (10.2) is also valid for electrons in the valence band, $E$ then being negative and measured downwards from our energy zero.

The probability of finding a positive hole at an energy $E$ in the valence band is clearly

$$f(p) = 1 - f(n) = 1 - \frac{1}{e^{(E-\mu)/k_B T} + 1}, \tag{10.4}$$

which, in the same low-temperature approximation, becomes

$$f(p) = e^{(E-\mu)/k_B T} = e^{-(|E|+\mu)/k_B T} \tag{10.5}$$

(remember that $E$ is negative in this case).

Now semiconductors are, as their name implies, very poor conductors and the concentrations of electrons and holes are small; the conduction band is almost empty and the valence band almost filled. We therefore assume that the band shapes are parabolic, i.e. of free electron form (Section 6.1.2), but associate effective masses $m_e$ and $m_h$ with the carriers. We therefore write for the band form

$$N(E) = \frac{V}{2\pi^2}\left(\frac{2m}{\hbar^2}\right)^{3/2} E^{1/2}, \tag{10.6}$$

whereas the densities of electron and hole states per unit volume are written

$$\frac{N_n(E)}{V} = \frac{1}{2\pi^2}\left(\frac{2m_e}{\hbar^2}\right)^{3/2} (E - E_g)^{1/2}, \tag{10.7}$$

$$\frac{N_p(E)}{V} = \frac{1}{2\pi^2}\left(\frac{2m_h}{\hbar^2}\right)^{3/2} |E|^{1/2}. \tag{10.8}$$

The electron concentration in the conduction band becomes, using (10.7) and (10.3),

$$n = \frac{1}{2\pi^2}\frac{2m_e}{\hbar^2}\left(\frac{2m_e}{\hbar^2}\right)^{3/2} \int_{E_g}^{\infty} (E - E_g)^{1/2} e^{-(E-\mu)/k_B T}\, dE, \tag{10.9}$$

which is readily transformed to give

$$n = \frac{1}{2\pi^2}\left(\frac{2m_e}{\hbar^2}\right)^{3/2} (k_B T)^{1/2} e^{(\mu - E_g)/k_B T} \int_{E_g}^{\infty} \left(\frac{E - E_g}{k_B T}\right)^{1/2} e^{-(E-E_g)/k_B T}\, dE. \tag{10.10}$$

We make the substitution

$$x = \frac{E - E_g}{k_B T}, \qquad dx = \frac{dE}{k_B T},$$

to obtain

$$n = \frac{1}{2\pi^2}\left(\frac{2m_e}{\hbar^2}\right)^{3/2} (k_B T)^{3/2} e^{(\mu - E_g)/k_B T} \int_0^{\infty} x^{1/2} e^{-x}\, dx. \tag{10.11}$$

The integral is of standard form,

$$\int_0^{\infty} x^{1/2} e^{-x}\, dx = \tfrac{1}{2}\pi^{1/2},$$

and our expression for $n$ finally becomes

$$\begin{aligned} n &= 2\left(\frac{m_e k_B T}{2\pi\hbar^2}\right)^{3/2} e^{(\mu - E_g)/k_B T} \qquad (10.12)\\ &= n_0\, e^{(\mu - E_g)/k_B T}. \end{aligned}$$

Similarly, we obtain the density of holes in the valence band as

$$p = \frac{1}{2\pi^2}\left(\frac{2m_h}{\hbar^2}\right)^{3/2} \int_0^{\infty} |E|^{1/2} e^{-(|E| + \mu)/k_B T}\, d|E|, \tag{10.13}$$

which may be rewritten

$$\frac{1}{2\pi^2}\left(\frac{2m_h}{\hbar^2}\right)^{3/2} (k_B T)^{1/2} e^{-\mu/k_B T} \int_0^{\infty} \left(\frac{|E|}{k_B T}\right)^{1/2} e^{-|E|/k_B T}\, d|E|. \tag{10.14}$$

The expression (10.14) may be evaluated in the same manner as (10.11), giving

$$p = 2\left(\frac{m_h k_B T}{2\pi\hbar^2}\right)^{3/2} e^{-\mu/k_B T} \quad (10.15)$$

$$= p_0 e^{-\mu/k_B T}.$$

For $T = 300$ K and $m_e = m_h = m$

$$n_0 = p_0 = 2.5 \times 10^{25} \text{ m}^{-3}.$$

*Expressions* (10.12) *and* (10.15) *for the equilibrium concentrations of electrons and holes are valid for any semiconductor irrespective of whether it is perfectly pure or whether it contains impurities. This is because only states in the conduction and valence bands are itinerant; the impurity concentrations are usually so small* ($\ll 0.1\%$) *that they can neither produce new bands nor affect the shape of those of the host material.* This statement is substantiated in the electronic theory of solid solutions, but we may appreciate its validity because normally the added impurity concentrations in semiconductors are smaller than the unavoidable impurity contents of many so-called pure metals.

If we form the product $np$, we find

$$np = 4\left(\frac{k_B T}{2\pi\hbar^2}\right)^3 (m_e m_h)^{3/2} e^{-E_g/k_B T}. \quad (10.16)$$

For any given semiconductor (i.e. fixed value of $E_g$) at a specific temperature, this product is constant. Thus if, in some way (e.g. small additions of foreign elements), we increase $n$, say, then this will automatically demand a reduction in $p$ to maintain the product constant. We shall discuss this aspect further a little later, but for the moment we return to the intrinsic situation where the excitation of an electron to the conduction band produces a hole in the valence band. The two types of charge carrier can only arise in pairs and, letting the subscript 'i' denote intrinsic concentration, of necessity we must have

$$n_i = p_i. \quad (10.17)$$

Using (10.12) and (10.15), we immediately find that

$$m_e^{3/2} e^{(\mu - E_g)/k_B T} = m_h^{3/2} e^{-\mu/k_B T},$$

whence

$$\mu = \tfrac{1}{2}E_g + \tfrac{3}{4}k_B T \ln \frac{m_h}{m_e}. \quad (10.18)$$

*Clearly at* $T = 0$ *an intrinsic semiconductor has its Fermi level in the middle of the band gap*

$$\mu = \tfrac{1}{2}E_g, \qquad T = 0. \quad (10.19)$$

If $m_e \approx m_h$ then this is true also for finite temperatures and *in practice we may usually assume the Fermi level to be at* $\frac{1}{2}E_g$ *for the intrinsic semiconductor.* However, for certain semiconductors $m_e$ and $m_h$ may differ markedly (see Table 10.3).

We can now appreciate that our earlier assumption $E - \mu \gg k_B T$ holds very well because the minimum value of $|E - \mu|$ is $\mu$, which, as we have seen, takes the value $\frac{1}{2}E_g \approx eV$ and the above condition is adequately fulfilled at all practical temperatures.

The intrinsic conductivity may be written

$$\sigma_i = \frac{n_i e^2 \tau_e}{m_e} + \frac{p_i e^2 \tau_h}{m_h},$$

but it is more usual to define a *mobility* $\mu_c$, which is the velocity (always positive by convention) that the carrier achieves in unit electric field strength. We write $\mu_e$ and $\mu_h$ for the electron and hole mobilities, or in general $\mu_c$ for the carrier mobility:

$$v = \left|\frac{e\tau}{m}\right| |\mathbf{E}|, \qquad \mu_c = \left|\frac{e\tau}{m}\right|.$$

Then

$$\sigma_i = n_i e\mu_e + p_i e\mu_h. \tag{10.20}$$

The mobility is temperature-dependent and remains to be determined by experiment or calculation, but we can expect that it will be of the form $\mu_c \propto T^n$, with $n$ somewhere in the interval $-2 < n < 2$, so that its variation is negligible compared with the exponential dependence on temperature of $n_i$ and $p_i$. We therefore approximate (10.20) by

$$\sigma_i = Ae^{-E_g/2k_B T}. \tag{10.21}$$

A knowledge of the intrinsic conductivity and its variation with temperature therefore allows the determination of the energy gap $E_g$. Data for $UO_2$ are shown in Fig. 10.3, and $E_g$ is determined as 2.6 eV for this substance. Germanium, on the other hand, has $E_g = 0.7$ eV, and we may calculate the concentration of electrons and holes for different temperatures from (10.12) and (10.15), assuming that $m_e = m_h = m$. These values are given in Table 10.2.

It is perhaps more instructive to calculate the number of carriers per atom; we can then compare this quantity with that for a typical metal such as Cu (1 carrier per atom). At room temperature the carrier concentration in Ge is about $10^{-9}$ that of a metal such as Cu.

## 10.2 Band Structure of Ge and Si

Ge and Si have a crystal structure built around the fcc lattice, the basis being 000, $\frac{1}{4}\frac{1}{4}\frac{1}{4}$ (Fig. 2.12). The fcc crystallographic cell contains eight atoms and the primitive

**Table 10.2** Carrier concentration in pure Ge ($n_i = n = p$, $E_g = 0.7$ eV, $m_e = m_h = m$)

| $T$ (K) | $n_i$ ($m^{-3}$) | $n_i$ (atom$^{-1}$) |
|---|---|---|
| 500 | $1.59 \times 10^{22}$ | $3.6 \times 10^{-7}$ |
| 400 | $1.49 \times 10^{21}$ | $3.4 \times 10^{-8}$ |
| 300 | $3.27 \times 10^{19}$ | $7.4 \times 10^{-10}$ |
| 200 | $2.05 \times 10^{16}$ | $4.6 \times 10^{-13}$ |
| 100 | $1.09 \times 10^{7}$ | $2.5 \times 10^{-22}$ |
| 0 | 0 | 0 |

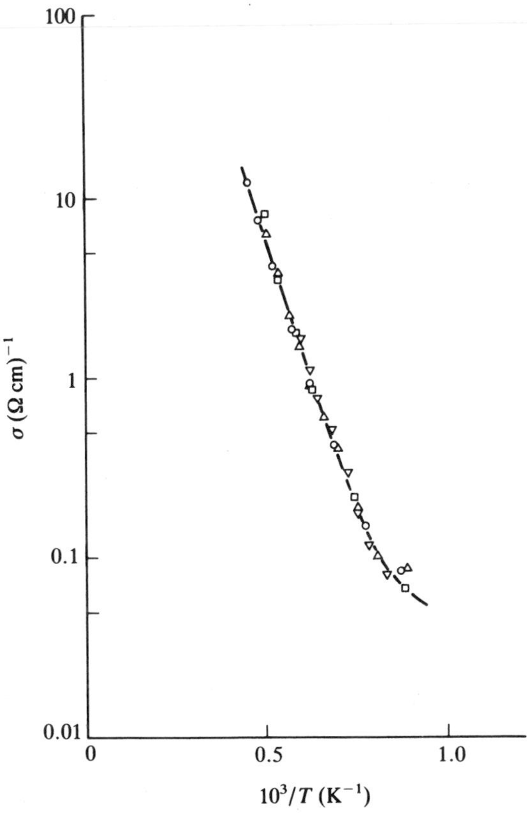


**Figure 10.3** The electrical conductivity of $UO_2$. The observed intrinsic conductivity corresponds to an energy gap of 2.6 eV between valence and conduction bands. (Reprinted with permission from Myers *et al.*, *Solid State Commun.*, **2**, 1964. Pergamon Press, Oxford.)

cell of the underlying Bravais lattice must be associated with the basis of two atoms. Each atom has eight sp electron states, four of which are occupied by electrons. Thus, all in all, our zone structure must accommodate 16 electron states, half of which are filled. This leads to a band structure of eight sub-bands, four completely filled and four completely empty at 0 K.

The bonding in Ge and Si is covalent in character, and in the first approximation we may consider the sp states in the free atoms as forming an eight-fold-degenerate level. Consider the formation of an Si dimer, i.e. a diatomic molecule. Just as in the formation of the $H_2$ molecule, we expect the electron states to interact and form bonding and antibonding levels (Fig. 10.4). In our simple degenerate sp approximation there will be room for eight electrons in each level. Together, the two Si atoms contain eight valence electrons, just sufficient to fill the bonding level. The separation into bonding and antibonding levels is the origin of the energy gap in the band structure of solid Si. It is of course a gross oversimplification to consider degenerate sp levels and, for the solid, detailed calculation shows that the sixteen sp states are to be associated with eight bands, each with its own dispersion pattern determined by the crystal potential. The lower four bands are filled and the upper four bands are empty. The band structure of Si is shown in Fig. 10.5, where the division into two groups of bands is clearly seen.

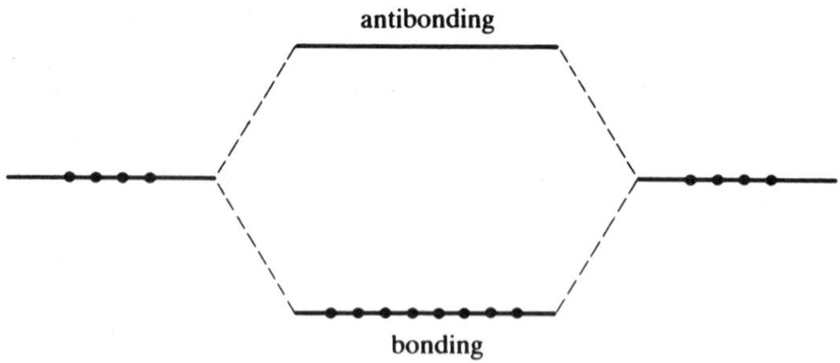


**Figure 10.4** A stylized picture of bonding and antibonding levels in a silicon dimer.

One should not worry about the detailed shapes of these bands but accept them as solutions of the Schrödinger equation in the assumed potential. The difference in energy between the highest occupied and lowest unoccupied state is indicated as 1.17 eV at 0 K. It is to be noted, however, that these limiting states lie at different points within the zone, and this minimum excitation of an electron from the valence to the conduction band demands that it be given crystal momentum equivalent to the difference in the **k** vectors; this is readily obtained in thermal excitation by the crystal momentum available from the phonons. Another feature of the band diagram shown in Fig. 10.5 is that the valence band has a composite structure. There are three uppermost bands degenerate at Γ together with a separate lower fourth valence band. These bands have different curvatures and therefore their holes have different effective masses. Thus, whereas diagrams such as Fig. 10.1 and equa-

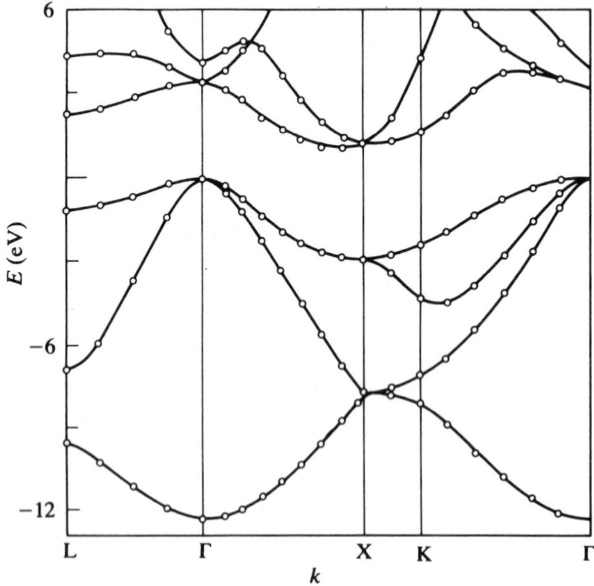


**Figure 10.5** The energy band structure of silicon. The highest occupied state lies at Γ but the lowest lying unoccupied state is close to the X point and the gap is said to be 'indirect'. Note the composite structure of the valence band. (After Chelikowsky and Cohen 1976. See also *Landolt–Börnstein* 1982.)

tions like (10.20) are convenient for discussing the principles of semiconductor behaviour, the actual properties of a specific semiconductor can be very complicated and only fully understood after much experimental and theoretical study.

The energy gap characterizing an intrinsic semiconductor may also be determined by optical absorption. At low temperatures there are very few carriers and the Drude contribution to the absorptivity is absent in an intrinsic semiconductor. Semiconductors are therefore transparent to infrared radiation and become absorbing only when interband transitions are excited. We might therefore expect the band gap to fix the threshold for interband transitions and in this way be readily determined by experiment. This would be the case if the band gap were determined by filled and empty states with the same reduced value of $\mathbf{k}$; the absorption of light would then cause a 'vertical' transition and give rise to a sharp edge at $\hbar\omega = E_g$. This is the case for many semiconductors, for example GaAs, whose band diagram and optical absorption are shown in Figs 10.6 and 10.7.

In Si (and many other semiconductors), we have seen that excitation across the minimum separation of filled and empty states demands a large change in wave vector, and such a transition cannot be initiated by a photon unless it has access to a source of crystal momentum. Again it is the phonons that provide the required momentum.

We write the conservation laws in the form

$$\left.\begin{aligned} E_f - E_i &= h\nu + \hbar\omega_f \\ \mathbf{k}_f - \mathbf{k}_i &= 0 + \mathbf{q}. \end{aligned}\right\} \qquad (10.22)$$

$\mathbf{q}$ and $\omega_f$ apply to the phonon involved in the transition. Now $E_f - E_i = E_g$ and it is clear that the inclusion of phonons produces an absorption edge at a somewhat

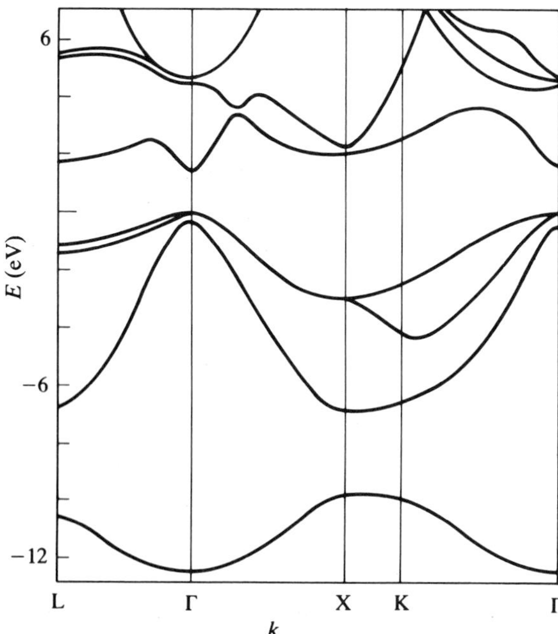


**Figure 10.6** The energy band structure of GaAs. This semiconductor has a direct energy band gap lying at the Γ point. (After Chelikowsky and Cohen 1976.)

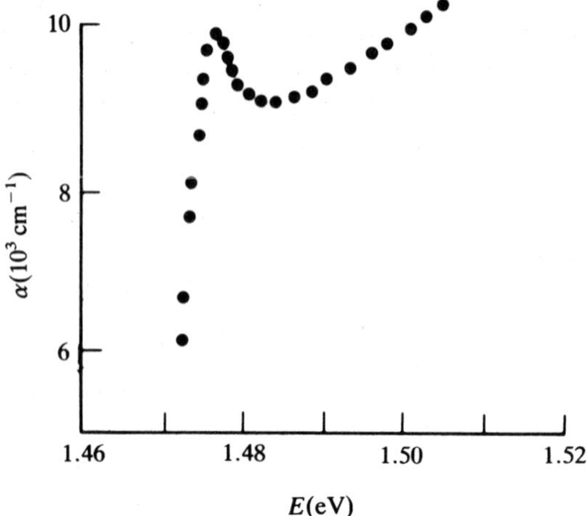


**Figure 10.7** The optical absorption for GaAs measured at 85 K. The absorption edge corresponds to transitions across the 'direct' band gap. The peaked appearance is attributed to the existence of an excitonic state (i.e. a metastable electron-positive hole molecule; see Section 9.8) lying just beneath the conduction band. (After Sturge 1962.)

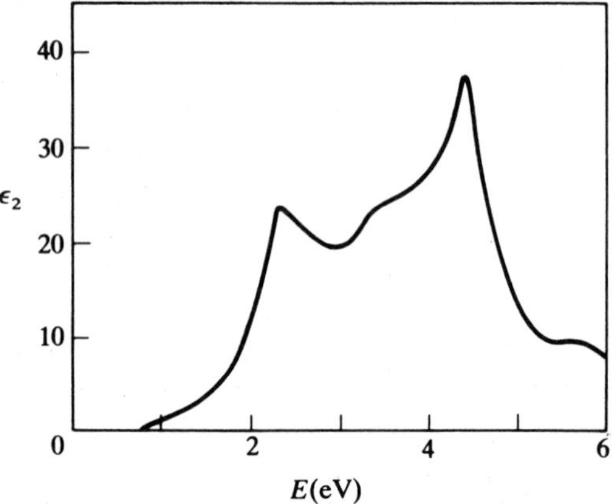


**Figure 10.8** The imaginary part of the dielectric constant of Ge. Below 0.7 eV Ge is transparent to light and $\varepsilon_2 \approx 0$; interband absorption begins near 0.7 eV but is weak until direct 'vertical' transitions start near 1.5 eV. (After Brust *et al.* 1962.)

lower energy, namely $E_g - \hbar\omega_f$. These indirect or phonon-assisted transitions produce only weak absorption compared with that associated with direct transitions. The optical absorption beyond the edge is therefore dominated by direct or vertical transitions and the spectrum has the same origins as the interband spectra of metals discussed earlier. Figure 10.8 shows the spectrum for Ge.

## 10.3 Doped Semiconductors

Early studies of semiconductors were plagued by the irreproducibility of results on what were purportedly identical samples. We now know that this situation arose because semiconductors are extremely susceptible to the presence of foreign elements and the only way to establish definitive properties is to obtain highly pure starting materials and then study the effect of impurity additions or 'dopants' in a systematic way. Germanium has now been obtained so pure that it contains only 1 part in $10^{12}$ impurity (the purest metal so far obtained is probably magnesium with 1 part in $10^8$ impurity, whereas many ordinary metals such as Mn or Co cannot be obtained purer than say 1 part in $10^4$). The handling of very pure semiconductor materials demands 'surgical cleanliness'.

Consider Ge containing a very small quantity of As; the latter is pentavalent and the bonding orbitals of the surrounding Ge atoms can be mated. But the arsenic has its fifth electron without a matching Ge bond. We think of this electron as attached to an $As^+$ ion, and electronically the arrangement simulates a 'one-electron atom' (Fig. 10.9). We might expect this electron to be bound to the $As^+$ ion with an energy

$$E = \frac{e^4 m}{2/(4\pi\varepsilon_r \varepsilon_0 \hbar)^2} \tag{10.23}$$

corresponding to the ionization energy of a hydrogen-like atom. Electrostatic neutrality demands that the fifth electron be bound to the parent arsenic ion. But it is embedded in a dielectric medium. Since germanium has a static dielectric constant $\varepsilon_r = 16$, if the extra electron experiences the full dielectric shielding of the medium then this reduces the ionization energy from 13.6 eV, typical of the hydrogen atom, to $13.6/256 = 0.053$ eV. If this full dielectric screening applies, we must find the fifth electron in a large orbit encompassing many Ge atoms, because the effective radius of the orbital is directly proportional to the dielectric constant $\varepsilon_r$. If the electron has an effective mass less than that of a free electron then this will further increase the orbital radius and decrease the binding energy. Experiment confirms that the fifth electron is weakly bound to the parent $As^+$ ion with an ionization energy 0.013 eV. Compared with the band gap of 0.7 eV, this is very small, and it is clear that the impurity atoms provide a much cheaper source of carriers than the intrinsic process can. On our band diagram we expect pentavalent atoms to possess their fifth electron in a weakly bound state lying just under the conduction band (Fig. 10.10). Thermal excitation of this state promotes the electron to the conduction band, where it becomes a free carrier. The impurity has given an electron to the matrix conduction band and it is therefore called a *donor* atom.

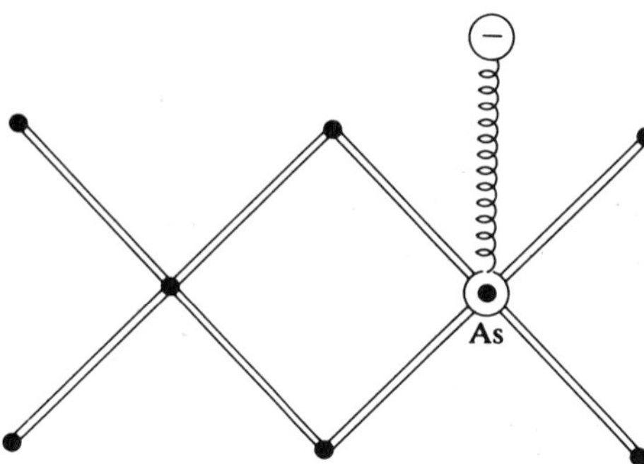


**Figure 10.9** The extra electron on an As impurity atom in Ge is loosely bound and occupies an orbital of large diameter.

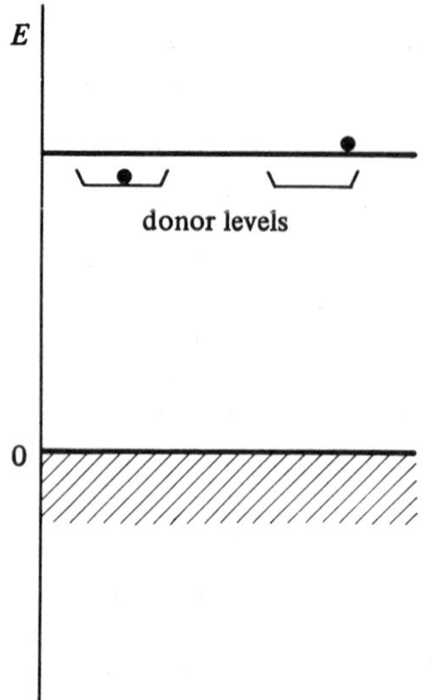

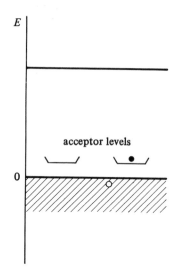


**Figure 10.10** Electrons in donor levels are weakly bound and lie just under the bottom of the conduction band. When they become ionized, electrons occupy states in the empty conduction band.

E
acceptor levels
0

**Figure 10.11** Acceptor levels lie just above the top of the valence band. If they become ionized, a free hole is introduced into the valence band.

Suppose our germanium sample contained not arsenic but boron as impurity. The latter is trivalent and can only mate with three of the four bonds of the germanium. The germanium matrix determines the band structure and, although the system is electrostatically neutral, the valence band lacks an electron; it therefore possesses a positive hole which we consider bound to the boron impurity. If, however, we create $B^-$, a negative boron ion, the boron then has four electrons and can satisfy the bonding requirements of the lattice. The situation can arise if an electron from a Ge atom is transferred to the boron. The positive hole is then found on a Ge atom and the hole is free to move on each Ge atom. The hole is no longer bound to the boron atom but belongs to the valence band – it has become a charge carrier. By accepting an electron from the matrix, the impurity boron atom has produced a free positive hole in the matrix valence band. We say that boron is an acceptor impurity. It costs energy to ionize the acceptor impurity atom, but again we find this energy to be a very small quantity (for B in Ge it is 0.010 eV) compared with the band gap. The acceptor states lie just above the top of the valence band (Fig. 10.11). The presence of donor or acceptor impurities is the origin of extrinsic semiconduction. Certain data for four important semiconductors are given in Tables 10.3 and 10.4.

## 10.4 Extrinsic Behaviour

We consider the presence of donor impurity atoms in an otherwise perfectly intrinsic semiconductor and attempt to determine the Fermi level (i.e. an expression

**Table 10.3** Data for certain semiconductors

| Material | $E_g$ (eV) | Mobility ($m^2\ V^{-1}\ s^{-1}$) n | Mobility ($m^2\ V^{-1}\ s^{-1}$) p | Effective mass $m^*/m$ n | Effective mass $m^*/m$ p | $\varepsilon_r$ |
|---|---|---|---|---|---|---|
| Si | 1.14 | 0.16 | 0.05 | 0.26 | 0.50 | 11.8 |
| Ge | 0.67 | 0.38 | 0.18 | 0.12 | 0.32 | 16 |
| GaAs | 1.40 | 0.85 | 0.04 | 0.07 | 0.68 | 13.5 |
| InSb | 0.18 | 8.0 | 0.14 | 0.016 | 0.4 | 16.5 |

equivalent to (10.19)). We use the following notation:

$$N_d = \text{concentration of donor atoms,}$$

$$N_{d+} = \text{concentration of ionized donors,}$$

$$N_{do} = \text{concentration of neutral donors.}$$

Some donor atoms are ionized, having provided electrons to the conduction band, others remain neutral, the relative proportions being governed by the statistical equilibrium appropriate to the particular temperature. We know that for Ge at 300 K the carrier concentration is about $5 \times 10^{13}$ $cm^{-3}$. Ge contains $4.4 \times 10^{22}$ atoms $cm^{-3}$, and 1 ppm impurity corresponds to $4.4 \times 10^{16}$ impurity atoms $cm^{-3}$. If only 1% of these impurity atoms were ionized at room temperature, they would still outnumber by a factor of 10 the intrinsic carrier concentration, so it is clear that the impurities dominate the behaviour. We therefore assume that $n \gg n_i$ because of this, and, by (10.16) the hole concentration is suppressed. We may think of the large concentration of electrons in the conduction band as causing a recombination of extrinsic electrons and intrinsic holes in the valence band, thereby suppressing the latter and maintaining (10.16). Effectively we may say that in the presence of a large donor electron concentration, the majority of the intrinsic electrons and holes recombine, leaving

$$n \approx N_{d+}. \tag{10.24}$$

Furthermore,

$$N_{do} = N_d f(E_g - E_d). \tag{10.25}$$

Clearly this is correct because the concentration of neutral donors must equal the concentration of electrons with energy $E_g - E_d$ (Fig. 10.12) (note that $E_d$ is measured from the bottom of the conduction band – it is the ionization energy of the donor level).

**Table 10.4** Ionization energies (meV) for impurity levels in Si and Ge

| | B | Al | Ga | In | P | As | Sb |
|---|---|---|---|---|---|---|---|
| Si | 45 | 57 | 65 | 160 | 45 | 49 | 39 |
| Ge | 10.4 | 10.2 | 10.8 | 11.2 | 12 | 12.7 | 1 |

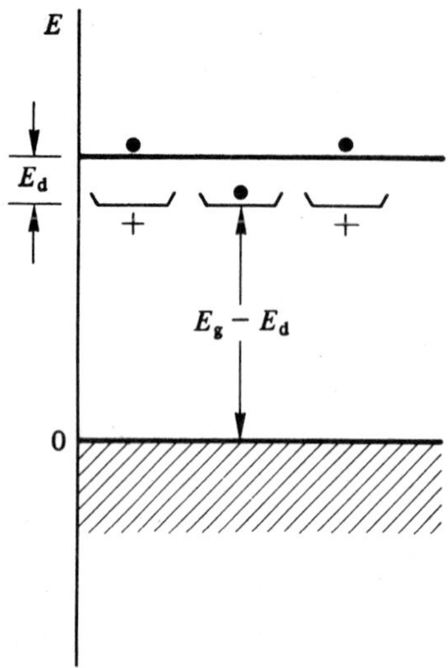


**Figure 10.12** The energy level diagram for an extrinsic n-type semiconductor.

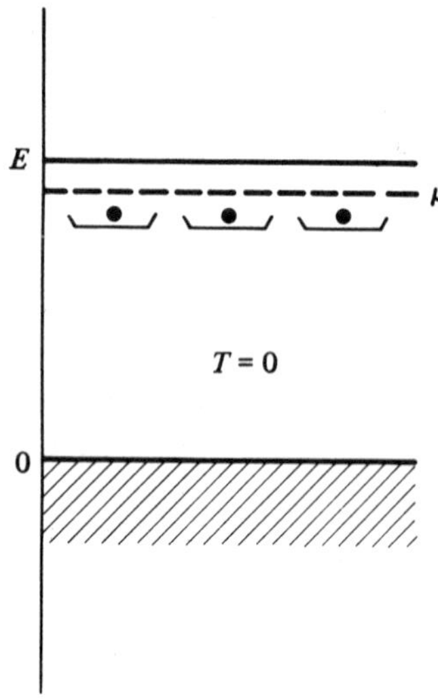


**Figure 10.13** At $T = 0$ K the Fermi level lies halfway between the source of electrons and their recipient; in the case of the n-type semiconductor this means halfway between the position of the donor levels and the bottom of the conduction band.

Thus

$$N_{d0} = \frac{N_d}{e^{(E_g - E_d - \mu)/k_B T} + 1}, \tag{10.26}$$

$$n = N_d^+ = N_d - N_{d0}, \tag{10.27}$$

which on evaluation using (10.26) becomes

$$n = \frac{N_d}{e^{(\mu - E_g + E_d)/k_B T} + 1}. \tag{10.28}$$

Our earlier relation (10.12) demands that

$$n = 2\left(\frac{m_e k_B T}{2\pi\hbar^2}\right)^{3/2} e^{(\mu - E_g)/k_B T},$$

which we write as

$$n = n_0 \, e^{(\mu - E_g)/k_B T},$$

where at room temperature $n_0 = 2.5 \times 10^{25}$ m$^{-3}$ (if $m_e = m$); so we have

$$n = n_0 \, e^{(\mu - E_g)/k_B T} = \frac{N_d}{e^{(\mu - E_g + E_d)/k_B T} + 1}. \tag{10.29}$$

Earlier we said that $\mu$ must lie between the source of electrons and the receiving levels. Now if the conductivity is impurity-dominated, the Fermi level must lie between the impurity donor levels and the bottom of the conduction band. Thus we expect $\mu > E_g - E_d$ and so $\mu - E_g + E_d > 0$. Therefore *at low temperatures*

$$n = n_0\, e^{(\mu - E_g)/k_B T} = N_d\, e^{-(\mu - E_g + E_d)/k_B T}, \tag{10.30}$$

which simplifies to

$$\mu = E_g - \tfrac{1}{2}E_d + \tfrac{1}{2}k_B T \ln \frac{N_d}{n_0}. \tag{10.31}$$

*When $T = 0$ we see that the Fermi level lies midway between the donor levels and the bottom of the conduction band* (Fig. 10.13). At low temperatures the impurity contribution dominates the behaviour and the Fermi level lies at $E_g - \frac{1}{2}E_d$. However, at high enough temperatures (the higher the larger $E_g$) the intrinsic behaviour must take over and then the Fermi level lies at $\frac{1}{2}E_g$. We may sketch the qualitative behaviour of $\mu$ with temperature (Fig. 10.14). If the impurities provide acceptor states then the valence band is the source of electrons and the acceptor levels are the recipients for them; the Fermi level is then found at $\frac{1}{2}E_a$, where $E_a$ is the energy of the acceptor level measured from the top of the valence band. We obtain an equivalent relation to (10.31) (see Problem 10.7) and the Fermi level varies with temperature as shown in Fig. 10.15.

Returning to (10.30), we find that at low temperatures

$$n = n_0\, e^{(\mu - E_g)/k_B T} = N_d\, e^{-(\mu - E_g)/k_B T} e^{-E_d/k_B T},$$

and we may write

$$e^{2(\mu - E_g)/k_B T} = \frac{N_d}{n_0}\, e^{-E_d/k_B T},$$

whence

$$n = n_0 \left(\frac{N_d}{n_0}\right)^{1/2} e^{-E_d/2k_B T} = (n_0 N_d)^{1/2} e^{-E_d/2k_B T}. \tag{10.32}$$

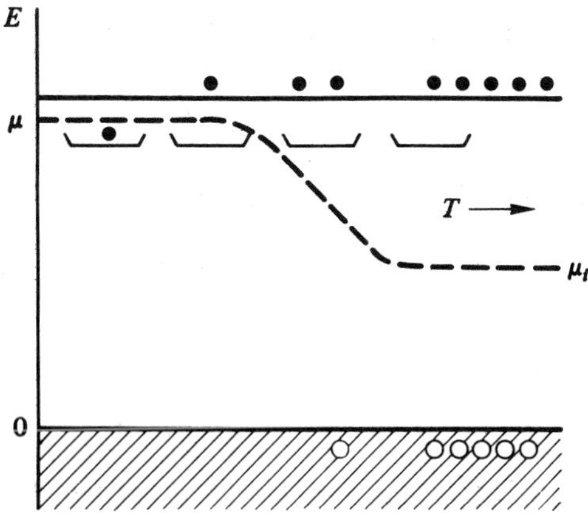


**Figure 10.14** This figure illustrates how the Fermi level in an n-doped semiconductor changes with temperature; the temperature increases from left to right in the diagram. At very low temperatures there are no free carriers and the first to arise come from ionized donor atoms, the Fermi level initially lying at $\frac{1}{2}E_d$; but at sufficiently high temperatures the intrinsic carrier concentrations become dominant and the Fermi level returns to $\frac{1}{2}E_g$. The lower the dopant concentration the lower the temperature at which the reversion to intrinsic behaviour occurs.

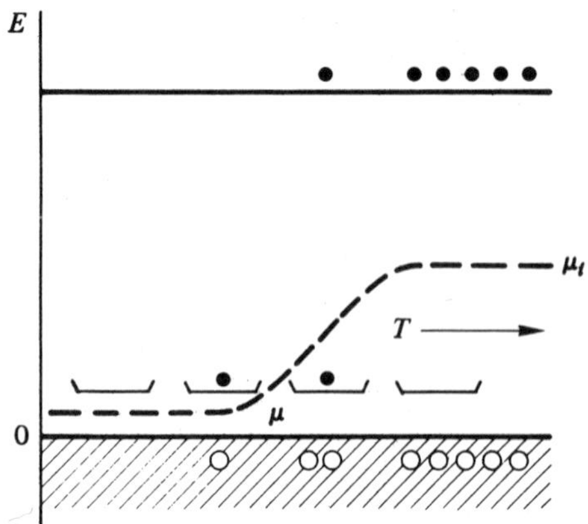


**Figure 10.15** Similar to Fig. 10.14 but for a p-doped semiconductor.

The conductivity is governed by

$$\sigma = ne\mu_e . \tag{10.33}$$

What happens if we have the simultaneous presence of acceptor and donor impurities? The answer to this question rests on the total electrostatic neutrality of the sample and we write an equation for the equality of positive and negative charge, namely

$$n + N_a^- = p + N_d^+ .$$

When we write out this equation in full we find that the Fermi functions complicate matters and it must be solved for the Fermi energy by iteration. However it is often possible to simplify the equation by neglecting an insignificantly small term. The presence of both acceptor and donor dopants produces compensation, which is to say that certain donor states contribute electrons to lower lying acceptor states; in this way the electrons and holes usually associated with these states are neutralized. Compensation may be partial or complete, dependent on the concentrations of the two types of impurity.

So far we have only considered dopant elements that have unit difference in valence from Ge or Si. There is no reason why impurities with greater difference in valence should not be incorporated in say Ge. Thus Ca and Zn may be considered as two electon acceptors. These and similar impurities provide a spectrum of shallow acceptor or donor levels. On the other hand there are elements like Cu, Au and many transition metals that may give rise to several bound electron states distributed throughout the band gap; Au in Ge for example produces five levels that act as acceptor or donor states. When such states arise close to the middle of the band gap they are known as 'deep' impurity levels. Such levels are detrimental in that they facilitate the recombination of electrons and holes by non-radiative processes, particularly in the indirect band gap elements Ge and Si. They thereby decrease the lifetime and mobility of the free charge carriers and particularly so for the minority carrier. In an indirect band gap semiconductor the electrons and holes recombine by the emission of photons with a very small efficiency $\sim 10^{-6}$ because the crystal momentum (i.e. the **k** vector) cannot be easily conserved, instead the recombination energy must be dissipated by the creation of a phonon. The deep impurity levels facilitate this process by providing places where an electron or hole may lie in wait for the appropriate carrier and also because the **k** conservation does not have to occur in a single event.

The carrier mobility $\mu_c$ is dependent on several factors that may vary with temperature. It is clear that the phonons and the presence of impurity dopant species must influence the movement of the charge carriers. At very low temperatures the dopants are neutral atoms, whereas at high temperatures they exist as ionized impurities. Impurity scattering is therefore temperature dependent, but it is not obvious what form this variation will take.

The mobility, particularly in lightly doped samples, may be measured by observing the transit of carriers in an applied electric field, the carriers being created at a particular place in a linear sample by illumination with light of suitable wavelength; such experiments provide a measure of the drift mobility. Another way is to make use of the Hall effect.

## 10.5 The Hall Mobility

The Hall effect is readily observed in semiconductors, being large on account of the small carrier concentrations. It immediately tells us the sign of the majority carrier. Together with the conductivity, it allows a direct measure of the mobility of the impurity carrier. Suppose we have Ge doped with As so that intrinsic conductivity is suppressed. The current is then carried wholly by electrons in the conduction band. Measurement of conductivity and Hall constant on the sample provide

$$\sigma = ne\mu_e, \qquad R = 1/ne,$$

whence

$$\mu_e = R\sigma. \tag{10.34}$$

*These two measurements give us the concentration and sign of the charge carrier and its mobility.* Experimentally one finds that for lightly doped samples $\mu_c \propto T^{3/2}$ at low temperatures (impurity scattering) and $\mu_c \propto T^{-3/2}$ at high temperatures (phonon scattering).

When both electrons and holes are present in the same sample the expression for $R$ becomes somewhat more complicated. As implied by (10.3), the charge carriers in a semiconductor (unless it is very heavily doped $\sim 10^{24}$ $\mathrm{m}^{-3}$) do not exist as a degenerate electron gas. They do not possess a Fermi velocity but one that is determined by the thermal energy 3/2 $(k_B T)$. Their velocity is therefore $\sim\sqrt{T}$. The density of the lattice vibrations above the Einstein temperature ($\sim$300 K) is $\sim T$. The carrier relaxation time depends inversely on the carrier velocity and the phonon density so the high temperature mobility $\sim T^{-3/2}$ in agreement with experiment.

## 10.6 Oxide Semiconductors

In semiconductor device technology, Ge and Si, together with the so-called III–V (e.g. GaAs) and II–VI (e.g. ZnS) compounds, occupy prominent places. The symbols III–V etc. denote the groups or columns of the periodic table from which the elements are selected. There are, however, other semiconducting substances, notably the oxides and sulphides (the so-called chalcogenides) as well as the pnictides (e.g. the nitrides and phosphides) of the transition and rare earth metals, which display a wide variety of electrical properties. In particular there are certain oxides such as VO and $V_2O_3$ that are semiconductors at low temperatures but become metallic at high temperatures; the transition is quite sharp. This phenomenon of the *semiconductor–metal transition* is a subject of considerable current interest. We shall now describe how semiconducting behaviour arises in a transition metal oxide. We might think that a perfectly stoichiometric transition metal oxide denoted by $M^{2+}\ O^{2-}$ would be an insulator; the common impression is that 'non-metals', e.g. oxides, are insulators. However an ion like $Mn^{2+}$ or any of its neighbours in the 3d series possesses an incomplete 3d shell. On the band model we should expect to find a partially filled narrow 3d band that would give rise to metallic properties. MnO is, however, in stoichiometric form, an insulator and a semiconductor when non-stoichiometric.

Energy bands arise when there is sufficient overlap between orbitals on neighbouring atoms to produce a system of closely spaced levels distributed over an appreciable energy range $\approx 1$–$10$ eV. In an oxide (or similar compound) with small d orbital overlap, i.e. tight binding conditions, the transition metal ion contains an integral number of electrons. If conduction is to arise in the d states an electron must be transferred from one ion (which then becomes $M^{3+}$) to an adjacent ion (becoming $M^{+}$). The extra electron on the now $M^{+}$ ion experiences considerable Coulomb repulsion from the electrons already on the former $M^{2+}$ ion and this repulsion may be large enough to prohibit the electron transfer; we say that there is a large correlation energy for this process. The substance at 0 K is then an insulator in the stoichiometric state. If the electron repulsion dominates (as is the case for the later 3d oxides owing to the small 3d shell radii) then our stoichiometric MO will be an insulator, whereas if the 3d orbital overlap is large (as in the oxides of Ti and V) metallic conductivity is found. The detailed behaviour of a particular compound is therefore dependent on the relative strengths of electron repulsion and orbital overlap. Those who might like to read further regarding the properties of specific substances are referred to Cox (1989).

Returning to our representative oxide MO we assume that in stoichiometric form it is an insulator. Deviations from stoichiometry are, however, more the rule than the exception and if the actual composition is $MO_{1+x}$ the structure contains extra oxygen ions and to every extra $O^{2-}$ we must find two balancing positive charges. This can only be achieved by converting two $M^{2+}$ ions into $M^{3+}$. Since we have removed an electron from an $M^{2+}$ ion to make the $M^{3+}$ ion we may consider the latter to be an $M^{2+}$ ion with an attached positive hole

$$M^{3+} \equiv M^{2+} + \text{positive hole} \qquad (10.35)$$

This positive hole is a latent charge carrier. Whether or not semiconducting properties arise depends on the mobility of the positive holes.

On the other hand, if the oxide possesses excess metal, if it has the formula $M^{2+}\ O^{2-}_{1-x}$, we may regard every excess metal atom as a divalent ion plus two attached electrons. If these electrons can be made free to move, we obtain n-type semiconducting properties. These deficit or excess metal compounds are, in principle at least, easy to understand, but the mechanism of conduction is significantly different from that of extrinsic conduction in doped Si or Ge. In the oxides, the nonstoichiometry fixes the number of charge carriers and the mobility is the decisive quantity. Thus if $MO_{1+x}$ is to conduct electricity, a positive hole must move from an $M^{3+}$ ion to a neighbouring $M^{2+}$, which in turn becomes $M^{3+}$. The extra charge associated with the $M^{3+}$ ion produces a polarization of the medium around it, and when the positive hole moves it must take the polarization and associated elastic strain field with it. Analysis shows that the conductivity is governed by

$$\sigma = Be^{-W/k_B T}, \tag{10.36}$$

where $W$ is an activation energy to move the charge carrier from one M site to an adjacent one. Although (10.36) is *similar* to (10.21), there is a significant difference because in (10.36) the exponential term applies to the mobility of the charge carrier. *We have a temperature-activated mobility.* Under the influence of an applied field and temperature, the positive hole hops from one metal ion to another and the process is in fact known as *hopping conduction.*

Clearly, just as departures from stoichiometry may produce semiconduction, so can the addition of a third element of different valence. If for example we dope NiO with Li then for every $Li^+$ ion we incorporate we must form an $Ni^{3+}$ ion to maintain charge neutrality. Each Li atom therefore introduces a positive hole:

$$Li_x Ni_{1-x} O \equiv Li_x^+ Ni_x^{3+} Ni_{1-2x}^{2+} O^{2-}. \tag{10.37}$$

Experiment shows that the presence of 0.05% Li in NiO causes a reduction in the resistivity by a factor of $10^4$. At sufficiently high temperature these extrinsic oxide semiconductors become intrinsic in behaviour. If we consider a typical oxide from the first long series of transition metals (i.e. the iron group), we find that the metal ions are small and the oxygen ions large. It is the latter that determine the size of the unit cell. Thus in general the $O^{2-}$ ions are in contact with one another and the 2p orbitals overlap, whereas the metal ions do not contact one another. This means that on the band model the outermost electrons of the oxygen ions, the 2p electrons, form a relatively broad filled energy band. The d electrons of the transition metal ions either form localized non-conducting states or a narrow energy band, and this means high effective electron mass and low mobility. They lie slightly above the oxygen 2p band.

## 10.7 Amorphous Semiconductors

In recent years, amorphous substances have attracted much attention, particularly with regard to their structural, electrical and magnetic properties. It is not so easy to visualize or illustrate the geometrical arrangement of atoms in an amorphous solid; in fact there is not one such arrangement but an infinite number. In keeping with our earlier discussion (Section 3.6), one is tempted to imagine a frozen liquid structure and in many instances this is probably a rather good approximation to the true situation, although the radial distribution functions for a true liquid and an amorphous solid do show distinct differences in detail. But such an approximation is not correct for pure Si or Ge. We have seen that as pure crystalline elements, these substances are intrinsic semiconductors; nevertheless, experiment shows that the liquids are true metals. Thus liquid Ge is a better conductor than liquid lead, and the Hall coefficient is in excellent agreement with the predictions of free electron theory. The semiconducting properties of crystalline Ge and Si are a result of the open and rigid tetrahedral structure. In liquid Ge or Si, these bonds are broken, and the coordination is greater and more typical of a close-packed metal like Pb. We presume that the $sp^3$ electron configuration typical of Ge in the solid form is replaced in the liquid by the configuration $s^2p^2$ as occurs in Pb. This is in keeping with our approximate description of a metal in terms of delocalized bonds.

Many substances, particularly alloys, may be obtained in the 'frozen liquid' amorphous state either by extremely rapid cooling from the melt or by evaporation onto a cooled substrate. Amorphous structures may also be formed by sputtering† the components on to a cooled substrate. In the case of Ge or Si, evaporation or sputtering onto substrates held below 300°C produces amorphous films, which are stable up to about 425°C. The density of such films is usually 15–20% less than their crystalline counterparts. The structure of these amorphous Ge and Si films is very different from that of the 'frozen liquid' – a result of the strong $sp^3$ covalent bonds that are so characteristic of these elements. In the amorphous state, these bonds still arise between nearest neighbours, but on account of the mechanical restraints placed on the movement of atoms at the relatively low temperatures at which the films are produced, the bonding is imperfect. Although there may be good four-fold coordination over small regions of the solid, there is no true tetrahedral symmetry and different regions are not linked to provide long-range order. The bond angles are severely distorted from the ideal value and many atoms may have only three neighbours. It is customary to describe the arrangement as a continuous random network of imperfectly bonded atoms in which even small voids may also be incorporated.

Suppose first that, although amorphous, each atom in the evaporated or sputtered film has four nearest neighbours, but that the bonds are distorted in both length and direction. The electron energy levels for such a structure might be expected to derive from the scheme typical for the crystalline phase. Although randomly arranged, the covalent bonds must give rise to some system of bonding and antibonding levels, leading to valence and conduction bands. We might expect the bands to be broader on account of the distortion of the bonds. The accepted picture is as shown in Fig. 10.16. Valence and conduction bands do arise, but the disorder produces pronounced 'tails' to the bands. Furthermore it is known that the electrons occupying these tail states are localized by the disorder in the potential; they therefore cannot take part in conduction. The energy required to give rise to a free electron-hole pair is as shown in the diagram; it is somewhat larger than that for the perfect crystal. However, the perfect four-fold-coordinated amorphous structure is an idealization. The introduction of three-fold-coordinated atoms produces significant changes to the picture of Fig. 10.16 because each such atom carries with it an uncompensated bond, a so-called 'dangling bond'. This is readily envisaged since ordinarily an Si atom needs four neighbours to mate with its four valence electrons. If there are only three neighbours then one bond must remain uncompensated or dangling. It might be thought that the electron associated with the dangling bond

† Sputtering is the process whereby atoms in a solid substance are knocked out by impinging energetic ions of a gas such as argon. The argon is at reduced pressure and is ionized by an electrical discharge. Atoms dislodged from the substance in question (which is usually an electrode or is in contact with such an electrode) condense onto a substrate placed above it. The method is convenient for the preparation of thin layers of most solids but has the disadvantage that argon and possibly oxygen are occluded in the film.

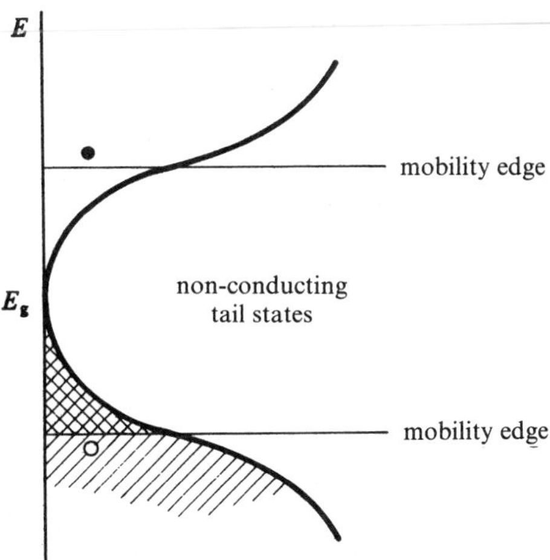


**Figure 10.16** In a fully coordinated but distorted tetrahedral semiconductor like Si, the disorder causes a spread of the energy levels into the region of the energy gap of the perfect crystal. These so-called tail states arise from the distorted bonds and are considered to be localized and non-conducting. The energy gap is replaced by a mobility gap.

would be a good charge carrier since the situation somewhat resembles that which we associate with a donor atom in crystalline Si.

This is not the case because the states associated with the dangling bond are located primarily in the middle of the band gap. We may liken the dangling bond to a 'deep' donor level. However the dangling bond is an uncompensated bond that can accommodate two electrons, so it may also be considered as a high lying acceptor level. Placing two electrons in the bond will invoke Coulomb repulsion so the donor and acceptor levels lie at different energies, the former centred somewhat below and the latter somewhat above the mid-gap position. Owing to the disordered structure the dangling bonds have different environments and the associated donor and acceptor levels each have their own distribution of energies. The two distributions overlap (implying that certain donor levels are empty and certain acceptor states filled) leading to a resultant distribution of dangling bond states similar to that shown in Fig. 10.17. Their concentration is so high, $\approx 10^{25}\ \mathrm{m}^{-3}$, that they control the position of the Fermi level.

In the pure but amorphous Si an occupied dangling bond state may be excited and an electron lifted over the upper mobility edge to become a free carrier, but the localized empty dangling bond state left behind does not provide a mobile hole. Only when electrons below the valence band mobility edge are excited to states above the conduction band mobility edge are electrons and holes created in equal numbers. This is difficult to observe in practice because the temperatures required cause the amorphous films to disintegrate. There is, however, another mechanism for electrical conduction between dangling bond states operative at low temperatures and known as 'variable range hopping' (Mott and Davis 1979, Mott 1985).

Consider a given atom possessing a dangling bond; as we have indicated, there are many such atoms, but they are distributed at random and furthermore the associated electrons and empty states have a distribution of energies. Our given atom certainly has similar atoms situated at various distances from it, and such an

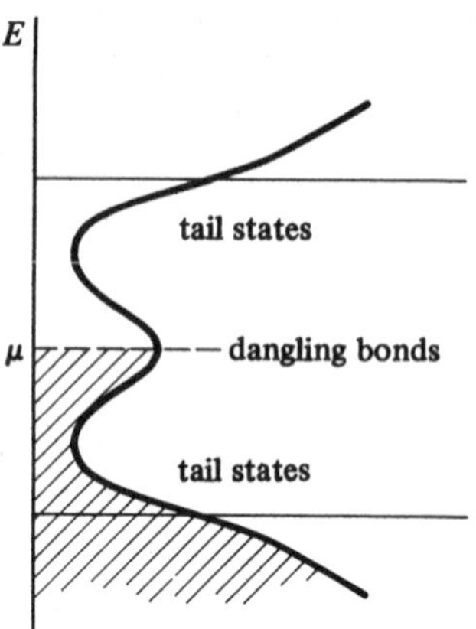


**Figure 10.17** In practice an amorphous tetrahedral semiconductor contains imperfectly coordinated atoms, leading to uncompensated or 'dangling' bonds. Each such bond produces a localized electron and a localized empty state. On account of their quasi-atomic character, these states are concentrated about the middle of the energy gap; they are sufficiently numerous to lock the Fermi level at $\frac{1}{2}E_g$. This has the consequence that the amorphous semiconductor becomes insensitive to doping.

atom may have an empty state at only very slightly higher energy than the occupied state on our chosen atom. Provided that the wave function of the electron extends to this empty state, since the activation energy for the electron to hop to the empty state is low, hopping occurs. Electrical conduction via hopping arises provided there are sufficient three-fold-coordinated atoms to provide continuous hopping paths through the solid. For concentrations of order $10^{25}\ \mathrm{m}^{-3}$ this is the case. Since the hopping distance varies owing to the random arrangement of atoms with suitable energy levels, the process is known as variable-range hopping. Mott has shown that under such circumstances the conductivity at low temperatures has the form

$$\sigma = A \exp\left[-\left(\frac{Q}{k_B T}\right)^{1/4}\right]. \tag{10.38}$$

Such behaviour has been observed in experiments, but there is still considerable discussion regarding the value of the power exponent (for further details see Mott 1985).

The large concentration of dangling bonds in evaporated or sputtered thin films of Ge and Si has another consequence – it makes the semiconducting property insensitive to impurity dopants. This is because the dangling bonds control the position of the Fermi level. Furthermore, it is clear that a boron atom in amorphous Si might well find itself in three-fold coordination, in which case there is no associated positive hole. Similarly, a phosphorus atom could have five, or more likely three, Si neighbours and thereby satisfy its quinquevalent or trivalent bonding requirements, eliminating any donor or acceptor function. But a phosphorus atom in fourfold coordination introduces a donor level just as it does in crystalline Si. The electron in this level, rather than be thermally excited to the conduction band, will fall into one of the unoccupied dangling bond states near the mid-gap position but its effect on the position of the Fermi level is insignificant on account of the already dominant density of levels arising from the dangling bonds in the amorphous Si. The conductivity of the amorphous Si is therefore unaffected by the presence of the phosphorus or any other conventional dopant.

In addition to evaporation or sputtering, amorphous Ge or Si films may also be prepared by the dissociation, in the vapour phase, of a chemical compound. Thus silane, $SiH_4$, can, with the aid of an electrical discharge, be converted into amorphous silicon. Such films may be doped and then give rise to impurity conduction. It is thought that this behaviour is a result of the trapping of atomic hydrogen in the silicon film. If we assume that the absorbed hydrogen leads to the formation of compensated Si–H bonds, then each such bond produces a system of two occupied bonding levels (that lie deep within the valence band) and two unoccupied antibonding states (that lie in the empty conduction band). The high density of dangling bond states in the mid-gap region is thereby markedly reduced by the presence of the hydrogen. If phosphine ($PH_3$) is added to the silane used to make the hydrogenated amorphous Si films, it is found that the conductivity is increased by several orders of magnitude; the films have become dopable. This is a direct result of the depletion of the mid-gap dangling bond states mentioned above. It is believed that doping also introduces dangling bond states, but in the presence of hydrogen they form a thinly distributed system of gap states that does not pin the Fermi level as is the case for unhydrogenated amorphous Si. The weakly bound electrons of the phosphorus donors fall into the vacant levels associated with the residual dangling bond acceptor states, leading to a Fermi level that becomes dependent on dopant concentration, whereby the conductivity becomes dependent upon a factor $\exp(\overline{E_c - E_F}/k_B T)$, $E_c$ being the energy of the conduction band mobility edge.

In the above very qualitative and brief account of amorphous semiconductors we have restricted our attention to the pure elements Si and Ge, but it will be appreciated that the infinite variety of amorphous structures causes the properties to be sensitively dependent upon the mode of preparation, temperature of preparation and subsequent heat treatement of the film, factors that have not been considered here. Furthermore, much of the present interest in disordered structures was stimulated by studies of the chalcogenide glasses, i.e. alloys or compounds containing one or more of the elements O, S, Se or Te for example, $As_2S_3$ is such a glass and has attracted interest because it exhibits a switching property whereby under the influence of an applied field it may be very quickly and reversibly converted from a high-resistance to a low-resistance state. It might thus find application as a binary device in computers.

### 10.7.1 Porous silicon

Silicon is perhaps the ideal semiconductor material. It is abundant, reasonably cheap, chemically and electronically well suited to device production. The only fly in the ointment is its indirect band gap which prevents its use as a light source (as light emitting diode or laser). It is expected that the ultimate computer design will use optical rather than electrical connections between circuit elements and even between larger units. Si that can be electroluminescent is therefore highly sought after. Hence the pronounced present day interest in porous Si.

Porous Si is formed by electrolytically etching the surface of a crystal wafer; it has fibrous structure (perpendicular to its planar extension) and the fibres can be made very narrow with diameters $\sim$1–5 nm and length several $\mu$m; the pores

**Figure 10.18** An impression of the structure of the porous Si layer on an electrolytically etched single crystal Si wafer. The individual filaments have diameter ~nm and length ~$\mu$m. (After Canham 1992.)

separating the fibres also have similar or larger size, Fig. 10.18. The fibrous structure has an enormous surface area, but is coated with hydrogen produced in the electrolytic process. All the dangling bonds that normally arise at the Si surface are therefore neutralized or passivated. It has been estimated that less than one surface state trap in more than $10^7$ surface atoms remains. By varying the experimental conditions such porous films have been made to luminesce at all the primary colours of the optical spectrum when irradiated by u-v light. Electroluminescence has also been demonstrated, although there are difficulties arranging the electrical contacts to the porous layer.

There is much discussion regarding the origin of the luminescent property of porous Si. It is however now known that the thin columns of Si are in fact crystalline Si and the most likely explanation seems to be that the fibres act as quantum wires or in certain cases quantum dots. These days there is much talk of 'quantum well' structures often formed in metallic films by photolithography. A quantum well is nothing other than a potential box with dimensions intermediate between atomic and $\mu$m sizes. The quantum well may be planar, linear like an ultra-thin wire or in the form of a dot in which confinement is achieved in all three dimensions. Current photolithographic technique is unable to produce quantum well structures much less than 100 nm in linear dimensions whereas in porous Si fibres with diameter $<5$ nm have been obtained.

In a quantum wire the electron wave function parallel to the wire axis has the usual Bloch form for a one-dimensional periodic potential, but in the transverse directions the electrons are confined to the sample in a manner similar to that for electrons in an atom. The available electron states form a spectrum of widely separated (compared with the ordinary solid) discrete levels. Each discrete level is associated with the itinerant states arising from the one-dimensional Bloch functions. The arrangement is a combination of discrete and band states and has quasi-one-dimensional character. There are energy gaps between the bands corresponding to the separation of the discrete states; these gaps are in the range 2 to 3 eV depending upon the fibre thickness, which is not necessarily the same at all parts of the fibre. The thicker portions have the smaller gaps and recombination of photo-excited electrons and holes occurs preferentially at these points leading to the emission of light. Unlike the case for bulk Si the interband transitions are believed to be direct thereby accounting for the pronounced efficiency of the luminescent process. However the experimentally observed decay times for the photoluminescence are ~ms and much too long for use in opto-electronic circuitry (~ns) so much further development is needed.

## 10.8 New Semiconducting Materials

Present-day semiconducting technology is based on Si, which has at least two advantages over Ge. First, the energy gap of 1.14 eV is significantly higher than that of Ge, 0.7 eV. This provides a greater stability with regard to temperature changes and the onset of intrinsic behaviour. Secondly, Si is much more convenient chemically than Ge because its oxide is stable and an excellent insulator ideally suited for the techniques of integrated circuit production. The basic starting material is the Si single crystal and it is now possible to produce these in the form of large ingots of up to 200 mm diameter. These ingots are then cut into thin slices, or wafers as they are called, which are then processed into 'chips'. Present-day development is directed to reducing the dimensions of the individual circuit elements and to increasing the packing density, leading to very large scale integration (VLSI).

The elemental semiconductors Si and Ge have indirect band gaps and the recombination of electrons and holes occurs via non-radiative processes; this means that they cannot be used as lasing materials. Non-radiative recombination is aided by the presence of extremely small concentrations of impurities (Cu, Au and certain other elements) that give rise to impurity levels near the middle of the energy gap, the so-called 'deep impurity levels'. Such levels must reduce the carrier lifetime and are therefore to be avoided in device production.

There is also a need to improve the performance of the basic semiconducting circuit element. Thus in computers the speed of operation of the logical element is directly dependent on the mobility of the charge carrier, which is better the greater the purity and structural perfection of the crystal. Similarly the generation of very high-frequency electromagnetic radiation ($\nu \approx 100$ GHz) is directly dependent on short carrier transit times, i.e. high carrier mobilities. Furthermore the rapid development of optical-fibre communication systems demands intense monochromatic light sources, which are most suitably obtained via semiconductor lasers. On account of these needs, much attention is now being devoted to GaAs and its various alloys with Al and In.

### 10.8.1 *GaAs and related III–V compounds*

Certain basic properties of some III–V compounds, all of which crystallize in the diamond cubic structure, are presented in Table 10.5. The data are for room temperature; an increase in temperature causes a reduction in the band gap energy – this can be attributed primarily (but not wholly) to the increase in atomic volume. At 0 K the band gaps are about 0.1 eV larger than those given in the table.

When two different metals crystallize with the same structure it is often possible to form a continuous series of solid solutions, provided that their lattice parameters do not differ by more than about 10% and provided that the metals do not differ radically in valence or electrochemical character. Thus Cu and Ni, Cu and Au (but not Cu and Ag) form such continuous solid solutions. The same goes for KCl-RbCl salts and many other systems. From Table 10.5, we should expect most combinations of the listed compounds to form mixed crystals, but attention has been directed mainly to those between AlAs and GaAs, i.e. alloy crystals of the form $Al_xGa_{1-x}As$. Why bother with such complication? The reason is that by alloying we can change the value of the band gap energy and thereby the wavelength of the

**Table 10.5**

| | Lattice parameter (Å) (RT) | Energy gap (eV) (RT) |
|---|---|---|
| AlP | 5.451 | 2.45 i |
| AlAs | 5.661 | 2.16 i |
| AlSb | 6.136 | 1.58 i |
| GaP | 5.451 | 2.26 i |
| GaAs | 5.653 | 1.42 d |
| GaSb | 6.096 | 0.73 d |
| InP | 5.869 | 1.35 d |
| InAs | 6.058 | 0.36 d |
| InSb | 6.479 | 0.17 d |

i ≡ indirect, d ≡ direct band gap.

recombination light. In pure GaAs the recombination radiation corresponding to $E_g = 1.42$ eV lies just in the infrared ($\lambda = 0.87$ $\mu$m). There can be good cause to produce light of shorter or longer wavelength.

Figure 10.19 shows how the energy gap varies in the AlGaAs system; note the complication that whereas in GaAs the gap is direct, it must at the higher Al concentrations become indirect, which is incompatible with laser operation. However, the main feature to emphasize is that there is a means of controlling the energy gap via the chemical composition of the crystal. These compounds may be made extrinsic semiconductors by doping with divalent elements. (p doping) or six-fold valence (n doping). Elements with valence four, like Si, are said to be amphoteric dopants because they may act as donor centres if they replace Ga or as acceptors if they replace As.

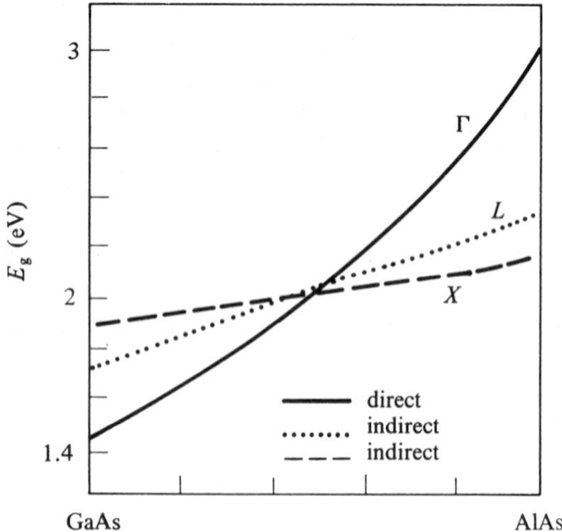


**Figure 10.19** The variation of the direct and indirect band gaps in the $Ga_xAl_{1-x}As$ compounds. (After Casey and Panish 1978.)

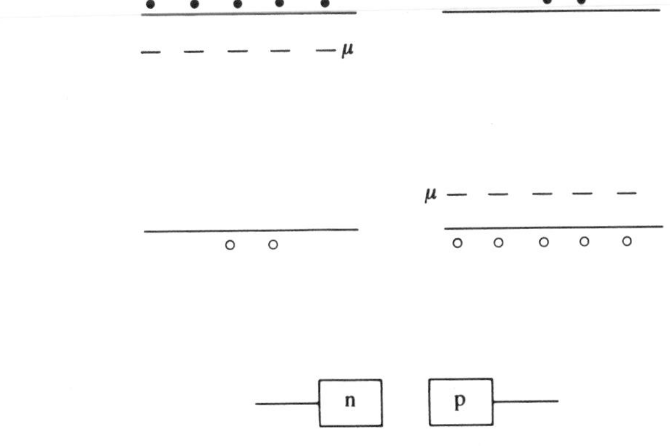


**Figure 10.20** The separate components of the pn junction.

Modern preparation techniques, particularly that known as molecular beam epitaxy,† allow the controlled production of layer structures, either intrinsic or doped, of the same or of different III–V compounds producing what are known as heterojunctions.

## 10.9 The pn Junction

Silicon may be doped with aluminium or arsenic to produce p- or n-type materials with equal concentrations of majority carriers. When such materials are placed into contact with one another, they form what is known as a pn junction, which is the basis for the development of many semiconducting devices. First of all, it should be realized that one cannot make a pn junction merely by placing p and n material into contact: the surfaces contain many defects that would impair the behaviour. The junction must be made 'inside' a piece of silicon by allowing the impurities to diffuse and form a barrier where they meet. There are many methods available to obtain pn junctions, but we shall not consider them.

Consider the separate p and n components – their energy level diagrams are as in Fig. 10.20. When p and n components form a closed circuit (without any source of e.m.f.) the Fermi level must be the same throughout the circuit. Recall that on our energy-level diagram, electrons tend to fall down into lower levels and holes tend to float upward. Because of the different electron and hole concentrations in the p and n components, a diffusion of charge carriers arises at the junction of the two materials. This causes a dipole layer to form and brings the Fermi levels into line.

† Molecular beam epitaxy (MBE) is a production method whereby the separate atoms Al, Ga, As, In, etc. are produced as pure molecular beams by heating the elements in constant-temperature ovens under ultra-high vacuum conditions. The atoms emerge as beams that condense onto a single-crystal substrate of GaAs. Arsenic only sticks in the presence of Ga, and by using excess As one can arrange the production of stoichiometric GaAs. If in addition an Al beam is used (with appropriate intensity) different concentrations of Al can be incorporated. The deposited atoms grow as layers (0.1–2 $\mu$m $h^{-1}$) and have the same single-crystal structure as the underlying GaAs crystal. However, generally in the epitaxial growth of single-crystal layers this is not necessarily the case. The overlayer crystal may have structure or orientation different from that of the substrate.

The charges in the dipole layer reside of course on the impurity or dopant atoms. Nevertheless, the bulk n and p materials are electrostatically neutral. At the barrier, electrons from the conduction band of the n material fall into the hole states of the p material. This leaves certain donor and acceptor states ionized, but without any compensating charge in the conduction or valence bands. We have a depletion of free carriers in the transition layer of the junction. The resultant energy diagram is as shown in Fig. 10.21.

Within the transition layer of the junction there are essentially no free carriers, but there is a distribution of ionized dopants that gives rise to a juction capacitance. The potential over the transition layer may be calculated using Poisson's equation

$$\nabla^2 V = -\rho/\varepsilon,$$

$\rho$ and $\varepsilon$ being the charge density and dielectric constant respectively. The thickness of the transition layer is smaller with larger dopant concentrations.

### The I(V) characteristic for the pn junction

Suppose we bias the n side of the junction positively. This means that we accentuate the barrier potential (reverse bias) (Fig. 10.22). If the p side is biased positively, we reduce the potential barrier across the junction (forward bias).

Note that when the barrier height decreases (increases), both electrons and holes can more (less) easily cross the barrier. The total current through the junction is composed of electron and hole components, both of which are affected in the same way by the bias potential. There are, however, *two* components to *each* carrier current. Let us consider the electron contributions at zero bias. We find exact compensation of the electron current thermally activated over the potential barrier from the n side by the current of electrons falling down the barrier from the p side.

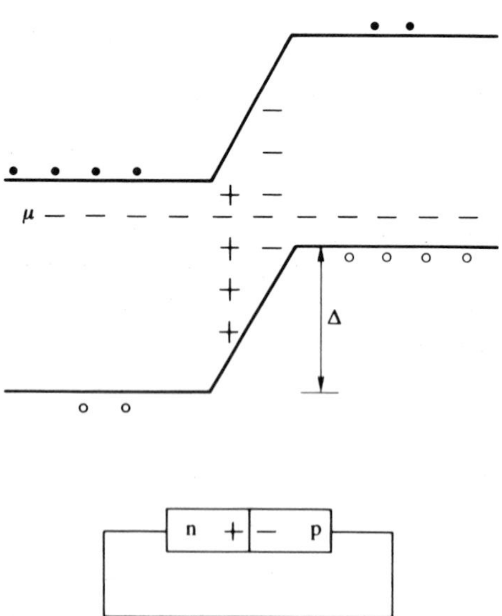


**Figure 10.21** The completed pn junction without external e.m.f. Carrier transfer across the junction creates a dipole layer, and the Fermi level assumes the same value throughout the circuit. The bands in the n and p parts attain a relative energy displacement Δ.

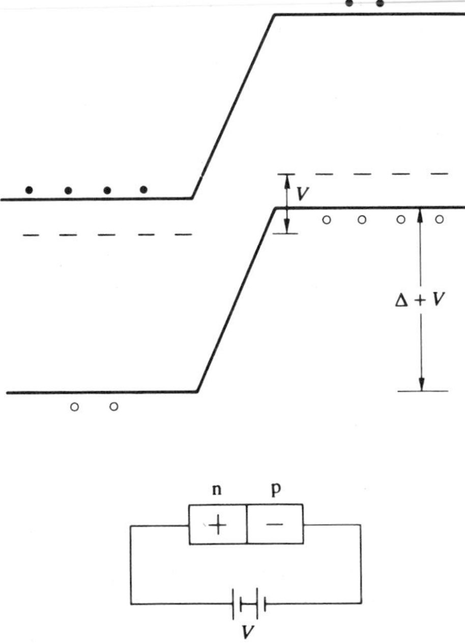


**Figure 10.22** An external potential biases the junction, leading to a barrier height $\Delta \pm V$. In this diagram, the n side has been given positive bias, thereby increasing the potential barrier, so-called reverse bias.

If at zero bias the potential barrier at the junction is $\Delta$ then the electron current from n to p material is

$$I_{\rm np} = Cn\mathrm{e}^{-\Delta/k_{\rm B}T}, \tag{10.39}$$

where $n$ is the concentration of electrons in the material. At zero bias we must have

$$I_{\rm np} = I_{\rm pn},$$

where $I_{\rm pn}$ is the electron current in the reverse direction caused by the electron concentration in the bulk p material. This latter current is independent of bias because it is always going 'downhill' and is determined by the small fixed electron concentration in the p semiconductor.

On applying a bias potential $\pm V$ to the p side ($+V$ corresponds to forward bias), we find

$$I_{\rm np} = Cn\mathrm{e}^{(-\Delta \pm V)/k_{\rm B}T}, \tag{10.40}$$

and $I_{\rm pn}$ remains unchanged. The net electron current through the junction becomes

$$I = I_{\rm np} - I_{\rm pn} = Cn\mathrm{e}^{-\Delta/k_{\rm B}T}(\mathrm{e}^{\pm V/k_{\rm B}T} - 1). \tag{10.41}$$

Negative bias does not affect the reverse current $I_{\rm pn}$, but forward bias causes $I_{\rm np}$ to increase exponentially. The hole current behaves in an exactly similar fashion and augments the electron current, leading to the $I(V)$ characteristic of Fig. 10.23. This characteristic is very non-linear and resembles that of a diode valve. The pn junction is a rectifying junction.

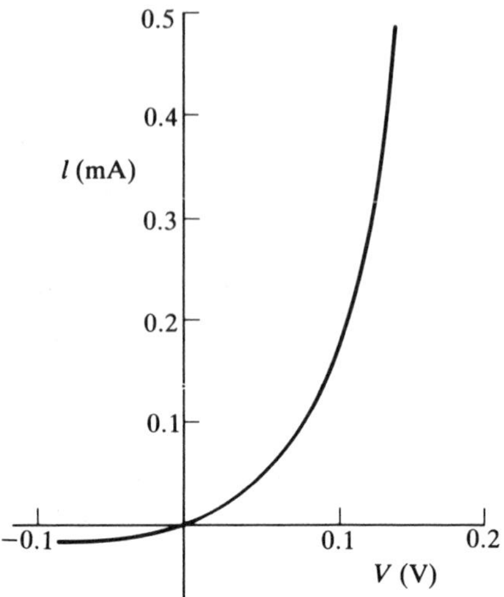


**Figure 10.23** The $I(V)$ characteristic for the pn junction, showing its rectifying properties.

## 10.10 The Semiconductor Laser

The operation of a laser demands

(a) population inversion;

(b) extremely low absorption losses;

(c) a Fabry–Pérot cavity with accurately parallel reflecting surfaces.

Initially the semiconductor laser was a thin active layer formed at a pn junction in heavily doped GaAs. The population inversion was produced by a forward bias potential across the junction and the parallel reflecting surfaces forming the optical cavity were obtained by cleaving the GaAs crystal. The advent of the MBE technique however now allows the growth of closely crystallographically matched and atomically juxtaposed single crystals of different chemical composition, and thereby different energy gaps, as well as different dopant concentrations. The composite layered structures are known as heterojunctions and permit the construction of efficient lasers with small dimensions.

The schematic form of a lasing heterojunction based on $In_{0.53}Ga_{0.47}As$, which has a band gap of 0.74 eV ($\equiv 1.67\ \mu m$), is shown in Fig. 10.24, whereas Fig. 10.25 describes the actual physical form of a practical laser. As Fig. 10.24(a) shows, the chemical composition of the active layer is adjusted to produce a smaller band gap than for the flanking InP ($E_g = 1.33$ eV) crystal layers. The discontinuity in the band edges at the interface between two semiconductors is known as the band offset, a difficult quantity to calculate; it is usually determined empirically. For the laser action the InP layers are heavily n and p doped respectively. Under closed circuit conditions a positive external bias on the p side causes electrons to be injected into the conduction band of the active layer and holes into its valence band, Fig. 10.24(b). The two carrier sorts collect in their respective bands because of the potential barriers that prevent their further transport through the junction. The

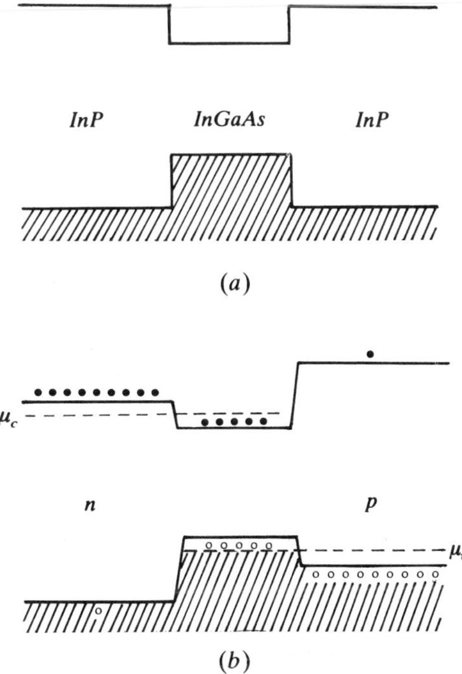


**Figure 10.24** The heterojunction laser consists of an active single crystal layer of InGaAs with a smaller band gap (0.74 eV) than the associated InP (1.33 eV) components of the junction; the conduction and valence band offsets are 0.25 and 0.34 eV respectively; (a), (in this diagram the band bending due to charge redistribution at the interfaces is not shown). When biased in the forward direction an inverse population of electrons arises because carriers from the heavily doped n and p InP components are injected into the potential wells that exist for both electrons and holes in the active layer, (b).

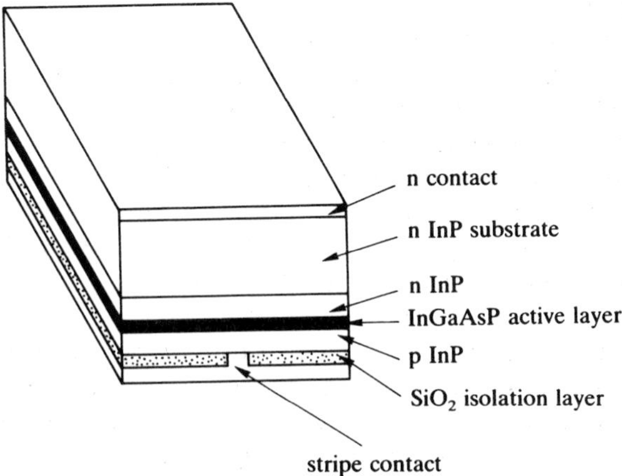


**Figure 10.25** The construction of an InGaAsP laser suitable for use in optical communication systems. The contact to the lasing junction is made in the form of a stripe to limit the size and divergence of the emergent beam and to promote coupling to the optical fibre. The overall size of the unit is about 300 $\mu$m × 200 $\mu$m × 100 $\mu$m. (After Smith 1982.)

electron–hole recombination time in the direct band gap active layer is sufficiently long to allow the electrons and holes to attain high concentrations in temperature equilibrium with the lattice; they each form a degenerate gas each with its own 'quasi Fermi level' that lines up with the appropriate Fermi level of the n or p contact. Eventually spontaneous recombination of the electrons and holes arises with the emission of band gap radiation at 1.67 $\mu$m. But the active layer forms part of an optical cavity and thereby supports stimulated recombination and stimulated emission of light, provided that the carrier injection rate is large enough. The injection rate is controlled by the current through the junction and the threshold value for laser action is $\sim 10$ mA (below the threshold current the system acts as a conventional light emitting diode). As is seen from Fig. 10.25 the light is emitted in the plane of the junction and along a narrow section to confine the beam and facilitate coupling to an optical fibre or other component. In spite of the small overall size of the laser shown in Fig. 10.25 it is considered too large and too power consuming for many purposes. Development is aimed at smaller cylindrical cavity lasers with diameters $\sim 10$ $\mu$m that emit light in the direction normal to the plane of the junction, an arrangement that allows the construction of planar arrays of lasers suitable for use in optoelectronic computer technology.

The need for semiconducting lasers is most acute in communication systems using optical fibres. Optimum design demands that the wavelength be matched to the performance of the fibre. One may require operation under conditions of minimum absorption, minimum dispersion, or preferably both.

Optical fibres for long-distance communication are made out of very pure $SiO_2$. Figure 10.26 shows the optical attenuation of this substance as a function of wavelength. Minima in attenuation arise at 1.3 $\mu$m and 1.55 $\mu$m. The strong peak at about 1.4 $\mu$m arises from contamination by OH radicals, which cause optical absorption through their vibrational spectrum. Minimum dispersion also arises near 1.3 $\mu$m, whereas the attenuation at 1.55 $\mu$m (about 0.2 d$B$ km$^{-1}$) is limited by inherent variations in the glassy structure of the $SiO_2$. Practical systems have been based on the multimode fibre with diameter $\phi \approx 50$ $\mu$m (Fig. 10.27), but optimum design then demands working at minimum dispersion. Single-mode fibres, $\phi \approx 10$ $\mu$m,

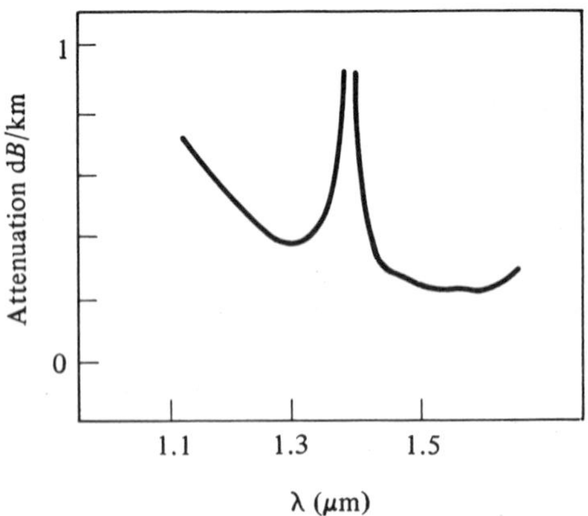


**Figure 10.26** The optical attentuation of ultrapure fused quartz as used in the construction of optical fibres. The marked absorption near 1.4 eV arises from the presence of $OH^-$ radicals. (After Smith 1982.)

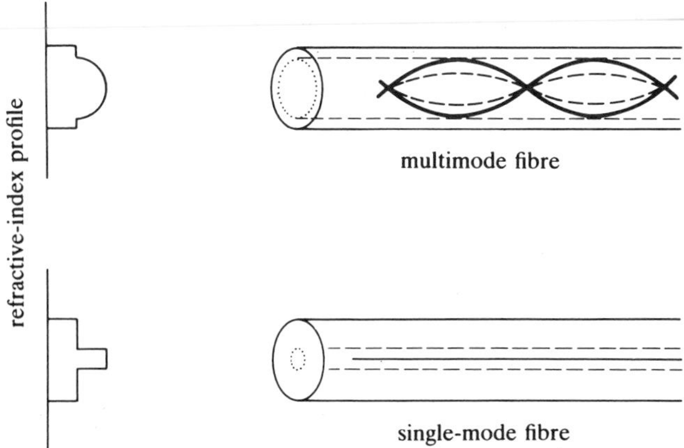


**Figure 10.27** Optical fibres use graded refractive-index profiles to guide the light along the fibre and to minimize signal degradation by dispersion. This is particularly important for the multimode fibre, which has a diameter of about 50 $\mu$m and a large numerical aperture, allowing several alternative paths (modes) for the passage of the light along the fibre. The single-mode fibre has diameter of about 10 $\mu$m and a small numerical aperture: there is essentially only one path for the light beam and only chromatic dispersion affects the performance (apart from attenuation). (After Smith 1982.)

are also available. Present-day optical fibre technology allows intercontinental communication in single fibre links that transmit data at a rate of 2.5 gigabits per second, equivalent to 32 000 simultaneous telephone channels. To obtain the necessary variation in refractive index, pure $SiO_2$, in the form of a hollow cylinder, is first doped on the inside with Ge to produce the 'high index' and then on the outside with P or F to give the 'low index'. The doping is done by oxidizing vapours containing these elements in a process known as 'modified chemical vapour deposition'. The cylinder is then collapsed and spun into fibre. Certain semiconducting lasers suitable for use near 1.5 $\mu$m are based on InGaAsP grown on a substrate of InP. The active lasing layer is made as a 'stripe' to promote coupling to the fibre (Fig. 10.25).

## 10.11 The Quantized Hall Effect

We now focus on mobile electrons that arise at the interface between an insulator and a p-doped semiconductor in the presence of a strong electrostatic field. We shall see that this leads to the realization of what is effectively a two-dimensional electron gas. Such as gas has been obtained in devices formed at the surface of Si or GaAs single crystals.

Consider the arrangement shown in Fig. 10.28. A piece of p-doped single-crystal Si has a surface with (100) or (111) orientation covered by a thin layer (about 200 nm) of insulating $SiO_2$, over which is evaporated a metal electrode called the gate. Independently of this gate, other electrodes may be attached to the semiconductor to provide the contacts shown in Fig. 10.28. In the absence of any gate potential, the energy-level diagram in simplified form is as in Fig. 10.29, but if the gate is biased positively then a strong electric field arises at the boundary between

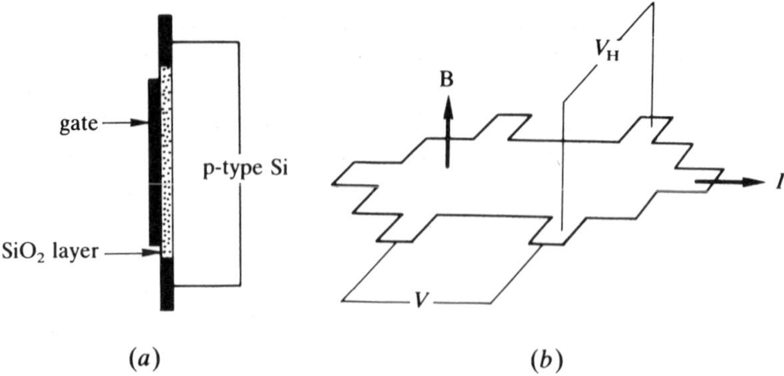


**Figure 10.28** (*a*) A metal-oxide-semiconductor structure. The p-doped Si is provided with metallic contacts and its surface is insulated from a gate electrode by an $SiO_2$ layer. (*b*) The disposition of magnetic field, device current and the potentials $V$ and $V_H$. The gate is not shown in this diagram.

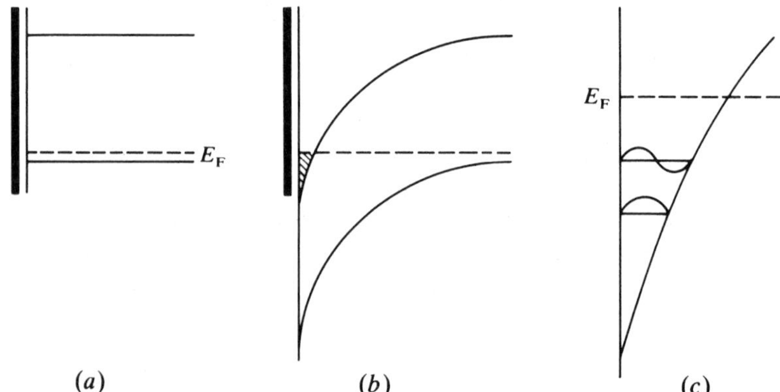


**Figure 10.29** In the absence of an external potential, the p-type Si has an energy diagram as shown in (*a*); the left-hand side of the diagram shows the presence of the gate (black) and the oxide layer (white). The Fermi level is controlled by the p doping. When the gate is biased positively relative to the p-type Si, a strong electric field is created at the surface of the semiconductor, leading to pronounced bending of the bands in this region. If the band bending is sufficiently strong, the conduction band is forced below the Fermi level and a dilute surface electron gas is created, (*b*). Perpendicular to this surface, this electron gas has very limited extension of a few Å and the electron energy levels have atomic character, (*c*).

the oxide and the semiconductor. The energy bands of the semiconductor become bent. This band bending near the surface of the semiconductor may be so pronounced that the conduction band, essentially empty in the presence of p doping, is depressed below the Fermi level, producing a triangular pocket of occupied electron states. These electrons, of order $10^{12}$ cm$^{-2}$, are confined to the surface in what is called an 'inversion layer'.† Within the plane of the surface their motion is unrestricted over macroscopic dimensions, but normal to the surface the layer is so thin

† If instead the field is reversed and holes are formed at the surface then an 'accumulation layer' is said to exist.

that the electron motion is quantized, producing discrete energy levels as would arise in a linear potential box of atomic dimensions.

The density of electron states in a two-dimensional free electron gas is independent of energy (Section 7.4), but because of the states available in the atomic like levels, the density of electron levels in the inversion layer appears as a series of surface sub-bands. As the gate potential is increased from zero, there comes a time when the first surface sub-band forms and, as the gate potential increases further, higher sub-bands will be captured. Thus the density of surface electron states varies in a stepwise fashion with gate potential. We shall again neglect the electron spin.

We now suppose that the gate potential is large enough to stabilize the first-sub-band of the two-dimensional electron gas in the inversion layer. In **k** space the states form a uniform quasicontinuum with a density $1/2\pi^2$ per unit area of the layer and the contours of constant energy are circles.

The properties of electrons in such inversion layers have been intensively studied during the past 15 years. Here we concentrate interest upon a particularly important aspect: the behaviour at very low temperatures, of order 0.1 K (to reduce scattering mechanisms to a minimum) and in the presence of an intense magnetic field directed normal to the inversion layer. This field, of order 10 T, is obtained using a superconducting magnet. Under these conditions, $\omega_c \tau \gg 1$ and $\hbar\omega_c \gg k_B T$, $\omega_c$ being the cyclotron frequency (see Section 9.2).

In real space the magnetic field causes electrons to follow helical paths owing to the transverse Lorentz force $-ev_x B_z$, but in **k** space the representative points move in circular orbits about an axis through the origin of **k** space and parallel to the field direction, i.e. an axis normal to the two-dimensional reciprocal lattice. All the points of **k** space have the same rotational frequency $\omega_c = eB/m$ about this axis. As we have described earlier, this regular rotation introduces discrete quantized Landau levels with energies

$$E_L = (L + \tfrac{1}{2})\hbar\omega_c. \qquad (10.42)$$

The previous quasicontinuous distribution of levels in the $(k_x, k_y)$ plane condenses into a series of rings corresponding to the discrete energies $E_L$. Each level has a degeneracy $p$ (spin included) per unit area of sample given by

$$p = \frac{2eB}{h}. \qquad (10.43)$$

Ideally these Landau levels are sharp; in practice broadening arises, but we neglect this. Initially, at low magnetic fields, the Landau levels are very closely spaced and nothing new happens, but as the field increases, $\omega_c$ increases and with it the separation and degeneracy of the levels. Eventually, at very high magnetic field strength the electrons occupy only the lowest, zero-point-energy, level. However, long before this limiting condition is reached, we find $\omega_c \tau \gg 1$ and $\hbar\omega_c \gg k_B T$. As the magnetic field increases and the Landau levels separate, they must pass through the Fermi level and eventually empty. In this process there arise situations where the Fermi level lies exactly between a completely filled level $L$ and a completely empty level $L + 1$. We must then have $N(E_F) = 0$. Thus, as the field increases, $N(E_F)$ periodically goes through zero (the period getting larger as the field increases). In practice, it is often more convenient to arrange this state of affairs by keeping the field constant and increasing the gate voltage, thereby introducing more and more carriers to fill the system of empty Landau levels.

Although it might appear difficult to arrange and to maintain the condition $N(E_F) = 0$, let us consider what might happen under such conditions. First, the electrical conductivity becomes thermally activated because no current can flow in a filled Landau level and the nearest empty level lies $\hbar\omega_c \gg k_B T$ higher in energy. Thus $\sigma_{xx} = 0$. Secondly, under the given conditions we expect $\rho_{xx} \approx \sigma_{xx} \approx 0$ (see Problem 10.12). Furthermore, owing to the degeneracy of the Landau levels, their large separation in energy and the very low temperatures used, there is, in this free electron model, no means for energy transfer between electrons (recall that in contrast with the same situation in three dimensions, the $z$ channel for energy exchange is completely closed). Under these conditions the electrons, in the presence of the crossed electric and magnetic fields $E_x$ and $B_z$ drift in the $y$ direction with a velocity $E_x/B_z$. If the specimen were unbounded, there would be no net movement of the electrons in the $x$ direction. The drift current in the $y$ direction is associated with the conductivity $\sigma_{yx}$. In an ordinary finite sample this current is zero and because of this a Hall voltage is produced. For the two-dimensional case the Hall resistivity is just the inverse of $-\sigma_{xy}$ (see Problem 10.12). Using the expressions determined earlier in our discussion of the Hall effect (Box 6.1), we find

$$\sigma_{xy} = -\frac{\sigma_0 \omega_c \tau}{1 + \omega_c^2 \tau^2} \rightarrow \omega_c \tau \gg 1 \rightarrow -\frac{ne}{B} \tag{10.44}$$

and

$$\rho_{xy} = \frac{B}{ne}.$$

When all the first $L$ levels are completely filled

$$\begin{aligned} n &= \sum p_L = (L+1)p \\ &= (L+1)\frac{2eB}{h} \end{aligned} \tag{10.45}$$

and

$$\rho_{xy} = \frac{h}{2e^2}\frac{1}{L+1}. \tag{10.46}$$

Thus, whenever the Fermi level lies exactly between two Landau levels, the Hall resistivity, which in two dimensions is identical with the Hall resistance, becomes quantized in terms of the quantum $h/2e^2$ (=129 05.5 Ω).

However, the chances of observing such a quantum effect appear to be very remote because the simple treatment given above assumes a perfectly free electron gas. The electrons in an inversion layer, as formed at a metal-oxide-semiconductor interface, are far from such conditions; many electrons are not free to move; they are localized (trapped) at imperfections in the crystal structure. It is therefore unlikely that, even if one could (against all expectation) arrange the singular situation where $N(E_F)$ lies exactly between two Landau levels, we could assume (10.46) to hold. Furthermore, in reality we do not expect the Landau levels to be perfectly sharp – they have an inherent width, reflecting thermal, impurity and disorder broadening.

It therefore came as a very great surprise when von Klitzing discovered the Hall plateaux that arise in certain inversion layers at very low temperatures and under very high magnetic field strengths. In practice the effects that one expects to be singular in the simple free electron picture are maintained even when $N(E_F)$ no longer lies exactly between two Landau levels. The data of Fig. 10.30 clearly show how, at certain values of $B$, $V_x$ falls to zero (because $\rho_{xx} = 0$) whereas the Hall voltage remains constant at certain plateau values associated with particular $L$ values. Still more surprising is the fact that the Hall resistivity, which can be measured to an accuracy of one part in $10^7$, is in excellent agreement with (10.46). There are two very important features of this result, one of fundamental, the other of practical significance. The quantized Hall resistance allows the ratio $h/e^2$ to be determined more accurately than was previously the case. The following relationship also holds:

$$h/e^2 = \tfrac{1}{2}\alpha^{-1}\mu_0 c.$$

$\mu_0$ and $c$ have their usual significance and $\alpha$ is the fine structure constant, which is of basic significance in all quantum physics. It is a measure of the ratio of the strengths of magnetic and electric dipoles because

$$\mu_B/ea_0 = \tfrac{1}{2}\alpha c,$$

$a_0$ being the Bohr radius.

The quantum Hall effect provides an independent measure of $\alpha$ that may be compared with values derived using very different techniques, in particular those of quantum electrodynamics.

The practical aspect of the quantum Hall effect lies in that it may be used to define a standard of resistance. The electrical properties of two-dimensional inversion layers are particularly suitable in this respect because they are independent of sample composition, band structure and geometrical size. The question nevertheless remains as to how the formulae of the simple free electron model can apply to real

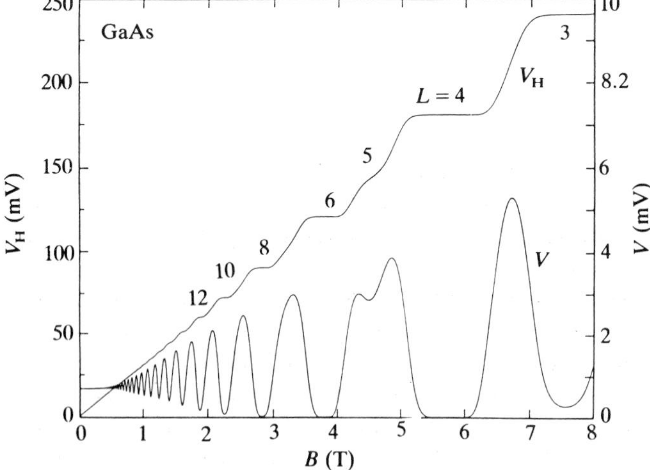


**Figure 10.30** The variation of the longitudinal ($V$) and Hall ($V_H$) voltages as functions of the magnetic field strength $B$. At low values of $B$ the behaviour is little affected; in high fields $V$ periodically becomes zero, but at the same time $V_H$ remains finite and constant. The plateaux in $V_H$ (and in $\rho_{xy}$) are associated with integral Landau quantum numbers as indicated. (From M. E. Cage *et al.*, IEEE, *IM*, **34**, 1985. © 1988 IEEE.)

highly complex specimens. The questions of whether electrical conduction arises and how it is dependent on crystalline order or disorder are very intricate and beyond the level of this text, but they have been satisfactorily solved (for a review see Pepper 1985).

## 10.12 Organic Semiconductors

One can envisage electrical conduction in an insulator (e.g. an inorganic salt like NaCl or an organic molecular crystal of the kind to be described later) by, in some way or another, adding electrons to the empty conduction band or removing them from the filled valence band.† However, owing to the dielectric properties of the medium, any charge carrier polarizes its immediate neighbourhood and in doing so may produce a potential well sufficient to localize it to a given site or to a limited region encompassing several lattice sites. In the former case we say that a small polaron has been formed and in the latter a large polaron. In different substances a whole range of polaron sizes is to be expected. The detailed properties of polarons are rather complex. Normally we consider electrons in a nearly empty band to be mobile, although they have a large band mass when the band is narrow; it is to be expected that the need to drag the polarization cloud along with it will enhance the effective mass of the polaron – increasingly, the smaller it is. Analysis shows that at low temperatures both large and small polarons move so that band conduction arises. The conductivity decreases with increasing temperature. Owing to the strong interaction with the lattice vibrations (particularly the longitudinal optical phonons, which have the same character as the polaron's displacements) the mean free path of the polaron at $T \approx \frac{1}{2}\theta_D$ is already reduced to a size comparable to the atomic spacing. This means that the polaron is effectively trapped at a particular lattice site and can then only move via a thermally activated hopping process. The mobility becomes exponentially dependent on temperature and we find semiconducting behaviour (see the previous discussion regarding oxide semiconductors in Section 10.6).

Polarons are considered important for electrical conduction in organic crystals. In contrast with inorganic insulators like $Al_2O_3$, MgO and NaCl, organic crystals usually have low melting points and are often mechanically weak. This is because such crystals are composed of molecules, often very large, that retain their molecular identity and interact only weakly through van der Waals forces. That this is the case is proved by the similarity of the optical absorption spectra of the individual molecules and the molecular crystal. The outer electrons of such molecules are usually tightly bound to the parent unit and occupy states in very narrow bands $<0.1$ eV wide.

Organic molecules and radicals having a planar structure lead to crystals built by the stacking of such units into sheets or columns, thereby forming anisotropic crystals that in their electrical and magnetic behaviour can, in certain cases, closely approximate one-dimensional systems. We shall confine our attention to certain organic substances that illustrate one-dimensional behaviour. Attention has been directed to these materials partly on account of the interest in the effect of dimensionality on behaviour and partly because it has been suggested that linear organic

† A brief discussion of the band structure of an alkali halide is given at the end of Chapter 12.

molecules may favour high-temperature superconductivity; although superconductivity has been observed in such structures, the transition temperatures found so far are between 1 and 10 K.

A one-dimensional metal, however, at low temperatures is expected to become unstable with respect to the insulating state. The reason is that if a new real-space periodicity of $\pi/k_F$ is in some way introduced then this will cause the creation of new zone boundaries at intervals of $k_F$ in reciprocal space. Energy band gaps then arise at the Fermi level, the Fermi surface disappears and the linear structure becomes an insulator. The presence of the new energy band gaps causes occupied states near $E_F$ to move to lower energies and this gain in binding energy stabilizes the transition. The new periodicity is caused by one of three possible processes:

(a) a regular variation in electrical charge density (static charge density waves);

(b) a regular variation in electron spin density (static spin density waves);

(c) a direct lattice distortion producing the new lattice periodicity (the Peierls' transition).

Linear metallic behaviour was first studied in inorganic salts like $K_2PtCl_6$, but current interest is directed more towards organic molecular crystals and polymers. A particular feature of these substances is that the electrons responsible for the conducting properties play only a weak role in the chemical bonding, so that wide variations in the electrical behaviour may be obtained without affecting the mechanical stability of the structures. We must also remember that in experiment, we only obtain approximate or quasi-one-dimensional samples. Real specimens are always three-dimensional and, although interactions between the separated chains of molecules may be weak, at sufficiently low temperatures they will become important and may then destroy the principal features of one-dimensional behaviour. These three-dimensional interactions are found to lead to either an insulating magnetic state or superconductivity.

Most organic substances, in pure crystalline form, have a very low electrical conductivity $<10^{-8}\ \Omega^{-1}\ m^{-1}$. This is because all the valence electrons are paired off in chemical bonds (but some less strongly than others) and there are no free charge carriers. It is, however, possible to form organic salts with either organic or inorganic radicals, and charge transfer may then introduce charge carriers into what were previously insulating units. In certain cases such salts (and certain polymers) may exhibit conductivities of order $10^7\ \Omega^{-1}\ m^{-1}$, a conductivity approaching that of metallic copper at room temperature.

One such radical (an anion, i.e. an electron acceptor) is tetracyanoquinodimethane, TCNQ (Fig. 10.31*a*). One of several salts incorporating TCNQ is TTF-TCNQ (where TTF is tetrathiafulvalene). It attracted much interest owing to a dramatic increase in electrical conductivity as the temperature is lowered below room temperature. However, at about 60 K this organic conductor transforms to the ferroelectric state (see Section 12.5). The radicals in TTF-TCNQ are large and planar and form the columnar structures mentioned previously. Usually the anions (TCNQ) and the cations (TTF), which are formed by charge transfer, are separated into individual columns, but they may also occur in an intermixed arrangement. The high electrical conductivity is only observed in the segregated structure and when charge transfer between columns is incomplete.

The conductivity arises in the quasi-one-dimensional stack of TCNQ radicals. If the charge transfer is complete, i.e. every TCNQ accepts one electron from the cor-

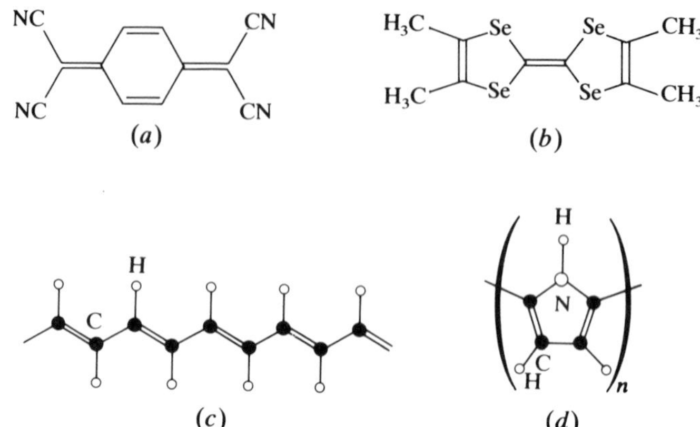


**Figure 10.31** The molecular structures of certain organic conductors: (*a*) tetracyano-quinodimethane (TCNQ); (*b*) tetramethyltetraselenafulvalene (TMTSF); (*c*) *trans*-polyacetylene: (*d*) polypyrrole. Not all the carbon atoms are depicted – their positions may be inferred from the bonds shown.

responding TTF, then this means that electrical conduction requires that two electrons be found on certain TCNQ units. The Coloumb repulsion between two electrons on the same radical, combined with the weak overlap between electrons on adjacent anions (remember the tight-binding aspect), makes this an unlikely situation; we say that we have a highly correlated system. Under these circumstances, electrical conduction only arises when the charge transfer between the organic radicals is incomplete. The details of the conduction process in these crystals and the role of polarons remain to be established.

Another organic radical that forms conducting salts, in this case with inorganic anions of the form $PF_6^-$, $ClO_4^-$ and similar complexes, is tetramethyltetraselenafulvalene, TMTSF (Fig. 10.31*b*). These salts, for example $(TMTSF)_2(PF_6)$, are somewhat simpler than those based on TCNQ because there is only one type of stack – that formed by the TMTSF molecules – and the electrical conductivity is associated wholly with these molecular chains, the anions remaining fully ionized and electrically inactive.

The covalent molecular orbitals that arise are denoted, according to their symmetry, as $\sigma$ or $\pi$ bonding orbitals and $\sigma^*$ and $\pi^*$ antibonding orbitals. The $\sigma$ bonds are axially symmetrical about the bond axis, whereas the $\pi$ bond has symmetry about a nodal plane containing the bond axis and has elongated form (see McWeeny 1979). In the neutral TMTSF molecules, all the valence electrons are paired in occupied bonding orbitals, which are separated from the empty antibonding orbitals by an energy gap. A $(TMTSF)_2^+$ cation may be formed by breaking a weak $\pi$ bond that links two TMTSF molecules in the chain structure. This bond is broken by charge transfer to form the electronegative anion. The broken $\pi$ bond on each pair of TMTSF molecules in the linear chain structure now produces a half-filled bonding energy band and electrical conduction arises along the molecular chain. At room temperature, the conductivity parallel to the chains is about 500 times greater than in the perpendicular direction, but this anisotropy decreases as the temperature is lowered whereas the conductivity shows a marked increase, attaining a value of order $10^7\ \Omega^{-1}\ m^{-1}$ near 10 K (Fig. 10.32). The behaviour is

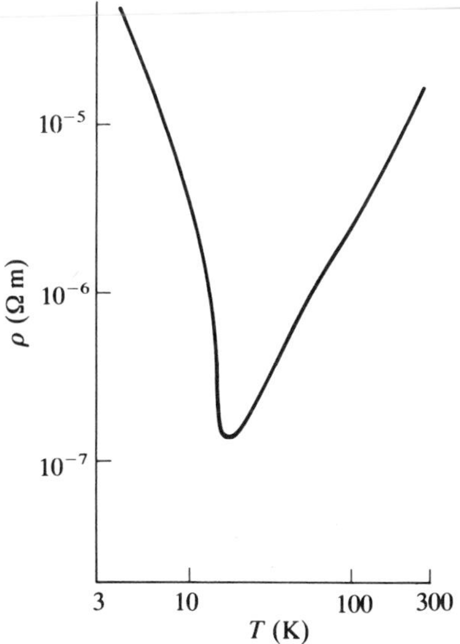


**Figure 10.32** The electrical resistance of $(TMTSF)_2(PF_6)$ as a function of temperature, showing the transition near 10 K. (Reprinted with permission from K. Bechgaard *et al.*, *Solid State Commun.*, **33**, 1980. Pergamon Press, Oxford.)

anion-dependent. At the lower temperature the weak interchain interaction becomes more important and these salts must then be considered as highly anisotropic three-dimensional structures rather than quasilinear ones. This leads to new properties, and the most stable electronic structure at the lowest temperatures and under ambient pressure becomes that of a magnetic insulator ($PF_6$) or a superconductor ($ClO_4$). In the case of the $PF_6$ salt, a superconducting state may also be obtained under a pressure of about 1 GPa. Using a range of pressure, temperature, magnetic field and choice of anion, these TMTSF salts display a wide range of properties (see Friedel and Jerome 1982).

Conducting polymers show much future promise. We restrict attention to what at present appears to be the most important example from the scientific point of view, namely polyacetylene. The simplest polymer is composed of molecules $X_n$ made by the repetition of $n$ identical units. In the case of polyacetylene the formula is $(CH)_n$. In its ordinary state the electrical conductivity is of order $10^{-6}\ \Omega^{-1}\ m^{-1}$, but the polymer may be doped (n or p character) and the conductivity increased by a factor as high as $10^{12}$. Such an organic material at room temperature may exhibit a conductivity approaching that of copper and could find use in circuit elements, plastic electrodes, display screens and plastic batteries.

The zig-zag linear structure of *trans*-polyacetylene is shown in Fig. 10.31($c$). There is a backbone of carbon atoms, each of which is linked to two other carbon atoms and a hydrogen atom. In this way, three electrons on each carbon atom are used to form three covalent single ($\sigma$) bonds, two of which bind the chain together. The remaining electron on each carbon atom is used to augment one of the bonds between two adjacent carbon atoms so that a double bond is formed: the latter then comprises a strong $\sigma$ bond and a weaker $\pi$ bond. The bonds between carbon atoms are therefore of alternating single ($\sigma$) and double ($\sigma$, $\pi$) character, they are also of slightly different length. It will be noticed that this arrangement leads to two possible bonding geometries: the double bond may point upwards to the left or upwards to the right. These are equivalent structures, but no resonant or hybridized structure (i.e. one in which both possibilities occur in equal amounts) arises because there is a considerable potential barrier to be overcome in order to change between the two. If this were not the case, polyacetylene would be a true metal.

The electrical properties of polyacetylene are associated wholly with the $\pi$ orbitals of the double bond. The occupied $\pi$ orbitals produce a completely filled valence band, separated from an empty conduction band formed from the antibonding $\pi^*$ orbitals. The energy gap is found optically to be of order 1 eV. Pure fully conjugated polyacetylene is therefore an insulator. The high electrical conductivity is achieved by the addition of electropositive or electronegative dopant ions; the sample may be immersed in a suitable solution of dopant, exposed to gaseous dopants or subjected to electrochemical doping. Dopant concentrations may be quite large, in the range 5–15%, causing the conductivity to change from of order $10^{-6}$ to $10^{6}\ \Omega^{-1}\ \mathrm{m}^{-1}$. The highest conductivity is observed in the best-oriented and best-ordered structures (Fig. 10.33). The polyacetylene is usually prepared as thin film with a density of about 0.5 g $\mathrm{cm}^{-3}$, and the structure is a fibrous conglomerate of small uniaxial crystallites which may be aligned by stretching the film. The conductivity and associated properties, like optical reflectance, are highly anisotropic, the ratio of the conductivities in the parallel and perpendicular directions to the fibre axis being of order 100, but it may be as large as 1000 in the best films. The high electrical conductivity associated with the low density and formability of polymers should make them suitable for technical applications. However, although it is an excellent model substance for scientific study, polyacetylene seems unlikely to find wide application because it is unstable in air. Alternative more stable synthetic conductors are therefore highly sought after; one such is polypyrrole PPy

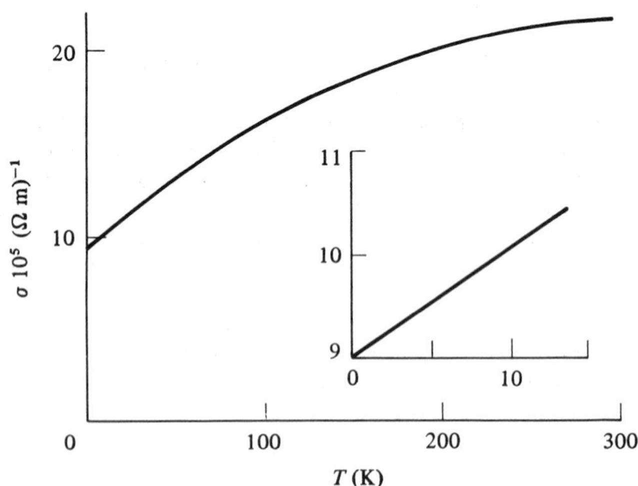


**Figure 10.33** The electrical conductivity of oriented *trans*-polyacetylene as a function of temperature under a pressure of 10 kbar. The inset shows the behaviour below 4 K in greater detail. (After Basescu *et al.* 1987. Reprinted by permission from *Nature*, **327**, p. 403. Copyright © 1987 Macmillan Magazines Ltd, London)

(Fig. 10.31*d*), which is, however, only obtainable in amorphous form, although this could be an advantage because of the associated isotropy.

Such polymers have great technological potential, but development is in its infancy – much study will be needed before the basic physics and chemistry of these substances are established.

## References


BASESCU, N., LIU, Z. X., MOSES, D., HEEGER, A. J., NAARMANN, H. and THEOPHILOU, N. (1987) *Nature* **327**, 403.

BECHGAARD, K., JACOBSEN, C. C., MORTENSEN, K., PEDERSEN, H. J. and THORUP, N. (1980) *Solid State Commun.* **33**, 119.

BRUST, D., PHILLIPS, J. C. and BASSANI, F. (1962) *Phys. Rev. Lett.* **9**, 94.

CAGE, M. E., DZIUBA, R. F. and FIELD, B. F. (1985) *IEEE Trans. Instrum. Meas.* **34**, 301.

CANHAM, L. (1992) *Phys. World* **5**, 41.

CASEY, H. H. and PANISH, M. B. (1978) *Heterostructure Lasers*, Part B. Academic Press, New York.

CHELIKOWSKY, J. P. and COHEN, M. L. (1976) *Phys. Rev.* **B14**, 556.

FRIEDEL, J. and JEROME, D. (1982) *Contemp. Phys.* **23**, 583.

LANDOLT–BÖRNSTEIN (1982) New Series, III, **17a**, Springer-Verlag, Berlin.

MCWEENY, R. (1979) *Coulson's Valence*, 3rd edn. Oxford University Press, Oxford.

MOTT, N. F. (1985) *Contemp. Phys.* **26**, 203.

MOTT, N. F. and DAVIS, E. A. (1979) *Electronic Processes in Non-Crystalline Materials*, Oxford University Press.

MYERS, H. P., JONSSON, T. and WESTIN, R. (1964) *Solid State Commun.* **2**, 321.

PEPPER, M. (1985) *Contemp. Phys.* **26**, 257.

SMITH, D. R. (1982) *Phys. Bull.* **33**, 401.

STURGE, M. D. (1962) *Phys. Rev.* **127**, 768.


## Further Reading


CHANG, L. L. and ESAKI, L. *Semiconductor Quantum Heterostructures*, in *Phys. Today* (1992) **45**, 10, pp. 36.

COX, P. A. *Transition Metal Oxides*, (1989) Oxford University Press, Oxford.

ELLIOTT, S. R. *Physics of Amorphous Materials*, (1983) Longman, Harlow.

PLOOG, K. *Fabrication of Custom-Designed Semiconductor Microstructures by Molecular Beam Epitaxy*. Proc. Int. Conf. Teaching Modern Physics, Condensed Matter, (1989) pp. 71. World Scientific Publishing Co., Singapore.

STREET, R. A. *Hydrogenated Amorphous Silicon*, (1993) Cambridge University Press, Cambridge.

WANG, S. *Fundamentals of Semiconductor Theory and Device Physics*, (1989) Prentice Hall International Editions, Eagle Wood Cliffs, N.J.

WOLFE, C. M., HULONYAK, N. and STILLMAN, G. E. *Physical Properties of Semiconductors*, (1989) Prentice Hall International Inc., Eagle Wood Cliffs, N.J.


## Problems

Use data presented in the text in the solution of numerical problems. Room temperature is 300 K (0.025 85 eV). The carrier mobilities may be assumed to be independent of temperature, but the electron and hole effective masses must be taken into account.

**10.1** When discussing the behaviour of either n- or p-type semiconductors, we always assume the conduction band and the valence band to have parabolic form, i.e. $N(E) \propto E^{1/2}$. How can this assumption be justified?

**10.2** A sample of Ge had the following values of resistance at the given temperatures:

| $T$(K) | 310 | 31 | 339 | 360 | 383 | 405 | 434 |
|---|---|---|---|---|---|---|---|
| $R(\Omega)$ | 13.5 | 9.10 | 4.95 | 2.41 | 1.22 | 0.74 | 0.37 |

Evaluate the energy gap.

**10.3** Continue the previous problem as follows: a concentration of 0.0001% As is added to the Ge; show that at room temperature all the donors are ionized. How will this affect the room-temperature conductivity? At approximately what temperature would the intrinsic conductivity equal that produced by the impurity atoms?

**10.4** Calculate the intrinsic conductivity of Si at room temperature. If $10^{20}$ atoms $m^{-3}$ phosphorus are added to the Si, what will be the new carrier concentration? Calculate the new position of the Fermi level.

**10.5** Calculate the intrinsic conductivity of InSb at room temperature. Some Sb is replaced by Te; calculate the following: (a) the ionization energy for the Te impurity; (b) the radius of the orbit of the weakest bound electron on the Te atom when immersed in InSb; (c) the concentration of Te required to cause overlapping of the Te impurity orbits; (d) the concentration of electrons and holes in the presence of the above concentration of Te and the position of the Fermi level.

**10.6** A semiconductor with a direct band gap is irradiated by light with photon energy $\hbar\omega(>E_g)$ and electron-hole pairs are formed. Determine expressions for the kinetic energies and wave vectors for these charge carriers. Photons of energy 1.6 eV are absorbed by GaAs, $E_g = 1.4$ eV. Calculate the appropriate energies and wave vectors for the electon-hole pair.

**10.7** Derive the relation equivalent to (10.29) for the case of strong p doping. What is the equation that holds when both donor and acceptor dopants are present?

**10.8** Si is doped with 1 ppm Al. Show that at room temperature the intrinsic carrier concentration is negligible compared with the extrinsic carrier concentration, and furthermore that >95% of the impurities are ionized. Then calculate the position of the Fermi level at 300 K and 100 K as well as the ratio of the conductivities at these temperatures.

**10.9** At 300 K a very pure sample of Ge has resistivity 3.9 Ω m. Calculate the band gap in Ge. The sample is then doped with $10^{22}$ $m^{-3}$ boron. What are the concentrations of electrons and holes and what is the new resistivity of the sample? Where does the Fermi level of the doped sample lie?

**10.10** A rectangular plate of semiconducting material has dimensions 10 mm × 4 mm × 1 mm, a current $I$ of 1.5 mA flows along its length and the associated potential drop is 78 mV. When a magnetic field **B** of strength 0.7 Wb $m^{-2}$ is applied normally to the major surface of the sample a potential difference of 6.8 mV appears across the width of the sample. Determine the character, concentration and mobility of the current carrier.

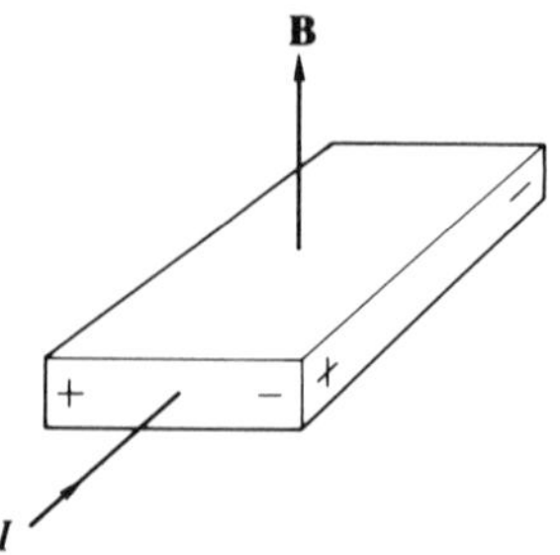

**10.11** The following Hall coefficient data were obtained for a certain semiconductor: Establish: (a) the character of doping; (b) the concentration of dopant.

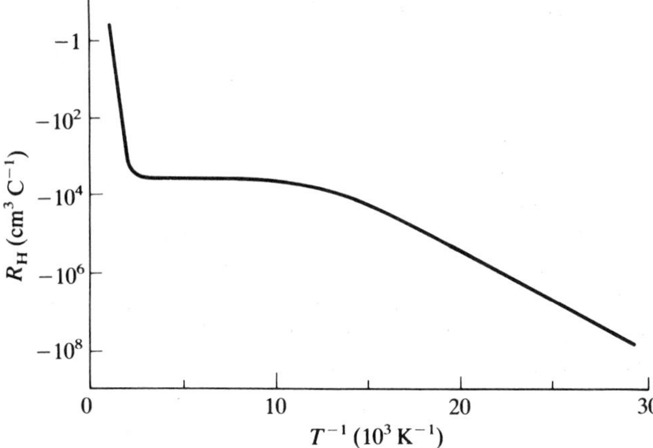


**10.12** Returning to the description of the Hall effect in Chapter 6, determine the conductivity matrix for the two-dimensional case ($\mathbf{B}_0$ parallel to the $z$ direction and electron motion confined to the $(x, y)$ plane). What form does the conductivity matrix take when $\omega_c\tau \gg 1$? Invert this matrix to obtain the resistivity matrix and show that when $\omega_c\tau \gg 1$

$$\sigma_{xx} = \frac{\rho_{xx}}{\rho_{xx}^2 + \rho_{xy}^2}, \qquad \sigma_{xy} = -\frac{1}{\rho_{xy}}.$$

These are the relations used in the discussion of the quantized Hall effect.

# 11

# Magnetism

Magnetic phenomena occupy a prominent position in solid state physics. The variety of magnetic properties exhibited in widely different classes of material offers much scope for experiment as well as theoretical speculation, whereas the practical applications of magnetism are of great technical and commercial importance. All forms of conventional matter, whether in the form of free atoms, ions, molecules or condensed aggregates, exhibit magnetism because in the presence of an external magnetic field they develop a magnetic dipole moment. However, it is also the case that certain atoms or ions may possess what is effectively a permanent magnetic dipole moment. See Fig. 11.1. The magnetization $M$ of a bulk sample is defined as the magnetic dipole moment with regard to either unit volume or unit mass. The magnetic susceptibility per unit mass is defined in terms of the magnetization per unit mass by

$$\chi = \mu_0 \frac{M}{B_0}, \qquad (11.1)\dagger$$

$B_0$ being the ambient magnetic field and $\mu_0$ the vacuum permeability.

Usually, since the mass of a sample is easier to determine than its volume, it is $\chi$, the mass susceptibility, that is of interest. One way to measure $\chi$ or $M$ is to determine the force $F$ on a sample of mass $m$ in a static or regularly varying magnetic field gradient.

$$F = (\mathrm{d}/\mathrm{d}z)(mMB_0) = (m\chi B_0\, \mathrm{d}B_0/\mathrm{d}z)(\mu_0)^{-1}$$

In the static method the gradient is directed perpendicularly to the field direction and, with proper design of the magnet, $B_0\, \mathrm{d}B_0/\mathrm{d}z$ may be arranged to be constant over the volume of the sample, usually a few $\mathrm{mm}^3$. The force is measured by weighing in a sensitive balance, usually an electromagnetic balance. In the oscillatory method the gradient is parallel to the field. The sample is attached to a stiff light rod and the vibrations induced in this rod are detected in a piezoelectric sensor. These

† $B_0 = \mu_0 H$; in this chapter we use $B_0$ to describe the ambient magnetic field in gauss or tesla. In SI units $\kappa$ is dimensionless whereas $\chi$ has dimensions $\mathrm{m}^3\ \mathrm{kg}^{-1}$; $\kappa = \rho\chi$.

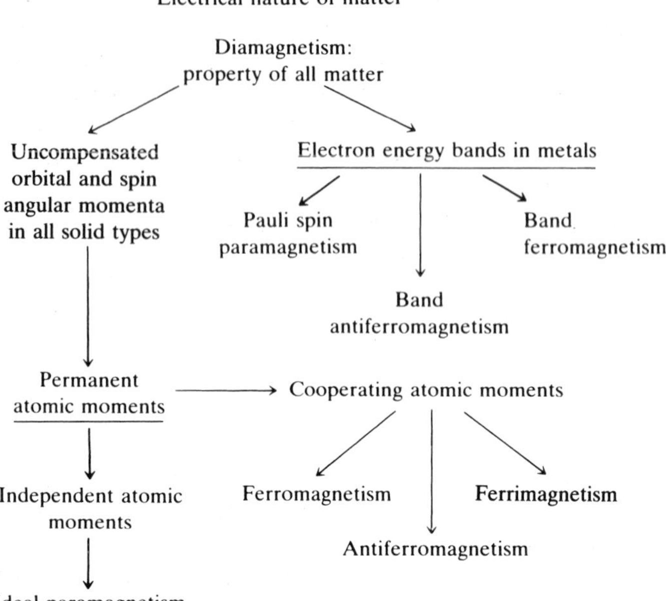


**Figure 11.1** The family tree of magnetism.

gradient methods require calibration against a known standard substance because of the difficulty in determining the field gradient.

Alternatively the sample may be passed through or vibrated in a system of pick up coils placed between the poles of a magnet. The signal voltage induced in the pick up coils may be measured and compared with that from a reference sample. The most sensitive example of this approach uses a highly uniform magnetic field from a superconducting magnet and a pick up coil forming part of a 'magnetic transformer', i.e. the pick up coil is in series with another loop that transfers a small fraction of the flux change to a carefully magnetically shielded microscopic superconducting flux detector called a SQUID (Superconducting QUantum Interference Device, see Sections 13.13 and 13.14). Such a detector can register changes corresponding to a fraction of the flux quantum $(h/2e)$ $2 \cdot 10^{-15}$ Wb.

## 11.1 Diamagnetism

The effect of an applied magnetic field on the orbital motion of the electrons in all forms of matter produces *diamagnetism*; in other words, there is always a *negative* contribution to the total susceptibility. We may appreciate this property in the following manner. A single electron in its atomic orbital resembles a classical resistanceless current loop and classically would possess an associated magnetic moment even in the absence of an external magnetic field. However, the application of a field causes a change in the magnetic flux threading the current loop orbital. Lenz's law applies and a back e.m.f. arises, leading to a reduction in the current and the associated moment. Since the current loop is resistanceless, this situation holds as long as the field remains applied. This is the negative diamagnetic contribution to the induced dipole moment arising in the core as well as the valence levels. It is always present and, although small, is of sufficient size that it represents a significant correction in studies of other forms of weak magnetic behaviour (such as the free-electron-like Pauli paramagnetism occurring in metals). The inert gases, whose atoms have closed electron shell structures, are, together with ions possessing similar electron configurations (e.g. $Cu^+$, $Zn^{2+}$, $Na^+$, $Cl^-$), examples of purely diamagnetic substances. The susceptibility is small and independent of both applied field strength and temperature (Fig. 11.2).

We expect the diamagnetic susceptibility to depend on an atom's electron content and the areas of the occupied orbitals, and thus on the orbital radii $R_i$. If $Z$ is the atomic number, calculation gives

$$\chi_{\text{dia}} = -\mu_0 \frac{Ne^2}{6m} \sum_{i=1}^{Z} \langle R_i^2 \rangle. \tag{11.2}$$

Normally in work with metals the corrections for diamagnetism are made empirically. Compared with other forms of magnetic behaviour, the study of diamagnetism has very restricted significance except for the special case of the de Haas–van Alphen effect. In what follows we shall assume that experimental data are always corrected for the diamagnetic contribution.

## 11.2 Atomic Magnetic Moments: Paramagnetism

In a *paramagnetic* substance, the total resultant dipole moment is increased by the application of an external magnetic field. The susceptibility is therefore positive; in normal field strengths, say $<1.5$ T, and at not too low a temperature, it is independent of the applied field (Fig. 11.2*a*). We have already encountered one form of paramagnetism in the free electron gas, where we found $\chi$ to be independent of

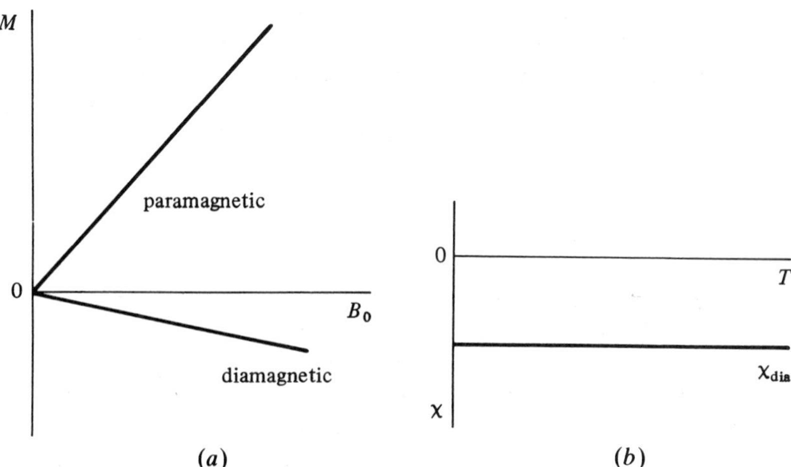


**Figure 11.2** For paramagnetic and diamagnetic substances the respective susceptibilities are independent of applied magnetic field strength, (*a*). The diamagnetic susceptibility is also independent of temperature, (*b*).

temperature (Section 6.3.4). We now consider other classes of material where the electron orbitals maintain their atomic character – for example free atoms (as occur in metallic vapours) and ions of salts like $FeCl_2$, as well as other forms of solid or liquid matter. We shall find that $\chi$ is usually strongly dependent on temperature.

Earlier we compared the orbital motion of an electron in an atom to a current loop, and to any electron with finite orbital angular momentum (quantum number $l$) may be attributed an associated $z$ component of magnetic moment $\mu_B l$.† The complete atom or ion possesses total orbital angular momentum appropriate to the quantum number $L$ and therefore carries a maximum $z$ component of orbital magnetic moment equal to $\mu_B L$. In addition, every electron has an associated spin magnetic moment equal to one Bohr magneton. If the atom has resultant spin quantum number $S$ then we write the total angular momentum in terms of the quantum number $J$, where

$$\mathbf{J} = \mathbf{L} + \mathbf{S}. \tag{11.3}$$

Different **J** values may be obtained from given **L** and **S**, producing multiplet structure. The lowest multiplet level is the ground state of the atom or ion, the appropriate $J$ determines the size of the associated $z$ component of magnetic moment, namely

$$\mu = g\mu_B J, \tag{11.4}$$

where $\mu$ is the maximum resultant $z$ component of the magnetic moment of the free atom. *This moment is permanent and indestructible.* The only changeable aspect is its resolved component in some specified direction such as that of an externally applied magnetic field. The quantity $g$ is the magnetomechanical ratio, and is also known as the Landé splitting factor:‡

$$g = 1 + \frac{J(J+1) + S(S+1) - L(L+1)}{2J(J+1)}. \tag{11.5}$$

We know from the Bohr–Stoner explanation of the periodic structure of the elements that electrons in atoms always occupy orbitals so as to produce the maximum resultant spin, subject of course to the Pauli principle. Closed shells of electrons have neither resultant orbital nor resultant spin angular momentum; they therefore have no resultant atomic magnetic moment. They have purely diamagnetic properties. On the other hand, any atom or ion with an incomplete shell or subshell of electrons (and here the transition metals and rare earth metals are of great importance) has an associated magnetic moment. In free atoms or ions the magnetic state is the preferred one because the electrons with similar spin, owing to the operation of Pauli's principle, avoid coming too close to one another, thereby reducing the Coulomb potential energy. We must emphasize that *for free atoms the magnetic state is an ordinary state of matter* (Fig. 11.3). On the other hand, when atoms form aggregates, the atomic moment is usually lost. We can readily appreciate why. Consider for example atoms of Na and Cl; both possess an atomic magnetic moment, but on combining to form $Na^+Cl^-$ an electron is transferred from the Na to the Cl and the ions so formed have inert gas electronic structures. The moment is lost. Similarly, the stable form of many elements is the molecule. Thus atomic hydrogen

† We assume that the $z$ direction is determined by that of a weak external magnetic field.

‡ See Appendices 11.1–11.3 for a more detailed description of these aspects.

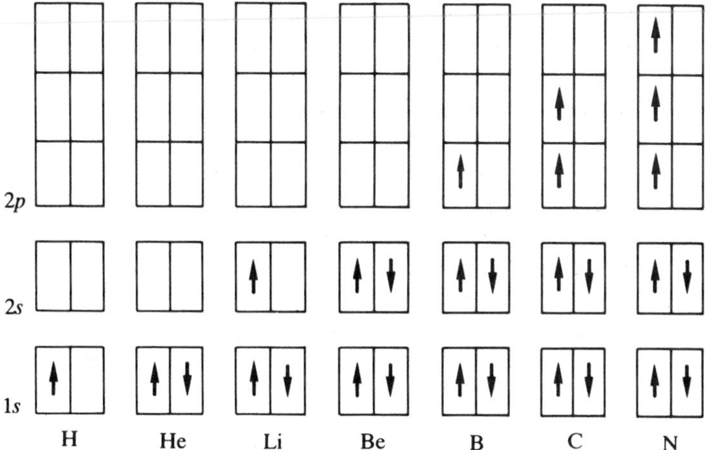


**Figure 11.3** The magnetic state is preferred by free atoms because the Pauli principle favours the occurrence of parallel spins. This is exemplified here in the electronic structures of the first seven elements of the periodic table.

has a magnetic moment of one Bohr magneton, but the stable form of hydrogen is the molecule and the electrons are paired and possess oppositely directed spins and zero orbital angular momentum. The hydrogen molecule is therefore a diamagnetic entity. The same is true for $N_2$ and in fact for most diatomic molecular gases, the important exceptions being $O_2$ and NO, which are both strongly paramagnetic (see Box 11.1, pp. 326, 327). Since the condensed forms of many elements like hydrogen and nitrogen are aggregates of the molecules, they also are purely diamagnetic (again with the exception of $O_2$ and NO).

Metallic atoms of odd valence possess atomic magnetic moments which are lost when the solid is formed. This, as we know, is because the valence electrons lose their atomic character and form an electron gas that gives rise to the characteristic Landau diamagnetism and the Pauli paramagnetism.

We conclude that atoms that have their magnetic moment associated with the valence electrons generally lose it when they form molecules, compounds or larger aggregates because the valence electrons become, in various ways (ionic, covalent or metallic bonds), shared with other atoms of the aggregate; they therefore lose their localized atomic character. Conversely, we can say that if the electrons that provide the atomic moment maintain their atomic character in the condensed state then the latter will be magnetic in a characteristic fashion. This situation is best illustrated by the rare earth metals. In these metals the atomic moment is located primarily in the 4f electrons: these, as we have already seen, lie deep within the atom and are little affected by the presence of surrounding atoms. The rare earth elements, whether in the form of the pure metals or as chemical compounds, are usually magnetic and present a rich diversity in their detailed behaviour. This is also true, but to a lesser extent, for the d transition metals, particularly when they form ionic compounds, because the essential atomic character of the d shell is preserved. However, the d electrons are among the outermost electrons of the transition metals and are strongly influenced by electrons on neighbouring atoms with which they interact, whereby they often lose their atomic character and associated moment. As we have already seen, the pure transition metals are magnetic in the manner of the electron gas. The strong magnetic properties of a minority of these elements (Mn, Fe, Co and Ni) must be considered exceptional. A true description of their magnetic behaviour remains a particularly difficult problem, particularly with regard to the temperature dependence of magnetization in both the ferromagnetic and paramagnetic states.

**Box 11.1 The $O_2$ Molecule**

The $O_2$ molecule is strongly paramagnetic, as are the molecules $S_2$ and NO, but $N_2$ is diamagnetic. The quantum mechanical approach to chemical bonding provides the explanation of the magnetic behaviour of these and other molecules. Molecule formation may be described in terms of molecular electron orbitals that concentrate the electronic charge between the nuclei (bonding orbitals) and those that distribute the charge at the extremities of the molecule (antibonding orbitals) (Fig. 11.4). If the molecular orbital has axial symmetry about the bond direction then we speak of $\sigma$ orbitals, whereas if the wave function has a nodal plane that contains the bond axis then we speak of $\pi$ orbitals.

In the molecular orbital approach to chemical bond formation each of the outer electrons (the 2s and 2p electrons in the case of $O_2$), in its orbit, is assumed to enclose both nuclei; each electron belongs to the molecule rather than to a particular atom. The molecular orbital is, however, described as a linear combination of overlapping atomic orbitals. The molecular states stem from the atomic orbitals and form a system of bonding ($\sigma$, $\pi$) and antibonding ($\sigma^*$, $\pi^*$) molecular orbitals as illustrated qualitatively in Fig. 11.5. The $\sigma$, $\sigma^*$, $\pi$ and $\pi^*$ orbitals arise irrespective of the initial symmetry of the associated level s, p, d etc. If the bond axis is in the z direction then we find a $\sigma(2p_z)$ bonding orbital normally associated

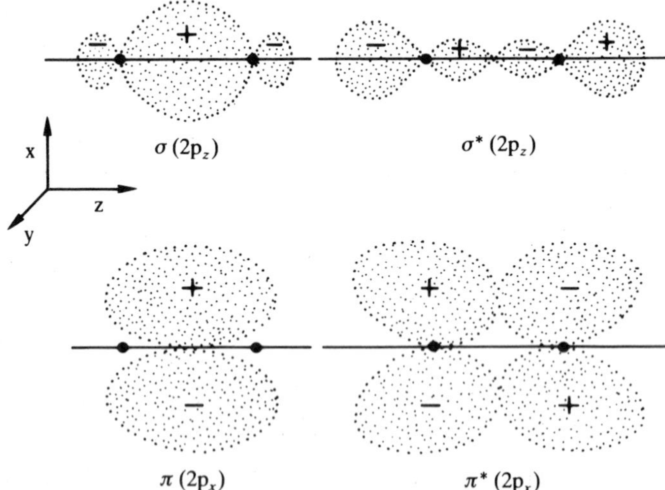


**Figure 11.4** A schematic illustration of sections of the bounding surfaces of the molecular orbitals formed from the atomic p orbitals in the oxygen atom. The oxygen ions are marked by the heavy black dots and the axis of the molecule is directed along the z axis. The $\sigma$ and $\sigma^*$ have rotational symmetry about the z axis whereas the $\pi$ and $\pi^*$ bonds have zero amplitude in the (y, z) plane. The corresponding $\pi(2p_y)$ and $\pi^*(2p_y)$ have the (x, z) nodal plane; the $\pm$ signs indicate the sign of the wave function amplitude.

*continued*

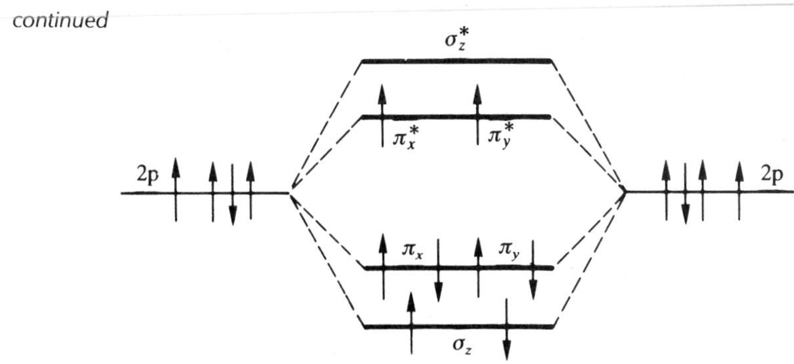


**Figure 11.5** On the left and right sides of this figure are shown, in schematic fashion, the p atomic orbitals of the free oxygen atom. The orbitals arising in the oxygen molecule are drawn in the centre of the diagram. The $\pi$ ($\pi^*$) orbitals are degenerate and occupied as shown (in practice the levels are not symmetrically arranged as shown because in $O_2$ the $\sigma_z$ orbital lies somewhat higher in energy than the $\pi_x$ and $\pi_y$ levels, but this is irrelevant here).

with the greatest overlap, whereas the $2p_x$ and $2p_y$ atomic orbitals produce a system of bonding and antibonding $\pi$ orbitals transverse to the bond axis. Owing to the equivalence of the $x$ and $y$ directions in the homonuclear linear molecule, these $\pi$ molecular orbitals are degenerate.

In the $O_2$ molecule there are eight 2p electrons (eight 3p in $S_2$) to be placed in the six available molecular orbitals originating from the atomic 2p states. The $\sigma(2p_z)$, $\pi(2p_x)$ and $\pi(2p_y)$ each take two electrons, leaving two over to occupy the degenerate $\pi^*$ antibonding levels. They will clearly do so with least electron electron Coulomb interaction energy if they occupy different orbitals, a situation guaranteed if they have parallel spins. Partially occupied degenerate antibonding orbitals (as may arise at the top of an energy band) are thus conducive to positive exchange interaction (see the later comment on the Heisenberg interaction, Section 11.7).

In the case of NO and $N_2$ we find a qualitatively similar distribution of molecular orbitals, but in NO there are seven 2p electrons to be accommodated and there is only one electron available for the system of degenerate $\pi^*$ orbitals. The NO molecule is magnetic, but less so than the $O_2$ molecule. In the $N_2$ molecule, however, the $\pi^*$ orbitals are empty and all orbitals of lower energy completely occupied; $N_2$ is therefore diamagnetic. For further details see Grey (1964) and McWeeny (1979).

In the following we shall, with the aid of very simple models, attempt to understand the magnetic behaviour of the following systems:

(a) an ideal magnetic gas defined as a system of non-interacting magnetic atoms;

(b) certain alloys in which transition metal atoms carry well defined and spatially localized atomic magnetic moments.

(c) aggregates of magnetic atoms that interact to produce, among other phenomena, that of ferromagnetism.

The word 'interaction' is used to represent the complex quantum-mechanical exchange and correlation forces that govern the behaviour of electrons and their magnetic properties. *The classical magnetostatic dipole-dipole interactions between atomic magnets are extremely small and completely insignificant; they do arise, but their effects would be observable only at ultra-low temperature, below 0.001 K. We therefore neglect them.*

The exchange interactions between the electron spins that carry the atomic moment are characterized by an exchange energy J. J may have positive or negative sign: the former is associated with the tendency for the parallel orientation of interacting spins, ferromagnetic interaction; the latter with an anti-parallel arrangement which is called antiferromagnetic interaction.

## 11.3 The Ideal Magnetic Gas: Classical Model

Although Faraday demonstrated that magnetic behaviour was a general property of matter, it was Curie who first made a systematic study of paramagnetism, particularly with regard to the influence of temperature. He found in 1895 that whereas the susceptibility of diamagnetic materials was independent of both the applied field and of temperature, that of paramagnetic non-metals, while independent of field strength, showed a characteristic temperature dependence represented by

$$\chi = \frac{C}{T}. \tag{11.6}$$

This is therefore known as Curie's law and $C$ as the Curie constant. In particular the above law is very well obeyed by oxygen gas, which we might consider a very good approximation to a magnetic gas. However, an ideal magnetic gas is not necessarily a gaseous substance (Fig. 11.6). Curie's law was given a simple theoretical explanation by Langevin, who in 1905, assumed that each atom of the gas carried a permanent moment $\mu$, although there was no knowledge of its true origin.† These moments become oriented in the direction of an externally applied field and it is assumed that this process is hindered by the thermal motion of the atoms.

Consider a volume of magnetic gas containing $N$ atoms, each with magnetic moment $\mu$. In the absence of an external field the moments are randomly distributed because any dipole-dipole interaction is vanishingly small, but the application of a field causes a reorientation leading to a resultant moment in the field direction. In Fig. 11.7 we represent the distribution of atomic moments according to the angle $\theta$ they make with the field direction. A moment $\mu$ inclined at an angle $\theta$ to the field $B_0$ leads to a magnetic contribution $-\mu B_0 \cos\theta$ to the total free energy of the atom. This magnetic potential energy causes a redistribution of atoms so that the number found with their moments oriented at an angle $\theta$ to the field $B_0$ is, according to Boltzmann,

$$\mathrm{d}N = A \exp\left(-\frac{-\mu B_0 \cos\theta}{k_B T}\right) \mathrm{d}\omega,$$

† It is not possible to explain the origin of magnetic phenomena classically; Langevin's assumption that there exists an atomic magnetic moment the same for all atoms and independent of temperature implies quantized energy levels.

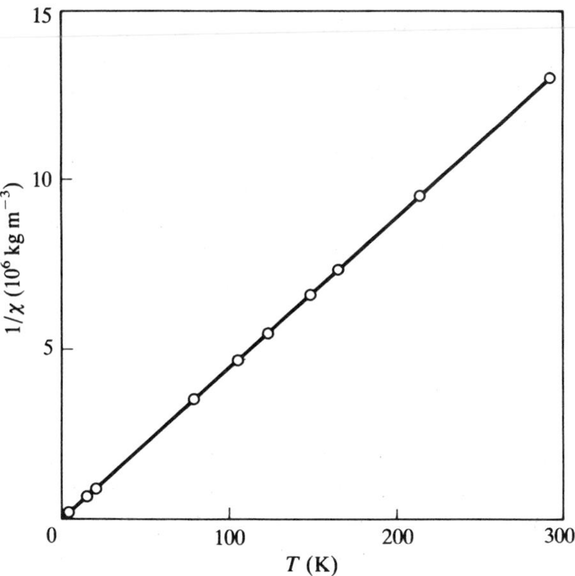


**Figure 11.6** Curie's law as observed in $CuSO_4 \cdot 5H_2O$. (After Crangle 1977.)

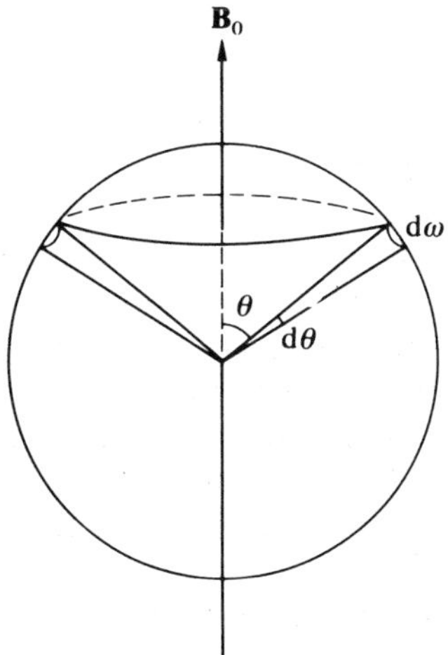


**Figure 11.7** Illustrating the angular distribution of atomic moments in the ideal paramagnetic gas in the presence of the field $\mathbf{B}_0$.

where $d\omega$ is an element of solid angle bounded by the cones $\theta$ and $\theta + d\theta$ and $A$ is constant. Writing

$$\alpha = \mu B_0/k_B T, \tag{11.8}$$

(11.7) becomes

$$dN = Ae^{\alpha \cos\theta}\, d\omega,$$

and

$$N = 2\pi \int_0^{\pi} Ae^{\alpha \cos\theta} \sin\theta\, d\theta = \frac{4\pi}{\alpha} A \sinh \alpha. \tag{11.9}$$

The total resolved moment in the field direction is

$$M = N\langle\mu\rangle = \int_0^{\pi} \mu \cos\theta \, \mathrm{d}N \tag{11.10}$$

$$= 2\pi A\mu \int_0^{\pi} \mathrm{e}^{\alpha\cos\theta} \sin\theta \cos\theta \, \mathrm{d}\theta.$$

Evaluating the integral and using (11.9) gives

$$M = N\langle\mu\rangle = N\mu\left(\coth\alpha - \frac{1}{\alpha}\right). \tag{11.11}$$

$L(\alpha) = \coth\alpha - 1/\alpha$ is known as the Langevin function.

Equation (11.11) governs the behaviour of a classical magnetic gas as a function of field strength and temperature. At constant temperature the magnetization varies with $B_0$ in the manner of Fig. 11.8. Initially the dependence is linear, but at high field strengths the moment slowly saturates. However, at room temperature there would be no marked departure from linearity until fields of order 100 T were achieved. We have already pointed out that magnetic energies are extremely small, and at normal temperatures $\mu_B B_0 \ll k_B T$. Therefore $\alpha$ is usually a very small quantity $< 10^{-2}$, and under such circumstances $L(\alpha)$ is close to $\frac{1}{3}\alpha$, so we can write

$$\frac{\langle\mu\rangle}{\mu} = \frac{\alpha}{3} = \frac{\mu B_0}{3k_B T} \tag{11.12}$$

or

$$M = N\langle\mu\rangle = \frac{N\mu^2 B_0}{3k_B T},$$

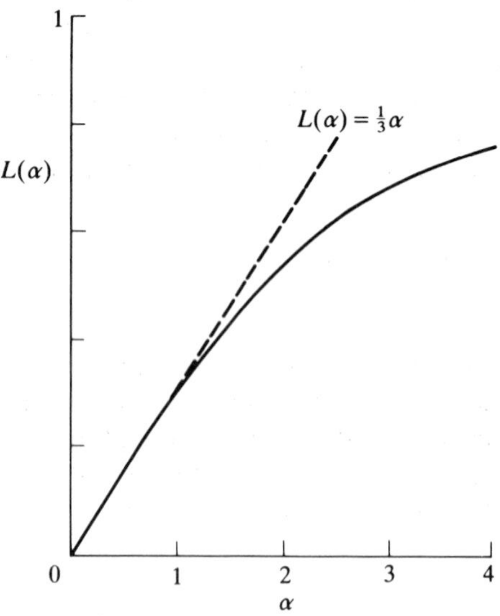


**Figure 11.8** The Langevin function.

and

$$\chi = \frac{\mu_0 M}{B_0} = \frac{\mu_0 N\mu^2}{3k_B T} = \frac{C}{T}. \tag{11.13}$$

If we consider a gram molecule of gas, $N$ becomes Avogadro's number and the molar susceptibility is

$$\chi_{\text{mol}} = \frac{\mu_0 N\mu^2}{3k_B T} = \frac{\mu_0 N^2\mu^2}{3RT}$$

$$= \frac{\mu_0 \sum^2}{3RT}, \tag{11.14}$$

where

$$\sum = N\mu$$

is the molar moment. We see immediately that (11.14) reproduces Curie's law and we associate the Curie constant with

$$C_{\text{mol}} = \frac{\mu_0 \sum^2}{3R}. \tag{11.15}$$

Thus a study of the susceptibility of a paramagnetic gas allows a direct determination of the atomic moment $\mu$.

## 11.4 The Ideal Magnetic Gas: Quantum Model

The Langevin model provides a good description of the temperature dependence of the susceptibility of an ideal paramagnetic gas but cannot explain the origin of the atomic moment; this only became possible with the advent of quantum mechanics. Confining attention to a free atom or ion, we associate the quantum number $J$ with the total resolved component of angular momentum, and then the atomic magnetic moment is $g\mu_B J$. In contrast with the classical theory, under spatial quantization, this moment can only assume discrete orientations in the presence of an external field. These orientations are those producing the following resolved components in the field direction:

$$-g\mu_B J,\ -g\mu_B(J-1),\ -g\mu_B(J-2), \ldots, 0, \ldots,$$
$$\ldots, +g\mu_B(J-2),\ +g\mu_B(J-1),\ +g\mu_B J.$$

In all there are $2J + 1$ possible orientations. We describe the resolved values of $J$ by the magnetic quantum number $m$, so that an arbitrary resolved moment becomes $g\mu_B m$. An atom with magnetic quantum number $m$ acquires a potential energy $-g\mu_B m B_0$ in the presence of the field $B_0$. We now apply Langevin's argument to this system; thus the number of atoms with moment $g\mu_B m$ in the field direction becomes

$$\mathrm{d}N = A \exp\left(-\frac{-g\mu_B m B_0}{k_B T}\right) \tag{11.16}$$

and

$$N = A \sum_{-J}^{+J} \exp\left(\frac{g\mu_B m B_0}{k_B T}\right), \tag{11.17}$$

so

$$dN = N \exp\left(\frac{g\mu_B m B_0}{k_B T}\right) \Bigg/ \sum_{-J}^{+J} \exp\left(\frac{g\mu_B m B_0}{k_B T}\right). \tag{11.18}$$

We write

$$\beta = \frac{g\mu_B B_0}{k_B T},$$

so that

$$dN = N e^{\beta m} \Bigg/ \sum_{-J}^{+J} e^{\beta m}.$$

The resolved magnetization in the field direction is readily seen to be

$$M = \sum_{-J}^{+J} g\mu_B m \, dN = N \sum_{-J}^{+J} g\mu_B m e^{\beta m} \Bigg/ \sum_{-J}^{+J} e^{\beta m}. \tag{11.19}$$

We assume that $\beta$ is small, and this gives us the high-temperature behaviour and allows us to write $e^{\beta m} = 1 + \beta m$, so that

$$M = N \sum_{-J}^{+J} g\mu_B m \,(1 + \beta m) \Bigg/ \sum_{-J}^{+J} (1 + \beta m). \tag{11.20}$$

We need the following sums:

$$\sum_{-J}^{+J} m = 0,$$

$$\sum_{-J}^{+J} 1 = \sum_{-J}^{+J} m^0 = 2J + 1,$$

$$\sum_{-J}^{+J} m^2 = \tfrac{1}{3} J(J+1)(2J+1).$$

On insertion in (11.20), we obtain

$$\frac{M}{B_0} = \frac{N g^2 \mu_B^2}{3 k_B T} J(J+1) \tag{11.21}$$

or

$$\chi = \frac{\mu_0 N g^2 \mu_B^2}{3 k_B T} J(J+1) = \frac{C}{T}. \tag{11.22}$$

If we compare (11.22) with the classical expression (11.13), we see that they are identical if we associate the classical moment $\mu$ with the quantity $\mu_B g[J(J+1)]^{1/2}$. This is therefore called the effective atomic magnetic moment:

$$\mu_{eff} = \mu_B g[J(J+1)]^{1/2} \tag{11.23}$$

or

$$p = g[J(J+1)]^{1/2}, \tag{11.24}$$

$p$ being known as the effective Bohr magneton number.

Implicit in the above derivation is the assumption that all the atoms are in the same spectroscopic state, i.e. all have the same $J$ value. A free atom or ion possesses definite values of $L$ and $S$, but these may combine vectorially to provide several possible $J$ values, $J$, $J'$, $J''$ .... There will usually be one value of $J$ associated with the lowest energy, and this is the ground state of the atom or ion. Our assumption has been that all other spectroscopic states are unoccupied and in particular that $J'$, which is that state closest to the ground state $J$, is sufficiently far away in energy that it has negligible population. In other words,

$$\hbar\omega(JJ') \gg k_B T. \tag{11.25}$$

This is usually, but not always, the case.

In very strong magnetic fields and at low temperatures, say about 4 K, we can no longer assume that $\beta$ is a small quantity, and (11.19) must be evaluated without approximations. We are content just to quote the result in the following form:

$$\begin{aligned} M &= N\langle\mu\rangle \\ &= Ng\mu_B J\left[\frac{2J+1}{2J}\coth\frac{(2J+1)\alpha}{2J} - \frac{1}{2J}\coth\frac{\alpha}{2J}\right], \end{aligned} \tag{11.26}$$

where

$$\alpha = \frac{g\mu_B J B_0}{k_B T}. \tag{11.27}$$

In other words,

$$\begin{aligned} M &= N\langle\mu\rangle \\ &= Ng\mu_B J B(J, \alpha), \end{aligned} \tag{11.28}$$

where $B(J, \alpha)$ represents the function in brackets in (11.26) and is known as the Brillouin function. It is the quantum equivalent of the Langevin function and determines the magnetization as a function of both temperature and applied field (Fig. 11.9). We note that in high magnetic fields saturation of the resolved moment is obtained at low temperatures, but at $M = Ng\mu_B J$ not $Ng\mu_B[J(J+1)]^{1/2}$. By fitting (11.28) to experimental data, we may attribute a specific $J$ to the magnetic atom.

However, a $J$ value may also be determined in the high-temperature approximation via Curie's law and application of (11.22). We conclude that a *knowledge of the magnetic behaviour of an ideal magnetic gas provides us with information about the spectroscopic state of the magnetic atom or ion.* One might object that the concept of the ideal magnetic gas is of very limited value since there are so few magnetic gases, but we do not need to have a gaseous substance to satisfy the criterion for an ideal magnetic gas. It is only necessary that the magnetic atoms be sufficiently separated to make any mutual exchange interaction insignificant. Thus a salt like $MnSO_4(NH_4)_2SO_4 \cdot 6H_2O$ is a good approximation to an ideal magnetic gas because the only ion possessing an atomic moment is the divalent Mn ion; these are dispersed in the mixed salt and further diluted by the water of crystallization. Such

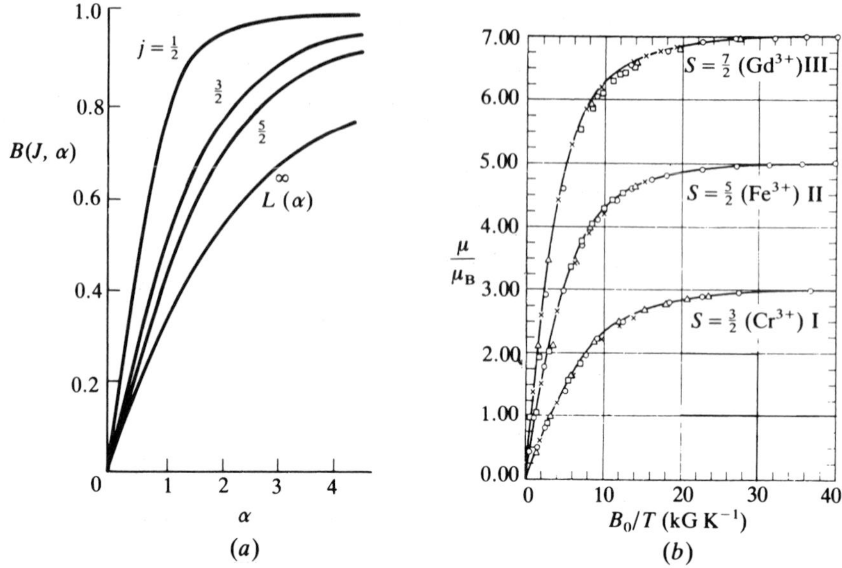


**Figure 11.9** The Brillouin function in theory (a) and experiment (b). ○, 1.30 K; △, 2.00 K; ×, 3.00 K; □, 4.21 K; ——, Brillouin functions. (Experimental data after Henry 1952.)

salts are known as dilute magnetic salts and have been the subject of intense study. Experiment shows that for these salts

$$\chi = \frac{C}{T + \Delta},$$

where $\Delta$ is a small quantity usually in the range $\pm 10$ K and often very close to zero. We may interpret $\Delta$ as a measure of residual magnetic exchange interactions. At low temperatures such salts have been shown to follow the Brillouin approach to saturation (11.26) accurately. Can we predict the spectroscopic ground state of a free atom or ion? In principle we must solve the Schrödinger equation, determine the eigenvalues corresponding to different spectroscopic states, and pick the one giving the lowest energy. With today's knowledge and computing facilities this is a practical proposition, but in 1927, when the Schrödinger equation was just one year old, the situation was very different. However, at that time Hund used the vector atom model to correlate empirical spectroscopic data and arrived at his well known rules which postulate that, always subject to the Pauli principle,

(a) $S$ is maximized;

(b) $L$ is maximized consistently with (a) above;

(c)

$$J = \begin{cases} |L - S| & \text{when a shell is less than half full,} \\ L + S & \text{when a shell is more than half full.} \end{cases}$$

When a shell is exactly half full $J = S$ since $\Sigma\, m = 0$.

Using these rules, it is a simple matter to establish the spectroscopic ground state of an ion (see Appendix 11.3).

**Table 11.1** The Bohr magneton values for the rare earth ions

| Ion | Electron configuration | Ground-state term | $g[J(J+1)]^{1/2}$ | $p$ (exp) |
|---|---|---|---|---|
| $Ce^{3+}$ | $4f^1 5s^2 p^6$ | $^2F_{5/2}$ | 2.54 | 2.4 |
| $Pr^{3+}$ | $4f^2 5s^2 p^6$ | $^3H_4$ | 3.58 | 3.5 |
| $Nd^{3+}$ | $4f^3 5s^2 p^6$ | $^4I_{9/2}$ | 3.62 | 3.5 |
| $Pm^{3+}$ | $4f^4 5s^2 p^6$ | $^5I_4$ | 2.68 | — |
| $Sm^{3+}$ | $4d^5 5s^2 p^6$ | $^6H_{5/2}$ | 0.84 | 1.5 |
| $Eu^{3+}$ | $4f^6 5s^2 p^6$ | $^7F_0$ | 0.00 | 3.4 |
| $Gd^{3+}$ | $4f^7 5s^2 p^6$ | $^8S_{7/2}$ | 7.94 | 8.0 |
| $Tb^{3+}$ | $4f^8 5s^2 p^6$ | $^7F_6$ | 9.72 | 9.5 |
| $Dy^{3+}$ | $4f^9 5s^2 p^6$ | $^6H_{15/2}$ | 10.63 | 10.6 |
| $Ho^{3+}$ | $4f^{10} 5s^2 p^6$ | $^5I_8$ | 10.60 | 10.4 |
| $Er^{3+}$ | $4f^{11} 5s^2 p^6$ | $^4I_{15/2}$ | 9.59 | 9.5 |
| $Tm^{3+}$ | $4f^{12} 5s^2 p^6$ | $^3H_6$ | 7.57 | 7.3 |
| $Yb^{3+}$ | $4f^{13} 5s^2 p^6$ | $^2F_{7/2}$ | 4.54 | 4.5 |

### 11.4.1 *Rare earth ions*

We compare $p$ values determined from magnetic measurements on dilute salts with the predicted ones. Table 11.1 provides details of the trivalent rare earth ions, and we note the very good agreement between theoretical and experimental $p$ values. Hund's rules work very well except for $Sm^{3+}$ and $Eu^{3+}$, where they fail completely – but it is not so much Hund's rules that are at fault as their application to these particular ions, which are now known to depart from the criterion (11.25). In fact, these ions have multiplet levels close to the ground state, and the occupancy of these levels varies with temperature. When these aspects are taken into consideration the anomalous behaviour of Sm and Eu can be accounted for. The dilute salts of rare earth ions therefore approximate the ideal paramagnetic gas very well indeed.

### 11.4.2 *Transition metal ions*

We consider only the first long series of transition elements. These exhibit primarily valences of two and three, although Mn and V are occasionally quadrivalent. They have been studied in dilute magnetic salts, often in the form of mixed ammonium sulphates. It is straightforward to apply Hund's rules and to calculate effective Bohr magneton numbers (Table 11.2). These are found to be in very poor accord with experiment, but agreement is restored if we calculate $p$ on the basis that only the spin moment reacts with the field so that

$$p = g[J(J+1)]^{1/2} \rightarrow 2[S(S+1)]^{1/2}. \tag{11.29}$$

Phenomenologically, we might say that the orbital angular momentum, although finite, is incapable of interacting with the external field: the orbital moments are said to be quenched, i.e. extinguished. Classically, we might think that the orbits are fixed in space and are unable to rotate into the direction of the external field.

**Table 11.2** The Bohr magneton values for certain 3d ions

| Ion | Configuration | Ground state | $g[J(J+1)]^{1/2}$ | $2[S(S+1)]^{1/2}$ | $p$ (exp) |
|---|---|---|---|---|---|
| $Ti^{3+}$, $V^{4+}$ | $3d^1$ | $^2D_{3/2}$ | 1.55 | 1.73 | 1.8 |
| $V^{3+}$ | $3d^2$ | $^3F_2$ | 1.63 | 2.83 | 2.8 |
| $Cr^{3+}$, $V^{2+}$ | $3d^3$ | $^4F_{3/2}$ | 0.77 | 3.87 | 3.8 |
| $Mn^{3+}$, $Cr^{2+}$ | $3d^4$ | $^5D_0$ | 0.00 | 4.90 | 4.9 |
| $Fe^{3+}$, $Mn^{2+}$ | $3d^5$ | $^6S_{5/2}$ | 5.92 | 5.92 | 5.9 |
| $Fe^{2+}$ | $3d^6$ | $^5D_4$ | 6.70 | 4.90 | 5.4 |
| $Co^{2+}$ | $3d^7$ | $^4F_{9/2}$ | 6.63 | 3.87 | 4.8 |
| $Ni^{2+}$ | $3d^8$ | $^3F_4$ | 5.59 | 2.83 | 3.2 |
| $Cu^{2+}$ | $3d^9$ | $^2D_{5/2}$ | 3.55 | 1.73 | 1.9 |

The causes of the quenched moments are the strong electrostatic fields that exist in the neighbourhood of any ion; these fields are usually called 'crystal fields'. They are significant for the transition metal ions because the magnetic electrons, the d electrons, are exposed to their full strength. The crystal field decouples the spin and orbital moments and the spectroscopic state is no longer characterized by a $J$ value. The crystal field, being electrostatic, causes a kind of Stark splitting of the originally degenerate $L$ level, and the sublevel separation becomes large compared with $k_B T$. The population of the sublevels cannot be altered either by temperature or still less by an external field; so the external field cannot influence the sublevels and create a resolved moment.

Normally the moment arises in the presence of an external field because the occupancy of the originally degenerate level is changed so that states with lower magnetic energy are occupied in preference to those with greater magnetic energy. If the crystal field removes this degeneracy and makes the sublevel splitting large compared with $g\mu_B mB_0$, the external field cannot change the occupancy of the levels of different $m$ and cannot produce an induced magnetic moment. On the other hand, the resultant spin angular momentum remains unaffected by the crystal field and the spin moment acts unabated. The d electrons are so vulnerable to crystal fields that in any metal, alloy or compound where atomic magnetic moments arise it is customary to assume that the contribution from the orbital angular momentum is quenched.

In the case of the rare earths the 4f electrons are protected by the surrounding $5s^2 5p^6$ electrons and the crystal field is not felt to anywhere near the same extent as for the d electrons. The crystal fields are significant for many properties of the rare earth atoms, but they are not sufficiently strong to quench the orbital moment.

Magnetic data obtained on dilute salts were very important for the characterization of the spectroscopic states of the rare earth and the transition metal ions. In recent years, they have played a similar role in the characterization of the 5f 'rare earths' known as the actinides (elements 89–96).

### 11.4.3 Paramagnetic dilute alloys

Earlier, in Section 9.5.2, the concept of the resonant bound state in dilute alloys was introduced. It is now convenient to present examples of dilute alloys which have been extensively studied during the post-war years. In the present connection 'dilute' means $\leqslant 1\%$ of solute. As earlier described Al is a poor solvent, but it does dissolve small amounts of a transition metal like Mn. Manganese, $3d^5 4s^2$, is strongly magnetic as a free atom and as an ion, but when dissolved in Al it does not exhibit a magnetic moment. The d electrons are believed to form a resonant bound state as illustrated in Fig. 9.10 and Fig. 11.10*a*, Box 11.2. Manganese dissolved in copper, however, forms a strongly paramagnetic alloy with a characteristic Curie-like temperature dependent susceptibility indicative of a Bohr magneton number $p = 4.9$ corresponding to $S = 2$. There is good reason to believe that the Mn acts as though it were almost divalent in these alloys so that its configuration approximates $3d^5 4s^2$. Again the d states arise in a resonant bound state, but now this state is spin split producing four unbalanced spins to agree with the experimentally determined magneton number. The arrangement is depicted in Fig. 11.10*b*. The Mn impurity atom is said to possess a localized magnetic moment, i.e. the moment is associated with the site of the Mn atom. The origins of the spin splitting are the Coulomb and exchange interactions which cause unfilled electron shells in free atoms or ions always to set their spins parallel to one another unless expressly forbidden by Pauli's principle. The occurrence of local moments in this kind of alloy is summarized in Fig. 11.11.

The concept of the magnetic resonant bound state was developed by Friedel and independently by Anderson. Friedel introduces a bare solute transition metal ion into an electron gas to which are added all the outer s and d shell electrons of the perturbing atom. The electrons in the gas are then allowed to interact with the bare ion and the behaviour is treated as a scattering problem. Anderson, on the other hand, submerges an idealized neutral transition metal atom in an electron gas. A single d orbital is assumed. Its wave function hybridizes with the plane wave functions of the free electrons. This causes the two initially sharp spin states of the d orbital to broaden and they assume a Lorentzian shape. An effective Coulomb repulsion energy exists between electrons of opposite spin in the two available spin states, which are only fractionally occupied because in this model there is only one electron in the d orbital. The problem then is to determine the conditions causing a separation in the energies of the two spin states and thereby the occurrence of a stable magnetic state. The two approaches are complementary and lead to the same conclusions. For our purposes the results of the Anderson approach are more readily summarized, see Box 11.2.

As Fig. 11.11 shows, local moments arise preferentially in solvents with low concentrations of conduction electrons implying small values of $E_F$ and $N(E_F)$. The 3d transition metal solute should possess a large magnetic moment in the free atom condition. It is particularly noteworthy that no local moment arises when Al is the solvent and Ni never carries a local moment in this kind of alloy (recall the earlier discussion of **Cu**Ni alloys). The observed behaviour is in good qualitative agreement with the ideas of Friedel and Anderson.

In the alloys where these local moments exist the high temperature behaviour ($> 50$ K) is characterized by a linear dependence of $1/\chi$ on temperature, because any coupling between impurity moments on different atoms is broken down by thermal agitation. At low temperatures and larger solute concentrations cooperative interactions between moments on different atoms arise, leading to long range ferromagnetic or antiferromagnetic order. In the earlier discussion it has been tacitly assumed that the conduction electrons serve only to broaden the d levels, but the s or con-

**Box 11.2. The Anderson Model of the Resonant Bound State**

Anderson (1961) considers a transition metal impurity atom, containing a single d orbital, embedded in a free electron gas with state density of given spin $N(E)$, and determines the conditions necessary for the occurrence of the magnetic as opposed to the non-magnetic state, Fig. 11.10.

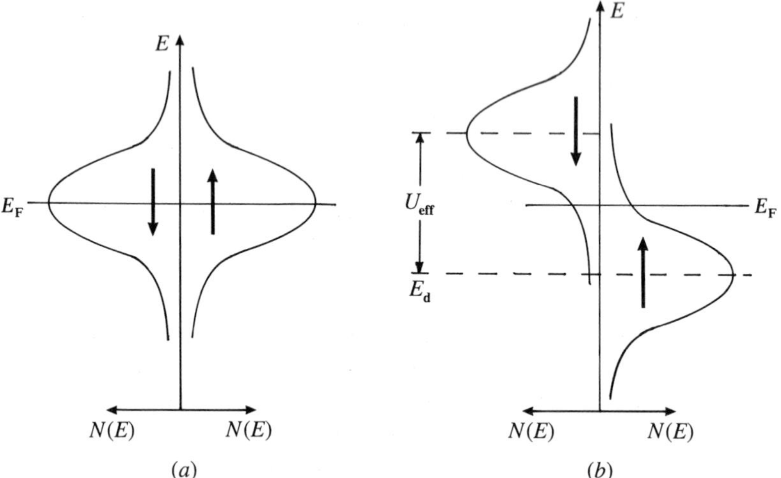


**Figure 11.10** The resonant bound d state in the non-magnetic (a) and magnetic (b) conditions.

The potential associated with the impurity ion is insufficient to capture the d orbital completely. It resonates (hybridizes) with the plane wave states $\psi_k$ of the electron gas according to a matrix element ${}_dV_k$ leading to a width $2\Delta$ to what in the free atom is a sharp orbital. This hybridization causes the fractional occupation of both spin states and complicates the problem in many body fashion. In the non-magnetic case there is no separation in energy of the two spin states in the orbital; this situation prevails in e.g. **Ag**Pd, **Cu**Ni and **Al**Mn. In **Cu**Mn and many other systems the impurity atom carries a magnetic moment, the spin up and spin down states having different energies, $E_d$ and $E_d + U_{eff}$ respectively. In the single orbital approximation $U_{eff}$ is the net Coulomb repulsion energy that a spin down electron experiences in the presence of a spin up electron. If $n_+$ and $n_-$ indicate the average occupations of the spin up and spin down states, this Coulomb contribution is $(n_+n_-)U_{eff}$. Since the two spin states are always partially occupied it is always present. The total energy of the system therefore comprises four components: the energies of the free electrons, the energy of the d orbital $E_d$, the effective Coulomb interaction energy between spin up and spin down d states $(n_+n_-)U_{eff}$ and the mixing parameter ${}_dV_k$. The problem is to determine the values of $n_+$ and $n_-$ that minimize the total energy of the system. Anderson's self-consistent treatment leads to the following results:

The density of states in each spin level is

$$N_d(E) = (\Delta/\pi) \cdot [(E - E_d)^2 + \Delta^2]^{-1}. \tag{11.30}$$

The width of the spin level is determined by

$$\Delta = \pi \langle {}_dV_k^2 \rangle N(E_F). \tag{11.31}$$

*continued*

The magnetic condition requires that

$$U_{\text{eff}}/\pi\Delta \geqslant 1. \tag{11.32}$$

Thus a narrow resonance coupled with a strong Coulomb interaction is needed for the magnetic state in this model. Having omitted the detailed treatment the results may be seen as plausible in that they demand that the resonance be sharp, i.e. the conditions should not depart too much from those characteristic of the free ion or atom. The host metal should have a low density of conduction band states at the Fermi level, which is the same as having a small band width ($E_F$).

We may also argue qualitatively as follows. In the non-magnetic case, Fig. 11.10*a*, the Coulomb interaction energy is already included together with the other contributions to the total energy and $n_+ = n_- = 1/2$ is apparently the stable state. Suppose however that the spin up and spin down states of Fig. 11.10*a* are each displaced (in opposite directions) an incremental energy $\delta E$ relative to the Fermi level so that

$$n_+ \to 1/2 + N_d(E_F) \cdot \delta E \quad \text{and} \quad n_- \to 1/2 - N_d(E_F) \cdot \delta E.$$

where $N_d(E)$ is the distribution of states associated with each spin level as given above. A spin up electron now has $N_d(E_F) \cdot \delta E$ fewer spin down electrons with which to interact and so we may imagine the spin up resonance to be lowered in energy by an amount $N_d(E_F) \cdot \delta E \cdot U_{\text{eff}}$. Conversely, a spin down electron now interacts with $N_d(E_F) \cdot \delta E \cdot U_{\text{eff}}$ more spin up electrons than before, effectively raising the spin down resonance by $N_d(E_F) \cdot \delta E \cdot U_{\text{eff}}$. Clearly if the net energy shift is greater than the original increment, $\delta E$, then the two spin states continue to separate until further change is no longer beneficial, thereby leading to a stable magnetic state. This condition, $U_{\text{eff}} N_d(E_F) \geqslant 1$, is, after evaluating $N_d(E_F)$, equivalent to that expressed above, $U_{\text{eff}}/\pi\Delta \geqslant 1$.

In experiment, however, d transition metals possess five orbitals and each spin level is five-fold degenerate. We can still use Fig. 11.10 to describe the situation, but now we associate five states with each resonant level of given spin. One immediate consequence of this is to include an exchange energy in $U_{\text{eff}}$ because if there is more than one electron in the spin up level, as is usually the case, the energy of these electrons is lowered by an amount J per pair of parallel spins. The exchange contribution is expected to be important for Cr and Mn impurity atoms (where $n_+$ is large and $n_-$ small), whereas its effect is reduced in Fe and Co impurity atoms because $n_-$ is also significant and exchange occurs between the spin down electrons too, thereby reducing $U_{\text{eff}}$. Ni or Pd as impurity atoms do not carry a local moment; $U_{\text{eff}}$ is too small.

Optical absorption and photoemission experiments on Ni, Pd and Mn dissolved in Ag or Cu have, assuming the above Lorentzian form for the resonant state, allowed estimates of the Anderson parameters; it is found that $\Delta \approx 0.5$ eV and ${}_dV_k \approx 1$ eV; these are about a factor of two smaller than those envisaged by Anderson. As for $U_{\text{eff}}$ this is estimated as $\approx 5 \pm 1$ eV for Mn dissolved in Ag or Cu; in the latter matrix it is estimated that $n_+ = 4.6$ and $n_- = 0.6$ leading to a resultant spin quantum number of 2, which is just that required to give the observed atomic magnetic moment.

The Mn atom has seven electrons in its outer unfilled shells so 1.8 electrons must be found in the conduction band. The distribution of electrons is not very different from that of the free atom in the $3d^5 4s^2$ configuration, but this is just the idea behind the Anderson approach. The local moments arise from magnetic atoms perturbed by their being dissolved in a free electron gas.

| Solvent \ Solute | Ti | V | Cr | Mn | Fe | Co | Ni |
|---|---|---|---|---|---|---|---|
| Cu | – | | + | + | + | + | – |
| Ag | | | + | + | + | | |
| Au | – | + | + | + | + | + | – |
| Zn | – | – | + | + | – | – | – |
| Al | – | – | – | – | – | – | – |

**Figure 11.11** The occurrence of local magnetic moments in certain alloy systems; occurrence (+), absence (−). A blank space indicates lack of data for metallurgical or other reason.

duction electrons also possess spin and a quantum mechanical s–d exchange† interaction exists. In simple terms, this s–d exchange causes the local moment to polarize the conduction electrons that are found in its vicinity. The net spin of the conduction electrons immediately surrounding the impurity atom is, however, oriented oppositely to its net d spin. This called an antiferromagnetic arrangement and is a prerequisite for the occurrence of the Kondo resistance minimum that arises at low temperatures in these alloys (see Fig. 9.41).

As already described (Section 6.8), owing to the sharpness of the Fermi surface this conduction electron polarisation has the RKKY oscillatory dependence on the distance $r$ from the site of the local moment of the form $(1/r^3) \cdot \cos 2k_F r$. The oscillatory decay, although small in amplitude, has considerable range and in more concentrated magnetic systems provides a means for coupling local magnetic moments on nearest and next nearest magnetic neighbours. This coupling may result in a parallel (ferromagnetic) or an anti-parallel (antiferromagnetic) setting of the moments dependent on the separation of the interacting atoms.

Although attention here has been directed to magnetism in certain dilute alloys, in some systems the transition metal may have extensive solubility as does Mn in Cu or Au for example. The local moment continues to arise, but the cooperative interaction between the moments on different solute atoms produces complex bulk magnetic behaviour and the Kondo effect is quenched. A particularly simple case is the occurrence of ferromagnetism in $Cu_2MnIn$, Fig. 11.12, which is discussed in more detail in a later section.

Having introduced the concept of the local magnetic moment, it is convenient to return to the subject of heavy Fermion alloys, Section 9.2, because they bear a certain similarity to a local moment Kondo alloy. There are now about thirty known heavy Fermion alloys and, apart from one or two exceptions, they have the common property that they all contain a relatively large amount (~25%) of either

† The s–d interaction that mixes s and d states and the s–d exchange that controls the spin settings are different and competing effects.

cerium (Xe) $5s^25p^6$ $4f^15d^16s^2$ or uranium (Rn) $5f^26d^17s^2$. Some examples have been given earlier and here we shall concentrate on $CeAl_3$, the alloy in which the behaviour was first observed. In Ce the 5d and 6s electrons are the valence electrons contributing to the conduction electron gas.

Recall that although the 4f levels in the rare earth metals have energies comparable to those of the itinerant valence states, their spatial extent is very limited; there is no $f_i$–$f_j$ resonance analogous to that for d electrons in, say, Mn. Furthermore, the interaction of the f states with the valence states is much weaker than the s–d resonance discussed earlier owing to the smaller overlap between the f and valence states. If, on this account, the single 4f electron in Ce maintains its atomic identity then it should carry an atomic moment similar to that observed on a $Ce^{3+}$ ion, and at high temperatures the alloy does in fact exhibit paramagnetic properties in agreement with a Curie law of the form $\chi = C/(T - \Delta)$. The heavy Fermion character becomes apparent below 1 K, in the mK range, where the electrical resistance becomes dominated by strong electron–electron scattering with $\rho = AT^2$. The coefficient $A$, which is proportional to $(m^*)^2$, $m^*$ being the effective band mass of the electron, is extremely large. Similarly, the electron heat coefficient $\gamma$ is 1.6 J $K^{-2}$ $mol^{-1}$. These results correspond to a charge carrier with an effective mass of about 1000 free electron masses. As mentioned earlier, de Haas–van Alphen measurements on this type of alloy confirm that these very large effective masses are associated with negative charge carriers on a Fermi surface.

The belief is that the Ce 4f electron, which at high temperatures behaves in normal fashion as a bound electron, becomes free to move at very low temperatures. There is, however (as in the case of the Kondo s–d exchange), a negative s–f exchange between the spin on the f electron and that of the valence electrons which causes the f electron to be surrounded by a cloud of valence electrons with a resultant equal and oppositely directed spin. The movement of the f electron and its polarization cloud corresponds to the movement of an electron-like quasiparticle of very large mass. Such a situation demands a radical change in the normal electronic band structure leading to an extremely narrow but incomplete band of quasiparticle states. Whereas the Kondo effect is a single impurity phenomenon, the heavy Fermion behaviour arises in concentrated local moment alloys: the latter are sometime called Kondo lattices on this account.

The heavy Fermion metals show several behaviour patterns in that at the lowest temperatures some are magnetic, others superconducting and still others 'normal' in their properties. Again it is noteworthy that these exceptional substances are based on cerium or uranium, elements that arise at the very beginning of the lanthanide and actinide series where the binding of the f electron to the atom is weakest.

Dilute salts, together with local moment alloys, form a minority group compared with magnetically concentrated matter such as Fe, Gd, $MnF_2$, $Fe_3O_4$, $Cu_2MnAl$, etc., i.e. all the conventional metals, alloys and compounds. The cooperative magnetic interactions in these substances lead to ferromagnetism, ferrimagnetism and antiferromagnetism.

## 11.5 Ferromagnetism

We normally associate ferromagnetism with the metals Fe, Co and Ni together with certain of their alloys and compounds, but the phenomenon is in fact much more widespread. A ferromagnetic substance is one that possesses a spontaneous magnetization, i.e. it is magnetized even in the absence of a magnetic field. The spontaneous magnetization varies with temperature in a characteristic fashion typical of a cooperative phenomenon (Fig. 11.12). It has a maximum value at 0 K and initially decreases very slowly with temperature; this can be understood when we consider that the moment on any atom is oriented in a particular direction by mutual interaction with surrounding neighbouring atoms. To reduce the moment, we must affect the behaviour not of a single atom but the coupled system of cooperating magnetic atoms, and this is difficult to achieve. Nevertheless, increasing the temperature inevi-

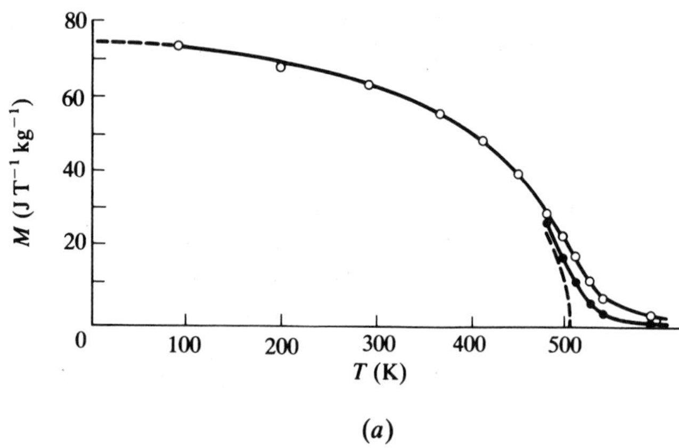


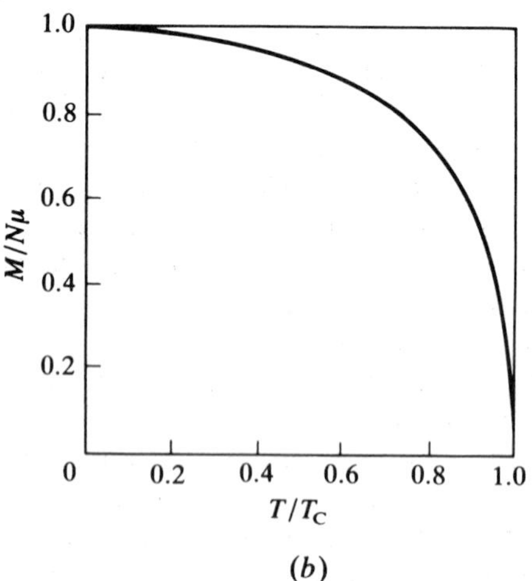


**Figure 11.12** The variation of the magnetization of a ferromagnetic material with temperature. Data for one of the Heusler alloys, $Cu_2MnIn$, is shown in (*a*) (after Coles *et al.* 1949). Near the Curie temperature the magnetization is sensitively dependent on the external field, but the spontaneous magnetization may be obtained by extrapolation procedures. It is customary to present data in reduced form as shown for Ni in (*b*).

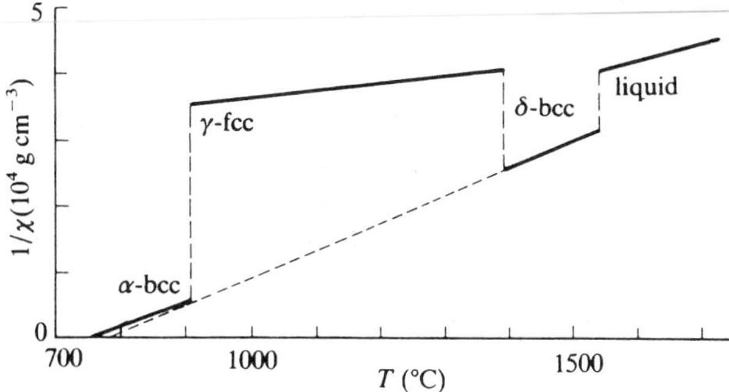


**Figure 11.13** Above the Curie temperature a ferromagnetic material becomes paramagnetic and follows the Curie–Weiss law. Here we present data for Fe showing the properties of the different phases of this element; we also see how magnetic analysis may be used to study phase transformations. (From *Landolt–Börnstein* 1986.)

tably causes a gradual decrease of the spontaneous magnetization. As the decrease proceeds beyond a certain point, a catastrophic breakdown of the coupling occurs and the spontaneous magnetization is reduced to zero in a rather narrow temperature interval; it finally becomes zero at a temperature $T_C$, the Curie temperature. Above this temperature the behaviour is paramagnetic; the susceptibility follows a modified Curie law known as the Curie-Weiss law (Fig. 11.13):

$$\chi = \frac{C}{T - T_C}. \qquad (11.33)$$

The reduction of the magnetic order as the Curie temperature is approached is most vividly appreciated from the behaviour of the heat capacity (Fig. 11.14). The variation of the magnetization in the neighbourhood of the Curie temperature depends sensitively on both the external magnetic field and the temperature. The presence of an external magnetic field causes the transition from the spontaneously magnetized to the paramagnetic state to broaden; the system is labile and the field may cause the temporary presence of clusters of magnetic moments, but these become less likely the lower the field strength and the more the temperature deviates from the ideal Curie temperature. Such 'critical fluctuations' arise in the vicinity of all phase transformations and are of great importance in statistical physics. The magnetic transition is an example of a second-order phase transition characterized by a change in the magnetic symmetry but which leaves the crystal structure unaffected (see Appendix 11.4).

A particular, and at one time puzzling, aspect of ferromagnetism is that a specimen of, for example, pure iron may appear to be unmagnetized and yet in the presence of an extremely small field, of order 100 $\mu$T, becomes magnetized to saturation corresponding to an induction $B \approx 2$ T. The explanation of this apparent paradox was provided by Weiss in 1907. He postulated that a ferromagnetic substance is always spontaneously magnetized appropriate to the given temperature, but that the material is divided into small regions called domains. Throughout any one domain the spontaneous magnetization has the same orientation. In the apparently unmagnetized state different domains are magnetized in randomly arranged

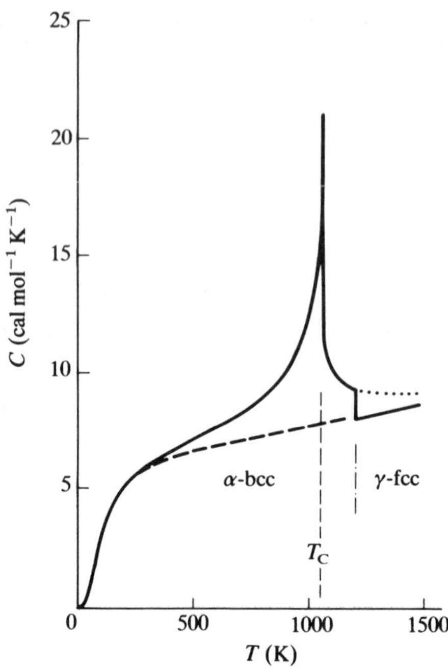


**Figure 11.14** The change in the internal order of a magnetic material (Fe) as it passes through the Curie temperature is clearly displayed in the heat capacity. Note that the α-γ transition only produces a discontinuity in the heat capacity (see Appendix 11.4). (After Austin 1932.)

orientations so that the resultant for the whole sample is zero or very nearly so. The application of a weak external field suffices to align the different domains so that the specimen exhibits the full spontaneous magnetization. The domain concept has been fully established, and has an elegance and importance far in excess of Weiss' early expectations. Domains are significant for the field-dependent properties of magnetic matter and therefore for many technical applications.

For the moment, we concentrate on the spontaneous magnetization and its temperature dependence, and ask what causes the ferromagnetic substance to become spontaneously magnetized. It is not easy to give a concise and convincing answer to this question. Admittedly, we can, just as for the atom, say that it arises as a result of quantum-mechanical exchange and correlation forces, but this statement explains very little. Therefore we postpone any attempt to explain the true origins of ferromagnetism and instead treat a simple model first introduced by Weiss but still very much alive today. We assume that the atoms carry an atomic moment as explained earlier and then ask how we can describe the fact that they cooperate and choose to align themselves in the same direction within a given domain. *Again we must emphasize that any magnetostatic interactions between the atomic moments are completely irrelevant*. On the other hand, we know from the behaviour of the paramagnetic gas that in sufficiently large external fields the magnetization becomes saturated. We therefore follow Weiss and suppose that the true interactions that cause ferromagnetism *may be represented by an inner molecular magnetic field* $B_i$ that is proportional to the spontaneous magnetization:

$$B_i = \lambda M, \tag{11.34}$$

the proportionality constant $\lambda$ (incorporating $\mu_0$) being known as the molecular field coefficient. To this, we must add any external field $B_0$ so that the total effective field $B_e$ is

$$B_e = B_0 + \lambda M. \tag{11.35}$$

All we need to do now is to replace $B_0$ in our earlier treatment of the paramagnetic gas by $B_e$. The interactions in the concentrated magnetic substance are represented by $\lambda M$.

Provided we can attain high enough temperatures – and this is always the case in practice – we can arrange that the high-temperature approximation is valid: this provides us with the paramagnetic solution. Recalling our previous treatment, we write

$$M = \frac{Ng^2\mu_B^2 J(J+1)}{3k_B T} B_e = \frac{C}{\mu_0 T} B_e, \tag{11.36}$$

or

$$M = \frac{C}{\mu_0 T}(B_0 + \lambda M). \tag{11.37}$$

We can solve this for $M$ to obtain

$$M = \frac{C}{\mu_0(T - T_C)} B_0, \qquad T_C = \frac{\lambda C}{\mu_0}, \tag{11.38}$$

or

$$\chi = \frac{\mu_0 M}{B_0} = \frac{C}{T - T_C}. \tag{11.39}$$

Equation (11.39) describes the paramagnetic behaviour above the Curie temperature correctly provided we associate $\lambda C/\mu_0$ with this temperature. If this is the case then we must show that a spontaneous magnetization arises for $\mu_0 T < \lambda C$.

The low-temperature behaviour of our concentrated magnetic substance is governed by the following two equations:

$$M = N\langle\mu\rangle = N\mu B(J, \alpha), \tag{11.40}$$

$$\alpha = \frac{\mu B_e}{k_B T} = \frac{\mu}{k_B T}(B_0 + \lambda M), \tag{11.41}$$

where $\mu$ has been written for $g\mu_B J$.

If spontaneous magnetization is to arise then there must be solutions for $M$ in (11.40) and (11.41) when $B_0 = 0$. The equations must be solved numerically or graphically. In Fig. 11.15 we illustrate these equations; they are plotted with $\alpha$ as abscissa. There are two classes of solution. Either the line (11.41) intersects the Brillouin function at two points or at the origin only. In the latter case $M = 0$ is the only solution, and this corresponds to the high-temperature paramagnetic regime. If there is a second intersection, corresponding to a finite $M$, then this corresponds to the stable spontaneous magnetization and we have a description of ferromagnetism. But we must show that this is indeed the case.

Suppose that the situation is as in Fig. 11.16 and $M = 0$; there is no spontaneous magnetization but each atom carries a moment. If a local fluctuation gives rise to a resultant small magnetization $\delta M$, this immediately produces an internal field $\lambda\,\delta M$,

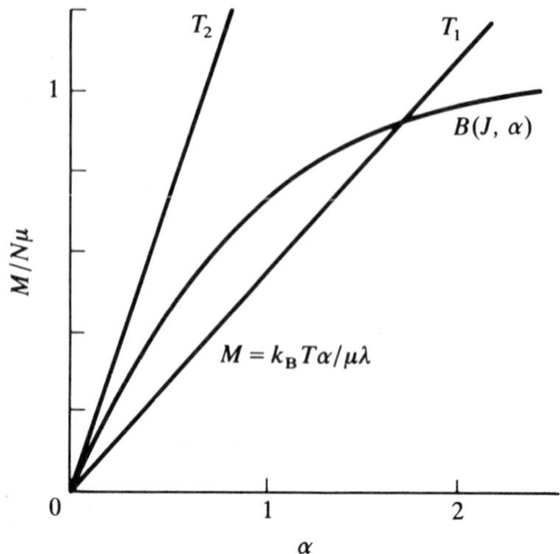


**Figure 11.15** The graphical solution for the spontaneous magnetization.

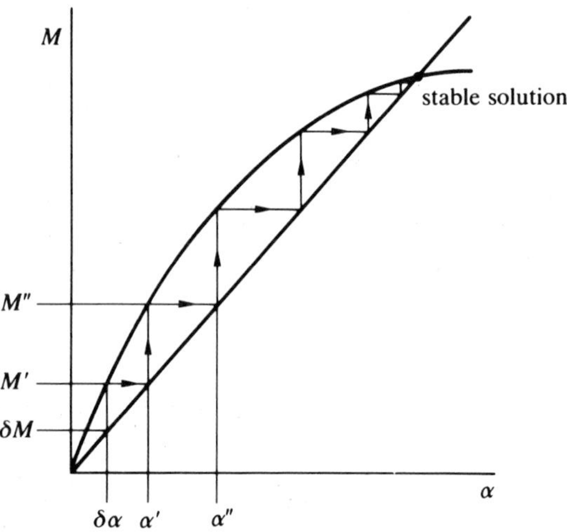


**Figure 11.16** The spontaneously magnetized state is the stable state below the Curie temperature.

which causes $\alpha$ to increase from 0 to $\delta\alpha$ in accordance with (11.41). But this $\delta\alpha$ corresponds, via (11.40), to an even higher value of magnetization $M'$, which in turn causes $\delta\alpha$ to become $\alpha'$, and this increases $M'$ to $M''$, and so on. The solution $M = 0$ is unstable in this case. Only when there is a magnetization corresponding to the second intersection is a stable state obtained; any fluctuation around this point causes changes that maintain the status quo. Now the limiting condition between the high-temperature paramagnetic and the low-temperature ferromagnetic solutions is when the line (11.41) is tangential to the Brillouin function at the origin. The temperature for which this occurs is the Curie temperature. Near the origin $\alpha$ is a small quantity and the Brillouin function may be written

$$B(J, \alpha) \rightarrow \frac{(J+1)\alpha}{3J} \text{ (small } \alpha) \tag{11.42}$$

and

$$M = N\mu \frac{(J+1)\alpha}{3J},$$

or

$$\frac{M}{\alpha} = N\mu \frac{J+1}{3J}, \tag{11.43}$$

whereas (11.38) gives

$$\frac{M}{\alpha} = \frac{k_B T}{\mu\lambda}.$$

The Curie temperature $T_C$ is therefore obtained from

$$\frac{kT_C}{\mu\lambda} = N\mu \frac{J+1}{3J},$$

$$T_C = \frac{N\mu^2}{k_B} \frac{(J+1)\lambda}{3J},$$

and, replacing $\mu$ by $g\mu_B J$, we obtain

$$T_C = \frac{Ng^2\mu_B^2 J(J+1)\lambda}{3k_B} = \frac{C\lambda}{\mu_0}.$$

This is the same as we obtained for the paramagnetic behaviour. The treatment is consistent. This is to be expected since at $T = T_C$, $\chi \to \infty$, which implies that the slightest fluctuation will produce a spontaneous magnetization. The solutions of (11.40) and (11.41) for different temperatures may be plotted as graphs of $M/N\mu$ as a function of $T/T_C$ for different $J$ (Fig. 11.17). Comparison with experimental results

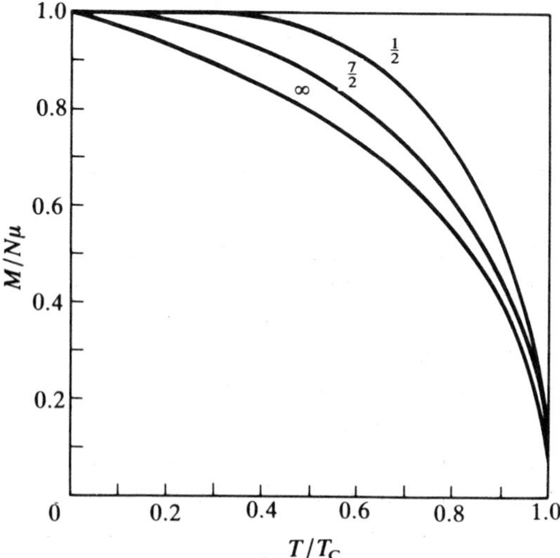


**Figure 11.17** The reduced magnetization plotted against reduced temperature for the molecular field model. Curves for three different $J$ values are shown. Compare with the experimental data for Ni in Fig. 11.12*b*.

for the principle ferromagnetic materials shows that the solutions with $J = \frac{1}{2}$ best represent the data.

### 11.5.1 Ferromagnetism: atomic moments

The atomic moment $g\mu_B J$ may be estimated accurately from measurements of the saturation magnetization at low temperatures. *For the 'd' transition metals it is assumed that the orbital moment is quenched and that the uncompensated spins act independently; their number is therefore a measure of the number of electrons contributing to the moment.* Some data for ferromagnetic substances are given in Table 11.3.

Note that the ferromagnetic Bohr magneton numbers, denoted by $p_B$, are non-integral. This is characteristic for the ferromagnetic 'd' metals and alloys and leads to the concept of band or itinerant magnetism in which the electrons carrying the magnetic moment are completely delocalized. This situation may be illustrated using the example of Ni. We have already discussed the similarities and differences in the electronic structure of Ni and Cu. The main difference is that for Ni the d band is not completely filled.

Magnetic data for pure Ni as well as for Ni alloys allow us to conclude that in Ni the d band contains 9.4 electrons. To obtain the necessary atomic moment of 0.6 $\mu_B$, we must assume that the d band is divided into spin up and spin down sub-bands; the latter are displaced in energy so that whereas one half band is filled, the other contains only 4.4 electrons. There is then a net difference of 0.6 electron spin per atom corresponding to the observed moment (Fig. 11.18).

The band model provides a convenient explanation of the non-integral moments; although we must always associate an integral number of electrons with each atom, the valence electrons may be distributed in non-integral fashion between different energy bands. But why should the spin up and spin down bands be displaced? Again we use our qualitative argument that the Coulomb repulsion energy is diminished between electrons of similar spin because they cannot approach one another as closely as can electrons of opposite spin. The quantum-mechanical exchange forces favour the development of parallel spins, but this effect is counter-balanced by the associated increase in the kinetic energy of the electrons (because the electrons must then have different **k** vectors). Now the d band is a narrow one and contains several sharp peaks in the density of states. This is particularly noticeable at the top

**Table 11.3** Values of $p_B$ and $T_C$ for certain ferromagnetic substances

| | $p_B$ per formula unit | $T_C$ (K) |
|---|---|---|
| Fe | 2.22 | 1043 |
| Co | 1.72 | 1394 |
| Ni | 0.6 | 627 |
| Gd | 7.6 | 290 |
| $Cu_2MnIn$ | 4.0 | 506 |
| $Fe_2B$ | 1.9 | 1015 |
| MnB | 1.92 | 578 |
| EuO | 6.8 | 69 |

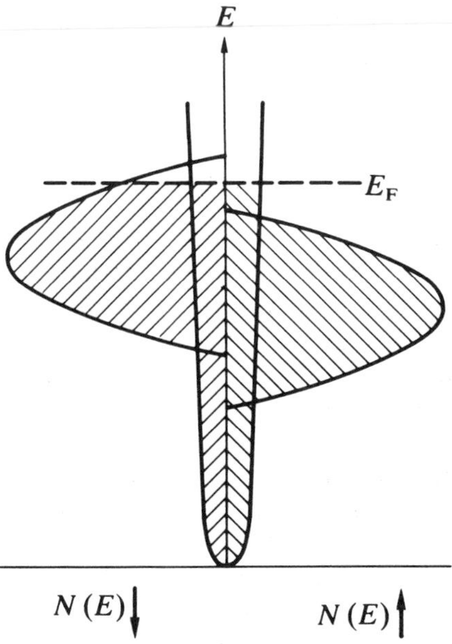


**Figure 11.18** The description of the non-integral magnetic moment of Ni in terms of the band model. The d states are said to be 'spin-split', leading to different occupations of the separate spin bands.

of the band, which is of interest in the case of Ni.† If $N(E_F)$ is large it does not cost very much kinetic energy to transfer electrons from the spin down to the spin up band, and if this increase in kinetic energy is compensated by the exchange energy then a stable magnetic state arises (Fig. 11.19). *Note that in this model the atomic magnetic moment does not exist other than as an average atomic moment.* The band of energy levels belongs to the crystal, and the crystal is magnetized as a whole. The occurrence of an average atomic moment and the cooperative interaction needed to produce the spontaneous magnetization are created at one and the same time. This then is band magnetism or collective electron ferromagnetism, as it is sometimes called.

The dependence of band ferromagnetism on the density of states at the Fermi level may be illustrated in the following manner. Suppose a d band possesses equal numbers of spin up and spin down electrons. An external magnetic field causes an imbalance as described earlier for free electrons, see Fig. 6.11. Introduce now an exchange interaction between electrons of similar spin so that the shift in the two sub-bands $\Delta E$, normally $2\mu_B B_0$, is increased by an amount $\Delta n\text{J}$, where $\Delta n = (n_+ - n_-)$ is the difference in occupation of the two bands and J is an effective exchange energy between parallel spins.

So

$$\Delta E = 2\mu_B B_0 + \Delta n\text{J}. \tag{11.44}$$

Since all the changes occur near the Fermi level

$$\Delta \text{n} = N(E_F)\Delta E/2.$$

† This can be seen in the band diagram for Cu (Fig. 7.25).

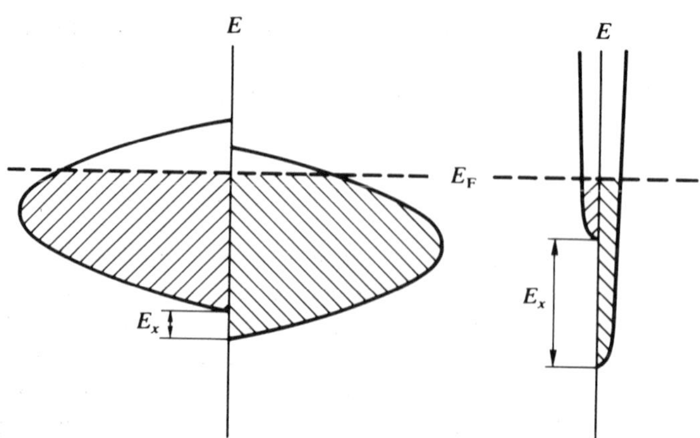


**Figure 11.19** Spin-split bands are increasingly probable the higher the density of states at the Fermi energy (i.e. the narrower the band). This is because the cost in kinetic energy for a given imbalance in spin is less when the state density at the Fermi energy is large.

$N(E_F)$ being the total density of states at the Fermi level. So the induced moment becomes

$$M = \Delta n\mu_B = (1/2)\{2\mu_B^2 B_0 + MJ\}N(E_F),$$

leading to a susceptibility per unit volume

$$\kappa = \mu_0 M/B_0 = \mu_0 \mu_B^2 N(E_F)/[1 - (1/2)JN(E_F)]$$
$$= \kappa_0/[1 - (1/2)JN(E_F)]. \qquad (11.45)$$

$\kappa_0$ being the 'bare' Pauli susceptibility. $\kappa$ is called the exchange enhanced susceptibility.

As already seen, a spontaneous magnetization arises when the susceptibility diverges, so a condition for its occurrence becomes $(1/2)JN(E_F) \geqslant 1$; it is known as the Stoner criterion. It provides a succinct summary of the band model, but the problem now is the evaluation of $N(E_F)$ and more particularly J.

At the present time the band model, in its most advanced form, is able to predict the magnetic ground state at 0 K for the 3d ferromagnetic elements, Table 11.4 (compare the predicted values with the measured ones in Table 11.3). It has also been calculated that the atoms in the outermost surface layer of a crystal possess appreciably larger atomic moments. Having fewer neighbours, the local d band breadth is smaller and $N(E_F)$ correspondingly larger for the surface atoms, so the

**Table 11.4** Predicted magnetic moments, $\mu_B$ (after Weinert and Freeman 1983)

| | | Bulk | Surface |
|---|---|---|---|
| bcc | Fe | 2.25 | 2.98 |
| hcp | Co | 1.64 | 1.76 |
| fcc | Ni | 0.56 | 0.68 |

effect of the exchange interaction is enhanced, leading to a larger atomic moment. It is not an easy task to demonstrate this in experiment, but measurements on very thin iron films with intentionally roughened surfaces have shown the magnetization to increase with increasing roughness.

As the temperature is increased above 0 K the spontaneous magnetization decreases, at first slowly, but as the Curie temperature is approached it rapidly falls to zero; thereafter the paramagnetic regime begins. The initial slow decrease is associated with the occurrence of spin waves. Thermal excitation causes the average resultant spin formally associated with an atom to precess about the direction of magnetization. The exchange interaction between neighbouring spins couples the precessions on different atoms so that travelling waves of spin precession arise. These are the quantized spin waves first introduced by Bloch in 1931 (see Section 11.8), and produce the characteristic $(1 - T^{3/2})$ dependence of the magnetization at low temperatures, which is observed in experiment.

Although the paramagnetic susceptibility above the Curie temperature is often described in terms of the Curie–Weiss relation (11.33), the fit is only approximate. Furthermore, the simplicity of the molecular field approach (and the implication of local atomic magnetic moments) requires that any interpretation in terms of the Curie–Weiss relation be considered phenomenological. On the other hand there is considerable low temperature experimental support for the band model (e.g. non-integral atomic moments and their changes on alloying and the presence of spin up and spin down sheets in Fermi surface studies). In spite of its success in predicting the magnetic state at the absolute zero of temperature, the band model, owing to the theoretical difficulties encountered, has not been able to predict the observed Curie temperatures of Fe, Co or Ni. There is uncertainty as to the form of the high temperature excitations that cause the continued decrease of the spontaneous magnetization and its disappearance at $T_C$. The paramagnetic range presents similar difficulties and the actual state cannot be likened to the random distribution of atomic magnetic moments as discussed in simple models, nor, for that matter, to a band model with zero exchange splitting between the spin up and spin down sub-bands.

We have given a schematic description of Ni in Fig. 11.18. Co is similar to Ni but lacks 1.7 electrons in the d band, whereas in Fe we must assume both spin sub-bands to be incompletely filled, otherwise there would be too few electrons in the sp conduction states. When one sub-band is completely filled and the other partially filled, as in Ni or Co, we speak of 'strong ferromagnetism', but when both sub-bands are only partially occupied, as in Fe, we speak of 'weak ferromagnetism'. It is the interaction that is described and not the size of the spontaneous moment. Of these three substances, Fe has the largest atomic moment and spontaneous magnetization, but because the exchange interaction cannot separate the sub-bands sufficiently for one of them to be completely filled, it is termed weak.

Exchange enhancement is not confined to the ferromagnetic d metals. It may arise but be insufficient to promote ferromagnetism. The Stoner condition is not fulfilled, but the Pauli susceptibility is increased since $J > 0$. This is the case for Pd, and a particularly striking demonstration of the enhancement is the existence of giant magnetic moments in **Pd**Fe alloys. It is surprising that an alloy of Pd containing only 0.15% Fe can be ferromagnetic. If, in this and similar alloys, the atomic moment is attributed to the iron atoms alone then one finds, in an alloy with 0.5% Fe, that it must be ascribed a moment of very nearly 12 $\mu_B$! It is clearly impossible that this so-called giant moment be associated with just the Fe atoms. Instead the solute Fe atom with its conventional local moment close to 4 $\mu_B$ must induce ferromagnetism in the itinerant 4d electrons of labile Pd atoms in its vicinity. Each iron atom is sufficient to produce a polarization cloud encompassing as many as 200 neighbouring Pd atoms with an average magnetic moment of 0.05 $\mu_B$. Similar behaviour is also observed in **Pd**Ni and **Pt**Fe alloys.

The rare earth metals show diverse and complex forms of magnetic behaviour. Gadolinium is relatively simple in that it is a ferromagnetic metal and the moment arises in the 4f shell, which is half filled, containing seven electrons. We should therefore expect an atomic moment of 7 $\mu_B$, but instead we find 7.63 $\mu_B$. There are two questions: where does the extra 0.63 $\mu_B$ come from and how do the atomic moments cooperate to produce the ferromagnetic state? We have already seen that the 4f electrons in Gd belong to the ion core and that no direct interatomic interaction between 4f electrons is possible. Thus the coupling of the atomic moments must be mediated through the conduction electrons, which, in the case of Gd, are the three outer electrons in the 5d and 6s states. These conduction electrons experience an exchange interaction with the half-filled 4f shell and as a result become magnetically polarized, leading to a moment of about 0.6 $\mu_B$ per atom, thereby increasing the total atomic moment to the required value. The conduction electrons occupy band states and extend throughout the crystal, and serve through their own polarization to align the 4f moments on the different atoms. This is the RKKY interaction.

This qualitative picture for Gd is more complete than that for Fe, say, because we know that the 4f electrons are completely localized and cannot possibly interact directly with those on an adjacent atom. Even so, the difficulties of calculation are very great and the theory has developed primarily in a phenomenological way. Nevertheless, the ferromagnetism of Gd arises through truly local atomic moments coupled via the conduction electron gas – this process is definitely established.

Another convincing example of the presence of the RKKY coupling mechanism is the cooperative magnetism of $Cu_2MnIn$ and similar alloys, Fig. 11.10. The spontaneous moment on the Mn atom at 0 K is 4 $\mu_B$ (Mn is the only realistic magnetic carrier). The alloy structure is highly ordered and the Mn–Mn separation is 8.75 Å, so any direct $d_i$–$d_j$ exchange interaction between manganese atoms is out of the question. The d states of each Mn atom therefore arise as a spin split resonant level with almost five electrons in the lower level and approximately one electron in the upper level, producing a local magnetic moment as described in Box 11.2. This provides the correct atomic moment and the coupling between the Mn moments is mediated via the weakly polarized conduction electron gas.

It should be pointed out, however, that the RKKY coupling mechanism may also cause atomic moments to align in antiparallel fashion. This is because the conduction electron polarization cloud surrounding a local moment oscillates in sign as it decays and the parallel or antiparallel coupling depends critically on the separation of the interacting atoms. Thus certain of the Heusler alloys display the antiparallel alignment.

Although the RKKY coupling is not invoked for the ferromagnetic behaviour of Fe, Co and Ni, the question of whether the conduction electrons in these elements are polarized by the strong magnetic moments of the d bands arises. According to neutron diffraction studies this is the case, but their contribution is a negative one in that their net spin ($\approx 5\%$ of the total moment) is in opposition to the moment from the d band. As in the case of the local moments discussed earlier, the s–d exchange leads to an antiferromagnetic interaction between s and d electrons in these elements.

## 11.6 Negative Interactions

In addition to pure ferromagnetic metals, simple alloys and paramagnetic dilute salts, there are more complex substances such as transition metal oxides MnO, mixed oxides $CoFe_2O_3$, halides $NiF_2$ and many other materials that exhibit magnetic behaviour different from that hitherto described. A particular feature of these substances is that one can usually distinguish two kinds of magnetic atom characterized by different chemical sort, valence or crystallographic arrangement. We therefore study the magnetic behaviour of a substance $AB$ where the two kinds of atom have atomic moments $\mu_A$ and $\mu_B$ and occupy specified lattice sites as illustrated in Fig. 11.20.

The presence of two (or more) different magnetic atoms leads to a complicated pattern of molecular field interactions that may extend beyond that between nearest neighbours, but we consider here only those interactions between the following nearest atom pairs $AA$, $BB$, $AB$. We speak of $A$ and $B$ lattices with associated magnetization $M_A$ and $M_B$ respectively.

We assume, following Weiss, that each atom experiences an effective field $\mathbf{B}_A$ or $\mathbf{B}_B$. The magnetic energy density of the substance $AB$ is then

$$U = -\tfrac{1}{2}(\mathbf{M}_A \cdot \mathbf{B}_A + \mathbf{M}_B \cdot \mathbf{B}_B), \tag{11.46}$$

where

$$\mathbf{B}_A = \lambda_A \mathbf{M}_A + v\mathbf{M}_B, \tag{11.47}$$

$$\mathbf{B}_B = \lambda_B \mathbf{M}_B + v\mathbf{M}_A. \tag{11.48}$$

This is a straightforward development of the Weiss approach to take account of the three interactions between pairs of atoms $AA$ ($\lambda_A$), $BB$ ($\lambda_B$) and $AB$ ($v$). Evaluation of

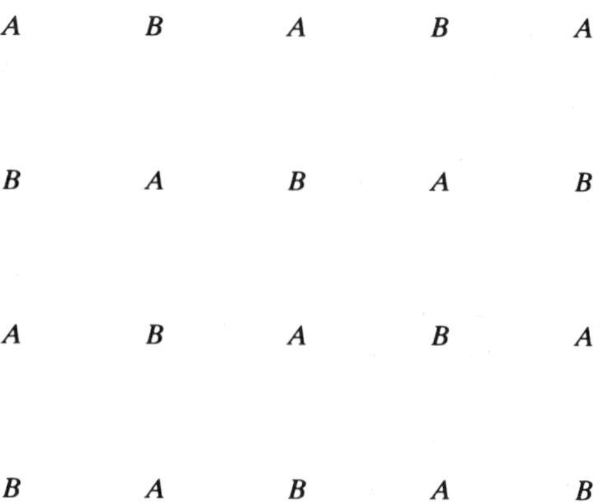


**Figure 11.20** A simple two-sublattice structure. The symbols $A$ and $B$ may be associated with different valence, different chemical type or different crystallographic site.

(11.46) using (11.47) and (11.48) gives

$$U = -\tfrac{1}{2}(\lambda_A M_A^2 + \lambda_B M_B^2 + 2v\mathbf{M}_A \cdot \mathbf{M}_B). \tag{11.49}$$

We may consider the following possibilities:

(a) $\lambda_A > 0$, $\lambda_B > 0$, $v > 0$; the energy is minimized when $\mathbf{M}_A$ and $\mathbf{M}_B$ are parallel; this leads to a ferromagnetic substance with two different magnetic components;

(b) $\lambda_A > \lambda_B > 0$, $v < 0$; the energy is minimized when $\mathbf{M}_A$ and $\mathbf{M}_B$ are oriented in opposite directions;

(c) $\lambda_A \approx \lambda_B < 0$; $v \ll 0$ the energy is minimized when $\mathbf{M}_A$ and $\mathbf{M}_B$ are oriented in opposite directions.

Provided $v < 0$ and $|v| \gg |\lambda|$, then, irrespective of the sign of $\lambda$, we expect the system to have least energy when $\mathbf{M}_A$ and $\mathbf{M}_B$ are antiparallel; and if

$$|\mathbf{M}_A| = |\mathbf{M}_B|, \qquad \mathbf{M}_A + \mathbf{M}_B = 0,$$

there is no resultant spontaneous magnetization and the system is said to exhibit *antiferromagnetism.* If, however,

$$|\mathbf{M}_A| > |\mathbf{M}_B|, \qquad \mathbf{M}_A + \mathbf{M}_B > 0,$$

a net spontaneous magnetization exists and the arrangement is said to exhibit *ferrimagnetism* (Fig. 11.21).

We now develop equations governing the properties of magnetic materials with negative interactions and we can simplify the treatment without losing the essence of the behaviour if we let $\lambda_A = \lambda_B = 0$ and write the former negative quantity $v$ explicitly as $-v$. As for the ferromagnetic case discussed earlier, we omit any consideration of fluctuations in the neighbourhood of transition temperatures.

In the paramagnetic regime the induced $A$ and $B$ magnetizations are parallel, we discard the vector notation and write

$$B_A = B_0 - vM_B, \qquad B_B = B_0 - vM_A. \tag{11.50}$$

At high temperatures any sublattice spontaneous magnetization is destroyed and the atomic moments are randomly oriented, but the negative interactions between $A$- and $B$-type atoms still exist. In the presence of an external field, an induced magnetization arises and we calculate the paramagnetic behaviour just as we did for

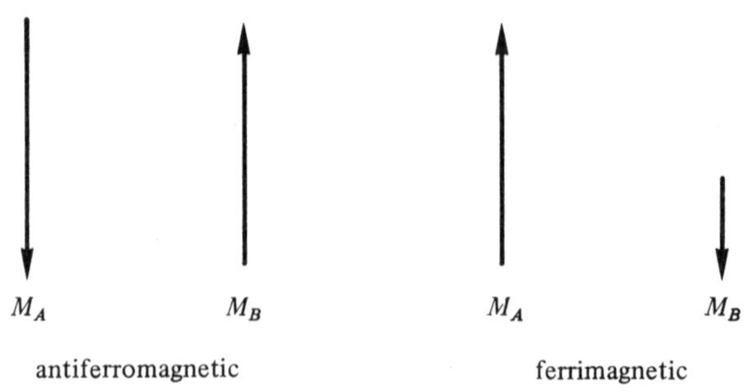


**Figure 11.21** A stylized presentation of the antiferromagnetic and ferrimagnetic states.

the paramagnetic gas. We use the formulae already obtained. Both sublattices act as paramagnetic gases producing, in an external field $B_0$, similarly directed magnetizations; the total magnetization is then $M_A + M_B$, but the $A$ and $B$ fields are still given by (11.50).

At high temperatures we have seen, (11.22), that

$$\mu_0 M_A = \frac{C_A}{T} B_A, \qquad \mu_0 M_B = \frac{C_B}{T} B_B. \tag{11.51}$$

Equations (11.50) and (11.51) are readily solved to give

$$\mu_0 M_A = C_A B_0 \frac{T - \nu C_B/\mu_0}{T^2 - \nu^2 C_A C_B/\mu_0^2} \tag{11.52}$$

with a similar expression for $M_B$, so the total magnetization becomes

$$\mu_0 M = \mu_0(M_A + M_B) = \frac{(C_A + C_B)T - 2\nu C_A C_B/\mu_0}{T^2 - \nu^2 C_A C_B/\mu_0^2} B_0. \tag{11.53}$$

This may be replaced by

$$\chi = \frac{(C_A + C_B)T - 2\nu C_A C_B/\mu_0}{T^2 - T_C^2}. \tag{11.54}$$

We have written

$$T_C \equiv \nu(C_A C_B)^{1/2}/\mu_0,$$

and note that for $T = T_C$, $\chi$ diverges and the sublattices become spontaneously magnetized. The sublattice magnetizations are calculable from the Brillouin function just as in the case of ferromagnetism.

### 11.6.1 *Antiferromagnetism*

When the sublattices $A$ and $B$ are identical we say that we have an *antiferromagnetic* material. This arises when the lattices are similar and occupied by the same ion, as occurs in MnO, $FeF_2$ and many other substances; (11.54) then becomes simply

$$\chi = \frac{2C}{T + \nu C/\mu_0} = \frac{2C}{T + T_C}, \tag{11.55}$$

where $C_A = C_B = C$ and $\mu_0 T_C = \nu C$. Equation (11.55) bears a certain similarity to the Curie–Weiss law, except for the sign of $T_C$, which, when $\chi^{-1}$ is plotted against $T$, leads to an intercept at a negative temperature (Fig. 11.22). This is borne out in practice, but observed negative intercepts vary from those predicted by (11.55) because our model is too simple and neglects $AA$, $BB$ interactions as well as those between more distant atoms. Nevertheless, the negative intercept is the hallmark of negative interactions. At temperatures below $+T_C$ the sublattices become spontaneously magnetized, but since their magnetizations are equal and oppositely directed, there is no resultant magnetization and the system remains paramagnetic. For this antiferromagnetic case $T_C = \nu C/\mu_0$ is known as the Néel temperature $T_N$ after L. Néel, who was the first person to make a thorough theoretical study of

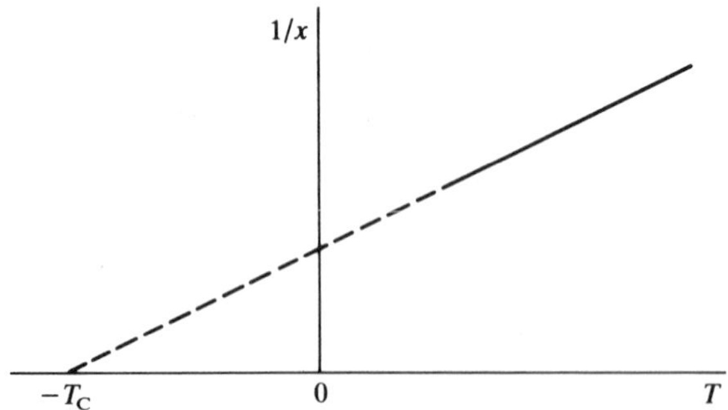


**Figure 11.22** The antiferromagnetic condition is characterized by a linear dependence of $1/\chi$ on temperature above the ordering temperature and a negative intercept on the temperature axis.

negative interactions. As the temperature is lowered, the sublattice magnetizations become more and more aligned antiparallel and the interactions restrict their possibility to orient themselves in the very much weaker applied field. The susceptibility of an antiferromagnetic substance therefore decreases as the temperature decreases below $T_N$. The Néel temperature is usually observed as a cusp in the variation of $\chi$ with $T$ (Fig. 11.23). In an antiferromagnetic single crystal the susceptibility is pronouncedly anisotropic. It is straightforward to show that $\chi$ is independent of temperature when measured perpendicular to the direction of the sublattice magnetizations, whereas when measured parallel to this direction it decreases monotonically as the temperature falls below $T_N$, and finally becomes zero at 0 K (see Problem 11.6).

Although an antiferromagnetic material is strongly magnetized below $T_N$, no resultant magnetization is apparent, but the variation of the heat capacity in the vicinity of the Néel point (Fig. 11.24) leaves us in no doubt as to the extent of the internal magnetic order.

### 11.6.2 *Ferrimagnetism*

Returning to (11.52), we see that if $C_A$ and $C_B$ are different then so are $M_A$ and $M_B$. This can occur if the kinds of atom are the same, but the lattices $A$ and $B$ are different or the lattices may be identical, but occupied by ions with different atomic moments. At low temperatures the antiparallel sublattice magnetizations do not cancel and a resultant magnetization $M_A - M_B$ exists. There is a strong similarity to a ferromagnetic substance. Detailed analysis shows, however, that a ferrimagnetic substance may have a resultant magnetization that varies with temperature in a manner either similar to or very different from that typical of a ferromagnetic substance, depending on the various parameters involved (Fig. 11.25). We find that ferromagnetic and ferrimagnetic substances are perhaps best distinguished (neutron diffraction excepted) by their high-temperature paramagnetic properties. For a ferrimagnetic substance, $\chi$ is given by (11.54); at temperatures well above $T_C$, $\chi^{-1}$ asymptotically approaches a linear variation with temperature (with negative intercept on the temperature axis) similar to that for an antiferromagnetic material, whereas in the range immediately above $T_C$ it has a pronounced and characteristic

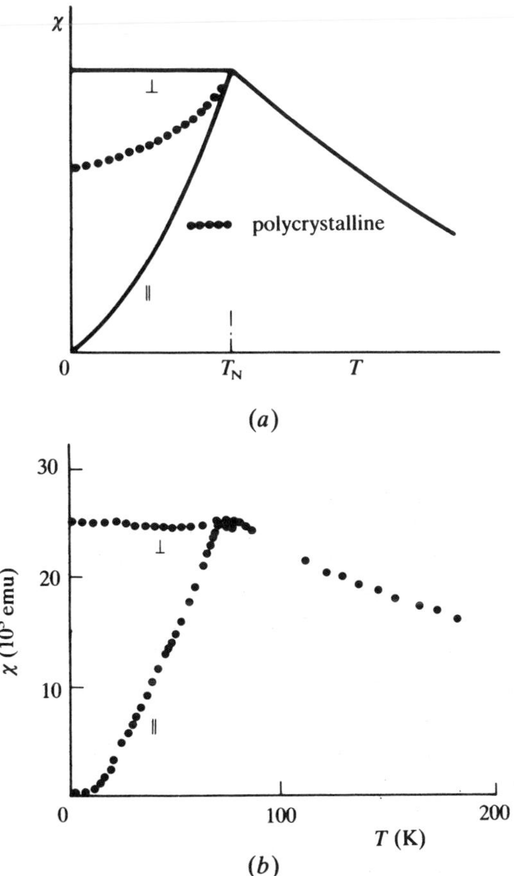


**Figure 11.23** A schematic presentation of the anisotropy in the temperature variation of the antiferromagnetic susceptibility (a), together with experimental data for a crystal of $MnF_2$ (b). (Experimental data after Trapp and Stout 1963.)

curvature concave to the temperature axis. This is in direct contrast with the behaviour of a ferromagnetic substance, for which $\chi^{-1}$ is in principle linear in $T$, although slight curvature does arise near $T_C$, but this is always convex to the temperature axis (Fig. 11.26). This is the feature that most surely distinguishes a ferrimagnetic from a ferromagnetic substance.

Magnetite, $Fe_3O_4$, is a ferrimagnetic substance and has provided the name 'ferrite' for a group of mixed oxides with the composition $TOFe_2O_3$, where T is a divalent transition metal ion, $Cu^{2+}$ or $Zn^{2+}$. The marked spontaneous magnetization of $Fe_3O_4$ and its poor conducting properties have made it useful in high-frequency transformers, aerials and formerly in computer elements. The oxides of the ferrite family crystallize in the spinel structure; the metallic ions are placed in the interstices available in a face-centred cubic arrangement of oxygen ions (see Fig. 1.7). Without going into details, we can say that the metal ions may occupy either $A$ sites (tetrahedral positions with four $O^{2-}$ neighbours) or $B$ sites (octahedral positions with six $O^{2-}$ neighbours). By varying the heat treatment during the preparation of these oxides, the divalent and trivalent ions arrange themselves in the $A$ and $B$ sites

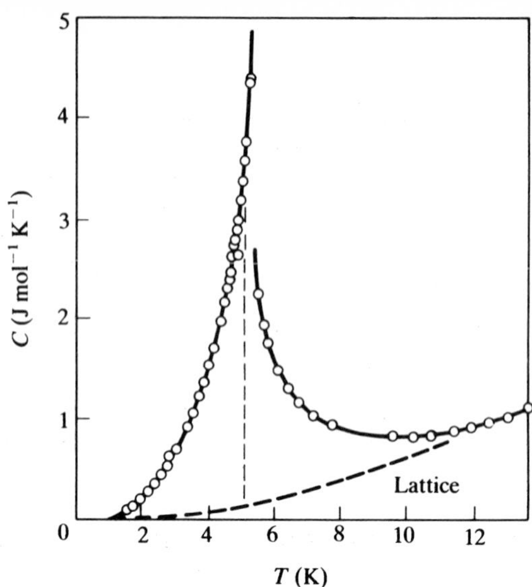


**Figure 11.24** Although no spontaneous magnetization is apparent in an antiferromagnetic substance ($NiCl_2 \cdot 6H_2O$), the internal order that arises is clearly evident in the heat capacity. (After Robinson and Friedberg 1960.)

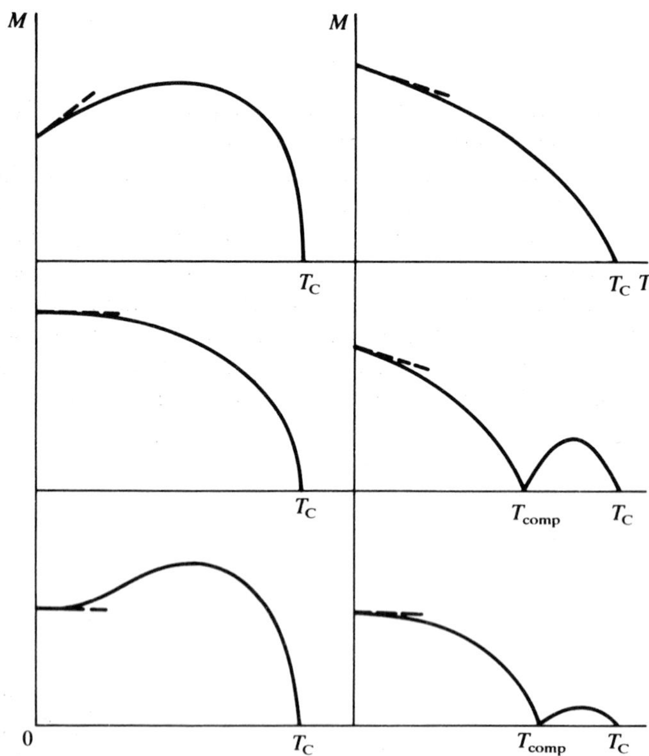


**Figure 11.25** A ferrimagnetic substance has a net spontaneous magnetization, and its variation with temperature may, as first shown by Néel, take one of several patterns dependent on the choice of molecular field coefficients. The possible patterns are illustrated in this figure, and it is seen that they may be similar to or very different from the pattern characteristic of a ferromagnetic material. Magnetic compensation is shown in Fig. 11.27. (After Néel 1948.)

**Table 11.5** Properties of the ferrites

| Spinel | Moment/$\mu_B$ A site | Moment/$\mu_B$ B site | Resultant moment/$\mu_B$ | Measured moment/$\mu_B$ | $T_C$ (K) |
|---|---|---|---|---|---|
| $ZnOFe_2O_3$ | −5 | +5 | 0 | 0 | — |
| $CuOFe_2O_3$ | −5 | 1 + 5 | 1 | 1.3 | 728 |
| $NiOFe_2O_3$ | −5 | 2 + 5 | 2 | 2.3 | 858 |
| $CuOFe_2O_3$ | −5 | 3 + 5 | 3 | 3.7 | 793 |
| $FeOFe_2O_3$ | −5 | 4 + 5 | 4 | 4.1 | 858 |
| $MnOFe_2O_3$ | −5 | 5 + 5 | 5 | 4.6 | 573 |

in several ways. One extreme arrangement, known as the inverse spinel structure, has the $Fe^{3+}$ equally distributed between *A* and *B* sites and the $T^{2+}$ ion only in the *B* sites. These *A* and *B* lattices become spontaneously magnetized in a ferrimagnetic fashion; the resulting moment per formula unit is then that provided by the $T^{2+}$ ions. If we assume the orbital magnetic moment is quenched then we expect the following magnetic moments per formula unit:

| T | Mn | Fe | Co | Ni | Cu | Zn |
|---|---|---|---|---|---|---|
| $\mu/\mu_B$ | 5 | 4 | 3 | 2 | 1 | 0 |

so Zn ferrite is a true antiferromagnetic material. Experiments with the inverse spinels have amply confirmed these moments, any slight disagreement being attributed to the presence of a parasitic orbital contribution (Table 11.5).

Another family of oxides that has received much attention is based on the mineral garnet. This usually occurs naturally as deep brownish-red crystals and has the composition $Al_2Ca_3\,(SiO_4)_3$. The structure is cubic, but the unit cell contains 160 atoms, so we shall not attempt to describe it. The rare earth magnetic garnets are based on the above mineral and have the composition $R_3Fe_5O_{12}$, where R is a trivalent rare earth atom – the Fe ions are also trivalent. The properties are described in terms of three sublattices, and the following diagram summarizes the

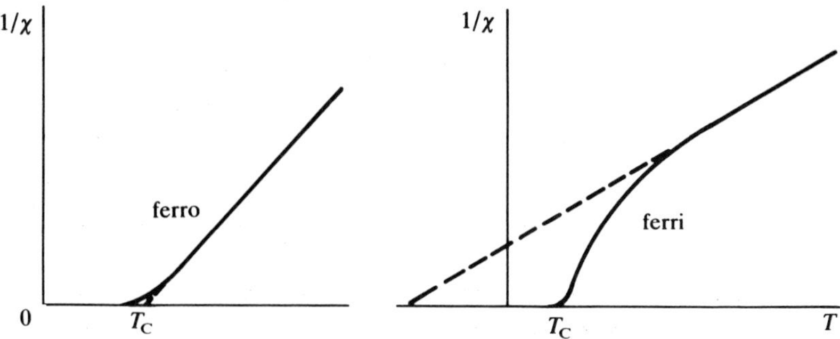


**Figure 11.26** Ferromagnetic substances may be distinguished from ferrimagnetic ones through their different temperature-dependent paramagnetic behaviour.

situation:

| | | | | | |
|---|---|---|---|---|---|
| lattice $A$ | $Fe^{3+}$ | $Fe^{3+}$ | | $10\mu_B$ down | $5\mu_B$ up |
| lattice $B$ | $Fe^{3+}$ | $Fe^{3+}$ | $Fe^{3+}$ | $15\mu_B$ up | |
| lattice $C$ | $R^{3+}$ | $R^{3+}$ | $R^{3+}$ | | $\mu_R$ down |

Strong negative interaction between lattices $A$ and $B$ leads to a resultant spontaneous moment equivalent to 5 $\mu_B$ per formula unit. The ions R on lattice $C$ couple in a weak negative manner to the resultant moment on the $B$ lattice. Since the rare earths may have large atomic moments, the resultant moment per formula unit may be as large as 15 $\mu_B$. These oxides show magnetic compensation, i.e. at a temperature well below $T_C$ the magnetizations of the sublattices have different temperature variations, and at $T_{comp}$ they are equal and opposite; the situation is shown in Fig. 11.27 and data for actual garnets are given in Fig. 11.28. These oxides contain only trivalent ions. Their insulating properties are therefore superior to those of the ferrites (where the mixture of $M^{2+}$ and $M^{3+}$ allows hopping conductivity). They are therefore better suited for microwave transformers; however, the high cost of the rare earth elements often prevents their practical application.

## 11.7 Exchange Interactions: Heisenberg Interaction

At the end of Chapter 6 we attempted a description of exchange and correlation. We shall not consider correlation again, other than as an influence that weakens the effect of exchange interaction. In free atoms and in the electron gas (whether it be ideally free or 'real') the exchange term arises only for electrons with parallel spins: it is always positive, leading to a lower total energy for the system. We say that exchange favours the occurrence of a magnetic moment in the electron gas as well as in free atoms. In the free atom the exchange is an intraatomic exchange, because the

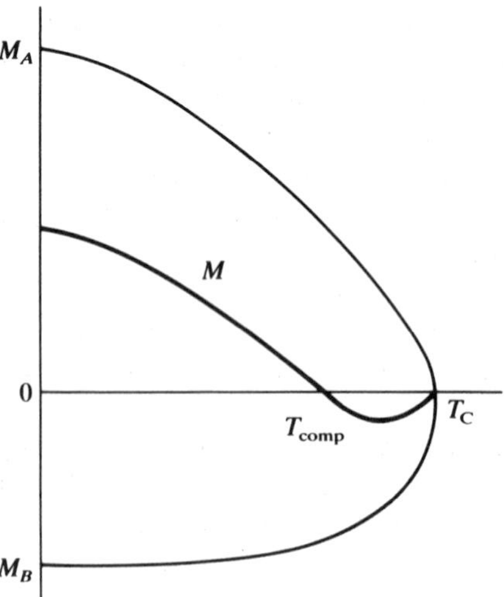


**Figure 11.27** Magnetic compensation arises when the two sublattice magnetizations are finite but equal at a certain temperature, leading to a change in sign of the resultant magnetization $M$.

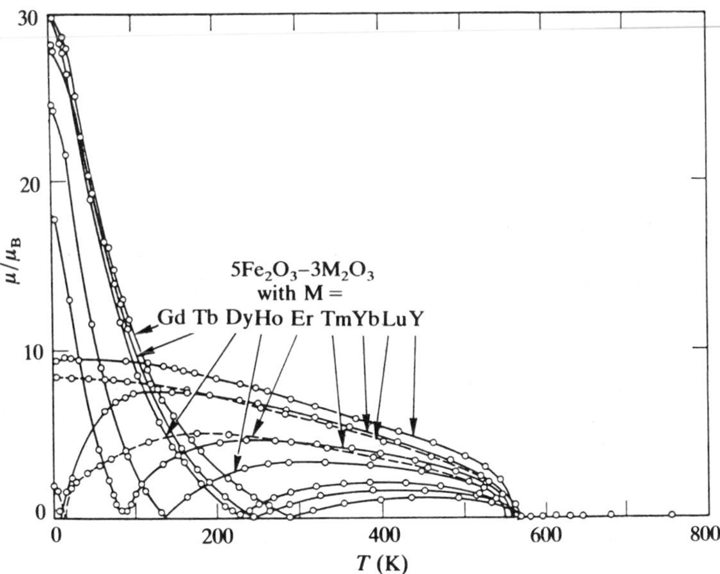


**Figure 11.28** Magnetic data for the rare earth garnets showing magnetic compensation. (After Pauthenet 1958.)

effect arises between electrons within the same shell or between electrons in different shells of the same atom. In aggregates of atoms exchange interactions also arise between electrons located on different atoms and we speak of interatomic exchange. In the case of the hydrogen molecule there are two separate atoms of hydrogen bound together by the indistinguishable valence electrons. In the Heitler–London approach to this problem, the electrons are assumed to be described by wave functions that are expressible in terms of atomic states associated with the individual atoms, i.e. a tight-binding picture is used. The overlap of the electron wave functions leads to the appearance of an interatomic exchange integral. Although the electrons are assumed tightly bound to the individual atoms, such overlap must occur, otherwise there would be no chemical bond formed. In contrast with the intraatomic or electron gas exchange described earlier, this interatomic exchange (also denoted by J may have either positive or negative sign, associated with parallel or antiparallel spin alignment respectively. In the hydrogen molecule, $J < 0$ and the electron spins are aligned antiparallel; the molecule does not carry a magnetic moment. It is generally held that the interatomic exchange favours chemical bonding and is not conducive to magnetic moment formation (however, see the earlier discussion of the $O_2$ molecule based on the molecular-orbital, as opposed to the Heitler–London, approach).

When discussing the origins of ferromagnetism, the major problem of principle arises with the 'd' elements, in particular Mn, Fe, Co, Ni and Pd. Whereas we can understand the origin of an atomic magnetic moment in terms of intraatomic exchange, it is difficult to account for the cooperative interaction that couples the moments on the different atoms (unless we adopt the band model of ferromagnetism – but then we delocalize the moment from the atom). On the other hand we might say that the occurrence of ferromagnetism in the 'd' elements is a relatively rare feature – only five of the 24 'd' elements are or become ferromagnetic. Furthermore, these five elements arise at the end of the series and the *d* electrons may be justifiably considered as tightly bound to the ions. Might it not be possible that we have an exceptional situation in these five elements and that the interatomic exchange is in fact such that $J > 0$, leading to a parallel alignment of the net spin on adjacent atoms?

Heisenberg proposed in 1928 that if the magnetic electrons were described by localized orbitals and if the exchange integral between electrons on adjacent atoms were positive (opposite to the case in $H_2$) then the resultant spins on the neighbouring atoms would align in parallel since this then minimizes the total energy. This is the direct interatomic exchange between d electrons. No calculations for real magnetic substances were made; it was postulated that J for the d electrons in Fe, say, was positive. Heisenberg's proposal allows the exchange energy to be written as

$$E_{ex} = -2J\mathbf{S}_i \cdot \mathbf{S}_j, \tag{11.56}$$

$\mathbf{S}_i$ and $\mathbf{S}_j$ being the resultant spins on adjacent atoms *i* and *j*. Clearly

if $J < 0$ then $E_{ex} < 0$ when $\mathbf{S}_i$ and $\mathbf{S}_j$ are antiparallel;

if $J > 0$ then $E_{ex} < 0$ when $\mathbf{S}_i$ and $\mathbf{S}_j$ are parallel.

The exchange integral was expected to depend sensitively on the degree of overlap between the orbitals on adjacent atoms, and it was *postulated* by Slater and Bethe that Fe, Co and Ni are ferromagnetic because in these elements there is a critical amount of d-orbital overlap that causes J to be positive. *There is, however, no theoretical calculation that justifies this postulate* (although it is often implied in many books). In fact there is very little quantitative theoretical explanation for magnetic phenomena. We have only the qualitative picture of an exchange interaction modified by correlation effects, and there are few reliable calculations of the quantities involved.

Nevertheless (11.56) is a convenient phenomenological or empirical description of the exchange energy, and the integral J is determined from experiment. We associate

$J > 0$ with positive interactions and ferromagnetism;

$J < 0$ with negative interactions and antiferromagnetism.

## 11.8 Neutrons and Magnetism

The study of magnetic behaviour permeates solid state physics. Initially interest was focused on bulk properties such as susceptibility and magnetization as well as their dependence on composition, crystal orientation and temperature. Perhaps the most important post-war developments in magnetism have been the application of magnetic resonance phenomena and the use of neutrons in diffraction and scattering experiments. Thus, although the concept of antiferromagnetism was introduced by Néel before the Second World War, it is only with the aid of neutron diffraction that we can determine the magnetic structures of elements and compounds in an unambiguous manner. The principles are exactly as in X-ray crystallography, but the scattering factor is dependent on the atomic moment, its orientation and its spatial distribution; the results of a complete neutron diffraction study therefore provide us with details of the size, crystallographic position and orientation of the atomic moments.

We have seen that a crystal as an assembly of point masses has a structure that may be determined by either X-ray or neutron diffraction, but in addition it has collective mechanical vibrations with specific frequency-wavenumber relationships that can be studied using inelastic neutron scattering (Section 5.11). In an analogous manner we find that arrays of magnetic moments, i.e. arrays of spins, are coupled by the quantum mechanical exchange interaction.

If the orientation of a given spin is altered in some way then this cannot leave the rest of the system unperturbed. Just as a lattice of coupled point masses has a spectrum of lattice vibrations, so a lattice of coupled spins has a spectrum of spin excitations with specific dispersion properties. For simplicity, we consider a linear ferromagnetic array of spins. At absolute zero the spins are aligned parallel to one another, but at a finite temperature the spins are no longer perfectly parallel; they precess around the direction of magnetization, and the resolved $z$ component is less than the maximum value obtainable at 0 K. Since the spins are coupled by the exchange interaction, the precession at each atom cannot be independent of that on its neighbours. This coupling of the precession gives rise to a travelling 'spin wave' just as the coupling of the atomic vibrations gives rise to a travelling displacement wave.

For a linear lattice of similar spins $S$ in which the Heisenberg interaction (11.56) is limited to nearest neighbours, the dispersion relation takes the form

$$\hbar\omega = 4JS(1 - \cos qa), \tag{11.57}$$

$\omega$ and $q$ being the frequency and wave vector of the spin wave. Just as for phonons, these spin wave excitations are quantized; they are called magnons. The magnons have energies similar to phonons and can exchange energy and momentum with neutrons similarly to phonons.

The inelastic scattering of neutrons by magnons allows us to determine the magnetic dispersion relationships for real substances, thereby providing direct knowledge of the exchange parameter J. At the present time, particular interest is directed towards studies of the rate earth elements as well as linear and two-dimensional magnetic lattices; the latter are approximated very well by substances in which the magnetic atoms form chains or nets and where the interaction between chains or nets is very much weaker than the interaction between atoms on the same chain or net. Thus the insulating compound $CsMnCl_3 \cdot 2H_2O$ is a very good approximation to a linear magnetic structure.

## 11.9 The Magnetization Curve: Hysteresis

We now consider the process whereby a ferromagnetic material like iron may be converted from an apparently non-magnetic condition to a highly magnetized state by the application of a very small external field of order 100 $\mu$T. Furthermore, we must explain how it is that some materials such as iron are easily magnetized and others, notably cobalt, are only magnetized with difficulty.

Experiment shows that the initial magnetization of a sample follows a characteristic pattern and although at any particular state of magnetization the process may be reversible for very small incremental changes in the external field, the cyclic

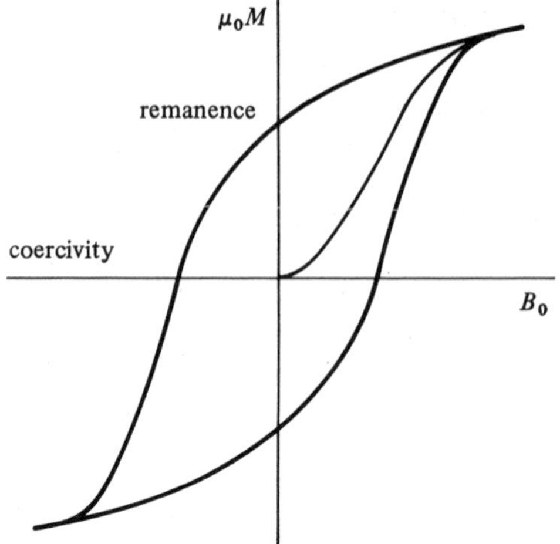


**Figure 11.29** The hysteresis curve.

variation of the external field leads to the well-known hysteresis curve (Fig. 11.29). In the following, we refer the magnetization $M$ to unit volume and use the volume susceptibility $\kappa = \rho\chi$, $\rho$ being the density.

We write

$$\kappa = \mu_0 M/B_0,$$

$$B = B_0 + \mu_0 M,$$

$$\mu_r = B/B_0 = 1 + \kappa.$$

We consider the behaviour at a constant temperature well below the Curie temperature; for many materials we may assume the behaviour is appropriate to room temperature. The magnetization process is dominated by the behaviour of magnetic domains. The concept of the magnetic domain, which was originally introduced by Weiss as an *ad hoc* hypothesis, has been substantiated completely. Domains can be seen, their shapes determined, and their movement in an applied field followed in detail. Domains are made visible by the use of magnetic colloids that delineate their boundaries, in electron diffraction as well as by the use of polarized light. Although domains were first seen in the early 1930s, the true development of experimental technique has taken place since the Second World War; a particular requirement is the preparation of accurately oriented strain-free single-crystal surfaces. Magnetic substances assume the domain structure because the magnetic contribution to the free energy is thereby minimized.

### 11.9.1 *Magnetic contributions to the free energy*

Normally a specimen of arbitrary shape is not uniformly magnetized unless it happens to be in the form of an ellipsoid of revolution. We therefore consider a single crystal in the form of a prolate ellipsoid.

### Magnetostatic energy

We assume the crystal to be isotropic in its microscopic behaviour (this is never true in practice, but the assumption may be made here). The ellipsoid is homogeneously magnetized along its major axis and is placed in field-free space; the specimen exists as a single domain (Fig. 11.30). The uniform volume magnetization produces a surface magnetization ('free poles') along the boundaries of the specimen. This surface magnetization in turn gives rise to a demagnetizing field $DM$. The geometrical demagnetizing factor $D$ is dependent on the ratio of length to breadth of the specimen, $l/b$; it varies approximately as $(l/b)^{-1}$. In the presence of an external field $B_0$ the true internal field producing the magnetization becomes

$$B_{0i} = B_0 - DM, \tag{11.58}$$

and in practice we must use $B_{0i}$ rather than $B_0$ when determining ferromagnetic behaviour. The magnetostatic energy of the sample in its own demagnetizing field is

$$U_s = \int_0^M DM \, \mathrm{d}M = \tfrac{1}{2}DM^2. \tag{11.59}$$

We call $U_s$ the magnetostatic self-energy density.

Suppose for the moment that our specimen can only exist as a single domain and that we attempt to establish the hysteresis behaviour along the major axis. The specimen is magnetized to saturation in the direction AB and the external field $B_0$ is then reduced, eventually becoming directed along BA. In the process of reversing the magnetization, the latter must rotate from the direction AB to that of BA. Since the specimen is isotropic, the rotation of the atomic moments involves no work (we neglect any eddy current losses or change the field very slowly), but the rotation is nevertheless hindered by the effects of specimen shape. Thus at some point in the demagnetizing process the bulk magnetization must be directed perpendicular to AB, i.e. in the direction of the minor axis. This situation is associated with a larger magnetostatic energy because the demagnetizing factor along the minor axis is much greater than that along the major axis. Calculations for a long thin ellipsoid give hysteresis curves of the form shown in Fig. 11.31. We conclude that single-domain particles possess a shape anisotropy arising from the demagnetizing factor and its dependence on length-to-diameter ratio. Ideally we might expect a coercive force of several thousand gauss in suitable single-domain particles of pure iron, and a good practical permanent magnet could be made from such particles. But can we expect to find particles as single domains? We shall see that we can, but the particles must be very thin, with diameters of the order of 1000 Å.

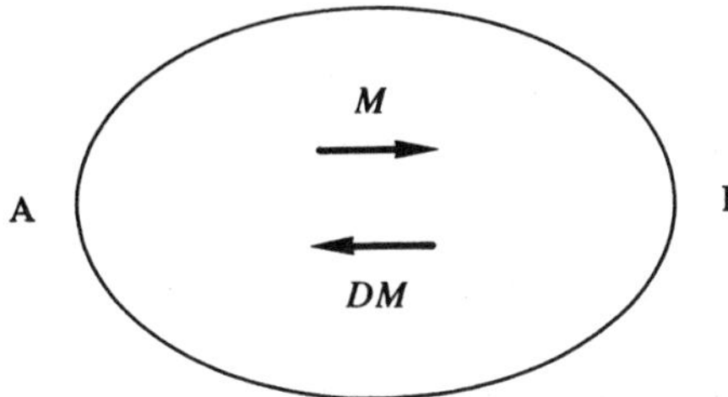


**Figure 11.30** The spontaneously magnetized single domain possesses a demagnetizing field that is governed by its shape.

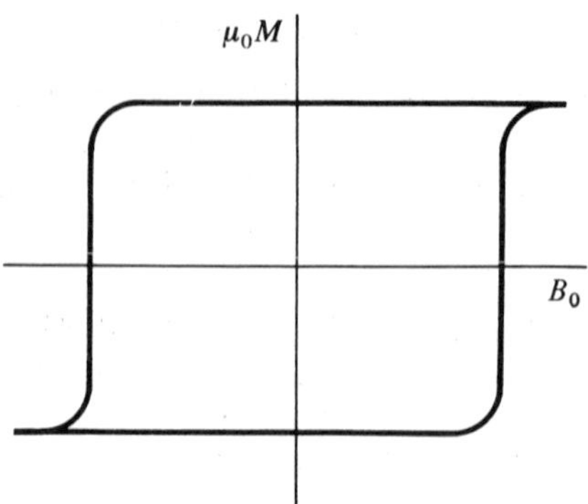


**Figure 11.31** Hysteresis may be caused by specimen shape.

A single-domain particle, provided it has a minor axis greater than a certain critical size, can always reduce the magnetostatic self-energy by assuming a multi-domain structure. Thus a long thin rectangular specimen (which may be considered a good approximation to an ellipsoid) can be divided into a number of domains, each magnetized along the major axis, but with alternate domains oppositely directed (Fig. 11.32). In this way, we reduce the net surface magnetization. Each individual domain has the same length but much smaller cross-section than the original specimen and therefore a much smaller self-energy. The multidomain structure has smaller magnetostatic self-energy, but the subdivision introduces boundary surfaces between domains with oppositely directed magnetizations. These domain boundaries contribute to the magnetic free energy because in the boundary, adjacent spins are antiparallel and their exchange energy is lost. Subdivision into a multidomain structure therefore continues until the additional decrease in magnetostatic self-energy is compensated by the associated increase in energy occasioned by the production of new domain boundaries.

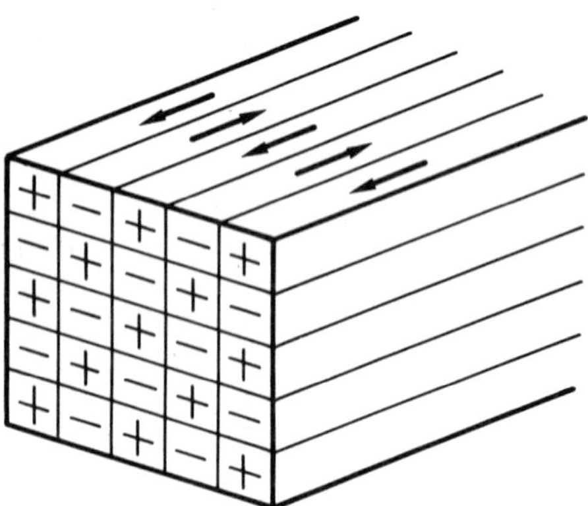

**Figure 11.32** A reduction in magnetostatic energy may be obtained in a multidomain structure, but this has its price in the energy associated with the domain boundaries.

*Energy of magnetocrystalline anisotropy*

As we have pointed out, crystals are not isotropic in their behaviour and the magnetization curves have different shapes in different crystal directions (Fig. 11.33). We speak of directions of 'easy' and 'difficult' magnetization. Hexagonal crystals, e.g. Co or $BaFe_{12}O_{19}$, are particularly difficult to magnetize in any direction other than along the *c* axis. This anisotropy is termed 'magnetocrystalline' and has atomic origins. The energy of a ferromagnetic substance is least when the atomic moments are directed in particular crystallographic directions and it requires work to rotate them into other orientations. This work is associated with changes in the electron charge distribution accompanying the rotation of the atomic moment from its

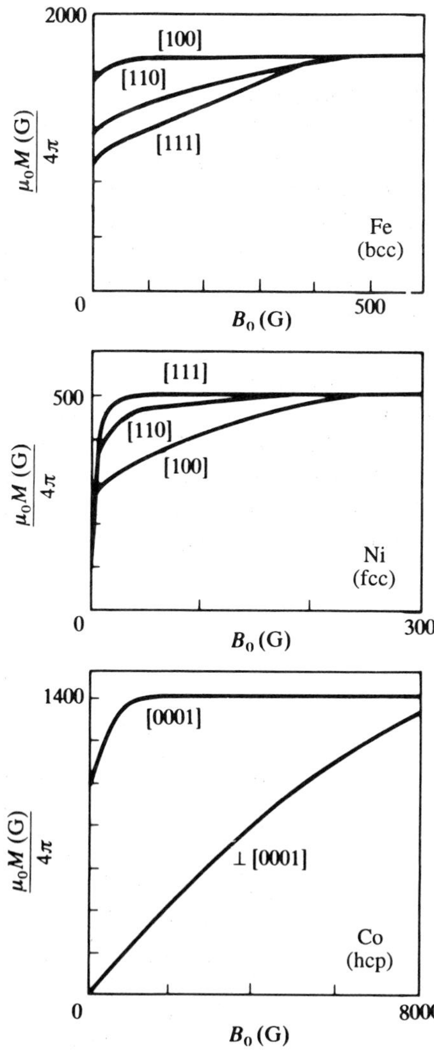


**Figure 11.33** Magnetocrystalline anisotropy as observed in Fe, Co and Ni. (After Honda and Kaya 1926.)

favoured direction. The spin–orbit interaction couples the spin magnetic moment with the orbital motion of the electrons. Rotating the spin moments from their preferred orientations (those of easy magnetization) requires that the orbital charge distribution assume a form less compatible with the demands of the crystal structure – the energy is thereby increased and work must be done to rotate the moment. It will be appreciated that this effect should be more pronounced in a uniaxial crystal like Co than for the more symmetrical and more isotropic cubic structures of Fe and Ni. It is customary to describe the magnetocrystalline energy $U_c$ in terms of empirical coefficients $K$:

$$U_c = \begin{cases} K_1 \sin^2\theta + K_2 \sin^4\theta & \text{(Co)}, \\ K_1(\alpha_1^2\alpha_2^2 + \alpha_1^2\alpha_3^2 + \alpha_2^2\alpha_3^2) + K_2\alpha_1^2\alpha_2^2\alpha_3^2 & \text{(Fe, Ni)}. \end{cases} \tag{11.60}$$

$\alpha_1$, $\alpha_2$, $\alpha_3$ are the direction cosines of the magnetization relative to the cube axes, and $\theta$ is the angle it makes with the $c$ axis in the case of a hexagonal structure.

We can readily see that a spherical single crystal of Co possessing no shape anisotropy would exhibit strong magnetocrystalline anisotropy leading to a pronounced coercivity. Clearly if we complement shape anisotropy with crystalline anisotropy then needle-like single-domain particles with their major axes coincident with the $c$ axis should have a large coercive force and be suitable for the preparation of permanent magnets.

### Domain boundary energy

Every domain is a region of homogeneous magnetization, but what happens at the boundary between two antiparallel domains? See Fig. 11.34. Is the boundary of

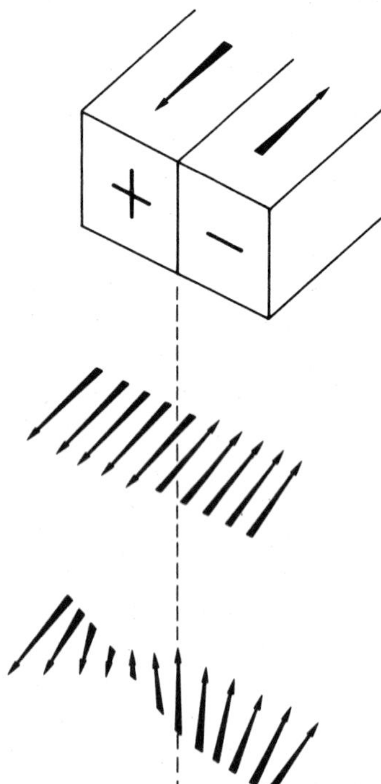


**Figure 11.34** The abrupt and gradual domain boundaries.

vanishingly small thickness as we implied in our discussion of the magnetostatic self-energy? Or does the boundary have finite thickness, the spins changing direction gradually? If we change the spin direction abruptly then, according to (11.51), we increase the energy by an amount $4JS^2$ per atom pair in the boundary. On the other hand, if the transition occurs in $N$ increments of $\pi/N$ then each atom pair within the boundary has an associated increase in energy

$$\Delta U_{\text{ex}} = -2JS^2 \cos\frac{\pi}{N} - (-2JS^2)$$

$$= 2JS^2\left(1 - \cos\frac{\pi}{N}\right) = 4JS^2 \sin^2\frac{\pi}{2N}.$$

Assuming $\pi/2N$ is small, $\Delta U$ becomes

$$\Delta U_{\text{ex}} \approx \frac{JS^2\pi^2}{N^2}. \tag{11.61}$$

For any line of atoms perpendicular to the boundary plane, there will be $N$ such increments and the boundary has a contribution from the exchange energy of

$$N\,\Delta U_{\text{ex}} = \frac{JS^2\pi^2}{N} = U_{\text{b}} \tag{11.62}$$

per line of atoms. Clearly, this energy is lower the larger the thickness $N$, which implies that a boundary should widen indefinitely. This neglects the influence of magnetocrystalline anisotropy. Normally a domain is magnetized in a direction of easy magnetization associated with minimum magnetocrystalline energy. In the domain boundary, the canting of the spins introduces a magnetization component in a less favourable direction associated with larger magnetocrystalline anisotropy energy. The wider the transition range, the larger the volume of material contributing to this energy. The boundary is therefore of limited thickness; this thickness is determined by the compromise between decreasing exchange and increasing magnetocrystalline contributions; see Problem 11.9.

Calculation shows that the boundary often has considerable thickness extended over about 300 atom layers. We now see that if a particle is to occur as a single domain then, if it has prolate ellipsoidal form, the diameter of the ellipsoid should not be much greater than a domain boundary thickness, i.e. of order 100 nm. The greater the magnetocrystalline anisotropy, the smaller the domain boundary thickness and therefore the smaller the critical size for a single-domain particle.

The wide domain boundary depicted in Fig. 11.34 is known as a 180° Bloch wall. There is a similar 90° variant. Bloch walls are the most common type of domain boundary in bulk magnetic matter. There are, however, other forms. In a Néel boundary the magnetization vector gradually rotates in fan-like fashion in the plane of magnetization. In films of thickness less than about 50 nm this boundary is calculated to have a lower energy than the Bloch wall and its occurrence has been demonstrated in experiment. For a thin film of a rather isotropic magnetic material like iron one would expect the magnetization to lie in the plane of the film because the demagnetizing factor is usually so much smaller in this direction than in that perpendicular to the film. This is in fact the case for films with thickness greater than about 20 nm. However, experiment shows that for thinner films the domains prefer the orientation perpendicular to the plane of the film. The reason for this preference is thought to lie in a surface anisotropy contribution to the free energy, a contribution that arises because surface atoms have a different environment from those in the bulk crystal. It becomes more significant the smaller the volume of the vertical domain.

#### Magnetoelastic energy

Changes in length are associated with the magnetization of ferromagnetic substances; the effect is known as magnetostriction, and a magnetostriction coefficient $\Lambda$ is defined as the fractional change in length as the magnetization increases from zero to its saturation value. $\Lambda$ may be positive or negative, with $|\Lambda| \approx 10^{-5}$. There is therefore an elastic strain energy $U_{el}$ associated with the magnetic state. Conversely elastic strain introduces additional magnetic anisotropy, and this is of particular significance in practice because solid substances usually have an internal strain distribution that can influence domain boundary movement.

### 11.9.2 Equilibrium domain structures

The total magnetic contribution to the free energy density of a ferromagnetic crystal may be written

$$U = U_s + U_c + U_b + U_{el}.$$

Crystalline matter is never perfect in its ordered structure and even so-called pure materials contain impurities, holes or small precipitates as well as internal stresses. The domain structure adopted by any ferromagnetic sample is therefore critically dependent on departures from perfection and the care with which the specimen is prepared and oriented. We cannot discuss other than very simple situations. Figure 11.35 shows the domain configuration first proposed by Landau and Lifshitz in 1935 for a single-crystal 'bar magnet' with its major axis parallel to an axis of easy magnetization, [100], in the case of Fe. This configuration eliminates the free poles at the ends of the sample by introducing triangular 'closure' domains. The magnetostatic self-energy is removed and the total energy is governed by the number of

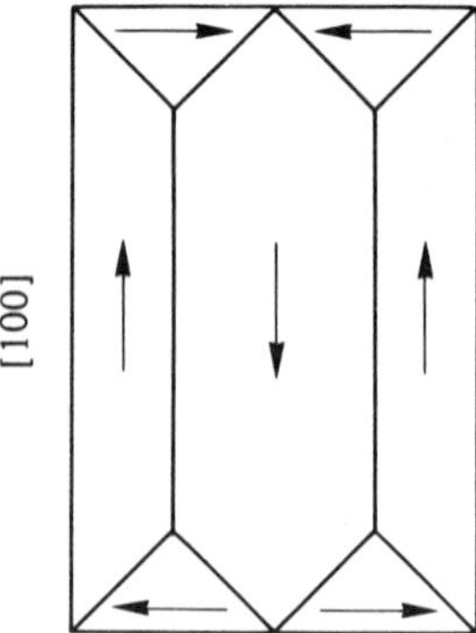


**Figure 11.35** An arrangement of closure domains that minimizes the magnetostatic energy of a single-crystal bar magnet of iron.

domain boundaries and the strain energy associated with the 90° boundaries. Ideally, such a specimen should not consist of more than four domains. This domain pattern has been observed in many experiments.

Consider a similar single-crystal bar magnet, but this time of Co, the axis of the bar magnet coinciding with the *c* axis of the crystal. Do we find the same pattern as in Fig. 11.35? Co is very much more anisotropic than Fe. Closure domains in this case must have their magnetization along a direction of difficult magnetization, and in Co this is associated with a very large magnetocrystalline energy, which prohibits their occurrence in the same way as for Fe.

Thus it may be preferable to accept a smaller magnetostatic self-energy rather than a larger magnetocrystalline energy. A possible equilibrium domain structure might consist of several domains parallel to the magnet axis and with free poles at the ends. The compromise now must be made between reduction of the self-energy, by introducing many antiparallel domains, and the associated increase in domain boundary energy that this occasions (Fig. 11.36). Another possibility is to use closure domains but of the least possible volume (Fig. 11.37). As with many other branches of solid state physics, the study of magnetic domains has become a specialist subject.

### 11.9.3 *The magnetization process*

A magnetic crystal becomes homogeneously magnetized by two processes:

(a) domain boundary movement;

(b) rotation of the magnetization within a domain.

Suppose that our single-crystal bar magnet of Fe is perfect, with no holes, no precipitates, no internal stresses. The domain configuration in the 'unmagnetized' condition is as shown in Fig. 11.35. In a suitably large external field, however, the crystal must become a single domain because the magnetostatic energy in the presence of the external field, $-\mathbf{B}_0 \cdot \mathbf{M}$, eventually becomes the dominant quantity. For

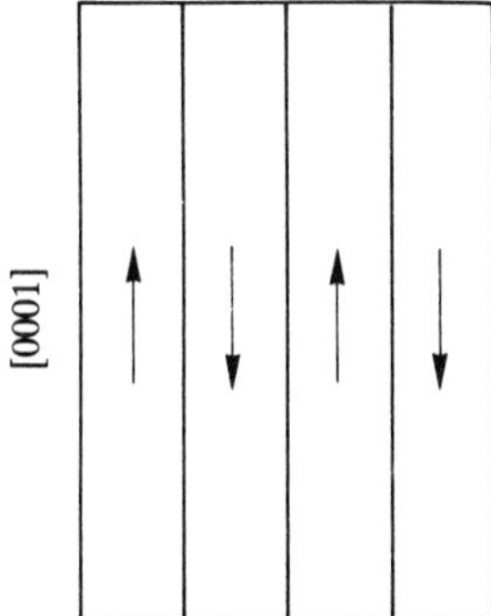


**Figure 11.36** In the highly anisotropic Co, a state of high magnetostatic energy may be preferable to one of higher energy of crystalline anisotropy.

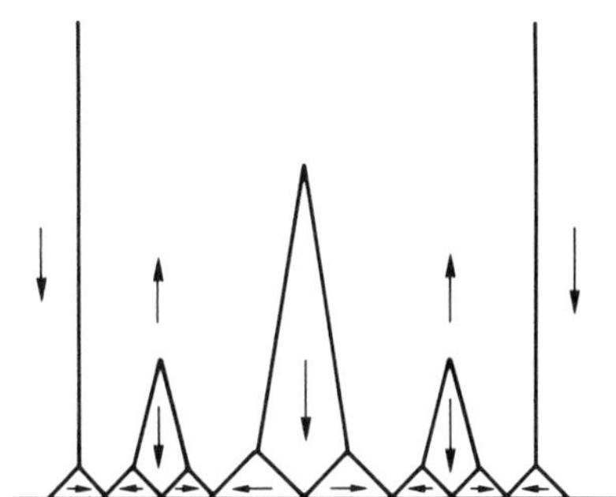

**Figure 11.37** If closure domains are to arise in Co, they must have the least possible size – the above arrangement has been suggested. (After Craik and Tebble 1965.)

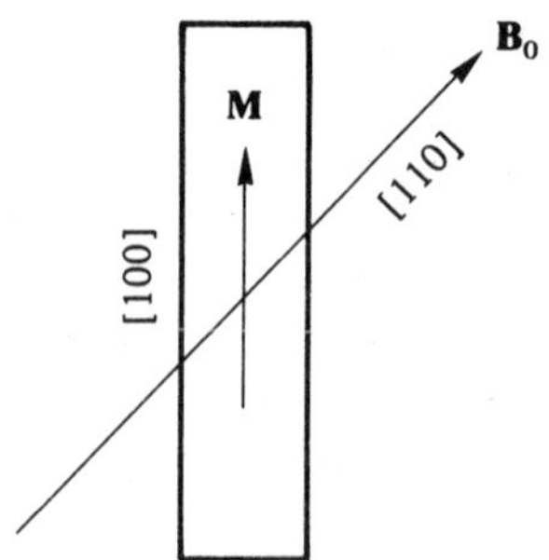


**Figure 11.38** When a single-crystal bar of Fe is magnetized in a 'difficult' direction, there is little hindrance to domain boundary movement to orient the domains in the most favourable [100] direction.

our iron bar magnet, only domain boundary movement is required. This demands very little work arising from the resistive losses associated with the induced electrical currents produced by domain boundary motion. Suppose now, however, that the crystal bar magnet is held fixed in space and an external field is gradually applied at an angle of 45° to its major axis (i.e. along [110]) (Fig. 11.38). As before, there is little hindrance to the production of a single-domain particle by domain boundary movement. We easily arrive at a magnetization in the field direction of $M \cos 45°$ or $M/\sqrt{2}$. Further magnetization in the direction of the field now demands rotation of the magnetization from its preferred direction to one of greater magnetocrystalline energy (and in this particular case greater magnetostatic self-energy); this occurs only with a certain difficulty and by performing work (Fig. 11.39). Iron has positive magnetostriction, so if we place our crystal magnet under tensile stress, as shown in Fig. 11.40, then it becomes even more difficult to rotate the magnetization from its preferred orientation.

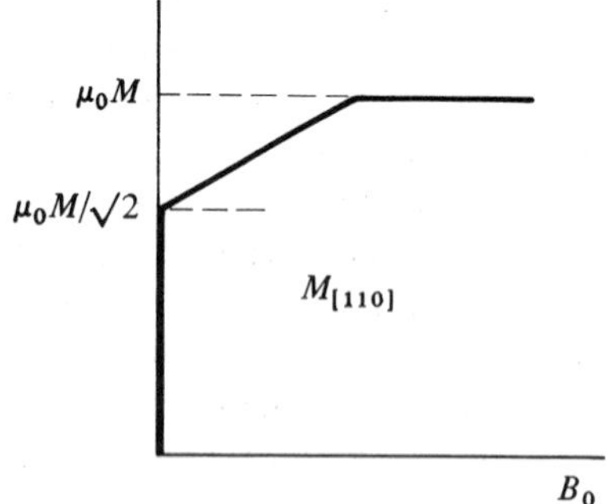


**Figure 11.39** Rotation of the magnetization within a domain requires that work be performed to increase the magnetocrystalline energy.

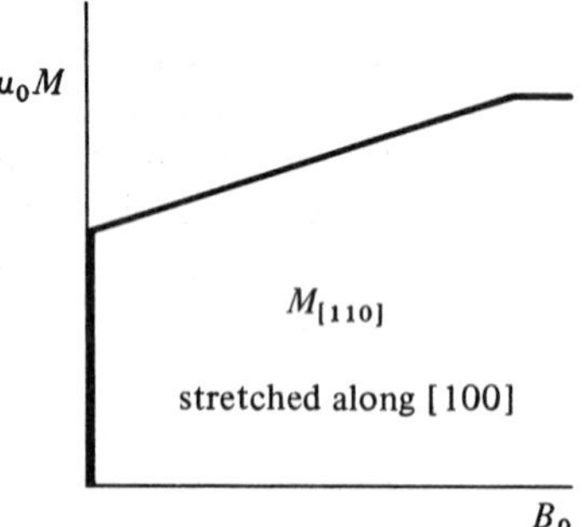


**Figure 11.40** If the sample has positive magnetostriction in the direction of easy magnetization then the application of a tensile stress in this direction makes it more difficult to rotate the magnetization into a less favourable direction.

### 11.9.4 Real magnetic specimens

The discussion of the previous section used an idealized specimen. Real samples can in fact come very close to the ideal situation. They must then be pure elements or highly ordered single-phase alloys without inner macroscopic faults (although grain boundaries are unavoidable in commercial materials) and in a strain-free condition produced by careful heat treatment. The magnetization process is easiest in such materials which therefore have high permeabilities and narrow hysteresis loops; the energy loss under cyclic magnetization is therefore small. These are the soft magnetic materials used in transformers and magnetic shields.

Commercial soft magnetic materials, as required for electronic components operating up to conventional radio frequencies, are tailored to particular requirements, e.g. maximum initial permeability, squareness of the hysteresis loop. These materials go under such names as 'Permalloy', 'Hymu' or 'Mumetal' and are invariably alloys of Fe and Ni near the composition $Ni_{80}Fe_{20}$. Careful production controls are required to obtain the optimum properties. Usually the coercive force is about 2 $\mu$T and the ratio of remanent to saturation magnetization about 0.8. To make maximum use of the magnetic material, the alloy is usually in the form of a ribbon of thickness 25, 50 or 100 $\mu$m since if thicker material is used, the oscillatory fields do not penetrate the bulk of the substance. A much cheaper core material is silicon-iron (3.2% Si). The silicon increases the electrical resistivity, thereby reducing eddy currents, but also promotes a coarse-grained structure, and by careful production techniques these grains can be oriented so that the direction of easy magnetization lies in the plane of the sheet. It is widely used in power and current transformers, magnetic amplifiers and saturable reactors. Since these core materials are electrical conductors, induced currents arise in them and a certain power loss is incurred. The objective in practice is to minimize this power loss by minimizing the area of the hysteresis loop and maximizing the resistivity of the core material.

In the past ten years much attention has been given to metallic glasses; these are amorphous metals formed by the rapid quenching of a thin stream of liquid metal. These glasses are usually produced as very thin ribbon. A typical metallic glass is $Fe_{80}B_{20}$ but there are very many other compositions. The glasses with large iron contents are ferromagnetic (as are the crystalline borides such as FeB, $Fe_2B$) and, although amorphous, they are very soft magnetically and show loss factors of three to four times less than that of silicon-iron. There is therefore considerable interest in the development of such amorphous alloys for 'soft' magnetic applications.

Homogeneity and isotropy are primary requirements for a soft magnetic substance, and this allows us, in part at least, to appreciate the properties of the metallic glass. On the other hand, when a magnetic material is neither homogeneous, owing to the presence of dispersed phases, nor isotropic, on account of internal strains and crystalline anisotropy, domain boundaries can only move reversibly over very small distances. The energy associated with the boundary becomes dependent on its position. Its movement under the action of an external field is discontinuous, the boundary remaining essentially fixed until it is compelled to move by the external forces; it then jumps rapidly to a new metastable position rather like the way a dislocation moves in a dispersion-hardened alloy. Permanent magnets should have large remanence and large coercive force. The first requirement determines the flux available and the second that the performance is not impeded by stray fields or the magnet's own demagnetizing field. These hard permanent magnet materials utilize the various contributions to magnetic anisotropy, shape, crystallinity and strain in order to confine the magnetization to a particular direction, from which it can be changed only with difficulty.

A closed ring of magnetic material possesses an induction $B$ proportional to the remanent magnetization $M_R$. Such a closed ring is of no use as a magnet because there is no working gap. As soon as we introduce a gap, we obtain a 'permanent' magnet, but at the same time we produce a demagnetizing field within the material of the magnet. This field is $-DM$, where $D$ is the demagnetizing coefficient and $M$ the new magnetization stable in the demagnetizing field. We write

$$B = B_0 + \mu_0 M,$$

$$B_0 = -DM.$$

Elimination of $M$ leads to

$$B = \frac{D - \mu_0}{D} B_0. \tag{11.63}$$

Now $D/\mu_0$ is always less than unity, so (11.63) is the equation of a straight line with negative gradient. The working conditions of the magnet are now determined not only by the hysteresis curve, which is a property of the magnetic material, but also by the line (11.63), which is determined by the geometrical shape of the magnet. The so-called 'working point' of the magnet is the intersection of the line (11.63) with the hysteresis curve: this intersection lies in the upper left-hand quadrant of the hysteresis curve (Fig. 11.41). The magnet designer attempts to make this working point coincide with that point on the hysteresis curve where the product $BB_0/\mu_0$ has its maximum value, the reason being that the minimum amount of material is then needed to obtain the desired performance (see any text on electromagnetism, e.g. Duffin 1965). The objective of magnet makers is to develop materials with larger values of $(BB_0/\mu_0)_{max}$; not so that higher magnetic fields may be produced (because they cannot), but because a given gap field may be obtained with a smaller and lighter magnet. We can readily estimate the maximum gap field that might be obtained with a permanent magnet. None of the modern permanent magnet

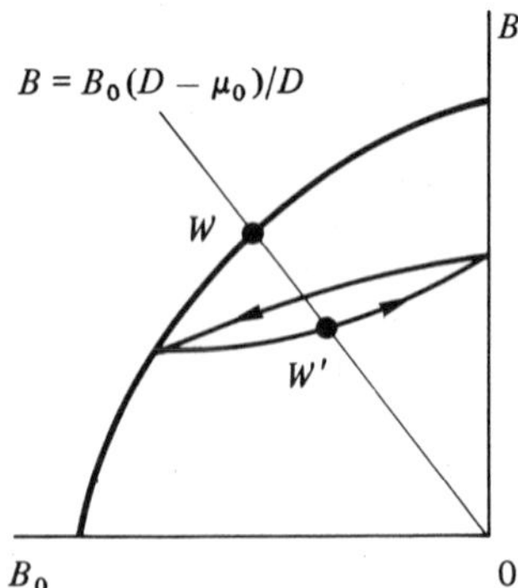


**Figure 11.41** The demagnetizing curve for a permanent magnet. $W$ is the working point determined by the demagnetizing factor of the magnet – it should coincide with the point of maximum $BB_0$. (After Crangle 1977.)

materials has a saturation induction greater than that of iron (about 2 T). Furthermore the remanent induction is usually about 0.75 of the saturation value. Any practical magnet with a reasonable working gap will have $|(D - \mu_0)/D| < 0.75$, so it is very unlikely that the gap field can be larger than of order 1 T. The data in Table 11.6 show that in practice the maximum gap field attainable is likely to be considerably less than this estimate because the remanent induction of available materials is usually $<1.2$ T.

In use, a permanent magnet may be subject to many influences, and it is often required that the magnet provide as reproducible a magnetic field as can be arranged. For this to be achieved, the magnet may be subjected to an initial demagnetizing field greater than its own intrinsic demagnetizing field so that it works on a minor hysteresis loop around the point $W'$. Although this causes a lower flux in the gap, the stability is improved, and if necessary the lower flux may be compensated by using a magnet of larger volume but similar gap size.

The significance of the requirement that $BB_0/\mu_0$ should be a maximum is clear: there is little point in using a material with large remanence if the coercivity is small, the magnet's intrinsic demagnetizing field would completely nullify its effect; nor is there any point in achieving a large coercive force in a material with a low saturation magnetization, since then there is little or nothing the coercive force can preserve. Large remanence must be combined with a large coercive force. Some data are shown in Table 11.6.

The many uses for permanent magnets create a large demand for them, so there is a constant search for improved or cheaper materials; the better the value of $(BB_0/\mu_0)_{max}$ the lesser the volume and associated weight of a magnet to produce the required flux. The best magnets are complex metallurgical objects; they are mechanically hard and brittle. Usually they are cast and then ground to shape. Every attempt is made to promote anisotropic behaviour, and heat treatment, often in the presence of a magnetic field, is used for this purpose. Sometimes the magnet material is used as a powder. The latter is mixed with a resin binder material, compressed and then heat-treated to provide the solid composite compact. Similarly, with the oxide magnet $BaFe_{12}O_{19}$, the particles are mixed with binder, cold-pressed

**Table 11.6**

| | Coercivity (T) | Remanence (T) | $(BB_0/\mu_0)_{max}$ (kJ m$^{-3}$) |
|---|---|---|---|
| $BaFe_{12}O_{19}$ | 0.36 | 0.36 | 25 |
| Alnico IV | 0.07 | 0.6 | 10.3 |
| Alnico V | 0.07 | 1.35 | 55 |
| Alcomax I | 0.05 | 1.2 | 27.8 |
| MnBi | 0.37 | 0.48 | 44 |
| $Ce(CuCo)_5$ | 0.45 | 0.7 | 92 |
| $SmCo_5$ | 1.0 | 0.83 | 160 |
| $Sm_2Co_{17}$ | 0.6 | 1.15 | 215 |
| $Nd_2Fe_{14}B$ | 1.2 | 1.2 | 260 |

*Note.* Alnico IV and Alcomax I have randomly oriented grains; the superior properties of Alnico V derive from the presence of directed columnar crystals, which enable better utilization of the crystalline anisotropy.

in the presence of a magnetic field to provide optimum alignment of the particles and then heat-treated to stabilize the arrangement. These powder magnets may be prepared in a variety of shapes.

Most conventional alloy magnets are based on iron to which important elements notably Al, Ni and Co are added – hence the trade names like Alnico and Alcomax. Small amounts of other elements, e.g. Cu, Ti and Nb, may also be used and there is a large number of variants around a given basic composition – the same material often has different names in different countries.

Nevertheless, it is clear from Table 11.6 that the magnets based on Sm and Co are far more powerful than conventional magnet materials, but their cost is also much greater. There is a certain production of $SmCo_5$ magnets, but their uses are restricted to situations where strong small magnets are needed and where cost is of secondary importance. There is yet to be large-scale use of rare earth magnets, but a significant advance is the recent discovery of magnets based on Nd, Fe and B alloys close to the composition $Nd_2Fe_{14}B$.

The performance of these alloys matches that of $SmCo_5$, allowing coercive field strengths of about 1.2 T to be obtained, but the cost of the raw alloy material is about one tenth that of the $SmCo_5$ starting material. On the other hand, magnet fabrication costs remain high – the raw alloy must be ground to powder, pressed to shape and then sintered at high temperature.

Now that the practical uses of magnetic materials have been described briefly, you might like to ponder over the following viewpoint. At the present time high temperature ($T_c = 130$ K) superconductivity is very much talked about. On the other hand one rarely, if ever, hears mention of high temperature magnetism: magnetism is, on the whole, a low temperature phenomenon. Most magnetic substances have $T_c$ below room temperature. A ferromagnetic or ferrimagnetic substance, if it is to be used at RT, must have a critical temperature of at least 600 K and preferably 900 K. Only iron and cobalt really meet this requirement. Although other elements are valuable for alloying purposes it is the high Curie temperature of Fe that allows transformers and similar equipment to operate at RT. One may very well wonder how technology would have developed if the Curie temperature of iron had been, say, 300 K and its exploitation required operation at liquid nitrogen temperatures.

## 11.10 Magnetism and Computers

### *11.10.1 Magnetic memories*

The rapid development of the electronic computer has been dependent on the production of semiconductor microprocessors, but equally so on the availability of large capacity magnetic memories. At present a hard disk magnetic memory has a maximum storage capacity of order 90 megabits per $cm^2$, and capacities of order 1.5 gigabits per $cm^2$ are projected for the future. The basic magnetic medium is usually a polycrystalline film of magnetic alloy based on cobalt and about 50 nm in thickness. The uniaxial crystalline anisotropy of the microcrystallites is similar to that of hexagonal cobalt. The 'bit' is the transition region between two oppositely magnetized regions less than 1 $\mu$m in length. The magnetization lies in the plane of the disk. There are some 50 000 bits per cm of track length. Although the magnetic domain structure is still dependent on the various contributions to the free energy discussed earlier, the magnetization–demagnetization process is very complex. One cannot assume, as in section 11.9.3, that coherent rotation of the magnetization vector arises in such processes. As the physical size of the magnetic bit is decreased to improve the density, the long range magnetostatic coupling between bits becomes more important. The computational problems in minimizing the free energy in order to model the domain structure are thereby considerably increased. Furthermore, the smaller the size of the bit the less stable it becomes with regard to thermal and demagnetizing effects, so the permanence of the storage is seriously affected. For this reason attention is directed to the perpendicularly oriented magnetic bit. In such a geometry it should be possible to combine a small in-plane storage site with a relatively large bit volume, because the thickness may be considered a free parameter. This would improve thermal stability. Demagnetizing effects are also weakened in this geometry. The writing and reading of the hard disk is arranged by 'heads' inductively coupled to the bits on the disk.

Perpendicularly oriented bits are the rule for magneto-optical memories and the bit is formed by thermomagnetic writing; a focused laser beam heats an elementary region to above the Curie point and a weak external field causes a reversal of magnetization on cooling. Reading is accomplished by detecting the variation in the rotation of the plane of polarization of a low intensity laser beam as it scans the disk.

### 11.10.2 *Giant magnetoresistance (GMR)*

The development of ultra-thin film deposition techniques as used in semiconductor research and technology has had repercussions in the study of thin films of magnetic metals. In addition to the study of surface properties attention has been directed to sandwich arrangements of the kind FNFNFN . . . where F represents a single crystal film of a ferromagnetic element with thickness of one or more atomic layers and N is an epitaxial non-magnetic film of similar order of thickness. The magnetization lies in the plane of the F film.

Consider first a three film arrangement, FNF. This is certainly a possible arrangement, but there is another, namely FNA, where A represents the same kind of ferromagnetic film but in the antiferromagnetic orientation relative to the first film. The non-magnetic layer, e.g. copper, is not completely magnetically inactive. Earlier we have seen that the polarization of the conduction electrons provides a means for the coupling of atomic moments, the RKKY mechanism. The conduction electrons of the N film are influenced by the atomic moments at the first FN interface leading to an oscillatory magnetic polarization that may extend throughout the N film, provided it is not too thick ($<2$ nm). The orientation of the second magnetic layer then depends on that of the conduction electron polarization at the outer surface of the N layer. Thus, dependent on the wavelength of the oscillatory polarization and the thickness of the N layer, this second magnetic film may have F or A character. By varying the thickness of the N layer it is possible to trace the oscillatory polarization of the conduction electrons. It has been found that the wavelength is considerably larger than that expected for the RKKY mechanism ($\pi/k_F$).

The electrons passing through the multilayer encounter F(A)/N interfaces where they are partly reflected and partly transmitted (as is light at a dielectric interface). The wave functions of the incident and transmitted electrons interfere with those of the electrons reflected at the interfaces. This leads to an oscillatory pattern of charge density within the multilayer, the wavelength being dependent on the thickness and electronic structure of the elements in the multilayer. However, the interaction at the magnetic interface is also spin dependent so the charge density oscillations are associated with oscillations in spin density, thereby producing the observed variation in magnetic polarization. Once the important parameters have been determined in experiment, it is possible to prepare multilayer sandwich arrangements with different magnetic structures: Co/Cu sandwiches have received such attention.

In addition to the concern regarding the fundamental origins and range of magnetic exchange coupling in composite magnetic thin films, there is considerable interest in their possible commercial exploitation as a result of the discovery of what is called 'giant magnetoresistance' (GMR). Ordinarily the effects of a magnetic field, applied longitudinally or transversely to the current flow, on a ferromagnetic metal are very limited in extent because the resistivity of the metal is already very large and external magnetic fields are usually insignificant relative to the Weiss effective field which represents the magnetic exchange coupling.

Consider a sandwich structure of the form FNANFNAN . . . , where F/A = Co and N = Cu. A current flows in the direction perpendicular to the plane of the films and a magnetic field is applied in the parallel direction, in say the spin up direction parallel to the magnetization in the F film. The resistance of the arrangement is measured when the magnetic field is increased in the up direction. As the magnetic field grows, the magnetization in the A films becomes less stable with regard to the up direction and eventually the A films must assume F character. In this transition the resistance of the composite film varies as shown in Fig. 11.42.

At room temperature a decrease in the resistance of up to 65% has been measured as a result of the transition from the FNANFNAN . . . pattern to that of FNFNFN . . . . At 4 K the change is 120%. These changes are very large, hence the name GMR. It is to be noted however that the 'saturation field' required to produce complete ferromagnetic alignment is quite high, approaching 0.5 T. The occurrence of GMR has caused considerable excitement because it offers the means to develop

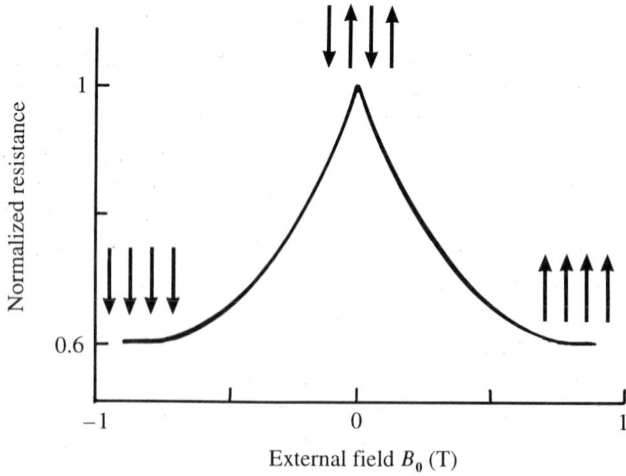


**Figure 11.42** Giant magnetoresistance as might arise in a layered FNAFNA . . . structure.

a magnetic sensor and possibly a magnetic transistor based on magneto-resistive effects (see Further Reading).

A qualitative explanation of GMR in Co/Cu composite films may be obtained as follows. Co is similar to Ni, it is a strong ferromagnetic material. The schematic band structure of Co is similar to that of Ni, Fig. 11.18. A spin split 3d band is assumed. The spin up 3d band lies below the Fermi level whereas the spin down sub-band cuts it so that it contains 3.3 electrons. This distribution produces the observed atomic moment of Co, 1.7 $\mu_B$. The sp bands may also be considered as spin split, but any small relative displacement is neglected here. Normally the electron spin does not enter the conduction process because the spin lifetime is much larger than the electron collision time for electrical conduction. When an electron suffers collisions with phonons or crystal defects the spin remains unchanged (the important exception is the Kondo effect described earlier). Although it has been suggested that d electrons may contribute to the conductivity of transition metals to a greater degree than hitherto proposed, we shall continue to assume that the conductivity of Ni and Co is to be attributed primarily to the sp conduction electrons and that s–d scattering is the major source of electrical resistance in these metals. If we consider pure Co (or Ni), since the spin remains unaffected, only spin down electrons can experience s–d scattering and the current is carried primarily by the spin up electrons because there are no empty spin up states in the d band that can cause scattering of these electrons, Fig. 11.18. The conduction process is to be considered as arising via two independent and parallel spin up and spin down channels.

In the ferromagnetic arrangement of the multilayer, FNFNFNFN, the resistance is essentially that of the separate Co layers. The spin down channel has the higher resistance due to s–d scattering possibly augmented by an interface contribution arising from the differences in crystal potential that arise at each interface. The spin up channel has a low resistance, and since it is in parallel with the spin down channel the total resistance is low. Other than coupling the magnetic layers the Cu films play no part, since their d sub-bands lie wholly below the Fermi level.

In the antiferromagnetic FNANFNAN . . . composite film the spin down electrons experience s–d scattering in the F layers as before, but in the A layers the situation is reversed: 'up' becomes 'down'. Therefore when a spin up electron enters an A layer, in addition to the interface resistance it will experience s–d scattering from the now incomplete spin up d band, whereas the spin down electron does not experience this scattering process because in the A layer it is the spin down sub-band that is completely filled. Thus in the antiferromagnetic arrangement both spin channels alternately experience high resistance paths whereas in the ferromagnetic arrangement the low resistance spin up channel effectively 'shorts out' the high resistance spin down channel. The different behaviours associated with the ferromagnetic and antiferromagnetic structures account for the GMR.

It is not necessary that the magnetic component, F, be a strong ferromagnet; the non-magnetic layer, N, may be a transition metal. Composite films containing alternate layers of Fe and Cr also display GMR. In this case the source of the effect is attributed to the very different values of $N(E_F)$ in these two metals (as indicated in Fig. 9.8). GMR has also been observed in sputtered granular films, a feature that would be important in future commercial use, but the major barrier to be overcome is to obtain lower values for the necessary 'switching magnetic field'.

## Appendix 11.1 The Atomic Magnetic Moment and the Landé Splitting Factor

Electrons in atoms possess both orbital and spin angular momenta. There are also resultant orbital and spin angular momenta associated with a complete atom or ion; these quantities are described in terms of the quantum numbers $L$ and $S$. In Russell-Saunders coupling, $L$ and $S$ may be combined vectorially to produce $J$, the total quantum number for the atom or ion. Given $L$ and $S$ may give rise to different $J$, thereby producing multiplet structure (see Appendix 11.3 on Hund's rules).

The eigenvalues for the atomic orbital, atomic spin and total atomic angular momenta are

$$\hbar[L(L+1)]^{1/2} \equiv \hbar L',$$

$$\hbar[S(S+1)]^{1/2} \equiv \hbar S',$$

$$\hbar[J(J+1)]^{1/2} \equiv \hbar J',$$

Consider the eigenvalue of the total angular momentum on the atom, $\hbar J'$. In the presence of a weak magnetic field (which provides a convenient reference direction along the $z$ axis), we find that the resolved components of $\hbar J'$ may take only the discrete values $-\hbar J$, $-\hbar(J-1)$, $-\hbar(J-2)$, . . . , $+\hbar(J-2)$, $+\hbar(J-1)$, $+\hbar J$. This is called spatial quantization (Fig. 11.43). The eigenvalues of the $z$ component of the angular momentum are written as $\hbar m$, where $m$ takes either integer or half-integer values in the range $-J \leqslant m \leqslant +J$.

When we wish to assess the magnetic moment of an atom or an ion, we must take account of the fact that the spin angular momentum is twice as effective as the orbital angular momentum. The magnetic moment parallel to $J'$ becomes

$$\mu' = \mu_B L' \cos(\mathbf{L}', \mathbf{J}') + 2\mu_B S' \cos(\mathbf{S}', \mathbf{J}'),$$

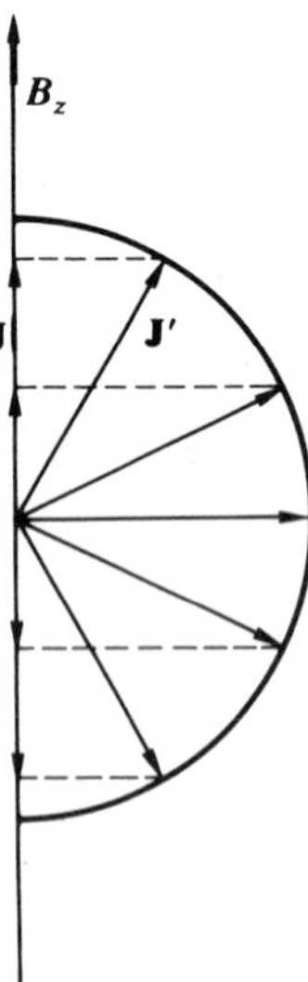


**Figure 11.43** In the presence of a magnetic field the component of the total angular momentum in the field direction may take only discrete values determined by the quantum number $m$, where $-J \leqslant m \leqslant +J$.

where (**a**, **b**) is the angle between the vectors **a** and **b**. However, it is preferable to express this magnetic moment directly in terms of $J'$, so that $\mu' = g\mu_B J'$. In spectroscopy $g$ is known as the Landé splitting factor controlling the Zeeman effect, whereas in the study of magnetism it is called the gyromagnetic ratio (but it is more correctly the magnetomechanical ratio). More often than not it is just called the $g$ factor.

An expression for $g$ is easily obtained from the triangle of vectors making use of the cosine rule, $L'$, $S'$ and $J'$ all being given (Fig. 11.44). In magnetic behaviour we are more interested in the resolved components of the expectation value $g\mu_B J'$ in the direction of the field (i.e. the $z$ direction). These resolved components are

$$-g\mu_B J, \; -g\mu_B(J-1), \ldots, 0, \ldots, +g\mu_B(J-1), \; +g\mu_B J,$$

which are most conveniently written as $g\mu_B m$, $-J \leqslant m \leqslant J$. The maximum value of the resolved magnetic moment is $g\mu_B J$ – we call this the atomic magnetic moment.

We perhaps should emphasize that in the absence of a magnetic field, an atom with $J > 0$ will be in a degenerate state, all the levels of different $m$ being equally represented in the ground state configuration. The application of an external magnetic field removes this degeneracy through Zeeman splitting. The different $m$ levels are now at different energies and they are occupied in accordance with Hund's rules to produce the magnetic moment as discussed here. Although the moment only arises in the presence of the external field, it does so however small we care to make the field, and we therefore speak of the permanent atomic moment. However, it will be appreciated that if some other non-magnetic agency (e.g. a strong electrostatic field) removes the orbital degeneracy then, if this agency is sufficiently strong, the

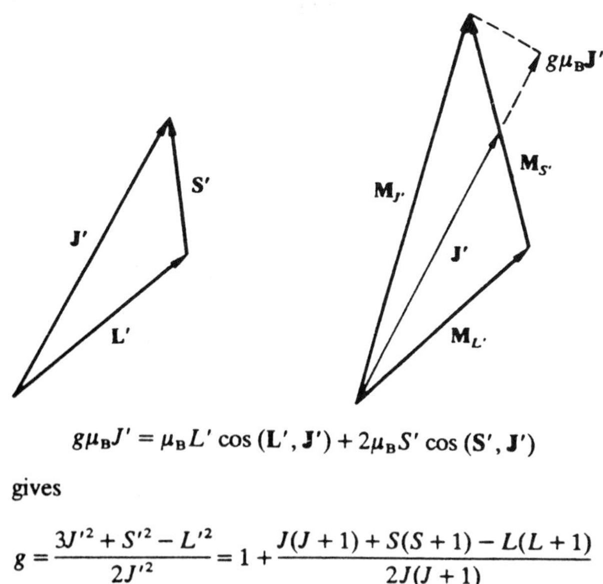


$$g\mu_B J' = \mu_B L' \cos(\mathbf{L}', \mathbf{J}') + 2\mu_B S' \cos(\mathbf{S}', \mathbf{J}')$$

gives

$$g = \frac{3J'^2 + S'^2 - L'^2}{2J'^2} = 1 + \frac{J(J+1) + S(S+1) - L(L+1)}{2J(J+1)}$$

**Figure 11.44** On the left is shown the vector composition of the total angular momentum while on the right is the corresponding geometry for the magnetic moment. The resultant moment $\mathbf{M}_{J'}$ is not parallel to $\mathbf{J}'$ and therefore precesses about $\mathbf{J}'$, leading to a resolved component of magnetic moment $g\mu_B J'$ parallel to $\mathbf{J}'$.

applied magnetic field cannot cause a redistribution of the electrons in the different $m$ levels, and there can be no orbital contribution to the magnetic behaviour.

It should be noted that the discussion in Section 11.4 leading to (11.23), $\mu_{\text{eff}} = g\mu_B[J(J+1)]^{1/2}$, has nothing to do with the discussion of eigenvalues given here. We consider as coincidental the fact that $\mu_{\text{eff}}$ (the effective classical moment) is found to be equal to the quantum-mechanical eigenvalue of the atomic moment.

## Appendix 11.2 The Atomic Moment and the Bohr Magneton

Consider an electron moving in a planar circular orbit of radius $r$. It rotates with frequency $\nu$ and simulates a current $i = -\nu e$. It therefore produces a magnetic moment

$$\mu = iA = -\nu e\pi r^2 = -\frac{\omega}{2\pi}e\pi r^2 = -\frac{\omega e r^2}{2}.$$

In vector notation this becomes

$$\boldsymbol{\mu} = -\frac{e}{2}\mathbf{r}\times\mathbf{v}$$

$$= -\frac{e}{2m}\mathbf{r}\times\mathbf{p} = -\frac{e\hbar}{2m}\mathbf{l},$$

$\mathbf{l}$ being a directed orbital quantum number. Because of the negative charge of the electron, $\boldsymbol{\mu}$ and $\mathbf{l}$ are oppositely directed; the quantity $e\hbar/2m$ is the Bohr magneton $\mu_B$.

In the presence of an external field the system acquires magnetic energy

$$-\boldsymbol{\mu}\cdot\mathbf{B}_0 = -\left(-\frac{e\hbar}{2m}\right)\mathbf{l}\cdot\mathbf{B}_0 = \frac{e\hbar}{2m}\mathbf{l}\cdot\mathbf{B}_0.$$

Spatial quantization allows the vector $\mathbf{l}$ to take $2l+1$ possible orientations between the limits $\pm l$; clearly the negative values of $l$ provide the lower energies.

When both orbital and spin momenta are present, we write

for a single electron $\boldsymbol{\mu} = g\mu_B\mathbf{j}$, $\mathbf{j} = \mathbf{l} + \mathbf{s}$;
for a whole atom $\boldsymbol{\mu} = g\mu_B\mathbf{J}$, $\mathbf{J} = \mathbf{L} + \mathbf{S}$.

$g$ is the magnetomechanical ratio.

## Appendix 11.3 Hund's Rules

The spectroscopic state of an atom is determined by the quantum numbers $L$, $S$ and $J$. Hund's rules state that the ground state is obtained when

(a) $S$ is maximized;

(b) $L$ is maximized subject to (a);

(c) $J = \begin{cases} |L-S| & \text{when a shell is less than half full,} \\ L+S & \text{when a shell is more than half full.} \end{cases}$

We shall determine the ground states of the ions $Pm^{3+}$ and $Dy^{3+}$ as examples.

| | | $Pm^{3+}$ | $Dy^{3+}$ |
|---|---|---|---|
| | | $4f^4 5s^2 5p^6$ | $4f^9 5s^2 5p^6$ |
| $l = 3,\quad m =$ | 3 | · · | · $-\frac{1}{2}$ · |
| | 2 | · · | · $-\frac{1}{2}$ · |
| | 1 | · · | · $-\frac{1}{2}$ · |
| | 0 | · $-\frac{1}{2}$ · | · $-\frac{1}{2}$ · |
| | $-1$ | · $-\frac{1}{2}$ · | · $-\frac{1}{2}$ · |
| | $-2$ | · $-\frac{1}{2}$ · | · $-\frac{1}{2}$ · $+\frac{1}{2}$ |
| | $-3$ | · $-\frac{1}{2}$ · | · $-\frac{1}{2}$ · $+\frac{1}{2}$ |

which gives

| $Pm^{3+}$ | $Dy^{3+}$ |
|---|---|
| $S = \sum s = 2$ | $S = \frac{5}{2}$ |
| $L = \sum m = 6$ | $L = 5$ |
| $J = \lvert L - S \rvert = 4$ | $J = L + S = \frac{15}{2}$ |

and the multiplicity is

| | |
|---|---|
| $2S + 1 = 5$ | $2S + 1 = 6$ |

The spectroscopic state is written in the form ${}^{2S+1}L_J$, where $L$ is the total orbital angular momentum. This may take integral values only, but is by custom denoted by a capital letter according to the following scheme:

| $L =$ 0 | 1 | 2 | 3 | 4 | 5 | 6 |
|---|---|---|---|---|---|---|
| S | P | D | F | G | H | I |

The two ions in question therefore have the spectroscopic ground states ${}^5I_4$ and ${}^6H_{15/2}$.

## Appendix 11.4 First- and Second-order Phase Transitions†

A phase is defined as a region of homogeneous matter confined within boundaries. Thus a bubble in water has well defined boundaries and contains a homogeneous quantity of water vapour – the vapour phase of water. Outside the bubble we find a homogeneous distribution of matter that we call the liquid phase of water.

A change of state, e.g. conversion of liquid water to water vapour, is a first-order phase transition (Fig. 11.45), and is always accompanied by a latent heat. Similarly

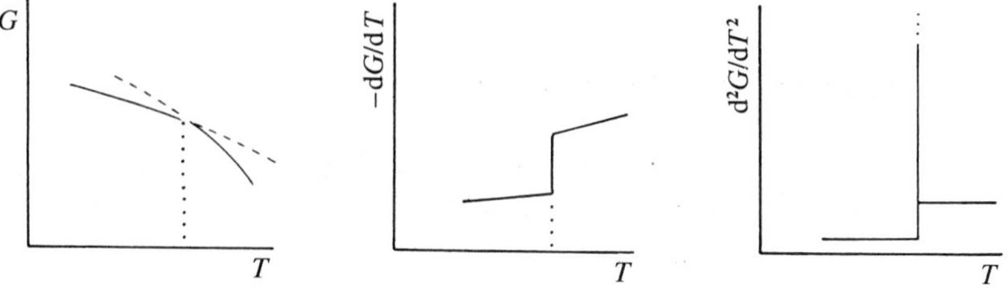


**Figure 11.45** $G$, $-\mathrm{d}G/\mathrm{d}T$ and $\mathrm{d}^2G/\mathrm{d}T^2$ for a first-order phase transition.

† For a discussion of first-, second- and higher-order phase transitions see Pippard (1960).

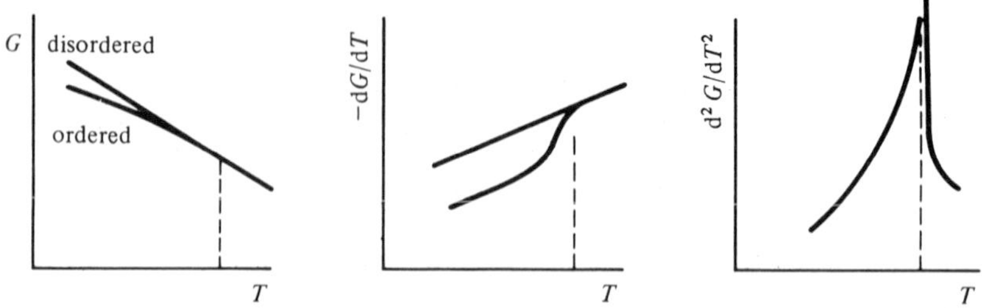


**Figure 11.46** $G$, $-\mathrm{d}G/\mathrm{d}T$ and $\mathrm{d}^2G/\mathrm{d}T^2$ for a second-order phase transition.

the bcc and fcc forms of iron are two distinct phases. We associate a Gibbs potential with each phase, and for each phase it exists over all temperatures and pressures. For any given conditions of temperature, pressure, magnetic field etc., the phase having the lowest Gibbs potential is the stable one. All changes of crystal structure are first-order phase transitions.

A second-order phase transition (Fig. 11.46) is not associated with changes in crystal structure, although slight distortions may arise in certain cases. Such a phase transition occurs when an internal symmetry is altered, as when a random arrangement of magnetic moments begins to order below the Curie or Néel temperatures, or when the change from normal to superconducting state occurs in the absence of an external magnetic field, or when in an alloy, say $AB$, a random distribution of the two kinds of atom undergoes a transition so that the different atoms occupy preferred positions. The characteristic feature is that above the ordering temperature there is a single value of Gibbs potential, whereas below this temperature the potential has two branches, corresponding to the ordered and disordered states. At the ordering temperature, these two branches merge continuously into the single branch characteristic of the high-temperature state.

## References

Anderson, P. W. (1961) *Phys. Rev.* **124**, 41.

Austin, J. B. (1932) *Indust. Engng Chem.* **24**, 1225, 1388.

Coles, B. R., Hume-Rothery, W. and Myers, H. P. (1949) *Proc. R. Soc. Lond.* **A196**, 125.

Craik, D. J. and Tebble, R. S. (1965) *Ferromagnetism and Ferromagnetic Domains.* North-Holland, Amsterdam.

Crangle, J. (1977) *The Magnetic Properties of Solids.* Edward Arnold, London.

Duffin, W. J. (1965) *Electricity and Magnetism.* McGraw-Hill, New York.

Grey, H. B. (1964) *Electrons and Chemical Bonding.* Benjamin, New York.

Henry, W. E. (1952) *Phys. Rev.* **88**, 559.

Honda, K. and Kaya, S. (1926) *Sci. Rep. Tohoku Univ.* **15**, 721.

*Landolt-Börnstein*, New Series, III, **19a**. Springer-Verlag, Berlin (1986).

McWeeny, R. (1979) *Coulson's Valence*, 3rd edn. Oxford University Press, Oxford.

Néel, L. (1948) *Ann. Physique* **3**, 137.

Pauthenet, R. (1958) *J. Appl. Phys.* **29**, 253.

Pippard, A. B. (1960) *Classical Thermodynamics.* Cambridge University Press, Cambridge.

Robinson, W. K. and Friedberg, S. A. (1960) *Phys. Rev.* **117**, 402.

Trapp, C. and Stout, J. W. (1963) *Phys. Rev. Lett.* **10**, 157.

Weinert, M. and Freeman, A. J. (1983) *Phys. Rev.* **B38**, 23.

## Further Reading

FRIEDEL, J. (1958) *Nuovo Cimento* **7**, 287.

FALICOV, L. Metallic Magnetic Superlattices, in *Physics Today* (1992) **45**, 10 pp. 46.

FREEMAN, A. J. and SCHNEIDER, K. A. (Editors) *Magnetism in the Nineties* (1991) North-Holland, Elsevier Science Publishers, Amsterdam.

HOWSON, M. A. (1994) *Contemporary Phys.*, **35**, 347.

JILES, D. *Introduction to Magnetism and Magnetic Materials* (1991) Chapman and Hall, London.

MATHON, J. (1991) *Contemporary Phys.*, **32**, 143.

*Physics Today* (1995) **48**, no. 4. Special issue on magnetoelectronics.

SMITS, J. (1992) *Phys. World* **5**, no. 11, 48.

## Problems

**11.1** The research literature on magnetism is to a great extent still based on the cgs (emu) system of units in which **B** is measured in gauss and **H** in oersteds, although these two units differ only in name. To get a feel for the cgs and SI units in magnetism, calculate: (a) the magnetization per unit volume and per unit mass in both cgs and SI systems for Fe, Co and Ni at 0 K given that the atomic moments are 2.22 $\mu_B$, 1.72 $\mu_B$ and 0.61 $\mu_B$ respectively; (b) the ratio of the Curie constants per mole in the two systems for any given substance.

| | cgs | SI |
|---|---|---|
| $\mu_B$ | $0.9273 \times 10^{-20}$ erg $G^{-1}$ | $0.9273 \times 10^{-23}$ J $T^{-1}$ |
| $k_B$ | $1.381 \times 10^{-16}$ erg $K^{-1}$ | $1.381 \times 10^{-23}$ J $K^{-1}$ |
| $\mu_0$ | 1 | $4\pi \times 10^{-7}$ H $m^{-1}$ |

**11.2** The $\alpha$ phase of iron above the Curie temperature has a paramagnetic susceptibility satisfying $\kappa = C/(T - T_C)$ where $C = 2.18$ K and $T_C = 1093$ K. Estimate the Weiss molecular field in iron at 0 K.

**11.3** The magnetic behaviour of $CuSO_4 \cdot 5H_2O$ is a very good approximation to that of an ideal paramagnetic gas. Express the susceptibility as a function of temperature in SI and cgs units. Give values with respect to unit volume and to unit mass as well as the molar value. The density of this copper compound is $2.284 \times 10^3$ kg $m^{-3}$.

**11.4** The magnetic susceptibility of Cu metal is $-9.63 \times 10^{-6}$ and the electronic heat capacity coefficient is 0.7 mJ $mol^{-1}$ $K^{-2}$. If the valence electron gas in Cu has a diamagnetic effect equal to one third of the Pauli paramagnetism, estimate the diamagnetic contribution of the ion cores in Cu.

**11.5** Au has a magnetic susceptibility of $-3.45 \times 10^{-5}$. Pt on the other hand has a susceptibility of $2.79 \times 10^{-4}$; estimate $\gamma$ for Pt. You may assume that the susceptibility is essentially independent of temperature. Look up the value of $\gamma$ for Pt and compare with your calculated values. Can you think of probable reasons why the two values differ?

**11.6** The Weiss molecular magnetic fields that are used to represent the true quantum-mechanical interactions that produce magnetism are very large – much larger than the laboratory fields produced by conventional electromagnets. With this in mind: (a) derive the susceptibility of an antiferromagnetic crystal in a direction perpendicular to the axis of antiferromagnetism; (b) when the external field is parallel to the axis of antiferromagnetism, show that the susceptibility at 0 K is zero and that it increases monotonically with temperature until the Néel point is reached (it is not necessary to evaluate exact expressions; instead produce a physical argument supported by sketches to illustrate what happens); (c) if the axis of antiferromagnetism is not strongly pinned to a particular crystallographic direction, what do you think will happen if the external field is progressively increased in the direction parallel to the antiferromagnetic axis?

**11.7** Describe the paramagnetic properties of the following substances: (a) $MnSO_4 \cdot (NH_4)_2SO_4 \cdot 8H_2O$; (b) Cu; (c) Fe; (d) $Fe_3O_4$.

**11.8** Determine the ground state and the Bohr magneton number for the following ions: $Mn^{3+}$, $Fe^{3+}$, $Sm^{3+}$, $Gd^{3+}$. Why, in certain cases, do the experimental values differ from those calculated?

**11.9** Equation (11.57) provides an expression for the exchange energy associated with an 180° domain boundary of width $N$ atomic diameters. If it were not for the magnetocrystalline anisotropy the boundary would widen indefinitely. If the magnetocrystalline anisotropy constant is $K$ J m$^{-3}$ obtain an expression for the equilibrium width of such a boundary. Estimate the width of this boundary in iron for which J $= 0.2\, k_B\, T_c$ J m$^{-3}$ and $K = 5.10^4$ J m$^{-3}$.

12

# Dielectric Media

'Insulators' are dielectric media, whose general characteristics usually comprise strong ionic or directed covalent bonds, brittle mechanical behaviour at ordinary temperatures, very high resistivities and in many cases transparency to visible and infrared light. A dielectric is a substance that becomes polarized in the presence of an electric field. The physical quantities of primary interest are the field vectors **E** and **D**, the polarization **P**, together with the electric susceptibility $\kappa$ and dielectric constant $\varepsilon_r$. If **E** denotes the macroscopic electric field within the medium then

$$\mathbf{D} = \varepsilon_r \varepsilon_0 \mathbf{E} = \varepsilon_0 \mathbf{E} + \mathbf{P}, \tag{12.1}$$

$$\mathbf{P} = \kappa \varepsilon_0 \mathbf{E}, \tag{12.2}$$

$$\varepsilon_r = 1 + \kappa. \tag{12.3}$$

An external field acting on a dielectric produces induced dipole moments, and we must consider the polarizabilities of the atoms; there is therefore some justification for beginning with a treatment of the isolated atom.

## 12.1 The Free Atom

In an isolated atom of Na or Xe, say, the electronic charge is distributed in a spherically symmetrical manner around the nucleus, but in the presence of an external field $\mathbf{E}_0$ the distribution is altered, leading to an induced dipole moment **p**; we say the atom becomes polarized (Fig. 12.1), and write

$$\mathbf{p} = \alpha \varepsilon_0 \mathbf{E}_{loc}, \tag{12.4}$$

where $\alpha$ is the atomic polarizability (assumed independent of the electric field) and $\mathbf{E}_{loc}$ the electric field acting at the site of the atom. We call this the local field. Since a dipole cannot act upon itself and because the atom is isolated, the local field is given in this case by the external field $\mathbf{E}_0$. The dipole moment arises as a result of separation of the centres of gravity of the electronic and nuclear charges, a separation that is determined by balancing the force exerted by the field against that

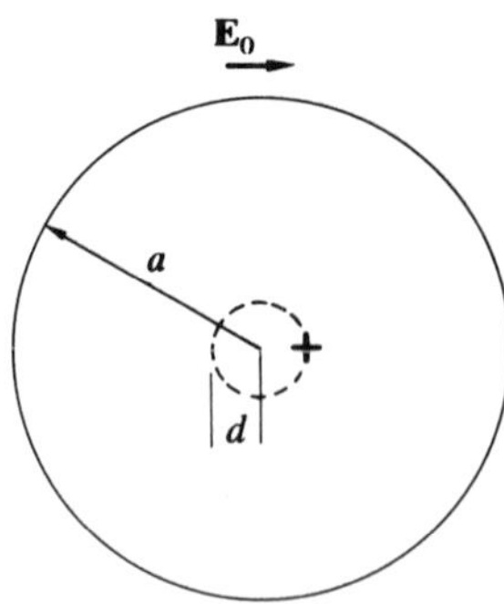


**Figure 12.1** The basis for an elementary calculation of the polarizability of the free atom. In the presence of the field $\mathbf{E}_0$ the electron charge cloud and the nucleus undergo a slight relative displacement.

between the charges themselves. Suppose the charge systems are displaced an amount **d**. Then, if the atom has radius $a$ and atomic number $Z$, and if we assume it to have uniform electron charge density, we find

$$E_0 Ze = \frac{d^3}{a^3} \frac{(Ze)^2}{4\pi\varepsilon_0 d^2}, \tag{12.5}$$

since Gauss' theorem implies that only the electronic charge within a sphere of radius $d$ produces a resultant force on the nucleus. Clearly,

$$d = \frac{4\pi\varepsilon_0 a^3}{Ze} E_0 \tag{12.6}$$

and

$$\mathbf{p} = 4\pi\varepsilon_0 a^3 \mathbf{E}_0, \tag{12.7}$$

giving

$$\alpha = 4\pi a^3. \tag{12.8}$$

Thus for a free atom $\alpha$ is determined by the atomic volume. We can now appreciate why the binding energies of the condensed noble gases increase with atomic number (Section 1.2). The free atom is characterized by an electronic polarizability, and this is also true for any molecule with a symmetrical distribution of charge. On the other hand, molecules formed of different atoms may present a large degree of ionicity when the centres of gravity of the nuclear and electronic charges do not coincide. This is the case for HCl and $H_2O$ (Fig. 12.2). Such molecules possess strong permanent dipole moments enormously larger than the induced dipole moment that can be produced by an ordinary electric field strength of order $10^6$ V m$^{-1}$.

Any discussion of microscopic dielectric behaviour is complicated by the concepts of macroscopic and local electric field strengths. When we consider condensed phases such as liquid and solid dielectrics (and even ordinary gases) we consider each atom or molecule to be polarizable, but we can no longer assume that the field acting on an atom is the external field that exists in the absence of the specimen. We must also consider the contributions to the field provided by all the other atomic dipoles, permanent or induced, that may act on any particular atom. Furthermore,

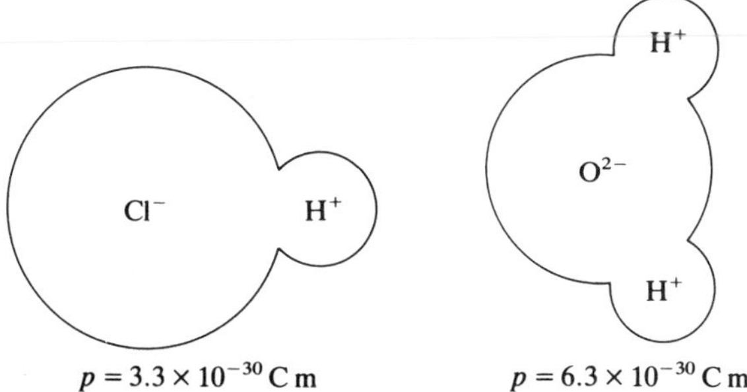


**Figure 12.2** Permanent electrical dipoles arise in molecules with non-centrosymmetric charge distributions.

we cannot assume that the macroscopic electric field within the dielectric (which takes account of the macroscopic polarization) is the same as the local field acting on a particular atom. The macroscopic field is a volume average over a region large compared with atomic dimensions, whereas the local field at an atom must take account of the discrete atomic structure and the associated rapidly varying atomic fields around a particular atom site.

## 12.2 The Local Electric Field

We begin by determining the macroscopic field within a dielectric that has prolate ellipsoidal form, since this leads to a uniform polarization of the medium. The polarization **P** produces a surface distribution of charge whose polarity gives rise to a field within the dielectric that is directed oppositely to the applied field $\mathbf{E}_0$ (Fig. 12.3). This is the depolarizing field analogous to the demagnetizing field already encountered in Section 11.9.1.

We write the macroscopic field within the dielectric as

$$\mathbf{E} = \mathbf{E}_0 - \frac{N'\mathbf{P}}{\varepsilon_0}, \tag{12.9}$$

$N'$ being the geometrical depolarizing factor, some important values of which are given in Fig. 12.4. The polarization is determined by

$$\mathbf{P} = \kappa\varepsilon_0 \mathbf{E}. \tag{12.10}$$

However, we should like to relate the macroscopic polarization **P** to the polarizabilities of the individual atoms or ions, and this requires that we can define the local

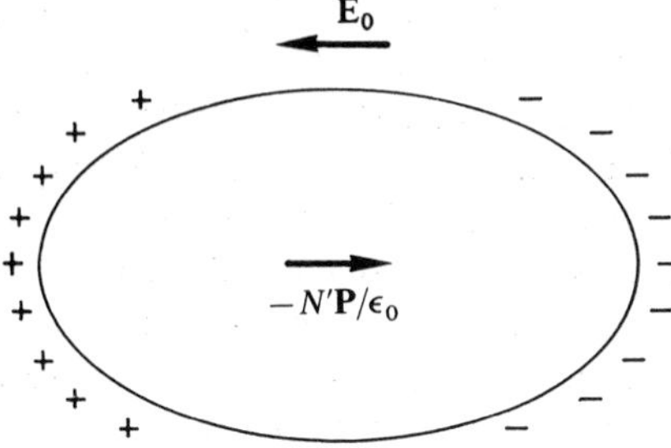


**Figure 12.3** The depolarizing field arises from the surface charge on a polarized substance.

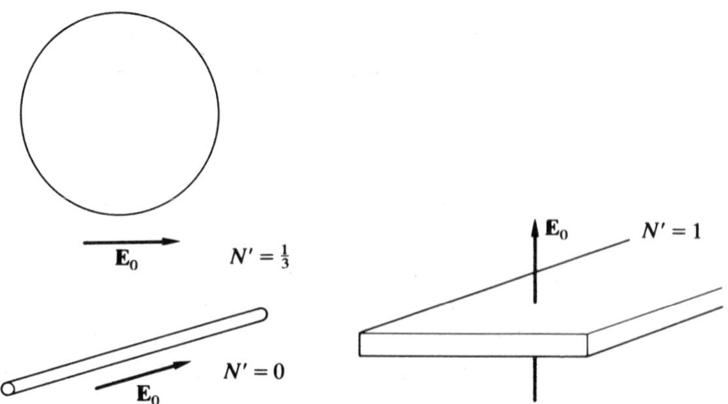


**Figure 12.4** Depolarizing factors for three important geometrical shapes.

field at any atom site. To do this, we first remove the atom in question at site $A$, say, so that we do not include its own dipole field, and then we ask whether we can define a near- and a far-field region. The far-field region is sufficiently remote that its effect is just equivalent to the macroscopic average field that exists there, whereas the near-field region is that volume surrounding the chosen atom site where we must take account of the individual dipole moments. We write

$$\mathbf{E}_{\text{loc}} = \mathbf{E}_{\text{far}} + \mathbf{E}_{\text{near}}. \tag{12.11}$$

To proceed, we assume the near-field region to extend over a radius $R$, and a sphere of material comprising the near-field region is imagined to be removed (Fig. 12.5). This produces a spherical cavity around our site $A$, and the field within this cavity is calculated to be (see Problem 12.1)

$$\mathbf{E}_{\text{cavity}} = \mathbf{E}_{\text{far}} + \frac{\mathbf{P}}{3\varepsilon_0} = \mathbf{E} + \frac{\mathbf{P}}{3\varepsilon_0}. \tag{12.12}$$

The field produced by the separate atomic dipoles (assumed frozen in the same positions they originally had before the sphere was removed) is dependent on their detailed crystallographic arrangement. *If the atomic dipoles are all oriented in the*

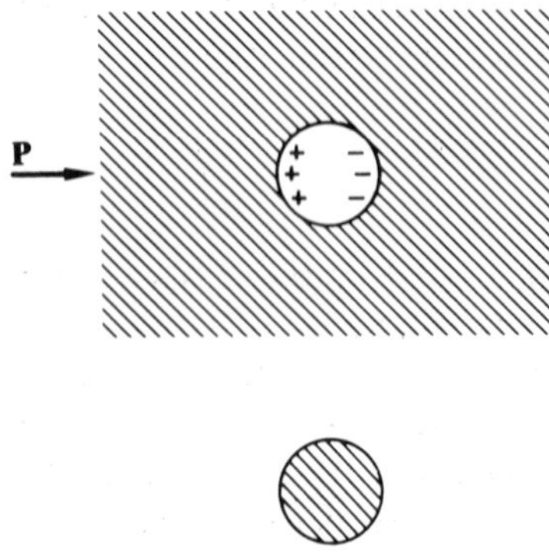


**Figure 12.5** The Lorentz cavity approximation for the calculation of the local electric field.

*same direction as* **P**, *and if the crystal has cubic symmetry, then this near field is zero.* The local field is therefore the same as the cavity field:

$$\mathbf{E}_{\mathrm{loc}} = \mathbf{E} + \frac{\mathbf{P}}{3\varepsilon_0}. \tag{12.13}$$

This approximation to the local field was first introduced by Lorentz and is usually known by his name.

A perusal of equations (12.9) and (12.12) shows that the Lorentz approximation to the local field in a cubic substance depends on the difference between the shape of the specimen (usually cylindrical or rectangular) and that of the cavity. The specimen shape determines the far field through the depolarization factor $N'$, whereas the cavity polarization field is determined by the spherical shape factor 1/3.

## 12.3 Clausius-Mossotti Formula

The polarization of a medium containing $N$ similar atoms per unit volume may be written

$$\mathbf{P} = N\alpha\varepsilon_0\,\mathbf{E}_{\mathrm{loc}},$$

which, on using (12.13), gives

$$\mathbf{P} = N\alpha\varepsilon_0\left(\mathbf{E} + \frac{\mathbf{P}}{3\varepsilon_0}\right),$$

leading to

$$\kappa = \frac{N\alpha}{1 - \frac{1}{3}N\alpha}, \tag{12.14}$$

and, by (12.3), this may be rewritten as

$$N\alpha = 3\,\frac{\varepsilon_{\mathrm{r}} - 1}{\varepsilon_{\mathrm{r}} + 2}. \tag{12.15}$$

If our dielectric medium is composite, containing different kinds of atom with concentrations $N_i$ and polarizabilities $\alpha_i$, and if these different kinds of atom act independently of one another, then

$$\sum_i N_i\alpha_i = 3\,\frac{\varepsilon_{\mathrm{r}} - 1}{\varepsilon_{\mathrm{r}} + 2}. \tag{12.16}$$

This is known as the Clausius–Mossotti formula relating the macroscopic quantity $\varepsilon_{\mathrm{r}}$ with the atomic quantities $\alpha_i$. So far we have implicitly assumed that the external field is a static field, but there is no reason why we should not consider time-dependent fields and in particular those associated with electromagnetic radiation. In the latter instance, we know that Maxwell's electromagnetic theory of light allows us to write the refractive index as

$$\mathcal{N} = \varepsilon_{\mathrm{r}}^{1/2}, \tag{12.17}$$

so that (12.16) may be written

$$\sum_i N_i \alpha_i = 3\frac{\mathcal{N}^2 - 1}{\mathcal{N}^2 + 2}, \tag{12.18}$$

which is known as the Lorenz–Lorentz formula.

If our dielectric medium possesses only electronic polarizability then we say it is non-polar and we do not expect the dielectric constant to change with frequency because electrons in atoms readily respond to optical frequencies. At X-ray frequencies, however, the dielectric constant is essentially unity (it is in fact slightly less than unity, which may be understood from (6.46) if we calculate a plasma frequency for the complete atom assuming $n = Ze/\Omega$) and there is no agreement between the static and X-ray dielectric constants. We shall also see that substances possessing permanent molecular dipole moments have very different low- and high-frequency behaviour, so that (12.16) and (12.18) cannot be expected to give the same atomic polarizabilities. This is readily evident for $H_2O$, which has a static dielectric constant of 81 but a refractive index of 1.33.

On the other hand, experiments with homopolar substances like the noble gases, in both gaseous and condensed forms, very pure Ge etc. show very good agreement with (12.16) and (12.18); in the case of the noble gases the same polarizabilities are found for the gaseous as for the condensed phases, demonstrating the applicability of the Lorentz field.

It is convenient to introduce the notation $\varepsilon_r(0)$ for the static dielectric constant and $\varepsilon_r(\infty)$ for the dielectric constant at optical frequencies (say the Na *D* lines). Note that $\varepsilon_r(\infty)$ does not signify the value of $\varepsilon_r$ at infinite frequency, which, as we have already indicated, is unity. We may rewrite (12.18) as

$$\sum_i N_i \alpha_i = 3\frac{\varepsilon_r(\infty) - 1}{\varepsilon_r(\infty) + 2}. \tag{12.19}$$

There are twenty alkali halides that can be formed from the five alkali metals and the four halogens, and it is possible to associate fixed polarizabilities with each ion and satisfy (12.19) to within 1% for all of them except the fluorides, which deviate somewhat more markedly. Thus the Lorentz approximation is quite satisfactory for these cubic materials (see Problem 12.3).

## 12.4 Frequency Dependence of ε

### 12.4.1 *Electronic polarizability*

We have already mentioned that we expect the electrons in atoms to be able to respond to rapid changes in the electric field, and this is borne out in experiment by the simultaneous satisfaction of (12.16) and (12.18) with the same atomic polarizabilities. There must, however, be some limiting frequency because eventually the applied field induces quantum transitions between occupied and unoccupied states of the atom, with a significant effect on the polarizability. Consider the classical picture of an electron bound to a centre of force and governed by an equation of simple harmonic motion

$$m\ddot{\mathbf{x}} + m\omega_i^2 \mathbf{x} = 0. \tag{12.20}$$

The force constant is $m\omega_i^2$, where $\omega_i$ represents the natural frequency of the oscillator. If a field $\mathbf{E} = \mathbf{E}_0 \, e^{i\omega t}$ is applied then we expect the electron to follow the field and be displaced according to

$$\mathbf{x} = \mathbf{x}_0 \, e^{i\omega t}.$$

The equation of motion becomes

$$m\ddot{\mathbf{x}} + m\omega_i^2 \mathbf{x} = -e\mathbf{E},$$

and is readily solved to give

$$\mathbf{x} = \frac{-e\mathbf{E}}{m(\omega_i^2 - \omega^2)},$$

$$\mathbf{p} = -e\mathbf{x} = \frac{e^2\mathbf{E}}{m(\omega_i^2 - \omega^2)},$$

leading to a polarizability

$$\alpha(\omega) = \frac{e^2}{\varepsilon_0 m(\omega_i^2 - \omega^2)}. \tag{12.21}$$

In a proper quantum-mechanical treatment of the problem (see e.g. Ziman 1972, Chap. 8) the form of (12.21) is almost the same:

$$\alpha(\omega) = \frac{e^2}{\varepsilon_0 m} \sum_i \frac{f_i}{\omega_i^2 - \omega^2}, \tag{12.22}$$

where $\omega_i$ now corresponds to any possible electronic transition and $f_i$, which is called an oscillator strength, is a measure of the probability of the specific transition associated with an energy $\hbar\omega_i$. We know from experience that many dielectric substances are highly transparent to ordinary visible light, so $\omega_i$ must lie in the ultraviolet region of the spectrum. If $\omega \ll \omega_i$, we expect to find little or no dependence of $\alpha$ on $\omega$. Furthermore, if there is no other source of polarization then (12.16) and (12.18) should hold simultaneously. We also see that the size of $\alpha$, and thereby $\varepsilon_r$, is governed by the $\omega_i$ and most strongly by the smallest. We illustrate this aspect by studying the values of $\varepsilon_r$ for the pure group **IV** elements (Table 12.1). The first electronic transition that is allowable in the pure elements is across the energy gap between the valence and conduction bands; we note that the larger this gap, the

**Table 12.1** Values of $\varepsilon_r(0)$ for the Group IV elements

| | $\varepsilon_r(0)$ | $E_g$(eV) |
|---|---|---|
| C | 5.7 | 5.4 |
| Si | 12.0 | 1.17 |
| Ge | 16.0 | 0.7 |
| $\alpha$-Sn | 23.8 | $\sim$0 |

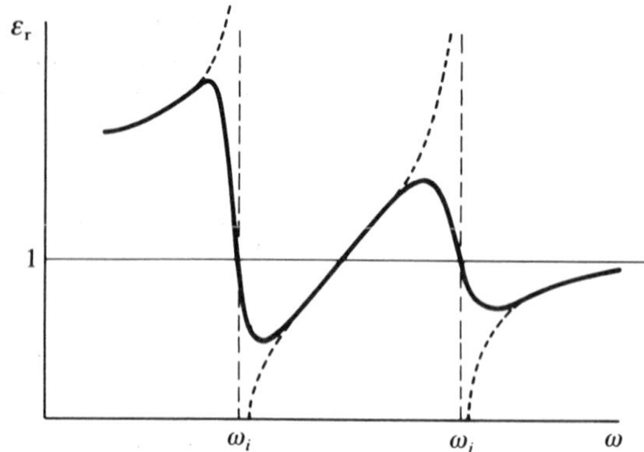


**Figure 12.6** Anomalous dispersion at resonant frequencies; in normal dispersion $\varepsilon_r$ increases with frequency.

lower the value of $\varepsilon_r$. Note that for Si and Ge $\omega_i$ lies in the near infrared, so these substances are not transparent in the optical range. $\alpha$-Sn is exceptional, but diamond is a typical transparent dielectric medium possessing only electronic polarizability. When the frequency of the applied field is equal to $\omega_i$ a resonance occurs leading to singular behaviour of $\alpha$. In practice, however, there is always some damping that causes the polarizability to remain finite, but it nevertheless undergoes rapid changes near $\omega_i$; this produces a variation in the refractive index known as anomalous dispersion. In 'normal' dispersion $\varepsilon_r$ increases with $\omega$. Since this phenomenon is a result of the quantized energy level structure, it arises in free atoms (i.e. gases) just as readily as in solids. We conclude that non-polar substances possessing only electronic polarizability display a simple variation of dielectric constant with frequency (Fig. 12.6).

### 12.4.2 Ionic polarizability

Although diamond and NaCl are both highly transparent to visible light, there is a significant difference in their dielectric behaviour. Diamond is non-polar whereas NaCl and similar polar substances contain ions with positive and negative charges. The application of an electric field to a polar substance causes a relative displacement of the ions leading to a lattice polarization described by an ionic polarizability. In the following we shall, for the moment, neglect the occurrence of any electronic polarizability and concentrate on the influence of the ions. Again we expect a varying electric field to produce an oscillatory motion of the ions, and, to the extent that they can follow the variations in the field, the polarizability will be independent of frequency. Ions are heavy particles and more sluggish in their motion than electrons. As the frequency of the applied field increases, there comes a time when the field changes so rapidly that the ions can no longer follow its variation; they experience the average local field as zero and there will then be no ionic contribution to the total polarizability. This state of affairs is reached in the conventional optical frequency range, a fact that we can appreciate by comparing $\varepsilon_r(0)$ and $\varepsilon_r(\infty)$ for a typical group of alkali halides such as the chlorides (Table 12.2). But then might we not expect some kind of resonance between the applied field and the ionic motion at

**Table 12.2** Values of $\varepsilon$ for the alkali metal chlorides

| | $\varepsilon_r(0)$ | $\varepsilon_r(\infty)$ |
|---|---|---|
| LiCl | 11.95 | 2.78 |
| NaCl | 5.90 | 2.34 |
| KCl | 4.84 | 2.19 |
| RbCl | 4.92 | 2.19 |
| CsCl | 7.20 | 2.62 |

a lower frequency? This is in fact the case; the ionic salt shows a resonant behaviour in the infrared region of the spectrum that does not arise in diamond.

### *Longitudinal (LO) and transverse (TO) optical modes*

In our discussion of lattice vibrations we found (Section 5.7) that we could define acoustic and optical modes, the latter being a contra-motion of different atoms (different either in sort or lattice site). In an alkali halide, the atomic positions are equivalent, but the atom types have different mass, and more importantly they are differently charged. Clearly, the LO and TO modes are associated with oscillatory polarizations, but in a cubic crystal LO modes have **E** and **P** parallel to **q** and TO modes have **E** and **P** perpendicular to **q**. Here **q** is of course the *phonon* wave vector.

*Furthermore, since light is a transverse motion, the* **E** *vector, being perpendicular to the wave vector, can couple only to the TO modes.* Nevertheless, the natural frequency of the LO mode near $\mathbf{q} = 0$ is a significant quantity, as we shall see shortly.

Our interest now lies in the interaction of light with the optical phonons. Since light has an insignificantly small wave vector, the phonons in question are also those with $\mathbf{q} \approx 0$, that is, they have very long wavelength. This means that we may assume that at any instant of time all ions of the same sign experience the same displacement from their equilibrium positions. We also assume that the displaced ions experience a restoring force proportional to the relative displacement of nearest-neighbour ions. We consider a sample of volume $V$ containing $N$ ion pairs with ionic masses $m_1$ and $m_2$ and charges $\pm n_0 e$, $n_0$ being the valence. At any given instant all the positive ions have displacement **u** and all the negative ions a displacement **v** from their equilibrium positions. These displacements clearly lead to a polarization

$$\mathbf{P} = \frac{\sum q\mathbf{r}}{V} = \frac{N n_0 e}{V}(\mathbf{u} - \mathbf{v}). \tag{12.23}$$

Suppose now that at each ion there exists a local electric field $\mathbf{E}_{\text{loc}}$; then the ions have the following equations of motion:

$$m_1 \frac{\mathrm{d}^2\mathbf{u}}{\mathrm{d}t^2} = n_0 e\mathbf{E}_{\text{loc}} - C(\mathbf{u} - \mathbf{v}), \tag{12.24}$$

$$m_2 \frac{\mathrm{d}^2\mathbf{v}}{\mathrm{d}t^2} = -n_0 e\mathbf{E}_{\text{loc}} - C(\mathbf{v} - \mathbf{u}), \tag{12.25}$$

$C$ being the force constant. If we subtract (12.25) from (12.24) and use (12.23) then it is a straightforward matter to obtain the following equation:

$$\frac{\mathrm{d}^2\mathbf{P}}{\mathrm{d}t^2} = \frac{Nn_0^2 e^2}{MV}\mathbf{E}_{\mathrm{loc}} - \frac{C\mathbf{P}}{M}, \tag{12.26}$$

where $M = m_1 m_2/(m_1 + m_2)$ is the reduced mass. Furthermore, we may write

$$\Omega_{\mathrm{p}}^2 = \frac{Nn_0^2 e^2}{MV\varepsilon_0} \tag{12.27}$$

and associate $\Omega_{\mathrm{p}}$ with a free ion plasma frequency. Thus (12.26) becomes

$$\frac{\mathrm{d}^2\mathbf{P}}{\mathrm{d}t^2} = \Omega_{\mathrm{p}}^2\,\varepsilon_0\,\mathbf{E}_{\mathrm{loc}} - \frac{C}{M}\mathbf{P}. \tag{12.28}$$

Both $\mathbf{E}_{\mathrm{loc}}$ and $\mathbf{P}$ are functions of the macroscopic field $\mathbf{E}$ within the medium, and if the latter has a time dependence $\mathrm{e}^{\mathrm{i}\omega t}$ then we assume that $\mathbf{E}_{\mathrm{loc}}$ and $\mathbf{P}$ have similar time dependences, allowing us to rewrite (12.28) as

$$-\omega^2\mathbf{P} = \Omega_{\mathrm{p}}^2\,\varepsilon_0\,\mathbf{E}_{\mathrm{loc}} - \frac{C}{M}\mathbf{P}. \tag{12.29}$$

We shall use (12.29) to illustrate properties of the LO and TO phonon modes in cubic polar crystals, in which $\mathbf{E}$ and $\mathbf{P}$ are always parallel vectors. We shall find that different behaviours arise because $\mathbf{E}_{\mathrm{loc}}$ takes different values in the two cases.

### The LO mode

Let us assume that our dielectric medium fills the space between the plates of a simple parallel-plate condenser. It is clear that if we apply a potential difference that, in the absence of the dielectric, causes a uniform field $\mathbf{E}_0$ then, when the medium is present, it will become uniformly polarized. The macroscopic field $\mathbf{E}$ depends upon the depolarization factor, which in this case is unity, so

$$\mathbf{E} = \mathbf{E}_0 - \mathbf{P}/\varepsilon_0. \tag{12.30}$$

If we now imagine that $\mathbf{E}_0$ fluctuates in time and space, causing similar fluctuations in $\mathbf{P}$, then it is clear that any nodal planes of $\mathbf{P}$ are perpendicular to the vectors $\mathbf{E}$, $\mathbf{P}$ and $\mathbf{q}$. We obtain longitudinal oscillations of $\mathbf{P}$ (Fig. 12.7). The alkali halides have cubic symmetry, so if the local field is given by the Lorentz approximation then

$$\mathbf{E}_{\mathrm{loc}} = \mathbf{E} + \frac{\mathbf{P}}{3\varepsilon_0} = \mathbf{E}_0 - \frac{2\mathbf{P}}{3\varepsilon_0}. \tag{12.31}$$

We cannot achieve such longitudinal $\mathbf{E}_0$ and $\mathbf{P}$ fields using light, but nevertheless the LO modes are natural vibrations of the crystal, so that, even in the absence of the external field $\mathbf{E}_0$, there is always a local field given by

$$\mathbf{E}_{\mathrm{loc}} = -\frac{2\mathbf{P}}{3\varepsilon_0} \tag{12.32}$$

whenever such a mode propagates.

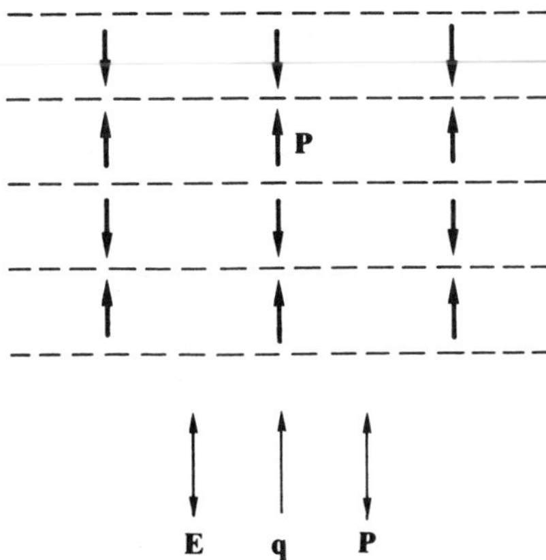


**Figure 12.7** The disposition of the vectors **E**, **P** and **q** in longitudinal optical modes.

Suppose that a long-wavelength LO mode is excited; then, from (12.29), the frequency $\omega$ is determined by

$$-\omega^2\mathbf{P} = \Omega_p^2\,\varepsilon_0\left(-\frac{2\mathbf{P}}{3\varepsilon_0}\right) - \frac{C\mathbf{P}}{M}. \tag{12.33}$$

In other words,

$$\omega^2 = \frac{C}{M} + \tfrac{2}{3}\Omega_p^2. \tag{12.33}$$

Since this is the natural frequency of the LO mode with $|\mathbf{q}| \approx 0$, we denote it by $\omega_L$:

$$\omega_L^2 = \frac{C}{M} + \tfrac{2}{3}\Omega_p^2. \tag{12.34}$$

### *The TO mode*

Suppose now that the dielectric medium is a cube with sides of length 1 cm. We shine light at normal incidence through a face of the cube and the transverse electrical field of the light wave becomes a driving force for the transverse optical lattice modes of vibration. Even a wavelength of order 1 $\mu$m corresponds to a very small value of wave vector on the scale of the Brillouin zone and may be considered appropriate to the situation $|\mathbf{q}| \approx 0$. Consider the pattern of polarization that the light field produces at a particular instant of time (Fig. 12.8). We find sheets of material with alternate polarizations aligned perpendicularly to the **q** vector of the light. These sheets are thin with thickness of order $\lambda$ and the driving field is parallel to the polarization sheets, which is consistent with a depolarizing factor of zero. If the field of the light wave in the medium is **E** and the polarization **P** then the Lorentz cavity approximation gives

$$\mathbf{E}_{loc} = \mathbf{E} + \frac{\mathbf{P}}{3\varepsilon_0}. \tag{12.35}$$

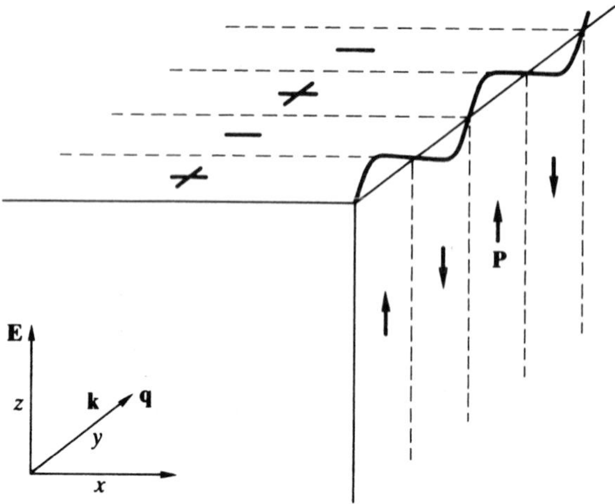


**Figure 12.8** The disposition of the vectors **E**, **P** and **q** in transverse optical modes.

If this field causes excitation of a TO vibrational mode then by (12.29) we must have

$$-\omega^2\mathbf{P} = \Omega_p^2\,\varepsilon_0\left(\mathbf{E} + \frac{\mathbf{P}}{3\varepsilon_0}\right) - \frac{C\mathbf{P}}{M}. \tag{12.36}$$

Just as we argued earlier, we may say that even in the absence of an external driving field, i.e. $\mathbf{E} = 0$, a natural TO mode with $|\mathbf{q}| \approx 0$ has a frequency $\omega_T$ determined by the above equation. Thus we find

$$\omega_T^2 = \frac{C}{M} - \tfrac{1}{3}\Omega_p^2. \tag{12.37}$$

$\omega_T^2$ is less than $\omega_L^2$ by $\Omega_p^2$.

Returning to (12.36), this equation may be rearranged to give the electric susceptibility

$$\kappa = \frac{\mathbf{P}}{\varepsilon_0\,\mathbf{E}} = \frac{\Omega_p^2}{C/M - \frac{1}{3}\Omega_p^2 - \omega^2}, \tag{12.38}$$

and it is clearly the case that

$$\varepsilon_r = \begin{cases} 1 + \kappa \to \infty & \text{when} \quad \omega = \omega_T, & (12.39) \\ 1 + \kappa = 0 & \text{when} \quad \omega = \omega_L. & (12.40) \end{cases}$$

In order to emphasize that the present discussion concerns transverse electromagnetic waves, we rephrase the above equations in terms of the refractive index:

$$\mathcal{N}^2 \to \infty \quad \text{when} \quad \omega = \omega_T, \tag{12.39a}$$

$$\mathcal{N}^2 = 0 \quad \text{when} \quad \omega = \omega_L. \tag{12.40a}$$

We say that there is a resonance when $\omega = \omega_T$, a frequency that we associate with the natural TO mode at $|\mathbf{q}| = 0$. $\varepsilon_r(\omega)$ is plotted in Fig. 12.9 as a function of fre-

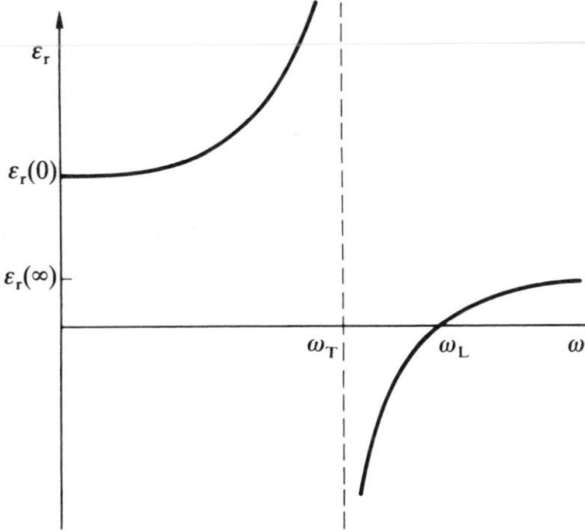


**Figure 12.9** In the interval $\omega_T < \omega < \omega_L$, $\varepsilon_r < 0$.

quency, and we see immediately that

$$\varepsilon_r < 0 \text{ in the interval } \omega_T < \omega < \omega_L. \tag{12.41}$$

This means that the refractive index is wholly imaginary, leading to perfect reflectivity, (6.47). Outside this frequency interval, the medium has the normal low reflectivity typical of a dielectric. In practice, damping rounds off the singular behaviour but the reflectivity remains high. Experiment confirms this pattern of behaviour (Fig. 12.10). If a continuous spectrum of light is subjected to repeated reflection from an ionic crystal then the light eventually becomes a narrow band of frequencies

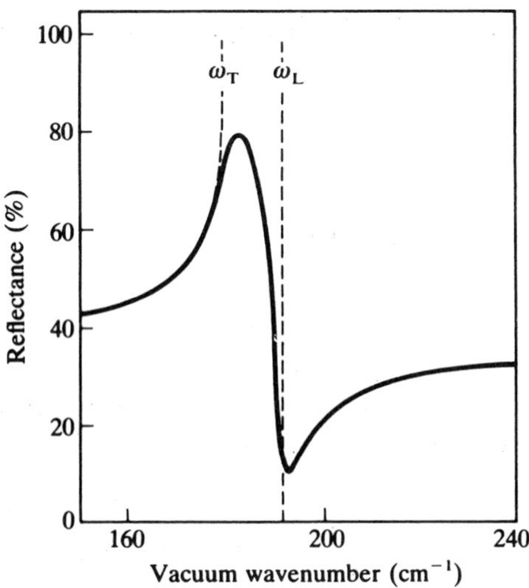


**Figure 12.10** An ionic crystal has a large reflectivity in the frequency range $\omega_T < \omega < \omega_L$. (Data for InSb from R. B. Saunderson, *J. Phys. Chem. Solids*, **26**, 1965. Pergamon Press, Oxford.)

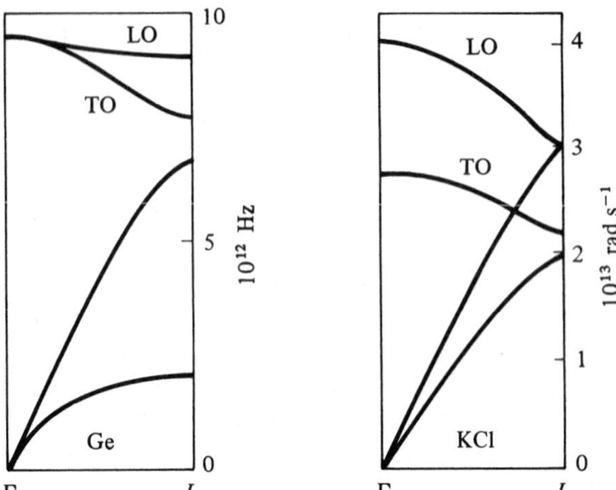


**Figure 12.11** The LO and TO modes in a non-polar crystal (Ge) and in a polar crystal (KCl). (Ge data after Nelin 1974; KCl data after Raunio and Rolandsson 1970.)

bounded by $\omega_T$ and $\omega_L$, and the reflected light is known as 'residual rays'. Both Ge and KCl have optical phonon branches, but the charged ions in the latter substance produce different local electric fields in the TO and LO vibrations, leading to different mechanical behaviour; the TO and LO modes are therefore different in KCl and similar polar compounds, whereas in the case of Ge there is no ionic polarizability so the TO and LO modes are degenerate at $|\mathbf{q}| = 0$ (Fig. 12.11).

### 12.4.3 *Longitudinal and transverse dielectric constants*

The dielectric constant enables us to describe the properties of polarizable media, and in principle we might expect the electric susceptibility to depend on both temporal ($\omega$) and spatial ($\mathbf{k}$) variations of the applied field; we ought therefore to write the dielectric constant as $\varepsilon(\omega, \mathbf{k})$.† In the detailed theories of solids, including metals, $\varepsilon(\omega, \mathbf{k})$ is one of the most significant quantities governing behaviour; this is not surprising since the properties of matter are controlled by electrical forces. However, as we have already indicated, we must distinguish between longitudinal and transverse fields; $\varepsilon_L(\omega, \mathbf{k})$ and $\varepsilon_T(\omega, \mathbf{k})$ are different quantities.

When discussing charge density variations in plasmas we need $\varepsilon_L(\omega, \mathbf{k})$, but the transverse optical fields interact with matter via $\varepsilon_T(\omega, \mathbf{k})$. The complete expressions for these quantities are very complicated, but in cubic materials and for $|\mathbf{k}| = 0$ they have simpler and identical form – however, we should not forget their basic difference.

When we have charge density fluctuations in a plasma, or excite an LO mode in a polar crystal, the overall system remains electrostatically neutral and the field

† In this section we omit the subscript 'r' on $\varepsilon_r$.

arises solely from the displacement of charges which, although of opposite sign, occur in equal numbers; there is no free charge. We may write

$$\nabla \cdot \mathbf{D} = \varepsilon_L \varepsilon_0 \nabla \cdot \mathbf{E} = 0. \tag{12.42}$$

On the other hand, as Fig. 12.7 shows, the nodal planes of polarization are perpendicular to both **k** and the polarization; therefore in a cubic material, where **E** and **P** are always parallel, $\nabla \cdot \mathbf{E}$ is certainly not zero and (12.42) can only hold if $\varepsilon_L = 0$, which thus becomes the condition for plasma oscillations.

In a transverse optical field, or a natural TO mode, **E** and **P** in a cubic material are again parallel to one another and to their nodal planes, which are again perpendicular to **k** (Fig. 12.8). The dielectric may be considered as being divided into infinitesimal sheets perpendicular to **k**, and in each sheet both $E_x$ and $P_x$ are constant, but the $y$ and $z$ components are zero. Clearly in this case both $\nabla \cdot \mathbf{E}$ and $\nabla \cdot \mathbf{P}$ are zero: (12.42) therefore holds.

Now Faraday's law in Maxwell's version says that

$$\nabla \times \mathbf{E} = -\frac{\mathrm{d}\mathbf{B}}{\mathrm{d}t}. \tag{12.43}$$

Although our discussion is restricted to vibrations of very long wavelength they are of high frequency and it is by no means evident that $\mathrm{d}\mathbf{B}/\mathrm{d}t$ is a negligible quantity equatable to zero. On the other hand, we have assumed that **P** faithfully follows the variations in **E** without any phase lag, which in fact means we have a 'quasi-electrostatic' situation. For truly static fields, the right-hand side of (12.43) is identically zero and we shall assume this to be so for our TO phonon. **E** and **P** have the forms $\mathbf{E}_0 \mathrm{e}^{\mathrm{i}(kz-\omega t)}$ and $\mathbf{P}_0 \mathrm{e}^{\mathrm{i}(kz-\omega t)}$; evaluation of $\nabla \cdot \mathbf{E}$ and $\nabla \times \mathbf{E}$ shows that, for a transverse wave, both can only be zero if $\mathbf{E} = 0$. But the polarization **P** is definitely finite, and if

$$\mathbf{P} = \kappa_T \varepsilon_0 \mathbf{E}$$

then this can only happen if $\kappa_T \to \infty$ and therefore $\varepsilon_T \to \infty$ also. Thus the natural transverse optical vibrations require that $\varepsilon_T \to \infty$. We can summarize our findings as follows:

$$\text{natural oscillations}\begin{cases}\text{longitudinal behaviour} & \varepsilon_L = 0,\\ \text{transverse behaviour} & \varepsilon_T \to \infty.\end{cases}$$

In quantum mechanics, natural oscillations become eigenstates. We may therefore appreciate the significance of the dielectric function in solid state physics because its zeros and poles define the eigenstates of a system.

Although optical behaviour is controlled by $\varepsilon_T$, the fact that in cubic structures, and for $\mathbf{k} = 0$, $\varepsilon_T$ and $\varepsilon_L$ have similar forms shows that if we determine the frequency at which $\varepsilon_T = 0$ by optical means then this frequency is also that for which $\varepsilon_L = 0$, and we can associate it with the natural frequency of plasma oscillations for $|\mathbf{k}| = 0$; but there is no interaction between the light and the plasma system. In our optical formula, $\omega_P$ (electrons) or $\omega_L$ (ions) are numbers that happen to be the same as the frequencies where $\varepsilon_L = 0$. In the case of longitudinal phonons, we see that the plasma frequency for $\mathbf{k} = 0$ is $\omega_L$ and not $\Omega_P$.

### 12.4.4 Lyddane–Sachs–Teller relation

If we assume that the electronic and ionic contributions to the total polarizability are additive then we may write

$$[\varepsilon_r(\omega) - 1] = [\varepsilon_r(\infty) - 1] + \frac{\Omega_p^2}{C/M - \frac{1}{3}\Omega_p^2 - \omega^2}, \tag{12.44}$$

which can be rearranged to read

$$\varepsilon_r(\omega) = \varepsilon_r(\infty) + \frac{\Omega_p^2}{\omega_T^2 - \omega^2}. \tag{12.45}$$

The latter equation may be used to obtain expressions for $\varepsilon_r(0)$ and $\varepsilon_r(\omega_L)$:

$$\varepsilon_r(0) = \varepsilon_r(\infty) + \frac{\Omega_p^2}{\omega_T^2}, \tag{12.46}$$

$$\varepsilon_r(\omega_L) = \varepsilon_r(\infty) + \frac{\Omega_p^2}{\omega_T^2 - \omega_L^2}. \tag{12.47}$$

But, as we have already seen, $\varepsilon_r(\omega_L) = 0$, so (12.46) and (12.47) may be combined to give

$$\frac{\varepsilon_r(0)}{\varepsilon_r(\infty)} = \frac{\omega_L^2}{\omega_T^2}. \tag{12.48}$$

This is known as the Lyddane–Sachs–Teller (LST) relation between the macroscopic quantities $\varepsilon_r(0)$ and $\varepsilon_r(\infty)$ on the one hand and the very-long-wavelength optical modes of the crystal on the other. The study of phonons by neutron scattering has enabled the LST relation to be tested. Although it is not possible to study the optical phonons with exactly zero wave vector by the neutron method, it has been found that for the alkali halides (provided the phonon wavelengths are small compared with the dimensions of the crystal but much larger than the ionic separation) the extrapolated data satisfy the LST relation. Some results are given in Table 12.3.

It is to be noted that although the overall agreement is good, the data for RbCl at 80 K deviate from the LST relation; the exact cause has not been established, but may well be due to instabilities associated with a change in structure from the conventional NaCl to CsCl form. One can also compare $\omega_T$ as measured in neutron

**Table 12.3** The Lyddane–Sachs–Teller relation for NaCl, KCl and RbCl

| | $T$(K) | $\varepsilon_r(0)$ | $\varepsilon_r(\infty)$ | $[\varepsilon_r(0)/\varepsilon_r(\infty)]^{1/2}$ | $\omega_L/\omega_T$ (neutron) |
|---|---|---|---|---|---|
| NaCl | 80 | 5.57 | 2.34 | 1.543 | 1.532 ± 0.03 |
| KCl | 80 | 4.59 | 2.19 | 1.448 | 1.448 ± 0.02 |
| | 300 | 4.84 | 2.17 | 1.493 | 1.487 ± 0.04 |
| RbCl | 80 | 4.67 | 2.19 | 1.461 | 1.389 ± 0.04 |
| | 300 | 4.89 | 2.18 | 1.498 | 1.49 ± 0.04 |

Courtesy G. Raunio and S. Rolandsson.

scattering experiments with values established by observation of residual rays, and excellent agreement is found.

Thus the general features of the dielectric behaviour of an ionic crystal like an alkali halide may be described as in Fig. 12.12.

### 12.4.5 Permanent dipoles, polar molecules

As we have already seen in Section 12.1, molecules may carry permanent dipole moments that in vapours (and under certain circumstances even in the condensed phases) can rotate in the presence of an external field and thereby make a large contribution to the total polarization. There is a close formal analogy between the behaviour of an ideal gas of electric dipoles and that of an ideal magnetic gas; we may repeat the analysis of Section 11.3 to obtain the electrical equivalent of Curie's law

$$\kappa = \frac{Np^2}{3k_B T}. \tag{12.49}$$

The presence of permanent dipoles is therefore apparent partly in the large susceptibility that they give rise to, but more significantly in its temperature dependence. The permanent dipole arises from the asymmetry of the charge distribution in the molecule, and this is associated with specific molecular geometries (Fig. 12.2); any reorientation of the dipole moment demands a change in orientation of the whole molecule. In a gas or vapour, this is readily visualized, but it can also occur in the molecular liquid and solid phases. Thus (12.49) is found to hold for solid polar substances. However, the larger the molecule and the closer the molecules are packed together, the more difficult the rotation. At sufficiently low temperatures the molecules lose their ability to rotate, they become frozen in position and can no longer turn into the direction of an applied field; thus, although the molecular dipole moment still exists, it becomes ineffective and the orientation polarizability is therefore lost (there is a formal similarity to the quenched orbital magnetic moment in the d metals). This usually occurs in a narrow temperature range, leading to a

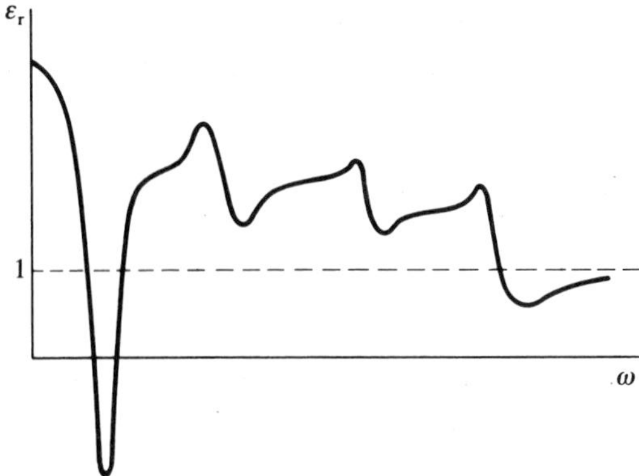


**Figure 12.12** A schematic representation of the frequency dependence of the dielectric constant of an ionic crystal.

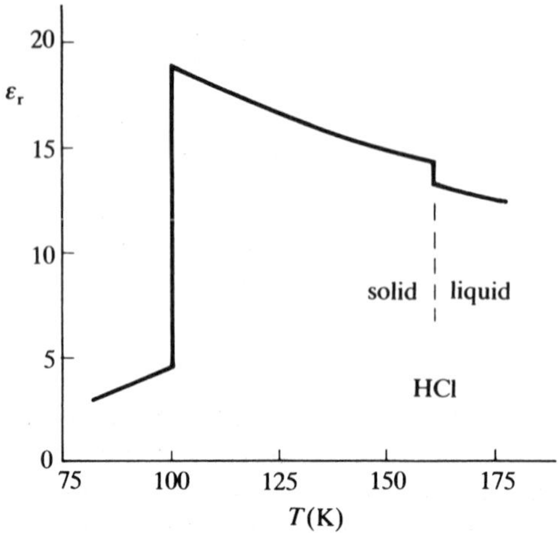


**Figure 12.13** Below a certain temperature, molecular crystals lose their orientational polarizability, which causes a marked decrease in the dielectric constant. (Reprinted with permission from C. P. Smyth and C. S. Hitchcock, *J. Am. Chem. Soc.* **55**, 1933. American Chemical Society, Washington D.C.)

sharp decrease in the dielectric constant as the temperature is lowered as shown in Fig. 12.13. Similarly, in the presence of alternating electric fields, there is a limiting high frequency above which the molecule cannot respond quickly enough and the permanent dipoles no longer contribute to the polarization. The frequency dependence of the dielectric containing permanent dipoles therefore appears as in Fig. 12.14.

## 12.5 Ferroelectrics

In principle it is possible to imagine that an ionic substance, although electrostatically neutral in that the ionic charges compensate one another exactly, may possess a macroscopic dipole moment even in the absence of an external electric field; we require that $\sum_i q_i r_i \neq 0$, which would be the case if ions of different sign were per-

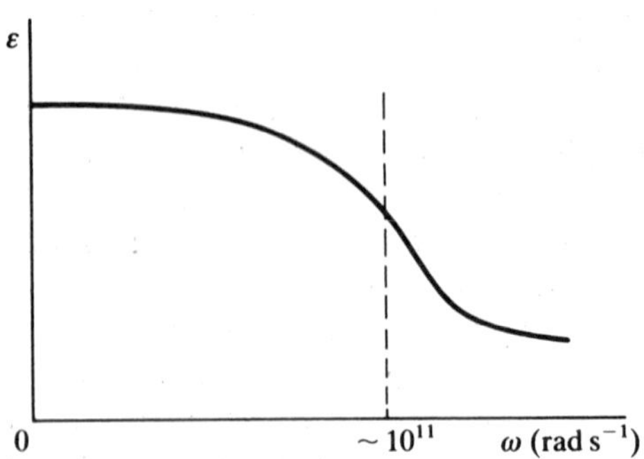


**Figure 12.14** At high frequencies orientational polarizability is lost because the dipolar molecules cannot respond quickly enough to follow the field.

**Table 12.4** Data for three titanates

| | $T_0$(°C) | $P_{spont}(T_0)$ ($\mu$C m$^{-2}$) |
|---|---|---|
| $BaTiO_3$ | 120 | $18 \times 10^{-4}$ |
| $PbTiO_3$ | 490 | $50 \times 10^{-4}$ |
| $KNbO_3$ | 430 | $26 \times 10^{-4}$ |

manently displaced so that each unit cell became polarized. This does happen spontaneously in certain substances known as ferroelectrics. Ferroelectrics are dielectric materials that possess a spontaneous polarization, and this places considerable restriction on the crystal symmetry. They are often of complex chemical composition, but of the approximately 100 known inorganic ferroelectric compounds some of the simpler ones are $NH_4HSO_4$ (monoclinic), $KH_2PO_4$ (orthorhombic) and $BaTiO_3$ (tetragonal); we shall use the last as an illustrative example – it is one member of a small class of titanates, zirconates and niobates, data for some of which are given in Table 12.4.

The ferroelectric state is stable only below a temperature $T_0$. At $T_0$, via either a first- or second-order transition, the material becomes 'para-electric' and the dielectric constant displays a typical 'Curie–Weiss' temperature dependence

$$\varepsilon = \frac{C}{T - T_C}. \tag{12.50}$$

$T_C$ is the Curie temperature, which in certain cases is identical with $T_0$. Below $T_0$ the dielectric constant decreases with temperature so that $T_0$ is characterized by a peak in $\varepsilon_r$, which may attain very high values ($>10^4$) (Fig. 12.15). In the case of $BaTiO_3$, $T_0$ characterizes a true first-order phase transition; above $T_0$ the structure is cubic and typical of the perovskites (Fig. 2.13), whereas below $T_0$ it distorts to a tetragonal form; the metal ions become displaced relative to the oxygen ions, leading to spontaneous permanent polarization.

From (12.50) we see that at $T_C$ ($T_0$), $\varepsilon_r \to \infty$ and $\varepsilon_r$ is $\varepsilon_r(0)$, so, by the LST relation, we expect $\omega_T \to 0$. Normally we associate a low oscillator frequency with a weak or 'soft' restoring force; an oscillatory mode that has a frequency near zero is therefore called a *soft mode*. Phenomenologically, the change of phase at $T_0$ is associated with $\omega_T \to 0$, and this has led to the idea that 'soft modes' indicate instability and are the precursors of certain phase transitions. It is perhaps not surprising that ferroelectrics are strongly piezoelectric (but the converse is not true; thus quartz is a common piezoelectric crystal but it is not ferroelectric), and furthermore they are non-linear with regard to the dependence of polarization on applied fields. The large values of $\varepsilon_r$ and piezoelectric coefficients at $T_0$, as well as their non-linear behaviour, are the source of many practical applications of these materials. Commercial development often aims at bringing $T_0$ as close to room temperature as possible, as well as at attempting to flatten the peak in $\varepsilon_r$ at $T_0$. It is clear why ferroelectrics find use in capacitors and transducers, while their non-linear properties are suitable for the construction of dielectric power amplifiers without the need for electron tubes or transistors. In the optical frequency range, the non-linearity is important because it

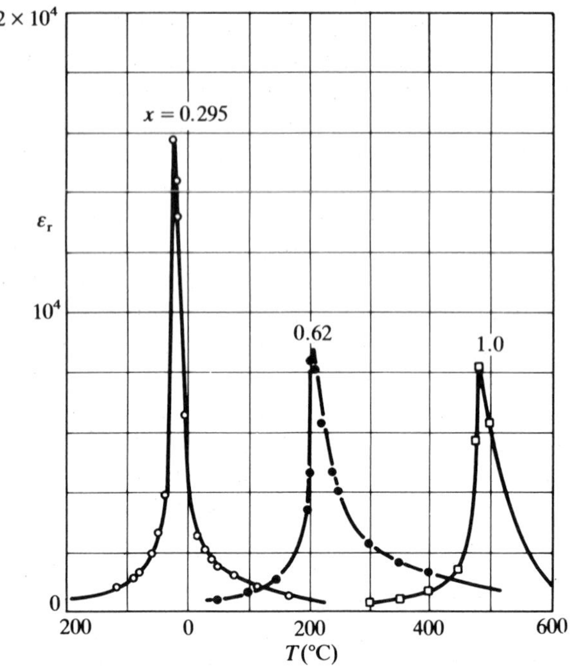


**Figure 12.15** The variation of the dielectric constant of a ferroelectric material ($Sr_{1-x}Pb_xTiO_3$) in the neighbourhood of the Curie temperature and its sensitivity to the chemical composition (From B. Dibenedetto and C. J. Cronan, *J. Am. Ceram. Soc.*, **51**, 1968, 364. Reprinted by permission of the American Ceramic Society, Westerville, Ohio.)

leads to a pronounced dependence of refractive index on external field and allows their use as electrooptical devices for the control and modulation of laser light. The reversal of the spontaneous polarization in static fields exhibits hysteresis and provides a switching property which has been considered for computer circuitry.

## 12.6 Band Structure of Ionic Compounds

When discussing the properties of metals and semiconductors we accentuate the band character of electron states, but this is normally unnecessary for a transparent insulator like NaCl. Nevertheless, we can still describe electron states in such substances in terms of bands, but these are either completely filled or completely empty and separated by large energies $>5$ eV. We present a short description of the main aspects of such a band structure because it illustrates certain new features not encountered earlier.

First of all, the pronounced ionic character of NaCl means that there must be a strong localization of electron charge density in space; the $Na^+$ ions lose their 3s electron, which is transferred to make the $Cl^-$ ion, thereby completing the chlorine 3p shell. We must also remember that the anions are large whereas the cations are small. Whether or not electron states form bands is dependent upon the strength of the mutual interaction between orbitals on different ions. For the small $Na^+$ ions, the 1s, 2s and 2p levels are so localized at the ion that at the equilibrium interatomic spacing no overlap arises and these levels therefore remain atomic-like and sharp. On the other hand, there is sufficient overlap of 3s states that these build a band of levels, but this is empty. Because the $Cl^-$ ions are large, there is much more tendency for their orbititals to overlap and, although the more tightly bound electrons remain in sharp atomic-like states, the 3s and 3p levels are spread into relatively narrow bands. In the occupied states there is little or no $Na^+ - Cl^-$ orbital interaction, partly because there is little or no overlap of the charge clouds and partly because the electron energies for the two types of ion do not overlap. This means that the bands of 3s and 3p occupied levels are confined wholly to the $Cl^-$ ions. Thus we can very well imagine an electron to be photo-emitted from the 3p band, thereby leaving behind a positive hole; conduction may then arise in this band, but the hole can only move via the $Cl^-$ ions. If we try to add another electron to an isolated $Cl^-$ ion (in an attempt to make $Cl^{2-}$), we know that the $Cl^{2-}$ ion is unstable and reverts to $Cl^-$ and a free electron. If such a situation were to hold in the solid NaCl then this would mean that there are no electron states between the filled 3p band and the vacuum level, above which occurs the continuum of levels appropriate to perfectly free electrons. However, in solid NaCl, the '$Cl^{2-}$ continuum' lies partly below the vacuum level and overlaps the empty band of Na 3s levels. Together, these overlapping bands compose the empty conduction band, and in this case, since there are levels on both $Cl^-$ and $Na^+$ ions, it extends throughout the whole crystal (Fig. 12.16).

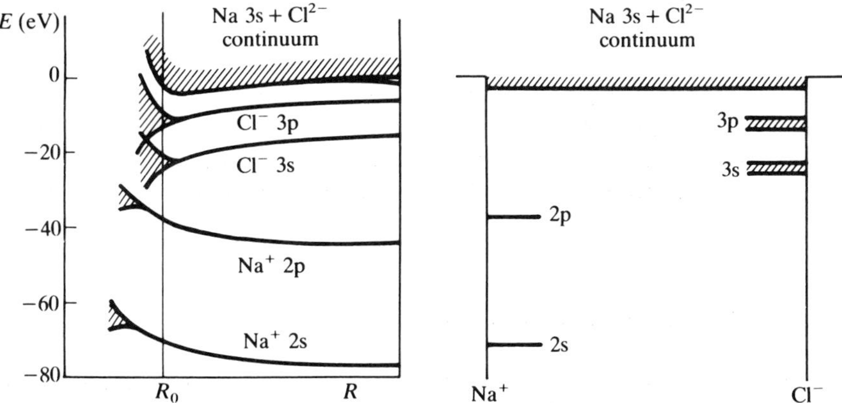


**Figure 12.16** The band structure of an alkali halide. The diagram on the left shows the energy levels associated with a lattice of $Na^+$ and $Cl^-$ ions for different interionic distance $R$. For large $R$ the levels shown retain their sharp atomic character, but for small $R$ orbital overlap causes resonance and electron energy bands arise, here indicated by the shaded areas. At the equilibrium interionic separation $R_0$, the 2s and 2p levels of the Na ion remain sharp whereas the 3s and 3p levels of the Cl ion (owing to their larger diameter) overlap and form narrow bands. The empty 3s ($Na^+$) and 4s ($Cl^-$) levels overlap to form an empty conduction band, and at $R_0$ the bottom of this band lies just below the vacuum level. The situation appropriate to $R_0$ is indicated in the conventional energy level diagram to the right, which shows how most levels remain confined to the parent ions; only the outermost levels extend over the whole crystal to form the empty shallow conduction band, which extends into the vacuum continuum.

The important difference between the band structure of a highly ordered ionic crystal like NaCl and that of a pure element such as Al or Ge is that there are two components and that these retain much of their individual character. This is an important feature even in metallic alloys, and we must be careful not to think of an alloy as a sort of 'pure metal' composed of atoms with a 'compositionally averaged electronic structure'. We often find that alloys have little or no ionicity and that to a good first approximation we may consider them to be composed of neutral atoms; thus in the case of $\beta$-brass ($Cu_{50}Zn_{50}$) we must expect to find an inhomogeneous distribution of valence electrons, since to maintain electrostatic neutrality there must be two 4s electrons around every Zn atom and one 4s electron on every Cu atom. The energies and orbitals of these different valence electrons interact strongly, leading to a common band of 4s levels, but nevertheless the wave functions must distribute them as described above. On the other hand, the d electrons in Cu and Zn lie at very different energies in the free atoms and in the CuZn alloy; there are therefore two separate d bands; one rather broad and confined to the Cu atoms (but narrower than in pure Cu because each Cu atom has fewer nearest Cu neighbours), the other narrow and lying at much lower energy, being confined to the Zn atoms (Fig. 12.17).

Thus there is a great deal of formal similarity between the band structure of an ionic salt and that of an alloy. Here we have only touched upon the most obvious feature, the inhomogeneity of the distribution of levels both in real and in energy space.

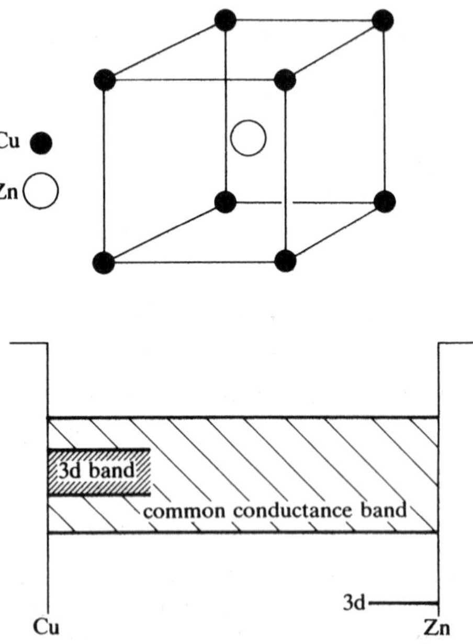


**Figure 12.17** The band structure diagram for the ordered alloy CuZn. There is a conduction band common to both Cu and Zn atoms, but the 3d levels of these atoms do not overlap one another. The Zn 3d levels are confined to the Zn atoms and form localized levels lying just below the bottom of the conduction band. The 3d electrons of the Cu atoms interact with both one another and the conduction electrons, thereby producing a 3d band in the usual manner, but this 3d band is confined to the Cu atoms in the alloy.

## References

DIBENEDETTO, B. and CRONAN, C. J. (1968) *J. Am. Ceram. Soc.* **51**, 364.
NELIN, G. (1974) *Phys. Rev.* **B10**, 4331.
RAUNIO, G. and ROLANDSSON, S. (1970) *Phys. Rev.* **B2**, 2098.
SAUNDERSON, R. B. (1965) *J. Phys. Chem. Solids* **26**, 803.
SMYTH, C. P. and HITCHCOCK, C. S. (1933) *J. Am. Chem. Soc.* **55**, 1830.
TESSMAN, J., KAHN, A. and SHOCKLEY, W. (1953) *Phys. Rev.* **92**, 980.
ZIMAN, J. M. (1972) *Principles of the Theory of Solids*, 2nd edn. Cambridge University Press, Cambridge.

## Problems

**12.1** Show that the Lorentz cavity field in a medium with uniform polarization **P** is

$$\mathbf{E}_c = \mathbf{P}/3\varepsilon_0 .$$

Show that this implies that the depolarization factor for a uniformly polarized sphere is $\frac{1}{3}$. (Hint: the polarization charge density at a point p on the spherical cavity is $-|\mathbf{P}|\cos\theta$, $\theta$ being the angle between the radius vector to the point p and the polarization **P**.

**12.2** Following the discussion of Section 11.4, introduce a damping term into the equation of motion for the electron so that

$$m\frac{d^2x}{dt^2} + \gamma m\frac{dx}{dt} + m\omega_i^2 x = -e\mathbf{E}.$$

(a) Obtain an expression for the polarizability of such an assembly of electrons, assuming that the local field is the same as the applied field. Show that if we allow $\omega_i$ to become zero (i.e. the electrons become free) then we obtain the Drude formula with $1/\gamma$ replacing $\tau$. (b) Obtain an expression for the dielectric constant, assuming that the local field is given by the Lorentz field.

**12.3** The following electronic polarizabilities appropriate to the sodium *D* line ($\lambda = 5893$ Å) were determined from the known molecular volumes and refractive indices of the twenty alkali halides (Tessman *et al.* 1953); the Lorentz local field was used and found to be the most appropriate approximation (the units are $10^{-30}m^3$):

| | | | |
|---|---|---|---|
| $Li^+$ | 0.364 | | |
| $Na^+$ | 4.580 | $F^-$ | 8.093 |
| $K^+$ | 16.76 | $Cl^-$ | 37.20 |
| $Rb^+$ | 24.87 | $Br^-$ | 52.25 |
| $Cs^+$ | 41.91 | $I^-$ | 80.81 |

(a) The refractive indices and molar volumes of three potassium halides are

| | $V_m$ $(10^{-30}\ m^3)$ | $\mathcal{N}$ |
|---|---|---|
| KCl | 61.86 | 1.4904 |
| KBr | 71.42 | 1.5594 |
| KI | 87.68 | 1.6670 |

Check how well the refractive indices of these three chlorides are given by the above polarizabilities. The molar volumes of AgCl and AgBr are 42.67 and 47.98 × $10^{-30}m^3$ and their refractive indices are 2.071 and 2.252 (Sodium D line) respectively. Determine the polarizability of the $Ag^+$ ion in these two salts. (b) Ionic diameters have been deduced from crystallographic data; for K, Cl and Br ions they are 1.33, 1.81 and 1.95 Å respectively. Are these in keeping with the empirical polarizabilities given above and our simple formula (12.8)? (c) Assuming that the electronic polarizabilities determined via the refractive index are valid under static conditions, what ionic displacements, in a local electric field of 10 kV $m^{-1}$, are to be expected in KCl? The static dielectric constant of KCl is $\varepsilon_r = 4.84$.

# 13

# Superconductivity

It has been established beyond doubt that certain substances, when cooled below a critical temperature $T_c$, completely lose all trace of electrical resistance in static electric fields: such materials are called superconductors. In a ring of superconducting material, induced currents persist for times as long as one has the patience to make measurements; the time constant of any decay of the current, which is controlled by $L/R$, where $L$ is the inductance and $R$ the resistance of the ring, is so large (of order $10^5$ years) that we are justified in assuming that the resistance is truly zero. First discovered in 1911 in the metal mercury, we now know that under ordinary conditions of equilibrium, which is to say the ordered crystalline state and normal pressure, some 28 pure metals are superconductors, and other elements become superconducting under particular circumstances of high pressure (e.g. Ge) or structural disorder (e.g. Bi) (Fig. 13.1).

For convenience, we shall use the letters N and S to denote the normal resistive and the superconducting states respectively. The N-S transition is, in pure strain-free single crystals, extremely sharp, occurring over $10^{-4}$ K in some cases (Fig. 13.2). The critical temperature $T_c$ that marks the onset of superconductivity in different pure metals ranges from near 0 to 9.2 K (Nb). Some 1000 intermetallic compounds and alloys become superconducting; certain of them have transition temperatures above 20 K and are used to produce powerful electromagnets. In 1986 a new class of superconducting material was discovered. These are mixed oxides of the form $La_{2-x}A_xCuO_4$; A being an alkaline earth metal (originally Ba or Sr). Transition temperatures up to 125 K have been measured (Table 13.1), and the associated critical magnetic fields are larger than hitherto observed. This discovery is promoting ever increasing activity in the development of such materials for new technical applications. Already phenomena discovered within the past twenty five years have been applied to the design and production of ultra-sensitive magnetometers, voltmeters as well as generators and detectors of ultra-high frequency microwaves (of order 100 GHz).

Although superconductivity was discovered in 1911, it was not until 1957 that its origins were correctly described – an interval that demonstrates the difficulties faced in developing an adequate theory. The experimental study of superconductivity

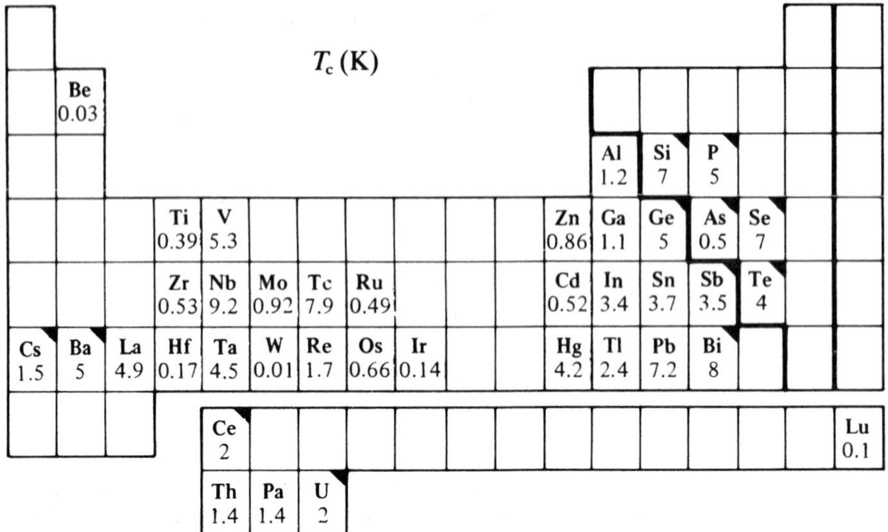


**Figure 13.1** The superconducting transition temperature in K for the naturally occurring elements. Those marked ◥ become superconducting only when subject to high pressures. Note that the 'd' transition metals are divided diagonally into two groups: the lower left group contains superconductors, the upper right group ferromagnetic or potentially magnetic metals. As we shall see, the presence of atomic moments is not advantageous for superconductivity because they exert a breaking influence on the Cooper pairs. The rare earth metals are strongly magnetic and are not therefore superconductors in the elemental state. Transition temperatures for a number of alloys are given in Table 13.1.

usually needs access to liquid helium to provide the low-temperature conditions. Today this is seldom a problem because liquid helium is available in large amounts in almost all solid state laboratories, but before about 1950 this was not the case and only a handful of laboratories had well-developed low-temperature techniques (nowadays it is those who work at or below 0.001 K who form the privileged few). It

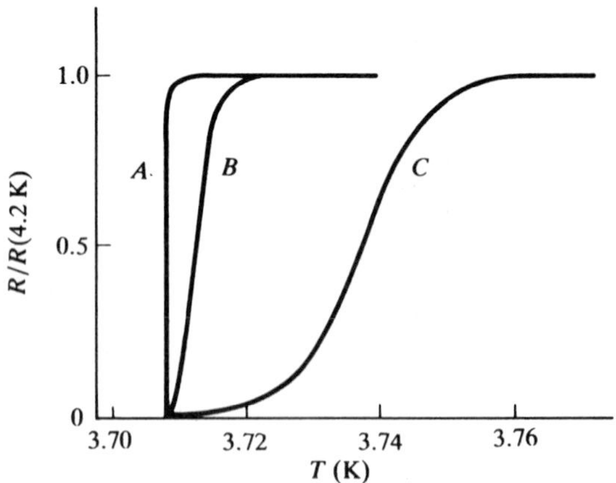


**Figure 13.2** The transition to the superconducting state may be extremely sharp, but decreasing perfection leads to a broader transition: (*A*) pure single-crystal Sn; (*B*) pure polycrystalline Sn; (*C*) impure polycrystalline Sn. (After de Haas and Voogdt 1931.)

**Table 13.1** Transition temperatures of some earlier and more recent high-$T_c$ superconductors

| | $T_c$ (K) |
|---|---|
| $Nb_3Sn$ | 18 |
| $Nb_3Ge$ | 23.2 |
| $V_3Si$ | 17.1 |
| $La_{1.8}Sr_{0.2}CuO_4$ | 35 |
| $Y_{0.6}Ba_{0.4}CuO_4$ | 90 |
| $YBa_2Cu_3O_7$ | 95 |
| $Tl_2Ba_2Ca_2Cu_2O_{10}$ | 125 |
| $Bi_{1-x}K_xBiO_{3-y}$ | 27 |

was not until about 1930 that a real experimental attack on superconductivity was begun. Furthermore, superconductivity did not receive the undivided attention of low-temperature physicists because liquid helium itself proved to have remarkable and exceptional properties and therefore attracted much attention (and still does).

Many physical properties of a metal are found to remain essentially or apparently unchanged in the N-S transition, e.g. crystal structure, lattice parameter and optical reflectivity; but others, namely the heat capacity and thermal conductivity, change in a radical way (Figs. 13.3 and 13.4) whereas the thermoelectric power disappears. Very soon after its initial discovery, experiment also showed that the S state could be destroyed by the application of a sufficiently strong magnetic field. This gave rise to a rather arbitrary classification into soft and hard superconductors, depending on whether the critical magnetic field strength $B_{c0}$ was low or high. Here $B_{c0}$ is the field required to quench superconductivity at 0 K (Fig. 13.5 and Table 13.2). This classification is no longer used, and we speak of type I and type II superconductors, which we shall define later. The existence of a critical field

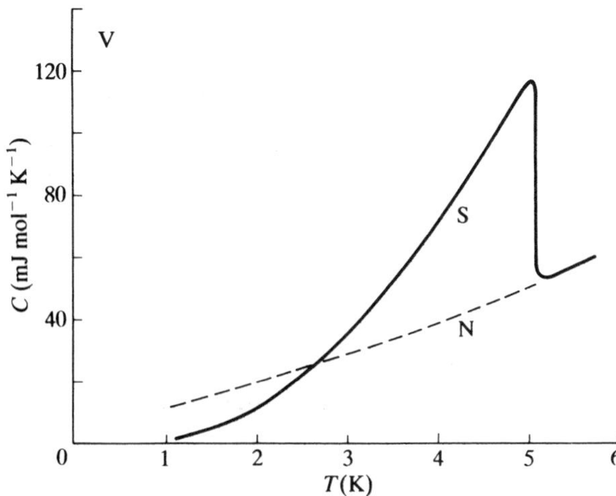


**Figure 13.3** The variation of the heat capacity through the superconducting transition. (After Corak *et al.* 1956.)

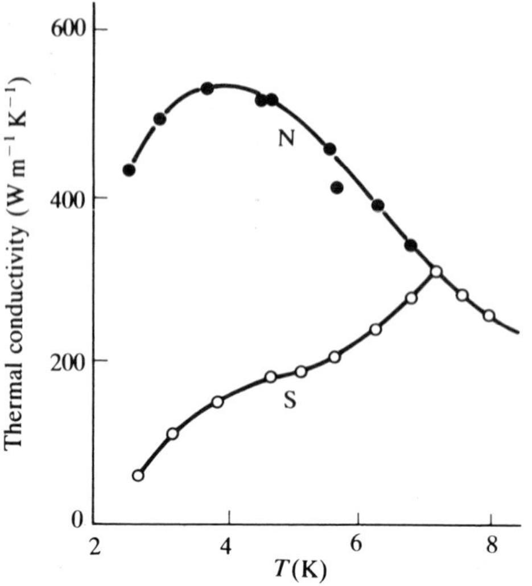


**Figure 13.4** The variation of the thermal conductivity in the normal and the superconducting states for lead. (After Olsen 1952.)

dashed any hopes that one might have had at that time regarding the construction of superconducting solenoids.

Perhaps the most significant experiment made before the war was that of Meissner and Ochsenfeld, who demonstrated in 1933 that an elemental superconductor always excludes all magnetic flux and is a perfect diamagnetic material. The experiment was very important because it showed that a superconductor, although having zero electrical resistance, is not the same as a perfect conductor.

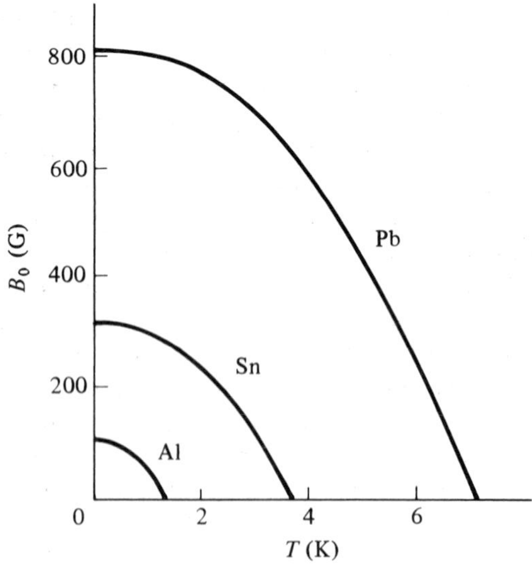


**Figure 13.5** An external magnetic field shifts $T_c$ to lower values; the S state is quenched at 0 K in the field $B_{c0}$. The critical field satisfies $B_c = B_{c0}(1 - T^2/T_c^2)$.

**Table 13.2** Some values of $B_{c0}$

| | $B_{c0}$ (G) |
|---|---|
| Al | 105 |
| Ga | 59 |
| In | 282 |
| Sn | 305 |
| Pb | 803 |
| Ta | 830 |
| V | 1410 |
| Nb | 2060 |

We may appreciate the difficulties in developing a theory of superconductivity in the following qualitative fashion. Experimentally the pure metals have $T_c \approx 5 \pm 5$ K, and this means that the energy stabilizing the S relative to the N state is of order $k_B T_c$ per atom and therefore extremely small relative to the average energy of the conduction electrons. We may well ask what kind of weak interaction can produce such a radical change in behaviour and what hope we can have of accurately calculating the effect of such a weak interaction for electrons with large energies. The surprising situation is that with the very successful theory of Bardeen, Cooper and Schrieffer (BCS) (1957), based on electron-phonon interaction, we now have a better quantitative appreciation of the S than of the N state! An important step in ascertaining the origins of the S state was the experimental discovery (1950) that different isotopes of the same element possess different transition temperatures; in fact, in many, but not all, cases

$$T_c M^{1/2} = \text{constant}. \tag{13.1}$$

This was a strong indication that electron interaction with the lattice is important for superconductivity. The BCS theory is based on electron-phonon interaction and describes how an electron, by polarizing the lattice through which it moves, can interact indirectly with another electron that experiences the polarization caused by the first. We can make a simple analogy with the indirect interaction between two boats when one travels through the wake of the other. This electron-phonon interaction leads to an attraction of the two electrons, which, at low temperatures, become bound together. The attraction is greatest if the two electrons have opposite spin and equal and opposite wave vectors. This 'electron molecule', as it were, is called a 'Cooper pair' – it is a doubly charged entity that can move through the metal lattice without being scattered so long as the electrons remain bound together. The superconducting particles are the pairs of electrons, but one should not think of the electrons as close neighbours because the interaction extends over large distance, of order $10^{-4}$ cm. The maximum distance of interaction is known as the coherence length $\xi$. Within this distance there are very many Cooper pairs – these are to be considered as mutually interacting quantities forming a cooperatively stable system. One might ask why we do not call them quasiparticles; the reason is that the interacting pairs are not an excited state of the system but form the ground state. The transition temperature is controlled by the binding energy $2\Delta$ of the Cooper pair in the system of interacting pairs called the 'condensate'.

## 13.1 The Meissner Effect

In Box 13.1 it is shown that a resistanceless conductor restricts the time dependence of the **B** field through the equation

$$\nabla^2 \frac{d\mathbf{B}}{dt} = \frac{ne^2\mu}{m}\frac{d\mathbf{B}}{dt}. \tag{13.2}$$

This equation may be rewritten as

$$\nabla^2 \frac{d\mathbf{B}}{dt} = \frac{\omega_s^2}{c^2}\frac{d\mathbf{B}}{dt} = \frac{1}{\lambda_L^2}\frac{d\mathbf{B}}{dt}. \tag{13.3}$$

We have written $\omega_s$ for a plasma frequency; $c$ is the velocity of light in the conductor and $\lambda_L$ is a new parameter called the penetration depth for reasons that will shortly become apparent. We have already seen that the magnetic behaviour is dependent on the shape of the object being magnetized; in what follows we shall always consider the simplest geometry and assume that our sample, whether it be normal, perfect or superconducting, is in the form of a long thin wire and that the external field is directed parallel to the wire. For this geometry the wire has zero demagnetizing coefficient. It should be understood, however, that different specimen-field geometries produce significantly different behaviour patterns, but we confine attention to the simplest case. Our coordinates are chosen so that the $z$ axis is perpendicular to the axis of the conductor and $z = 0$ at the surface of the conductor.

Now a particular solution of (13.3) is

$$\frac{d\mathbf{B}(z)}{dt} = \frac{d\mathbf{B}(0)}{dt} e^{-z/\lambda_L}, \tag{13.4}$$

which says that for a perfect conductor $d\mathbf{B}/dt$ decays exponentially, with decay constant $\lambda_L$, as we pass from the surface into the body of the conductor. We can readily estimate $\lambda_L$, see (13.3), and find that it is of order $10^{-6}$ cm; this means that, apart from a very thin skin, the perfect conductor is characterized by

$$\frac{d\mathbf{B}(z)}{dt} = 0 \tag{13.5}$$

This implies that the magnetic state of a perfect conductor is dependent on the order in which the field is applied and the state of perfect conductivity achieved (Fig. 13.6).

If the field is first switched on after the material is in the perfectly conducting state (Fig. 13.6*a*) then it cannot penetrate more than a thickness of order $\lambda_L$ of the specimen; on the other hand, if the material is transformed to the state of perfect conductivity after the field is applied (Fig. 13.6*b*) then the field penetrates the specimen and remains frozen there even if the source of the external field is removed. This behaviour pattern is very different from that observed for a superconductor, as exemplified in the original experiments of Meissner and Ochsenfeld. The 'Meissner effect', as it is now called, shows that a superconducting pure metal, in the chosen geometry, *always expels the magnetic flux irrespective of the sequence in which the field is applied or the material becomes superconducting. Thus a superconductor is not a perfect conductor but a perfect diamagnetic material with zero electrical resistance.*

**Box 13.1 The Perfect Conductor**

If we assume that the absence of electrical resistance implies perfect conductivity, so that certain electrons, under the influence of an applied field **E**, are continuously accelerated, then

$$\frac{d\mathbf{v}}{dt} = \frac{-e\mathbf{E}}{m},$$

$$\mathbf{J} = -ne\mathbf{v},$$

implying that

$$\mathbf{E} = \frac{m}{ne^2}\frac{d\mathbf{J}}{dt},$$

$n$ being the density of 'superelectrons'.

Maxwell's version of Ampere's circuit law requires that

$$\nabla \times \mathbf{B} = \mu\mathbf{J},$$

which leads to

$$\mathbf{E} = \frac{m}{ne^2\mu}\nabla \times \frac{d\mathbf{B}}{dt},$$

so

$$\nabla \times \mathbf{E} = \frac{m}{ne^2\mu}\nabla \times \nabla \times \frac{d\mathbf{B}}{dt}.$$

Remembering that $\nabla \cdot \mathbf{B} = 0$ and using the vector identity

$$\nabla \times (\nabla \times \quad) = \nabla(\nabla \cdot \quad) - \nabla^2(\quad),$$

we obtain the equation

$$\nabla^2 \frac{d\mathbf{B}}{dt} = \frac{\mu n e^2}{m}\frac{d\mathbf{B}}{dt}.$$

This may be rewritten as

$$\nabla^2 \frac{d\mathbf{B}}{dt} = \frac{\omega_s^2}{c^2}\frac{d\mathbf{B}}{dt},$$

where

$$\omega_s^2 = \frac{ne^2}{\varepsilon m}, \qquad c = \frac{1}{(\mu\varepsilon)^{1/2}}.$$

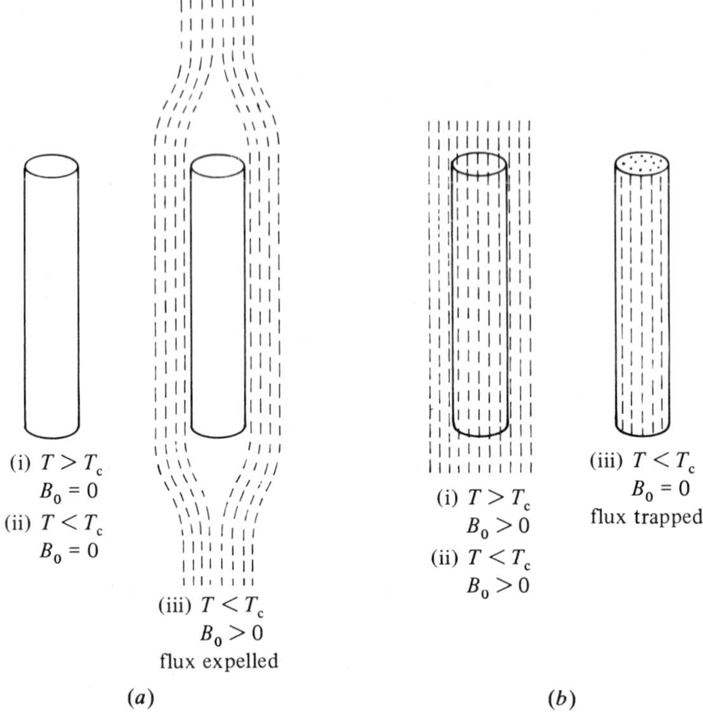


**Figure 13.6** The behaviour of a perfect conductor in a magnetic field depends not on the final-state conditions but on the sequence of events leading to the final state. Although not shown, the lines of B in (iii)b are continuous as for a bar magnet.

## 13.2 Perfect Diamagnetism and Stability of the S State

A perfectly diamagnetic substance is characterized by

$$\mu_r = 0$$

and

$$\mathbf{B} = \mathbf{B}_0 + \mu_0 \mathbf{M} = 0, \tag{13.6}$$

whence the volume susceptibility $\kappa$ becomes

$$\kappa = -\frac{\mu_0 |\mathbf{M}|}{|\mathbf{B}_0|} = -1. \tag{13.7}$$

The diamagnetic magnetization $\mathbf{M} = -\mu_0^{-1}\mathbf{B}_0$ may be imagined as arising from a network of induced Ampèrian currents spread throughout the superconductor. Within the body of the superconductor these cancel, leaving only a finite surface current.

We now consider the free energies of the following substances:

(a) a normal metal, e.g. Cu, that is weakly diamagnetic – so weak that we can safely assume $\kappa \approx 0$;
(b) a normal but perfectly diamagnetic metal, $\kappa = -1$;
(c) a superconductor, $\kappa = -1$.

In case (a) the normal metal with $\kappa \approx 0$ acts like a vacuum; there is no interaction with the applied field and therefore no magnetic contribution to the free energy; we take this state as our zero of energy. Phase stability is governed by the Gibbs free energy, which we write as $G(B_0, T)$, i.e. we have $G(0, 0)$ when both $B_0$ and $T$ are zero and $G(0, T)$ when $B_0$ is zero at finite temperature. For the N state we write

$$G_N(B_0, T) = G_N(0, T), \tag{13.8}$$

because the field does not interact with the sample. Magnetically, the perfectly diamagnetic normal metal and the superconductor are identical; both possess diamagnetic magnetizations leading to an energy $B_0^2/2\mu_0$ per unit volume.

However, we must not forget that, in contrast with the normal but perfectly diamagnetic metal, the superconductor is in a much more highly ordered state (in a manner yet to be determined). The heat capacity (Fig. 13.3) and the derived entropy (Fig. 13.7) clearly show this to be the case. Nevertheless, the diamagnetic magnetization causes an increase in the free energy density of the S state and we write

$$G_S(B_0, T) = G_S(0, T) + \frac{B_0^2}{2\mu_0}, \tag{13.9}$$

and when the external field attains the critical value $B_c$

$$G_S(B_c, T) = G_S(0, T) + \frac{B_c^2}{2\mu_0}.$$

But when $B_0 = B_c$ the transition from the S to the N state occurs, i.e. the S and N states have the same free energies. In other words

$$G_S(B_c, T) = G_S(0, T) + \frac{B_c^2}{2\mu_0} = G_N(B_c, T) = G_N(0, T), \tag{13.10}$$

whence

$$G_S(0, T) = G_N(0, T) - \frac{B_c^2}{2\mu_0}, \tag{13.11}$$

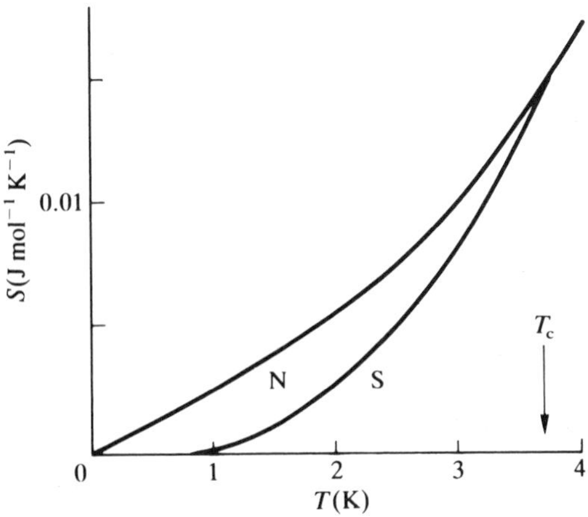


**Figure 13.7** As shown here for Sn, the superconducting state has lower entropy than the normal state and is therefore more highly ordered than the latter. (After Keesom and van Laer 1938.)

and at absolute zero

$$G_S(0, 0) = G_N(0, 0) - \frac{B_{c0}^2}{2\mu_0}. \tag{13.12}$$

We see that the critical field at a given temperature $T$ provides a measure of the stability of the S state. Furthermore, in any field $B_0 < B_c$ the S state is always more stable than the normal but perfectly diamagnetic state by the energy $B_c^2/2\mu_0$.

From the above discussion, it follows that if the superconductor is in a magnetic field $B_0 < B_c$ then

$$G_S(B_0, T) = G_N(0, T) - \frac{B_c^2 - B_0^2}{2\mu_0}. \tag{13.13}$$

## 13.3 The Heat Capacity

For a magnetic substance we write the Gibbs free energy as

$$G = U - TS + PV - \mathbf{B}_0 \cdot \mathbf{M}, \tag{13.14}$$

where the symbols have their usual thermodynamic meanings. Furthermore, since work must be performed on the system to create an element of magnetization $\mathrm{d}\mathbf{M}$ in the external field $\mathbf{B}_0$, we write

$$dU = T\,\mathrm{d}S - P\,\mathrm{d}V + \mathbf{B}_0 \cdot \mathrm{d}\mathbf{M}, \tag{13.15}$$

which, together with (13.14), gives

$$\mathrm{d}G = -S\,\mathrm{d}T + V\,\mathrm{d}P - \mathbf{M} \cdot \mathrm{d}\mathbf{B}_0. \tag{13.16}$$

If $P$ and $\mathbf{B}_0$ are held constant then the entropy is

$$S = -\left(\frac{\partial G}{\partial T}\right)_{P,\,B_0},$$

and, using (13.13), we find that

$$\begin{aligned} S_S - S_N &= \frac{\mathrm{d}}{\mathrm{d}T}\left(\frac{B_c^2 - B_0^2}{2\mu_0}\right) \\ &= \frac{B_c}{\mu_0}\frac{\mathrm{d}B_c}{\mathrm{d}T}. \end{aligned} \tag{13.17}$$

The heat capacity is given by

$$C = T\frac{\mathrm{d}S}{\mathrm{d}T},$$

so we obtain

$$\begin{aligned} C_S - C_N &= T\frac{\mathrm{d}}{\mathrm{d}T}\left(\frac{B_c}{\mu_0}\frac{\mathrm{d}B_c}{\mathrm{d}T}\right) \\ &= \frac{T}{\mu_0}\left[\left(\frac{\mathrm{d}B_c}{\mathrm{d}T}\right)^2 + B_c\frac{\mathrm{d}^2 B_c}{\mathrm{d}T^2}\right]. \end{aligned} \tag{13.18}$$

Now when $T = T_c$ the critical field $B_c$ is zero, so

$$(C_S - C_N)_{T_c} = \frac{T_c}{\mu_0}\left(\frac{dB_c}{dT}\right)^2. \tag{13.19}$$

This means that we can predict the discontinuity in the heat capacity at $T_c$ and in zero external field if we know how $B_c$ varies with $T$ near $T_c$. Although we have no detailed knowledge of the character of the superconducting state, thermodynamic reasoning provides a valuable means of correlating experimental data.

In the absence of an external magnetic field, the transition from the N to the S state is characterized by a continuous variation of both $G$ and $dG/dT$, but $d^2G/dT^2$ is discontinuous at $T_c$, leading to the jump in the heat capacity. Because $dG/dT$ is continuous at $T_c$, there is no latent heat associated with the N-S transition in zero external field. Such a transition is usually called a 'second-order phase transition' and is analogous to the magnetic transitions at the Curie or Néel temperatures. A second-order transition is associated with a change in the internal order or symmetry but leaves the lattice structure unaffected. (See Appendix 11.4.)

On the other hand, a conventional phase change, such as that associated with freezing or melting, is accompanied by a latent heat because, although $G$ is continuous through the melting point, $dG/dT$ is discontinuous, and therefore so is the entropy. The change in entropy at a fixed temperature, in this case the melting point, corresponds to a change in the internal energy of the system associated with the latent heat. If our superconducting wire is in an external field $B_0 < B_{c0}$ then the N-to-S transition occurs at a temperature below $T_c$, the value in zero external field. Under these conditions $G$ again varies continuously, but $dG/dT$ is discontinuous, leading to a discontinuous change in the entropy given by (13.17) (remember that $B_c$ is no longer zero but is determined by the strength of the applied field $B_0$). The superconducting transition is now accompanied by a latent heat and is a true first-order phase transition.

## 13.4 The London Equation: $\mathbf{J} = -(ne^2/m)\mathbf{A}$

In early phenomenological theories it was found useful to assume that certain electrons, the 'superelectrons', are responsible for superconductivity. Now a perfect conductor is characterized by $d\mathbf{B}/dt = 0$, but experiment shows that in a superconductor $\mathbf{B} = 0$. A possible description of a superconductor may be obtained if in addition to (13.4) we demand that

$$\frac{ne^2\mathbf{E}}{m} = -ne\frac{d\mathbf{v}}{dt} = \frac{d\mathbf{J}}{dt} \tag{13.20}$$

and

$$\nabla \times \mathbf{J} = -\frac{ne^2}{m}\mathbf{B}. \tag{13.21}$$

Equation (13.20) expresses the acceleration of the charge carriers, whereas (13.21) is obtained, with the help of (13.20), by the integration of Faraday's law and the assumption that the integration constant is zero.

We now use Maxwell's form of Ampère's law,

$$\nabla \times \mathbf{B} = \mu \mathbf{J} \left( \frac{\mathrm{d}\mathbf{D}}{\mathrm{d}t} = 0 \right), \tag{13.22}$$

and may substitute for either **J** or **B** in (13.21). Substituting for **J**, we obtain

$$\nabla \times \nabla \times \mathbf{B} = -\frac{ne^2\mu}{m}\mathbf{B},$$

which, using the vector identity $\nabla \times (\nabla \times \ ) = \nabla(\nabla \cdot \ ) - \nabla^2(\ )$, leads to

$$\left.\begin{aligned} \nabla^2\mathbf{B} &= \frac{ne^2\mu}{m}\mathbf{B} = \frac{\mathbf{B}}{\lambda_L^2} \\ \text{and} \quad \mathbf{B}(z) &= \mathbf{B}(0)\mathrm{e}^{-z/\lambda_L} \end{aligned}\right\} \tag{13.23}$$

If, on the other hand, we express (13.21) in terms of **J**, we find

$$\left.\begin{aligned} \nabla^2\mathbf{J} &= \frac{\mathbf{J}}{\lambda_L^2} \\ \text{and} \quad \mathbf{J}(z) &= J(0)\mathrm{e}^{-z/\lambda_L} \end{aligned}\right\} \tag{13.24}$$

The last expression shows that any supercurrents associated with the presence of the external field $\mathbf{B}_0$ are confined to a surface sheath. It is also clear that within the bulk of a superconductor carrying only steady current (i.e. $\mathrm{d}\mathbf{J}/\mathrm{d}t = 0$), by (13.20) and (13.23), both **E** and **B** are zero so there is no acceleration of the ordinary electrons.

If we introduce the vector potential **A** and write

$$\mathbf{B} = \nabla \times \mathbf{A}, \quad \text{with} \quad \nabla \cdot \mathbf{A} = 0$$

then (13.21) may be rewritten as

$$\mathbf{J} = -\frac{ne^2}{m}\mathbf{A}. \tag{13.25}$$

This equation shows that the supercurrent is directly related to the value of the vector potential. The equation (13.21), more usually expressed as (13.25), is known as the London equation and explains the Meissner effect because the magnetic flux is kept out of the superconductor.

The London equation was postulated in an *ad hoc* fashion and provided a phenomenological description of the electromagnetic behaviour of a superconductor. It did not provide any insight regarding the microscopic origins of superconductivity, but in the hands of F. London it was important for the development of theory and led to the following important conclusions.

(a) An external magnetic field penetrates a distance of order $\lambda_L$ into the superconductor (Fig. 13.8).

(b) A supercurrent carried by 'superelectrons' and a normal current carried by ordinary electrons coexist. However, when a superconductor carries a constant surface current, by (13.20) there can be no electric field within it and the normal

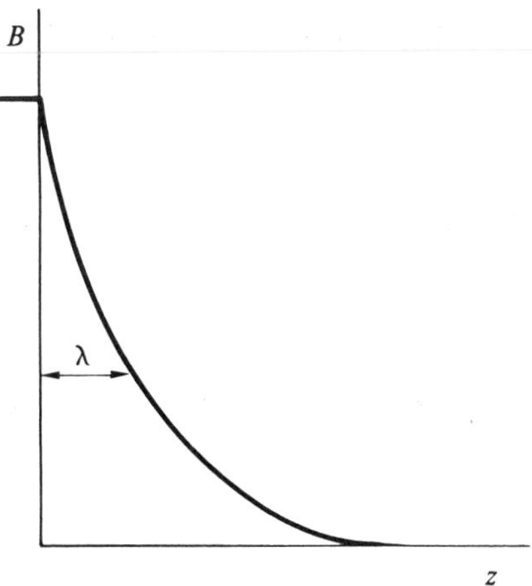


**Figure 13.8** A magnetic field external to the superconductor decays exponentially as it penetrates the superconductor – the decay constant is called the penetration depth $\lambda$. The vacuum-superconducting interface is at $z = 0$. The supercurrent shows a similar variation.

electrons are not accelerated. The density of superelectrons is zero at $T_c$, rising to a maximum value at 0 K, and because of this the London penetration depth $\lambda_L$ is very large near $T_c$, becoming smaller at lower temperatures. The presence of the penetration field means that a superconductor cannot have perfect diamagnetic susceptibility since the flux is not completely expelled; normally this is of no consequence, but in very fine particles or very thin films the deviations from perfection become measurable and allow direct estimates of the penetration depth via susceptibility measurements. The experimentally observed penetration depth $\lambda_e$ is larger than the London value $\lambda_L$.

(c) The theory predicted that superconductivity would arise if a fraction of electrons – the 'superelectrons' – were in a condensed state of minimum momentum, and if in fact the momentum were zero then by the Heisenberg uncertainty principle the carrier wave function would have unlimited extent. On this basis London was able to show that if the wave function of these superelectrons remains unchanged in the presence of a weak magnetic field then superconductivity arises. This led to the concept of the 'rigid' wave function, and in London's picture the rigidity extended throughout the superconductor.

(d) The quantization of magnetic flux: if a superconductor contains a hole, e.g. if it is in the form of a ring or a hollow cylinder, then the wave function of the superconducting electrons, being rigidly fixed in space, is single-valued, and any line integral of the wave function along a closed path surrounding the hole is also single-valued, with the exception of integral phase factors of $2\pi$. This, as described in Section 13.13, has the consequence that any magnetic flux threading the hole is quantized in integral multiples of $\phi_0$, where

$$\phi_0 \equiv h/q, \tag{13.26}$$

$q$ being the carrier charge. This quantization of flux was confirmed by experiment, but first in 1961, ten years after the original prediction was made and four years after the BCS theory was established. It did not therefore come as a surprise when the quantity $q$ was found to be equal to twice the charge on the electron.

## 13.5 The Coherence Length

The London theory is called a local theory because the supercurrent at a particular point is assumed to be controlled by the magnetic vector potential at the same position, (13.25). For a superconductor in an external magnetic field the theory introduces the concept of the penetration depth and suggests a skin of transitional material enclosing the main body of the superconductor. For pure elemental superconductors this transitional region may be considered associated with a positive surface energy that prevents the division of the material into a sequence of domains with alternating S and N character. Such a division would otherwise allow the superconducting condition to exist with a minimum cost in diamagnetic energy.

In 1950 the London theory was superseded by a new phenomenological approach developed by Ginzburg and Landau; again, in its original form at least, it was a local theory. According to Ginzburg and Landau, the S state is presumed to be different from the N state in the presence of internal order (as yet undefined) that is characterized by an order parameter. The latter may be regarded as an effective wave function for the superelectrons, and its square is a measure of their density. Ordinarily we associate the spatial variation of the wave function with the momentum (and thereby with the energy) of a particle; this is clearly valid because the rapid spatial variation of the wave function demands the inclusion of high spatial frequencies that are associated with large momentum. Thus, if the density of superelectrons varies throughout a sample then this is associated with an increase in the total energy. In the absence of an external magnetic field Ginzburg and Landau assumed that the order parameter is the same throughout the whole sample – there is complete rigidity of the wave function. When an external field is applied the order parameter varies within the surface layer, leading to a surface energy with two components: one arising from the conventional penetration of the magnetic field and the other from the variation of the order parameter. We shall not present the details of this theory, but make the following comments. Since the theory is a phenomenological one, significant parameters are expressed in terms of experimentally measured quantities. In particular there is a parameter $\kappa$ (not to be confused with the magnetic susceptibility):

$$\kappa = \frac{2e^2}{h^2} B_c^2 \lambda_e^4,$$

$\lambda_e$ being the experimental penetration depth in the limit of very weak external fields. This quantity $\kappa$ is important because it controls the surface energy of a superconductor.

It is found that when $\kappa < 2^{-1/2}$ there is a positive surface energy between normal and superconducting regions and when $\kappa > 2^{-1/2}$ there is a negative surface energy.

This leads to a classification of superconductors into type I or type II according to whether the surface energy is positive or negative. The importance lies not only in the distinct classification but also because the type II superconductor may have an extremely large critical field, allowing the construction of superconducting magnets

That the Ginzburg-Landau theory is very successful may be appreciated from the fact that it is consistent with the microscopic BCS theory. However, for our purpose, it is simpler to attempt a qualitative appreciation of the surface energy using the concept of coherence length.

It is clear that the penetration depth is a significant feature of a superconductor, but is not an easy one to measure using magnetic methods. Normal metals in the presence of very high-frequency electromagnetic waves display the 'skin effect'; the field penetrates only a small distance $\delta$ into the conductor. In an analogous manner, a superconductor exhibits a surface impedance controlled by the penetration depth. When a microwave field enters the superconductor, the supercurrent becomes oscillatory and gives rise to a finite electric field (13.20). This influences the normal electrons, thereby producing resistive effects, but only within the region $\lambda_e$. A microwave cavity containing a metal wire therefore has a different resonant frequency depending on whether the wire sample is in the S or the N state. The effect is directly related to $\delta - \lambda_e$. On this basis, Pippard (*circa* 1950) made an extensive study of the field penetration depth as a function of temperature, external field and, most significantly, additions of alloying elements.

It was found that the penetration depth is larger than as predicted by the London formula (13.3), and furthermore, in contrast with $B_c$ or $T_c$, it is very sensitive to the presence of alloying elements. This implies that $\lambda_e$ is sensitive to the mean free path of electrons in the normal state because the latter is the only quantity that a small amount of alloying element can affect in a radical manner. A particular feature of these results is that the increase in $\lambda_e$ due to the added impurity is most marked when the electron mean free path is smaller than a certain size.

These results led Pippard to propose that the interactions producing the perfect superconducting order that is encountered within the pure bulk superconductor have a range characterized by a length $\xi_0$. The implication is that the superelectrons respond to the value of the magnetic vector potential averaged over the region of extent $\xi_0$ because within this region the superelectrons act as a unit, i.e. coherently. The parameter $\xi_0$ is therefore called the coherence length and we say that a non-local relationship exists between **J** and **A**.

The coherence length in a superconducting alloy, $\xi$, is less than that for the pure metal, $\xi_0$. $\xi$ is a function of the electron mean free mean path $l$ with limiting value $\xi_0$. The penetration depth is written as

$$\lambda_e = \lambda_L \frac{\xi_0}{\xi}, \qquad \frac{1}{\xi} \approx \frac{1}{\xi_0} + \frac{1}{l}. \tag{13.27}$$

In an elemental superconductor $\xi_0$ is of order $10^{-4}$ cm, very much larger than the penetration depth. Since it is possible to obtain thin films and colloidal particles that remain superconducting even when their linear dimensions are less than $\xi_0$, Pippard concluded that $\xi_0$ is not a critical length of the existence of superconductivity, but rather the maximum range for coherence within the bulk superconductor. Further support for the concept of coherence came from thermodynamic arguments. Thus, amongst other things, Pippard pointed out that the extremely sharp N-S transition, which in certain cases occurs in a temperature interval $<10^{-3}$ K, precludes fluctuations of superconducting order within regions less than a certain size, i.e. of linear dimension $\xi_0$.

The above ideas have been substantiated in the microscopic theory. We may associate the coherence length with the linear extension of a Cooper pair. The concept of coherence may also be grafted onto the Ginzburg-Landau theory, and it is then found that $\kappa$ and $\xi_0$ are related by

$$\kappa \approx \lambda_e/\xi_0 .$$

Also,

$$\xi \approx \hbar v_F/\Delta$$

($\Delta$ is defined in Section 13.9), i.e. the distance travelled by an electron at the Fermi level and in a time characteristic of superconductivity. If the superconductor acts coherently over regions of linear extent $\xi_0$, the order parameter has its maximum value over these regions. Furthermore, if this parameter is compelled to change (e.g. by the presence of a boundary) then it always assumes the slowest possible variation since this minimizes the free energy. On the other hand there is nothing to be gained by variations over a distance larger than $\xi$. So at any surface separating a superconductor from a vacuum or normal material we expect the order parameter, even in the absence of a magnetic field, to start near zero at the surface atoms and grow smoothly to the bulk value within a distance $\xi_0$ from the surface. Normally in a pure metal superconductor $\xi_0 > \lambda$,† but by alloying we can make significant reductions to the coherence length so that $\xi < \lambda$; these situations correspond to the type I and II superconductors respectively.

## 13.6 The Surface Energy

Normally in solid state physics we avoid the surfaces that arise in real specimens unless we have a specific interest in them. We tend to calculate the properties of infinite single crystals. Nevertheless, whenever a surface is created we find an associated surface energy. Thus the physical surface of a crystal contains atoms with uncompensated bonds producing the surface energy and associated surface tension. Similarly for the superconductor, the material near the surface is in a different physical state from an infinite bulk superconducting medium, and it therefore possesses a characteristic surface energy. This energy depends upon the relative sizes of $\xi$ and $\lambda$ and allows a classification of superconductors.

*Type I*: $\xi > \lambda$ (Fig. 13.9*a*). The surface energy is positive and the surface area is therefore minimized; in our long-wire geometry the surface is restricted to the unavoidable geometrical surface of the wire. Such a superconductor exhibits a complete Meissner effect, which is to say that, apart from the penetration depth, all flux is excluded at the superconducting transition, and the magnetization curve shows a characteristic triangular appearance (Fig. 13.10).

*Type II*; $\xi < \lambda$ (Fig. 13.9*b*). The surface energy is negative – the more surface, the lower the energy of the system. The material adopts a composite S/N structure with alternate domains of S and N character, which in the limiting condition becomes a fibrous 'vortex state' characterized by a geometrically regular distribution of lines of flux quanta. Each line is a single flux quantum produced by a supercurrent vortex. The magnetization curve shows two critical field strengths: $B_{c1}$, at which the flux first penetrates the sample, and $B_{c2}$, when the transition to the N state is complete (Fig. 13.11). The condition appropriate to the interval bounded by $B_{c1}$ and $B_{c2}$ is

† Henceforth we dispense with the subscript on $\lambda$.

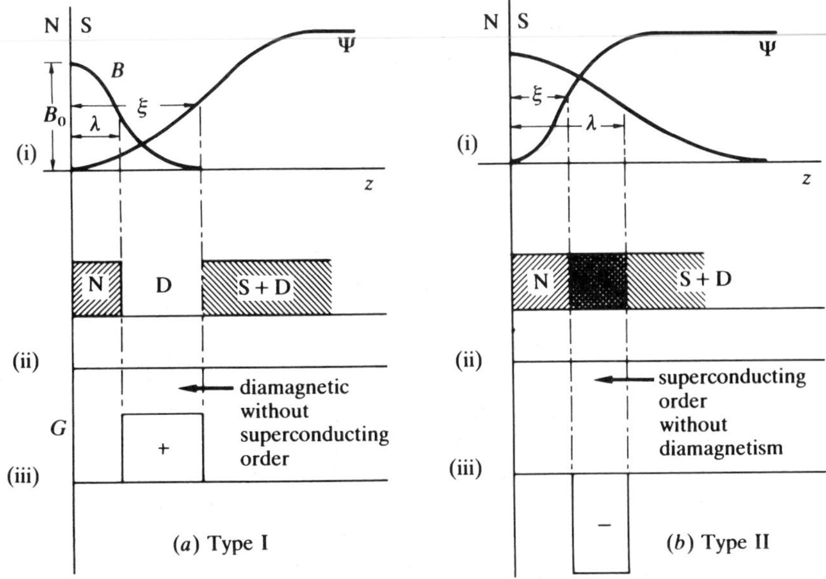


**Figure 13.9** We imagine a superconductor placed in an external field $B_0$: (a) type I; (b) type II. On the left there is a vacuum, V, and on the right superconducting material, S. The field penetrates according to the parameter $\lambda$, and the density of charge capable of resistanceless motion increases over a distance controlled by $\xi$. In type I, $\xi > \lambda$, whereas in type II, $\xi < \lambda$. The upper diagrams (i) show the smooth variations that arise in practice, but for ease of demonstration diagrams (ii) show the same situation simplified in terms of step thresholds. In type I we see that the regions of field penetration and superconducting order do not overlap, they are separated by a purely diamagnetic region; the magnetic energy is not compensated by the order characteristic of a superconductor, so the boundary energy is positive, (iii). In type II substances, however, the regions of field penetration and superconducting order overlap. This region has the benefit of superconducting order, which lowers the energy, without the associated diamagnetism that in the presence of an external field leads to a positive contribution to the energy (see (13.9)); the result is therefore a negative surface energy, (iii).

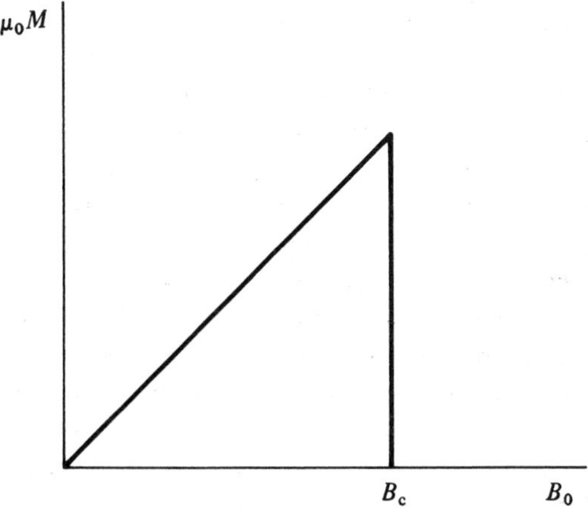


**Figure 13.10** For type I superconductors the diamagnetization grows linearly with magnetic field until the critical value is reached. At this point the flux penetrates the sample and the diamagnetic susceptibility assumes the normal low value.

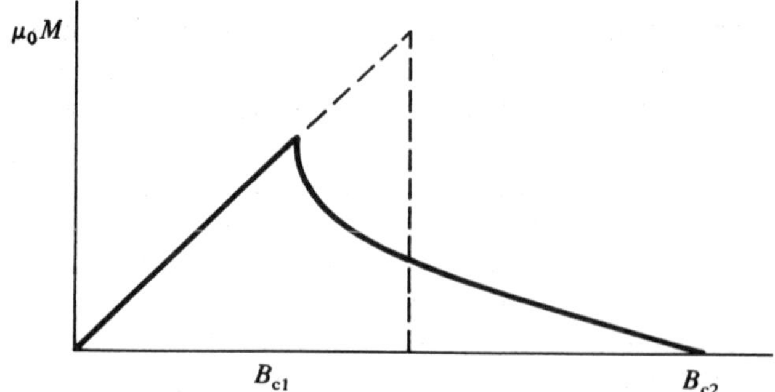


**Figure 13.11** For type II superconductors the initial behaviour of the diamagnetization is linear with external field, but at a certain field $B_{c1}$ flux begins to penetrate the sample and the magnetization decreases, initially rapidly, but then slowly, until in a field $B_{c2}$ the sample becomes saturated with lines of flux quanta.

often referred to as the 'mixed state'. Values of $B_{c2}$ can be very large indeed, approaching 50 T in certain materials; this enables type II materials to be used in superconducting magnets to produce very intense static magnetic fields. A common commercial magnet material is $Nb_3Sn$, one of the $\beta$-tungsten structures (Fig. 2.18), with a $B_{c2}$ of about 25 T, but the working field of a superconducting magnet is always considerably less than the ideal value. Type II superconductors are preferably alloys of transition metals; the reason is that the Type II behaviour demands a small coherence length. We have already seen that the coherence length decreases as the conventional mean free path decreases. Since the transition metals are poor conductors to start with, and alloying can only make them worse by introducing inhomogeneities, it is understandable that they form type II superconductors. It is of course possible to convert a type I superconductor such as Pb into type II by alloying, but extremely high $B_{c2}$ values are only obtained using the transition metals (exceptions are the new ceramic high-$T_c$ materials).

## 13.7 The BCS Theory

The difference in energy between the S and N states is exceedingly small, of order $10^{-6}$ eV per valence electron,† which is very much smaller than the uncertainty in the calculation of the energy of an electron in the normal state. The only hope for any quantitative atomic theory of superconductivity rests in the belief that one specific mechanism is at work and that the N and S states have one unique difference. If in all other respects they are identical then, however imperfect our understanding of electronic behaviour, it will apply equally to both states and we may hope for a cancellation of errors. In what follows we shall first assume the temperature to be at or very close to absolute zero.

† In general terms a superconducting metal contains of order $4 \times 10^{22}$ atoms cm$^{-3}$, corresponding to a valence electron concentration of order $10^{23}$ cm$^{-3}$. At 0 K some $10^{20}$ electrons cm$^{-3}$ form Cooper pairs, leading to a condensation energy of $10^{20}\,\Delta$ eV cm$^{-3}$. Since $\Delta \approx 10^{-3}$ eV, the S state has energy lower than the N state by an amount of order $10^{-3}\,\Delta$ eV per valence electron, i.e. of order $10^{-6}$ eV per valence electron.

The question is what mechanism is at work and what kind of internal order characterizes the superconducting state. Experiment and theory point to an electron-phonon interaction. This is the mechanism responsible for the major portion of the electrical resistance of metals at ordinary temperatures, so it may appear surprising that the same mechanism should favour the occurrence of superconductivity. On the other hand, experiment also shows that most of the pure metal superconductors are poor conductors ordinarily. The best metallic conductors, the monovalent metals, have so far not been observed to be superconducting at the lowest temperatures at present available (of order $10^{-3}$ K).

As far as the internal order in a superconductor is concerned, Cooper's demonstration that two electrons excited to levels above an otherwise fully occupied calm Fermi sea would, in the presence of an attractive interaction, no matter how weak, form a bound pair, was a most important result. The discrete pair state has zero total spin and is most stable when its centre of mass is at rest, i.e. when the two electrons forming the pair have equal and opposite wave vectors. Nevertheless, the pair state can be given momentum and move through the lattice, but if its kinetic energy approaches the pair state binding energy then it dissociates into two independent electrons. The spatial extension of the pair wave function is of order $10^{-4}$ cm when the binding energy is of order $k_B T_c$, so there is already a hint of a significant dimension similar to the coherence length.

How can the electron-phonon interaction result in an indirect attractive force between electrons? The electrons and the ions that form the lattice are oppositely charged. Any electron in its motion through the lattice polarizes the lattice along its path, and the polarization remains during the time it takes for the pulse of phonons making up the ion displacements to disperse. This time for dispersion is sufficiently long to allow another electron to experience the local polarization caused by the first electron. Its potential energy is thereby reduced, so the process may be considered as an effective attraction between electrons. The attraction is greatest for electrons with opposite spins and equal and opposite wave vectors. For strong electron-phonon coupling and large values of $N(E_F)$ the attraction is sufficient to overcome the ordinary screened Coulomb repulsion between the electrons involved (which are far from one another), so that there arises a net attractive force that allows the creation of the Cooper pair.

Bardeen, Cooper and Schrieffer developed their theory on the premise that the electron–photon interaction provides a means for creating Cooper pairs, but in principle the development of the general theory is independent of the origin of the interaction, provided that it leads to a net attraction. However, all the evidence, experimental and theoretical, that has accumulated since the publication of the theory in 1957 has confirmed the initial assumption of electron-phonon interaction as the true source of the attractive force between electrons in ordinary metallic superconductors. The heavy Fermion and high temperature cuprate superconductors are exceptional and at present there is no consensus regarding the origins of superconductivity in these materials.

## 13.8 Interacting Pairs

In the case of ferromagnetism the occurrence of an atomic moment is a prerequisite, but it is not sufficient in itself to produce ferromagnetism; we need a mechanism for the coupling of the atomic moments on different atoms. Similarly, the creation of Cooper pairs is a necessary but not a sufficient condition for the creation of superconductivity. The pairs must interact in a cooperative fashion to produce a new stable electronic structure completely different from that of the ordinary metal. This new structure is the ground state of the superconductor – the condensate.

The wave function of a given Cooper pair is so extended in space that it overlaps those of some $10^6$ similar stationary pairs. The interactions between these overlapping pair states produce the new superconducting ground state. The 'interaction' may be described in terms of a transition between two *degenerate* electron configurations that differ only in that a pair state $\uparrow k_+$, $\downarrow k_-$ changes to $\uparrow k'_+$, $\downarrow k'_-$; clearly for this to happen the states $k$ must initially be occupied and the states $k'$ empty.

The wave function of a many-electron system is ordinarily constructed as a superposition of antisymmetrical products of the individual single-particle wave functions. Each product corresponds to a particular configuration of electron coordinates. Because of the indistinguishability of electrons, the different configurational products are degenerate and equally probable, and each product must, according to the Pauli principle, be antisymmetric. This means that any arbitrary configuration of coordinates is just as likely to produce a product wave function with negative as with positive sign. When the electrons behave as independent particles, the two-particle interactions involve the products of two different wave functions (one for each electron configuration) and, because of the equal probabilities of positive and negative sign, the sum of two-particle interactions averages to zero. On the other hand, if the interactions arise between electron configurations that differ only with respect to the occupancy of pairs of states then the appropriate products of wave functions are always positive. In the event of an attractive two-particle potential this produces a pairing interaction that leads to a reduction in energy that becomes larger the greater the number of pair transitions that can occur.

Which electrons form Cooper pairs? Since the attractive interaction is mediated by the phonons, there must be an exchange of energy and momentum. It is usual to represent the interaction in the manner of Fig. 13.12. To form a Cooper pair, electrons must be excited above the Fermi energy. Because of the Pauli principle, only electrons very close to the Fermi energy may receive small energy additions. Since, in a Debye model, the phonons are limited to energies $\leqslant \hbar\omega_D$, Bardeen, Cooper and Schrieffer assumed that only electrons within a surface layer $E_F \pm \frac{1}{2}\hbar\omega_D$ could form pairs ($\frac{1}{2}\hbar\omega_D$ is an approximate average phonon energy). This implies that $\frac{1}{2}N(E_F)\hbar\omega_D$ electrons have access to $N(E_F)\hbar\omega_D$ states that may be used for pair interactions. Another description is to say that electrons with excitation energy $>\hbar\omega_D$ have effective excitation frequencies lying outside the vibrational response of the lattice and cannot therefore produce the ion displacements necessary for the positive interaction.

We arrive at the apparently paradoxical situation whereby an increase in the kinetic energy of electrons near the Fermi level leads to the formation of Cooper pairs, thereby causing a reduction in potential energy that more than compensates for the initial excitation and any screened Coulomb repulsion between the electrons. The total energy is thereby reduced relative to that of the conventional one-electron description of states in a metal. Under such conditions and at low enough temperatures the Cooper pairs form spontaneously. The pair wave functions all have the same form, and the superposition of the pair wave functions describes what is known as the 'condensate'. The condensate or aggregate of interacting pairs represents a new homogeneous entity in the electronic structure.

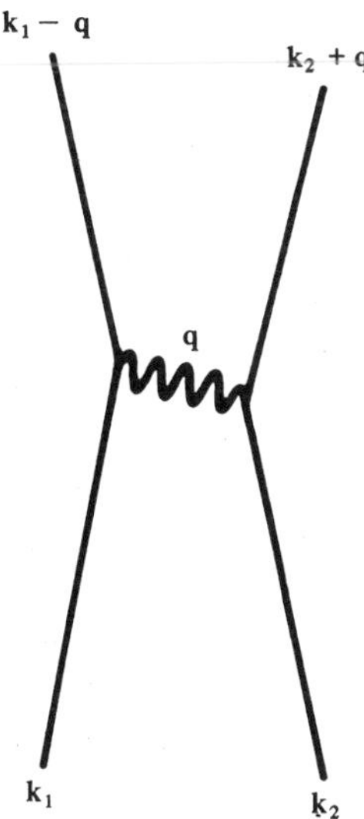


**Figure 13.12** A schematic picture of phonon-mediated electron interaction. An electron with wave vector $\mathbf{k}_1$ polarizes the lattice, creating a phonon with wave vector $\mathbf{q}$. Another electron with wave vector $\mathbf{k}_2$ absorbs the phonon. The end result is two electrons with wave vectors $\mathbf{k}_2 + \mathbf{q}$ and $\mathbf{k}_1 - \mathbf{q}$.

## 13.9 The Condensate

A superconductor consists of electrons occupying conventional one-electron states and the condensate, which is a single quantum state available only to overlapping interacting Cooper pairs (Fig. 13.13). The separate pair wave functions have the same frequency (i.e. energy), the same wavelength (i.e. momentum) and an invariant phase relationship. The condensate is described by the product of the pair wave functions, and in the absence of electric and magnetic fields the centre of mass is stationary. The complete system of interacting pairs is phase-locked to form a unique entity. If a pair is removed from the condensate then it must be 'broken' into two independent electrons; this demands an energy $2\Delta$, the binding energy of a Cooper pair within the condensate.

But why should this give rise to superconductivity? The answer lies in the unique character of the condensate, its coherence and the associated energy gap. Within the condensate all the Cooper pairs must have the same energy and be phase-locked. If the condensate moves, it must move coherently without altering the energy or the phase-locking of the pairs. The only way in which a pair can behave differently from any other pair in the condensate is by its ceasing to be a pair, and this demands an energy $2\Delta$. If a uniform electric field is applied to a superconductor, all the pairs in the condensate experience the same force and move in the same direction – they constitute a current. If the condensate is to maintain its unique character, all the pairs must maintain the same energy and phase relationship.

Suppose a voltage difference $V$ existed across the two ends of the superconductor, then any Cooper pair traversing the superconductor would gain energy $2eV$. If this were to happen then different parts of the condensate would have different energies. Similarly, after a time $t$ there would arise a phase shift of $2eVt/\hbar$ between

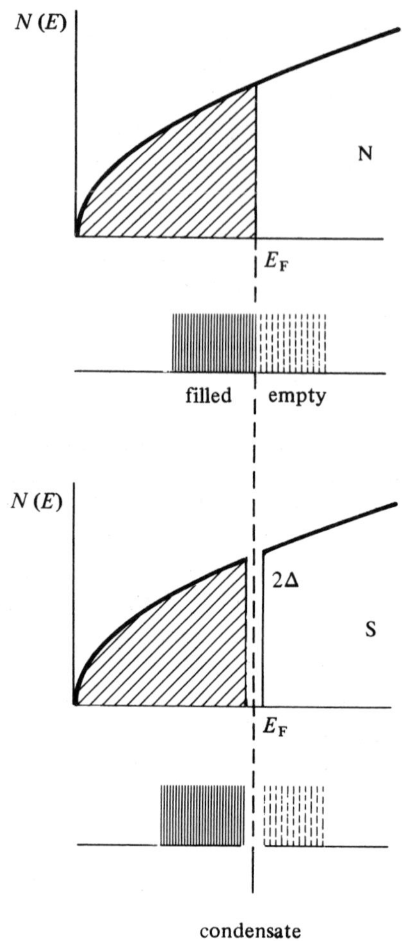


**Figure 13.13** Electron states, filled and empty, as they arise in the normal and superconducting states. The condensate exists only in the superconducting state and it is drawn at the Fermi level – but being a collective ground state it cannot be represented on the conventional band diagram. $2\Delta$ is the energy gap associated with the presence of the condensate (see Table 13.3).

**Table 13.3** Some values of the energy gap

| Element | $2\Delta(0)$ (meV) |
|---|---|
| Al | 0.34 |
| Sn | 1.15 |
| Pb | 2.73 |
| V | 1.6 |
| Nb | 3.0 |
| Ta | 1.4 |

pairs at the two ends of the superconductor. The overlapping pairs would no longer have the same energy or the same phase relationship and the condensate would be lost. To maintain its existence, the condensate must therefore move and carry current without producing a potential difference. It should also be noted that in the absence of magnetic impurity atoms there are no scattering mechanisms for the pairs.

In simple terms we may say that the condensate exists as a homogeneous phase-locked entity, or it does not exist at all. If it exists, it may carry current; and, provided the energy associated with its motion is insufficient to destroy it, there will be no voltage drop associated with its motion, because it cannot receive energy in amounts less than $2\Delta$, the energy needed to break a single pair from the condensate. At absolute zero the condensate is destroyed by an external magnetic field of strength $B_{c0}$ or a supercurrent that is sufficient to generate this field.

What happens if we increase the temperature above zero? An increase in temperature produces a certain thermal excitation of the condensate and there is a finite probability that pairs be broken. The density of pairs within the condensate decreases, thereby causing it to become less stable. The pair binding energy decreases and the critical field becomes smaller. Otherwise the condensate maintains its unique properties. Owing to the very large density of Cooper pairs at absolute zero,† the initial reduction in numbers has little effect on $2\Delta$, but as the temperature is gradually increased the cooperative stability of the condensate is slowly diminished, leading to a progressive reduction in $2\Delta$. This process becomes increasingly rapid as $T$ approaches $T_c$ (Fig. 13.14). At $T_c$, $2\Delta = 0$, the density of pairs is zero; the condensate ceases to exist; but so long as it occurs there is a complete absence of electrical resistance.

In its simplest form the BCS theory leads to the following quantitative formulae:

(a)
$$T_c = \theta_D \exp\left[\frac{-1}{VN(E_F)}\right], \tag{13.28}$$

where $V$ represents the effective electron-photon interaction and $N(E_F)$ the density of states of a given spin at the Fermi level – it is understood that $VN(E_F) \ll 1$, the so called weak coupling limit.‡

(b)
$$W = -\frac{B_{0c}^2}{2\mu_0} = -\frac{1}{2}N(E_F)\Delta^2, \tag{13.29}$$

where $W$ is the condensation energy per unit volume;

(c)
$$2\Delta = 3.53 k_B T_c \quad \text{for} \quad T = 0; \tag{13.30}$$

(d)
$$N(E') = \begin{cases} N(E_F)\dfrac{|E'|}{(E'^2 - \Delta^2)^{1/2}} & (E' \geqslant \Delta), \\ 0 & (E' < \Delta), \end{cases} \tag{13.31}$$

where $N(E')$ is the density of independent electron states excited from the condensate, $E'$ being the excitation energy measured relative to the Fermi energy.

† Density of pairs in lead at 0 K: $N(E_F) \approx 0.3$ eV$^{-1}$ at$^{-1}$, $2\Delta(0) \approx 2.7 \times 10^{-3}$ eV; therefore $N(E_F)\Delta(0) \approx 4 \times 10^{-4}$ at$^{-1}$ $\approx 1.3 \times 10^{19}$ cm$^{-3}$, say of order $10^{19}$ pairs cm$^{-3}$.

‡ When this condition does not apply strong coupling holds as in the $\beta$ tungsten compounds; the BCS theory then requires modification.

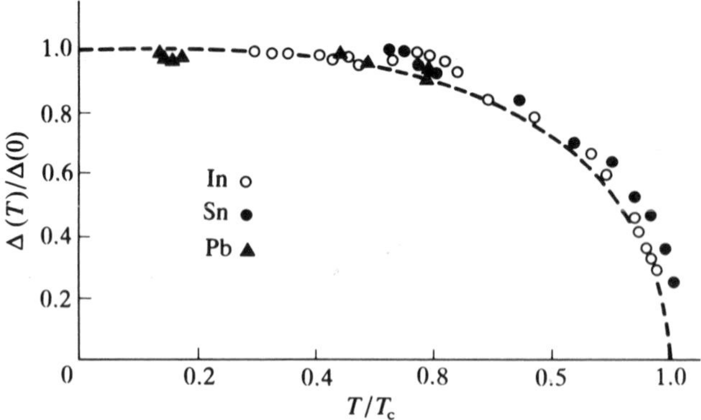


**Figure 13.14** The variation of the gap energy as a function of temperature. The dotted curve is the variation predicted by the BCS theory, whereas the data points were obtained in tunnelling experiments as described in Section 13.11. (After Giaever and Megerle 1961.)

## 13.10 Energy Spectrum for Single-Particle Excitations (Quasiparticles)†

Near absolute zero a superconductor consists of the condensate and ordinary electrons; the latter occupy single-particle states with energies below $E_F - \Delta$ whereas states with energies greater with $E_F + \Delta$ are empty (Fig. 13.13). If we add electrons from an external source, or create them by breaking pairs, then we say that single-particle excitations, often called quasiparticles, are produced. The density of states of these excitations, $N(E')$, is calculated to have the form (13.31) as illustrated in Fig. 13.15. In this figure the upper horizontal line represents the density of electron states at the Fermi energy for the normal material, whereas the peaked curves describe $N(E')$. It is readily calculated from (13.31) that there are $N(E_F)\Delta$ of one-electron states contained within each of the two peaks symmetrically placed about $E_F$.

At absolute zero there are no free single-particle-like excitations (Fig. 13.13), and we can, purely for convenience, represent this situation in a manner analogous to the perfectly insulating semiconductor state. Thus a representation of Fig. 13.13, so far as single-particle excitations are concerned, is that shown in Fig. 13.15. We consider the lower band of $N(E')$ to be completely filled and the upper counterpart completely empty. An increase in temperature causes pair breaking, leading to the occurrence of unpaired electrons. This, in terms of Fig. 13.15, is pictured as an excitation of electrons across the gap $2\Delta$, thereby producing free single-particle states in both the lower and the upper bands. As the temperature increases, $\Delta$ becomes smaller, the condensate becomes more and more dilute and when $T = T_c$ we regain the conventional normal-state band form (Fig. 13.16). We emphasize that this 'semiconductor' model for single-particle excitations is used only as a convenient analogy to facilitate a pictorial description of the formation of single quasiparticles from the condensate. In particular, we should always remember that the energy gap $2\Delta$ is tied to the Fermi surface and has no connection with the energy

† The single particles (i.e. electrons) are called quasiparticles because they represent an excitation from the ground state of the system. There is no phase coherence of the wave functions of the different quasiparticles and they cannot give rise to those properties peculiar to the pair condensate.

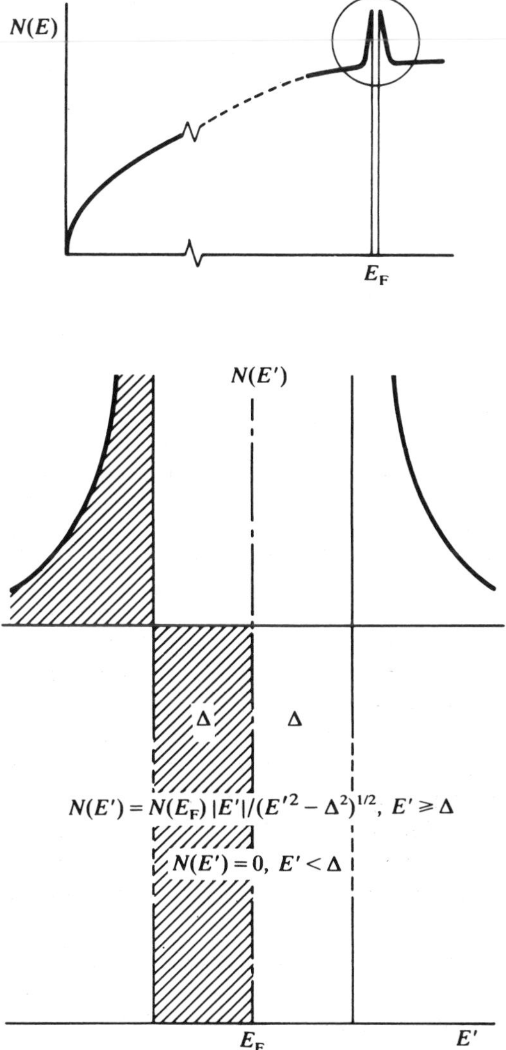


**Figure 13.15** The density of quasiparticle (i.e. electron) excitations $N(E')$ in the S state at 0 K. $E'$ is measured relative to the Fermi level. At 0 K the situation is similar to that of an intrinsic semiconductor. No current can be carried by the electrons because all occupied states are separated from the empty states by the energy gap $2\Delta$. The shaded areas are equal. Single-particle states normally occupied in the absence of the gap have been accommodated below the gap; similarly, the empty states just above the Fermi level have been pushed to higher energy. The quasiparticle states are restricted to a narrow energy interval near $E_F$ and are insignificant in size on the scale of the conventional energy band structure.

gaps at the Brillouin zone boundaries that are so significant for the behaviour of ordinary metals and semiconductors.

Nevertheless, the above description is useful in a discussion of tunnelling between two superconductors separated by a very thin insulating layer, a process to be treated in the next section, as well as providing a qualitative understanding of the heat capacity and thermal conductivity of a superconductor.

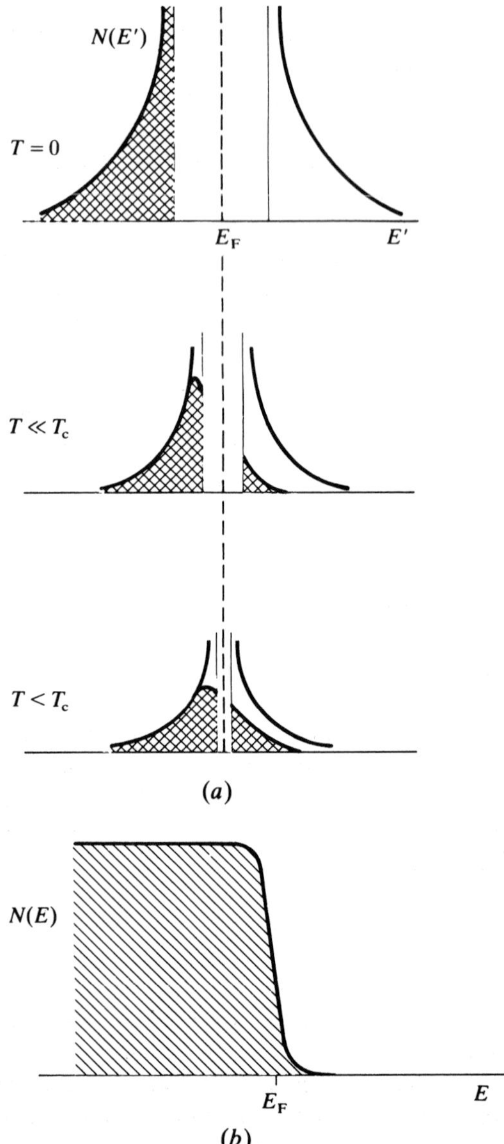


**Figure 13.16** (a) The effect of temperature on the density of quasiparticle states; because the energy gap decreases with temperature, the number of quasiparticle states also decreases. The base line is $N(E_F)$, as is illustrated in Fig. 13.15. When $T = T_c$, as in (b), the energy gap has disappeared and with it the quasiparticle states – the normal state band structure is regained.

At very low temperatures the electronic component of the heat capacity of a superconductor is found to be exponentially dependent on temperature; this may be understood in terms of the energy gap $2\Delta$ separating the single-particle excitations from the condensate. At low temperatures $2\Delta$ varies slowly with temperature (Fig. 13.14), thereby leading to a fixed excitation energy and the exponential dependence of the heat capacity. For a pure metal like Cu, Sn or Pb the conduction of heat in the N state depends strongly on the thermal excitation of electrons in states near the Fermi level. By interacting with the phonon gas, the excited electrons revert to empty states immediately below the Fermi level and energy is conveyed to the lattice. However, before this inelastic collision takes place, the electrons travel a distance $\mathbf{v}_F \tau$, $\tau$ being the relaxation time for electron-phonon collisions. Thus in the presence of a temperature gradient energy is transferred from the hotter to the colder regions of the sample. There is no net flow of charge because cold electrons, with a similar mean free path $\mathbf{v}_F \tau$, flow into the hotter region.

Clearly the superconducting condensate cannot conduct heat well because it cannot receive energy in amounts less than $2\Delta$. Furthermore, the creation of the condensate removes the single-particle states ordinarily most useful in thermal conduction, i.e. those near $E_F$. So the marked reduction in the thermal conductivity, which is one characteristic of the N–S transition, is immediately understandable on the basis of the BCS superconducting ground state. Admittedly, for temperatures in the range $0 < T < T_c$ there always exist broken pairs, leading to adjacent empty and filled single-particle states in both upper and lower bands (Fig. 13.16), but their contribution to the thermal conductivity is very limited until $T$ is of order $T_c$.

## 13.11 Giaever Tunnelling

Tunnelling is the name given to the quantum-mechanical penetration of potential barriers that in classical terms would be insurmountable. Here the barrier is a thin insulating layer of oxide separating either an ordinary metal and a superconductor or two superconductors; we consider the latter case. Experimentally, the 'tunnel junctions', as they are usually called, are formed by first evaporating, through a screen or 'mask', an aluminium strip, say, approximately 1 mm wide and 200 nm thick, onto a clean glass microscope slide. This strip is then exposed to the atmosphere (or to a partial atmosphere of oxygen) until an oxide layer some 1–1.5 nm has formed. A second strip of similar dimensions, but now of the other metal, let us say tin, is evaporated diagonally across the first. The junction is then a small 1 mm$^2$ sandwich of Al and Sn, the 'filling' being the aluminium oxide insulating layer. Electrodes to which leads may be soldered are also evaporated onto the slide. Such a junction may present a resistance of 50 Ω. The oxide layer must be sufficiently thin to allow the wave function of an electron in the Al to penetrate the oxide layer and have finite amplitude in the other metal, in our case Sn. The junction (Fig. 13.17) is used in a simple series circuit. With zero applied voltage, the Fermi levels of the two metals coincide and no electron current flows. We now bias the junction so that the Al is positive with respect to the Sn (Fig. 13.18); a small electron current then flows from the Sn to the Al. It is small because the current arises from the small density of electrons excited to levels above the energy gap of the Sn that can tunnel through the oxide and find available empty states in the Al. Remember that tunnelling demands that an electron in an occupied state has an empty state available at the same energy on the opposite side of the barrier.

As the bias is increased, the lower limit of the Al energy gap approaches that of the Sn, and the current is augmented by electrons flowing from the nearly filled Sn band below the energy gap to the empty levels in the corresponding band of the Al. The current attains a subsidiary maximum value when these lower gap energies coincide, and ceases to increase (it may in fact decrease owing to the decrease in available state density) with further increase in the bias because the almost-filled Sn

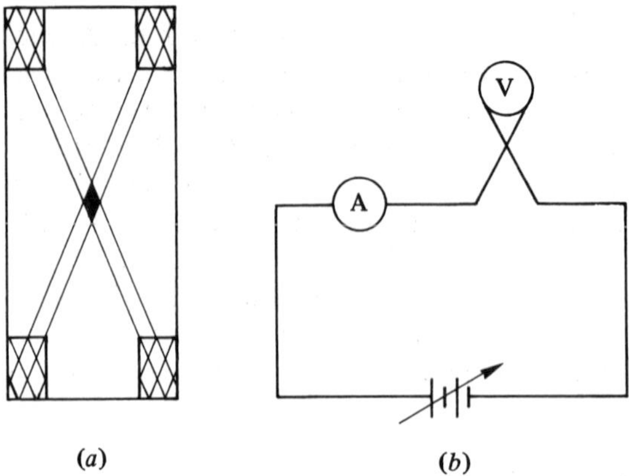


**Figure 13.17** (a) A tunnelling junction formed at the crossover of two thin metal strips evaporated onto a microscope slide. (b) A simple DC tunnelling circuit.

**Figure 13.18** The quasiparticle states in a $Sn/Al_2O_3/Al$ junction. As the bias voltage changes, the Al states slide past those of the Sn in the manner shown, leading to the current-voltage characteristic presented in the lower diagram. In this way one may determine $\Delta(Sn) - \Delta(Al)$ and $2\Delta(Al)$. Points marked $V_1$, $V_2$ and $V_3$ correspond to the situations shown in the upper part of the figure.

band begins to overlap the lower half of the Al energy gap and there are no states in the Al to receive electrons from the Sn. This situation holds until the upper limit of the Al energy gap coincides with the lower boundary of the Sn gap. With a further increase in the bias we now find the 'filled' Sn band opposite the 'empty' Al band; this is an ideal arrangement and the current increases markedly with increased bias.

The complete tunnelling characteristic has the form shown in Fig. 13.18; it allows direct measurement of $\Delta(\mathrm{Sn}) - \Delta(\mathrm{Al})$ and $2\Delta(\mathrm{Al})$. A series of experiments at different temperatures provides information regarding the variation of $\Delta$ with temperature. Such results were first obtained by Giaever in 1960. They have vindicated the BCS theory in great detail and furthermore have led to new problems of physical interest.

## 13.12 Phase and Momentum†

### 13.12.1 *Kinematic and canonical variables*

In classical mechanics Hamilton's equations are expressed in terms of the canonical or 'ruling' variables $p_i$, $x_i$. In the absence of charged particles and magnetic fields, the quantity $\mathbf{p}$ is the same as the ordinary or kinematic momentum, i.e. $\mathbf{p} = m\mathbf{v}$. However, when a particle carries charge $q$ and moves in a magnetic field associated with the vector potential $\mathbf{A}$, the canonical ($\mathbf{p}$) and kinematic ($m\mathbf{v}$) momenta are no longer the same but are related by

$$\mathbf{p} = m\mathbf{v} + q\mathbf{A}, \tag{13.32}$$

so

$$m\mathbf{v} = \mathbf{p} - q\mathbf{A} \tag{13.33}$$

and

$$\tfrac{1}{2}mv^2 = \frac{1}{2m}(\mathbf{p} - q\mathbf{A})^2. \tag{13.34}$$

### 13.12.2 *Quantum mechanics*

In quantum mechanics the distinction between kinematic and canonical momenta is maintained and the *canonical* momentum is replaced by an operator:

$$\mathbf{p} \to -i\hbar\nabla. \tag{13.35}$$

For a charged particle moving in a vector potential $\mathbf{A}$ we find the corresponding expressions for the kinematic momentum and the kinetic energy to be

$$m\mathbf{v} \to -i\hbar\nabla - q\mathbf{A}, \tag{13.36}$$

$$\tfrac{1}{2}mv^2 \to \frac{1}{2m}(-i\hbar\nabla - q\mathbf{A})^2. \tag{13.37}$$

† The content of the following sections is based on the treatment given in Feynman *et al.* (1965), Chap. 21.

### 13.12.3 *Particle currents*

The flow of particles constitutes a material current, which may be described by a current density vector

$$\mathbf{J} = \frac{\hbar}{2mi}(\psi^*\nabla\psi - \psi\nabla\psi^*), \tag{13.38}$$

or, in the case of charged particles moving in a vector potential **A**,

$$\mathbf{J} = \frac{1}{2m}\{[(-i\hbar\nabla - q\mathbf{A})\psi]^*\psi + \psi^*(-i\hbar\nabla - q\mathbf{A})\psi\}. \tag{13.39}$$

In the latter case, to obtain the electrical current vector, we must multiply **J** by the charge $q$ on each particle.

### 13.12.4 *The condensate wave function*

As we have already seen, the condensate contains Cooper pairs with a density of order $10^{19}\ \mathrm{cm}^{-3}$, and these pairs are all phase-locked, producing a unique electrical fluid. The energy of the condensate is extremely low. The density must be uniform, otherwise there would be rapid variation of the wavelength, leading to higher-energy components. We write a pair wave function

$$\psi = n^{1/2}e^{i\theta}, \tag{13.40}$$

where $n$ is the pair density and $\theta$ a phase. Since all the Cooper pairs are phase-locked, $\theta$ becomes a macroscopic quantity characterizing the condensate. We insert (13.40) into (13.39) to obtain

$$\mathbf{J} = \frac{\hbar}{m^*}\left(\nabla\theta - \frac{q\mathbf{A}}{\hbar}\right)n, \tag{13.41}$$

where $m^*$ is the mass of the Cooper pair. Since $\mathbf{J} = n\mathbf{v}$, we readily see that (13.41) is equivalent to

$$m^*\mathbf{v} = \hbar\nabla\theta - q\mathbf{A}; \tag{13.42}$$

in other words, the canonical momentum is associated with the spatial gradient of the phase.

## 13.13 Flux Quantization

In equation (13.26) we have an expression for London's flux quantum. We are now in a position to derive this expression using (13.42). In a hollow superconducting cylinder, currents flow only in the outer skin of material; there is no current in the body of the superconductor, so there can be no resultant kinematic momentum of the pairs. Thus $m\mathbf{v} = 0$ and we obtain

$$\nabla\theta = \frac{q}{\hbar}\mathbf{A}. \tag{13.43}$$

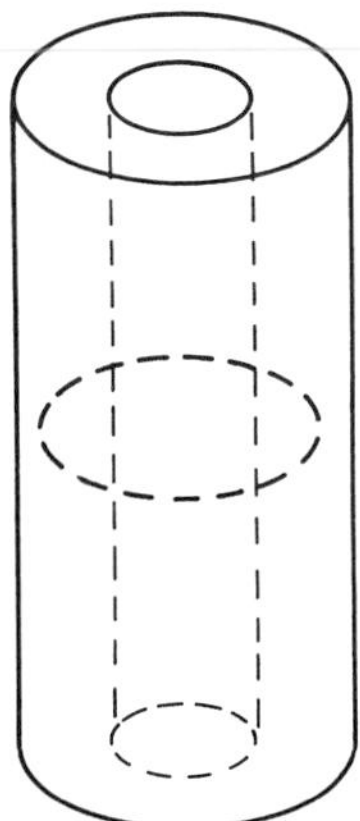

**Figure 13.19** In a hollow thick-walled superconducting cylinder any current is confined to the outer surface skin of thickness of order $\lambda$; the line integral around a contour within the body of the superconductor is in 'current-free' space.

If we now take the line integral around a closed path within the superconductor (Fig. 13.19) then

$$\oint \nabla\theta \cdot \mathbf{dl} = \oint \frac{q}{\hbar} \mathbf{A} \cdot \mathbf{dl}. \tag{13.44}$$

But the phase at our common starting and finishing point of the line integral must be uniquely determined, so the phase integral must be equal to zero or an integral multiple of $2\pi$. Thus

$$\oint \frac{q}{\hbar} \mathbf{A} \cdot \mathbf{dl} = 2\pi p. \tag{13.45}$$

But $\oint \mathbf{A} \cdot \mathbf{dl}$ is a measure of the magnetic flux $\Phi$ threading the hollow superconductor, so

$$\Phi = \frac{2\pi\hbar}{q} p = p \frac{h}{2e}. \tag{13.46}$$

The flux trapped within the superconductor is therefore quantized in units

$$\frac{h}{2e} = 2 \times 10^{-15}\ \text{Wb}$$
$$= 2 \times 10^{-7}\ \text{maxwell}.$$

Experiment has verified London's prediction.

## 13.14 Josephson Tunnelling

When two metals, either in the normal or the superconducting state, are in close, but not perfect, contact with one another on account of the presence of a thin oxide layer, we know that electrons, independently, may pass through the barrier by quantum-mechanical tunnelling. The probability for tunnelling is small, of order

$10^{-10}$. One may ask whether Cooper pairs may also tunnel through such thin layers. At first sight this seems very unlikely because if two electrons are simultaneously to cross the barrier then the probability becomes of order $10^{-20}$ and the process undetectable. However, the electrons comprising a Cooper pair are no longer to be considered as independent particles but components of a new entity the pair. The electrons are inextricably coupled and behave in a coherent fashion. So the probability of tunnelling remains that of a single particle. The tunnelling of pairs is named after Josephson, who, in a very short theoretical paper, described the behaviour to be expected and pointed out many possible associated applications.

Two superconductors separated by a thick insulating layer, say 10 nm, are essentially two independent superconductors without any joint properties. On the other, hand two superconductors in perfect contact are, electromagnetically speaking, a single superconductor; but when they are separated by a thin insulating layer (1 nm thick) they become a system of coupled superconductors in a sense analogous to that used when we speak of coupled circuits or coupled oscillators. Because of this, the tunnelling barrier is often known as a 'weak link'. It is customary to discuss Josephson tunnelling in terms of the 'DC' and 'AC' aspects.

### 13.14.1 The DC effect

To begin with, we assume that no magnetic field is present and that any fields caused by supercurrents are negligibly small. The DC effect is then most simply described by saying that the weak link (i.e. the oxide layer) separating the two superconductors itself becomes a superconductor. Thus, if such a junction is joined to a current generator, a current flows through the junction without any voltage drop. The condensates in the two superconductors are no longer independent; they have leaked into the oxide barrier, joined up and formed a continuous fluid. Of course the order parameter and thereby the concentration of pairs is very much smaller in the link than in the flanking superconductors. There is a common Fermi level for the whole system, and this means that pairs can tunnel from one side of the junction to the other without transfer of energy. This makes it possible to have current flow in the absence of a voltage difference. The phases characterizing the two bulk superconductors are neither the same (as when perfectly joined) nor independent of one another (as when completely separated). There arises a distinct phase difference $\Delta\theta$ between them. Josephson showed that the DC current flowing through the junction is related to this phase difference according to the equation

$$I = I_0 \sin \Delta\theta. \tag{13.47}$$

Thus if we have a perfect current generator and couple it to the weak link then the phase difference across the link is determined by the amount of current we push through it. Similarly, if by any means (particularly magnetic flux) we change the phase difference then we alter the current flowing. If we try to push current greater than $I_0$ through the link then we obtain normal resistive behaviour. The current $I_0$ is a characteristic parameter for a given junction; it is related to the tunnelling conductance between the junction metals in their normal states. We shall not concern ourselves with the detailed composition of $I_0$; for us, it is merely a junction parameter.

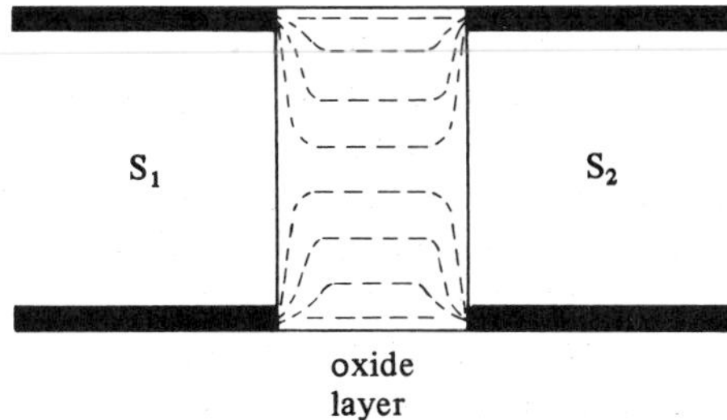


**Figure 13.20** In the oxide layer of a junction the supercurrent is uniformly distributed because $\lambda$ is very large.

Let us assume that we have a junction formed by a narrow thin oxide layer between two pieces of the same superconducting metal, e.g. a Pb-PbO-Pb junction. Because of the lower concentration of pairs in the oxide layer, the penetration depth $\lambda$ is much larger than in bulk superconductors (see (13.23)). Thus, although in the bulk superconductor the current is confined to the surface layers, in the junction itself it is essentially uniformly distributed (Fig. 13.20).

We first discuss the behaviour of a perfectly symmetrical circuit containing two identical junctions. For convenience we assume these junctions to be 'point junctions' because their thicknesses and cross-sections are so very much smaller than the dimensions of the complete circuit loop (Fig. 13.21). We now ask how the DC Josephson current varies when a magnetic field threads the circuit. We recall that for a bulk superconductor (13.42) holds; that is,

$$\begin{aligned} \nabla\theta &= \frac{m^*\mathbf{v}}{\hbar} + \frac{q\mathbf{A}}{\hbar} \\ &= \frac{q}{\hbar}\left(\mathbf{A} + \frac{m^*}{q^2 n}\mathbf{j}\right), \end{aligned} \tag{13.48}$$

where

$$\mathbf{j} = nq\mathbf{v}$$

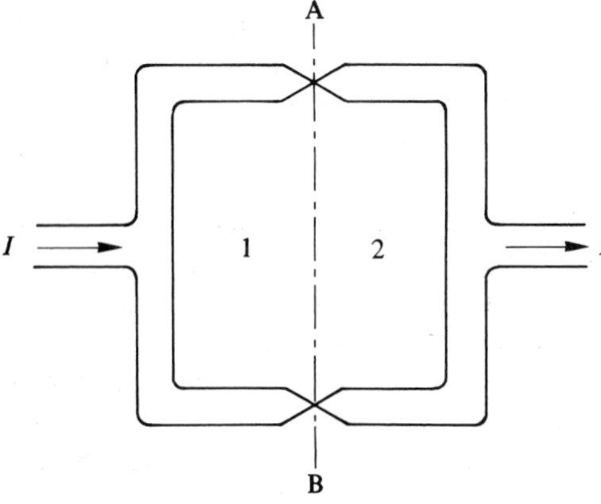


**Figure 13.21** The superconducting analogue of Young's double-slit experiment in optics.

is the electrical current density. We have written $q$ as the charge in the pair, so $q = -2e$. The gradient in the phase depends upon the magnetic vector potential and the current density. We have already used this equation to determine the quantum of magnetic flux and found that by a suitable choice of line integral we could avoid the term in **j**. This is also possible in other circumstances, so usually we can assume that $\nabla\theta$ depends only on **A**.

In the absence of a magnetic field, our symmetrical circuit contains two identical junctions assembled in parallel; let us call them A and B. The phase difference $\Delta\theta$ is the same for both links, so we write

$$I = 2I_0 \sin \Delta\theta. \tag{13.49}$$

We now apply a uniform magnetic field $B_0$ perpendicular to the plane of the circuit loop. In the absence of this field, the phases on the left-hand sides of the two junctions are the same; so are those on the right-hand sides. However, this is not the case in the presence of the field; then $(\Delta\theta)_A \neq (\Delta\theta)_B$. The difference, which we denote by $\delta(\Delta\theta)$, is found using (13.48). Denoting the left- and right-hand sides of the junctions by subscripts 1 and 2 respectively, we have

$$\delta(\Delta\theta)_1 = \frac{q}{\hbar}\int_{B_1}^{A_1} \mathbf{A}\cdot d\mathbf{s}.$$

The path **s** lies in the body of the left superconductor, where $\mathbf{j} = 0$ on account of the small penetration depth. Similarly,

$$\delta(\Delta\theta)_2 = \frac{-q}{\hbar}\int_{B_2}^{A_2} \mathbf{A}\cdot d\mathbf{s}.$$

Thus the total difference in phase difference between junctions A and B becomes

$$\begin{aligned} {}_A\delta_B(\Delta\theta) &= \frac{q}{\hbar}\left(\int_{B_1}^{A_1} \mathbf{A}\cdot d\mathbf{s} + \int_{A_2}^{B_2} \mathbf{A}\cdot d\mathbf{s}\right) \\ &= \frac{q}{\hbar}\oint \mathbf{A}\cdot d\mathbf{s} = \frac{q\Phi}{\hbar}, \end{aligned} \tag{13.50}$$

where $\Phi = \oint \mathbf{A}\cdot d\mathbf{s}$ is the total flux threading the circuit loop.

Thus the current $I$ in the presence of the field $B_0$ becomes

$$I = I_0 \sin \Delta\theta + I_0 \sin\left(\Delta\theta + \frac{q\Phi}{\hbar}\right).$$

Now $\Delta\theta$, the value in the absence of the field, may have any principal value in the interval $0 \leqslant \Delta\theta \leqslant \frac{1}{2}\pi$ and, without loss of generality, we can assume that it takes the value $\frac{1}{2}\pi$ and write

$$\begin{aligned} I &= I_0\left(1 + \cos\frac{q\Phi}{\hbar}\right) \\ &= 2I_0 \cos^2\frac{q\Phi}{2\hbar}. \end{aligned} \tag{13.51}$$

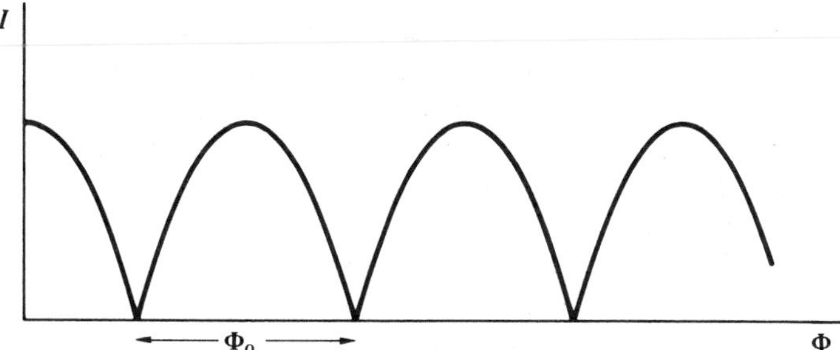


**Figure 13.22** The superconducting analogue of Young's fringes, in the absence of Fraunhofer diffraction.

Clearly the current varies in a periodic manner with $\Phi$ and therefore with $B_0$ (Fig. 13.22). The maxima in current appear when

$$\frac{q\Phi}{2\hbar} = m\pi \tag{13.52}$$

where $m$ is an integer.

The interval between two adjacent maxima, i.e. the period of the oscillations, is therefore

$$\delta\Phi = \pi\,\frac{2\hbar}{q} = \frac{h}{q}, \tag{13.53}$$

and, since $q$ is the charge of the Cooper pair, $\delta\Phi$ is none other than the quantum of magnetic flux, $\phi_0$.

Clearly the periodic variation of current arising as a result of the oscillatory changes in the difference of phase difference across the two junctions is a straightforward effect of interference. The circuit of Fig. 13.21 is the superconducting analogue of Young's two-slit experiment in optics. It represents a superconducting interferometer.

However, in the case of the optical experiment we know that the interference pattern is modulated by the diffraction pattern appropriate to a single slit. This diffraction pattern occurs because even in a single slit the phase of the light leaving a given element of the slit in a specific direction varies in a linear manner across the breadth of the slit. Can we expect analogous behaviour in the case of the single Josephson junction? In our discussion of interference the two junctions were considered to be truly point contacts, but they must have a finite area and thickness. In an external magnetic field, flux penetrates the thickness of each junction and the quantum phase varies (transversely to the current flow) across the thickness in a manner fully analogous to the variation of the phase of a parallel light beam incident on a single slit in optics. If the two junctions are geometrically identical then each causes the current through it to vary with the flux, producing the characteristic Fraunhofer pattern. The interfering currents discussed earlier are therefore modulated by the Fraunhofer envelope in exactly the same way as Young's fringes are.

A device such as the two-junction interferometer is known as a SQUID, an acronym for Superconducting QUantum Interference Device. We have seen that the two-junction interferometer produces current oscillations of period $\phi_0$. In principle this gives us a means of measuring $e/h$, but this requires that we can measure the flux to high accuracy. We shall soon see that other superconducting phenomena allow more accurate measurements of $e/h$, so for the present it is more appropriate to assume a value for this ratio and use our interferometer to measure magnetic flux; this can be done to an astounding sensitivity, of order $10^{-11}$ G. Commercial magnetometers using these principles are now available.

### 13.14.2 The AC effect

The previous discussion concerned the behaviour of the direct current arising in a weak line in the absence of an applied voltage. What happens if we apply a DC voltage and attempt to determine a current-voltage characteristic for the junction? There is of course no physical difference between a Josephson junction and one used for Giaever tunnelling; it is just that in the latter attention is directed to quasi-particle (i.e electron) currents whereas in the former the current carried by the pairs is the central feature. Thus our $I(V)$ curve appears as in Fig. 13.23. We find a quasi-particle threshold at $V = 2\Delta$ (assuming a symmetrical junction). There is also the DC Josephson current at $V = 0$. It would appear that there is nothing new to be learned, but that is because the simple DC characteristic does not reveal all that is going on. It was not until Josephson pointed out what to look for that strikingly new phenomena were observed. In the following we shall always assume the absence of any magnetic flux.

When the current through a superconducting junction exceeds the critical value $I_0$ the induced superconducting property of the oxide layer is destroyed and we obtain the normal state. The critical current is small for the oxide layer of a junction, and when the oxide enters the normal state electrons as well as Cooper pairs carry the current. The ordinary electrons cause a potential difference $V$ to appear across the junction, and the pairs also experience the associated electric field and are accelerated. On crossing the junction, the pairs acquire energy $qV$ and their phase

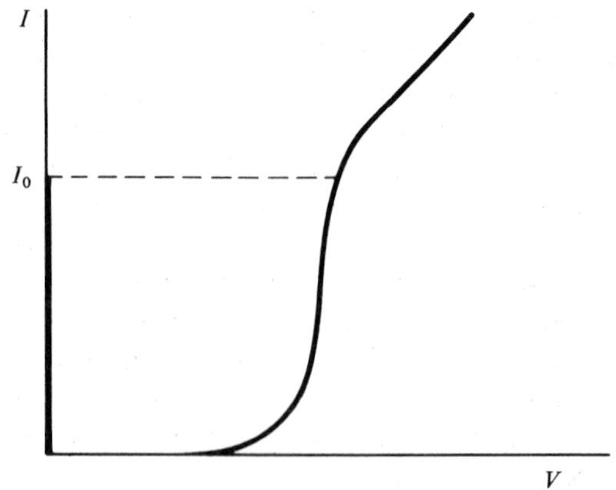


**Figure 13.23** The $I(V)$ characteristic for a Josephson junction. When current greater than the critical current of the oxide layer flows, this layer becomes normal and the applied voltage that arises is able to accelerate the Cooper pairs. This leads to the production of radio-frequency current not registered on the above characteristic.

varies with time according to

$$\frac{\hbar \cdot \mathrm{d}\theta}{\mathrm{d}t} = qV. \tag{13.54}$$

Thus, given an initial phase difference $\Delta\theta$, the application of a voltage $V$ across the junction causes $\Delta\theta$ to change to

$$\Delta\theta + \frac{qVt}{\hbar},$$

and the pair current according to (13.47) becomes

$$I = I_0 \sin\left(\Delta\theta + \frac{qVt}{\hbar}\right). \tag{13.55}$$

The pair current oscillates with angular frequency

$$\omega = \frac{qV}{\hbar}, \tag{13.56}$$

corresponding to a frequency

$$\nu = 483.6 \text{ MHz } \mu V^{-1}. \tag{13.57}$$

The oscillating pair current has an associated radiation field so the junction acts as an antenna. The power generated in the junction is very small, of order $10^{-10}$ W, and furthermore, because of the difficulties of impedance matching, it is very difficult to extract the radiation from the junction. Nevertheless, using resonant waveguide techniques, the existence of this radiation has been fully confirmed and the AC Josephson effect has been extensively studied – not least because there are many possible applications. The behaviour can be very complex, so we only describe certain principal features.

Equation (13.57) shows that very small voltages of order $10^{-4}$ V bring the frequency of the junction radiation into the region of 50 GHz ($\lambda = 6$ mm). Furthermore, the junction is tunable by the adjustment of the DC voltage difference. In various refined experiments, often making use of additional modulating microwave fields, the frequency of this radiation for a specific DC voltage has been accurately measured. If we know $v$ and $V$ then we may derive a value for $e/h$. The accuracy of the measurement depends primarily on that with which the voltage can be determined; nevertheless, the method gives $e/h$ to within 1 ppm, which is the most accurate determination of this ratio so far and has led to a reassessment of several of the fundamental constants. It is also clear that the AC Josephson effect provides a relatively straightforward and highly accurate method for the comparison of voltage standards, and many national laboratories now make use of this possibility.

Although the radio-frequency power developed in a Josephson junction is very small, the radiation arises in the interval between the far infrared and the microwave regions – one for which sources are scarce. Considerable effort has therefore been devoted to improving the power output. One possible way that has been tried, with some success, is to arrange the coherent oscillation of a series of perhaps as many as 40 or more junctions. $N$ junctions, each of power $w$, acting independently produce a total output $Nw$. If, however, the junctions operate coherently, we must add the separate currents before we square to determine the output. Thus coherent oscillation gives an output $N^2w$.

Under the action of DC and AC voltages (not to mention a magnetic field), the Josephson junction is a complex non-linear circuit element. This leads to further applications as a parametric amplifier, mixer or detector of very high-frequency radiation. Much effort is devoted to the construction of efficient high-sensitivity low-noise amplifiers and detectors in the 50–100 GHz region, particularly for use in radio astronomy. The Josephson junction is also a good detector for far-infrared radiation. The non-linear characteristic of the Josephson junction also makes it an excellent medium for the study of chaos.

Perhaps the most widespread use of Josephson junctions will be in computers. The basic unit of a computer is a bipolar element, one with two stable arrangements, that can be used as a switch. It has long been realized that a superconductor itself may form such a switch because it may easily be converted from the S state (producing no potential drop) to the N state when a voltage appears across it. The switching is arranged with the aid of another superconductor (say Pb), which when actuated produces a magnetic field larger than the critical field of the switch metal (say Sn). Such a unit is called a *cryotron*.

Similarly, a Josephson junction may be switched from the superconducting state, when current flows without a voltage drop, to the quasiparticle tunnelling state, when a voltage equal to the gap energy, typically of a few mV, appears. Again the switching is carried out with the aid of an auxiliary element, the so-called 'gate'. Such junctions may be incorporated in large-scale integrated circuits just as for semiconducting elements. There are, however, more complicated Josephson devices than the simple switch (e.g. SQUID), but our objective is merely to demonstrate the feasibility of superconductive switching. Computer development is directed to cheaper, smaller, more reliable, less power-consuming apparatus. But perhaps from the point of view of improving absolute computer performance, the most important goal is faster operation. The shorter the switching time, the more quickly complicated analyses may be made. The Josephson junction is superior to the ordinary cryotron in switching speed (of order 1 ns). However, the present opinion is that superconducting elements will only become economical in very large computers, larger than have hitherto been made. It is also the case that *semiconducting* materials are undergoing constant improvement.

## 13.15 High-Temperature Superconductivity

Prior to 1986 'high-temperature' superconductivity implied a $T_c$ in the interval of 15–23 K. The most important materials were the $\beta$-tungsten compounds (see Fig. 2.18), of which one important example is $Nb_3Sn$; this compound has $T_c = 18$ K and remains superconducting in a magnetic field $>20$ T. It finds commercial use, as either tape or multifilamentary cable, in the manufacture of high-performance superconducting magnets providing working fields up to 15 T at 4 K. The intense magnetic field subjects the windings to considerable mechanical stress, and the superconducting films or filaments must be bonded to copper to provide mechanical stability as well as good ordinary electrical and thermal conductivity as a safeguard against local failure of the superconductor.

Although there is much discussion of the technical application of superconductivity, high performance magnets and ultra-sensitive magnetometers are as yet the only major commercial developments; they have found use primarily in the research laboratory. An exception is the use of superconducting magnets in applications of nuclear magnetic resonance (see Section 15.3) to chemical analysis and to magnetic resonance imaging (MRI). Arrays of SQUIDS are also being developed to enable real time imaging of the brain and its response to external stimulus, a technique that is becoming known as magnetic source imaging (MSI). It is usually stated that the need to use large amounts of expensive liquid helium as refrigerant is the major obstacle to the technical development of superconducting equipment, for example in superconducting power transmission lines or rotating machinery.

The discovery of what are now called high-temperature superconductors, namely the ceramics based on copper oxide, may permit wide practical application of superconductivity since they allow helium to be replaced by the considerably cheaper and more effective coolant liquid nitrogen. Superconducting oxides were known prior to the recent discoveries, but their transition temperatures were invariably much lower than the best $\beta$-tungsten compounds; the highest $T_c$ values were found in $LiTi_2O_4$ (14 K) and $BaPb_{1-x}Bi_xO_3$ (13 K). Before 1986, the highest $T_c$ ever observed was 23.2 K, in $Nb_3Ge$. In fact, between the discovery of superconductivity in 1911 and the year 1973 the average rate of increase of $T_c$, in the persistent attempt to find new and better materials, was 0.3 K per annum, and by 1975 the rate had become virtually zero. The general opinion was that, as far as ordinary metals and alloys were concerned, the limit had been reached.

Bednorz and Müller's discovery of a $T_c \approx 30$ K in the system $La_{2-x}Ba_xCuO_4$ was therefore a major event in the study of superconductivity. As is always the case it did not take long before other workers reproduced their data and began to develop substances with still better properties leading to the discovery of a $T_c$ of 90 K in $YBa_2Cu_3O_7$ (this substance is usually called YBCO). The past decade has seen an enormous outburst of activity leading to the establishment of many national research and development programmes. Much attention has been directed to $YBa_2Cu_3O_7$ and its derivatives. At the time of writing the systems providing the highest confirmed transition temperatures are the cuprates containing thallium or mercury namely: $TlBa_2Ca_{n-1}Cu_nO_{2n+2+\delta}$, (Tl-12$(n-1)n$) and $HgBa_2Ca_{n-1}Cu_nO_{2n+3+\delta}$, (Hg-12$(n-1)n$). There is also an analogous BiSrCa cuprate system Bi-12$(n-1)n$. The bracketed symbols are a shortened representation of the chemical formulas where only the metallic ions are listed. Tl-2223 has $T_c =$ 125 K whereas that for Hg-1223 is, at atmospheric pressure, believed to be just above 130 K, but so far the pure phase has not been obtained. We shall return to these substances later. Here we shall attempt to describe principal features of these superconducting materials by reference mainly to the compounds $La_2CuO_4$ and $La_{2-x}Ba_xCuO_4$.

In stoichiometric form $La_2CuO_4$ is an antiferromagnetic insulator with Néel temperature 240 K. La has valence 3 and Cu valence 2, so there is perfect charge compensation between the electropositive and electronegative atoms. The only ion possessing an incomplete shell is the $Cu^{2+}$ ion which has nine 3d electrons or, more conveniently, one 3d hole; the ion also carries a magnetic moment. Recalling previous discussion of conduction in oxide semiconductors the insulating properties of $La_2CuO_4$ are attributed to the strong correlation effects arising from the charge and magnetic moment associated with the Cu holes.

The structure of $La_2CuO_4$ is orthorhombic and the bonding between the Cu and O ions predominantly covalent. When La (which is always trivalent) is replaced by Ba the charge balance is disturbed and an electron deficient structure is formed. In other words each Ba atom that replaces La introduces a new hole. We may look upon Ba as a hole dopant. However, it is thought that these extra holes are associated primarily with the oxygen sites in the Cu-O planes and furthermore, in contrast to the 3d holes, they are mobile. The Néel temperature decreases with increasing Ba content and at sufficiently large Ba concentrations the compound ceases to be antiferromagnetic. It adopts a tetragonal structure, becomes metallic and superconductivity arises at low enough temperature (however, if doped beyond a certain limit the system ceases to display superconductivity). Hall effect measurements confirm that the charge carriers are holes; their concentrations are relatively small ($\sim 5 \times 10^{21}$ cm$^{-3}$), but larger than in many oxide superconductors discovered earlier. It is also the case that measurement of the flux quantum in these substances shows that the superconducting charge carriers have charge 2e so hole pairing arises. Furthermore nuclear magnetic resonance Knight shift measurements (see Section 15.3) provide evidence for the anti-parallel alignment of the carrier spins. There seems to be no doubt that Cooper pairs are active in these materials.

The tetragonal structure of $(LaBa)_2CuO_4$ is shown in Fig. 13.24; it has $c/a > 1$ so the Cu-O separation is larger in the c direction than that in the Cu-O planes perpendicular to c. The particular feature to notice is the layered arrangement and that all the layers or planes contain O atoms, but the Cu, on the one hand, and the La and Ba, on the other, lie on separate planes. The arrangement is considered to be a collection of Cu-O planes that are separated by the other atoms. This is a feature common to all these high-$T_c$ materials; the metallic atoms other than Cu serve to separate the Cu-O planes and, provided the tetragonal structure is maintained, the chemical character of the separator atoms is not so critical, hence the large number of substitutions that can be made.

When there is only one Cu atom per formula unit it always lies at the centre of an octahedron formed by six surrounding oxygen atoms; this is readily seen in the middle portion of the unit cell depicted in Fig. 13.24. In certain compositions with higher copper content the octahedron splits into two pyramids, one upright the

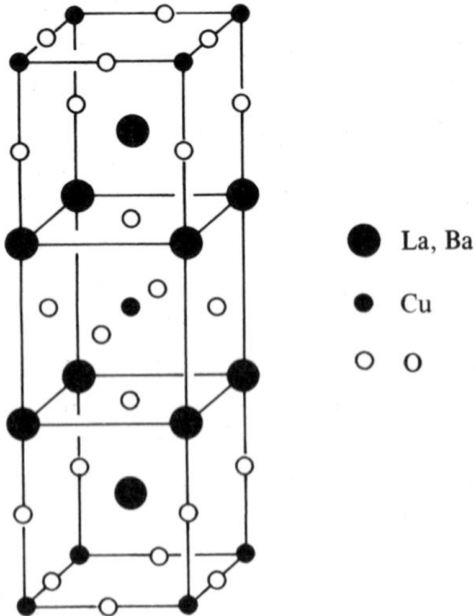


**Figure 13.24** The atomic structure of the high-temperature superconductor $La_{1-x}Ba_xCuO_4$. The Cu atoms lie at the centres of octahedra formed of oxygen atoms, this is most clearly seen in the mid portion of the figure.

other inverted; these pyramids are separated by planes of, for example, yttrium (YBCO) or calcium (as in Hg-1222 or Hg-1223). In the base of each pyramid lies a copper atom, thus each split octahedron produces two so called adjacent Cu-O planes. In certain systems a third Cu-O layer, flanked by Ca atoms, can be inserted between these pyramids; there are then three adjacent Cu-O layers per structure unit. The coefficient $n$ for the number of Cu atoms in the basic formula of the Bi, Tl or Hg cuprate tells us how many Cu-O planes are to be found in the crystallographic unit cell and therefore the number of adjacent Cu-O planes separated only by planes of Ca atom, Empirically there is a strong correlation between the size of the transition temperature and the number of adjacent Cu-O planes. Thus in the Tl-cuprate family $T_c$ is respectively 43 K, 102 K and 125 K for one, two or three Cu-O planes per unit cell. This has led to the search for cuprates with $n > 3$ and the attempt to make artificial structures by controlled epitaxial growth of thin films.

In spite of the experimental effort devoted to the subject the properties of both the normal and superconducting states of these cuprates are not thoroughly characterized nor fully understood. The relatively small carrier concentrations result in a small occupied volume of k space and a small Fermi velocity. On the other hand a high transition temperature implies a large energy gap parameter. The coherence length, $\xi \sim v_F/\Delta$, must therefore be smaller than for conventional metallic superconductors. In fact experimental estimates indicate that it is $<30$ Å parallel to the Cu-O planes and only 5 Å in the perpendicular direction. The cuprates are therefore extreme examples of type II superconductors with upper critical fields as high as 100 T. The critical currents required to destroy the S state differ by a factor $\sim 100$ in the parallel and perpendicular directions.

The N and S states are markedly anisotropic because the electrical properties of both states are associated with charge carriers confined to the Cu-O planes. In the normal state metallic conductivity is found parallel to these planes but the behaviour in the perpendicular direction is that of a semiconductor. Even so the normal state metallic behaviour differs markedly in its temperature variation from that encountered in an ordinary metal because it appears that processes involving other than pure phonon scattering are responsible for the temperature variation of the resistance. A proper understanding of the cuprates demands a treatment of a two-dimensional system in which the carriers, which must have marked polaron character, experience both strong charge and magnetic correlation and as yet this treatment is not forthcoming (for a short review see Fulde and Horsch 1993).

There is a lack of definitive experimental data. Different workers obtain different results on what, on paper at least, are similar samples. The ceramic cuprates are, however, with their many components, need for high temperature processing and tendency to degrade, very difficult to prepare in a reproducible manner. They are often difficult to obtain in single crystal form and they twin† readily. So it is perhaps not surprising that, as yet, many important features of their behaviour, e.g. the variation of penetration depth with temperature, are not known with certainty. The best samples appear to be those grown epitaxially as single crystal films, thickness

† A twinned crystal is one containing a grain boundary that divides the crystals into two regions, the lattice arrangement in each being the mirror image of the other in the grain boundary (twin boundary) plane. A twinned 'crystal' may contain any number of such 'twins'. Twinning is particularly prevalent in fcc structures.

< 0.5 μm, on a suitable substrate ($SrTiO_3$ or MgO). Such specimens show muc higher critical current densities than bulk samples. The epitaxial growth process als permits the control of the composition of the individual layers as they are deposite and it is possible to create metastable and maybe stable artificial structures. In th way samples with more than three adjacent Cu-O layers have been built.

It was reported recently (Lagues *et al.*, 1993) that such an artifically prepare 30 nm thick sample of BiSrCaCuO with a structural repeat unit containing eigh adjacent CuO layers showed a superconducting transition at 250 K, more tha 100 K higher than any transition observed previously. Still more surprising perhap is another communication (Tholence *et al.*, 1994) describing the onset of supercon ductivity, also at 250 K, in bulk Hg-1223 and Hg-1245 both synthesized under hig pressure. However these results have not been confirmed.

The principal theoretical question concerns the coupling mechanism for super conductivity in the cuprates. The critical temperature as expressed in the BC theory, (13.28), depends on three quantities: $\theta_D$, $N(E_F)$ and $V$ the effective electron phonon coupling strength. It is considered unlikely that, even with an optimizatio of all three parameters, the BCS model could produce a $T_c$ in excess of 50 K however, this does not exclude the possibility that electron–phonon interaction aug mented by another, as yet undetermined, mechanism could produce the high tran sition temperatures that are observed. Charkravarty *et al.* (1993) have in fact show that, although charge carriers are not transported perpendicularly to the Cu-O layers, Cooper pairs may tunnel between them; the coherence length and the inter planar spacing are of similar size. They have calculated that, irrespective of it origins, an in-plane attractive carrier coupling, enhanced by inter-plane Josephso tunnelling, produces transition temperatures of the right size and that also depen on the number of adjacent Cu-O planes. The pair wave function has s wave charac ter, i.e. the pair state possesses no orbital angular momentum, just as in the BC model. This has the consequence that the order parameter amplitude has every where the same sign and that any anisotropy in the superconducting properties i associated with crystallographic origins; there is no inherent anisotropy associate with the pair state.

Recall that the progenitors of these superconducting oxides are anti ferromagnetic. Although there is no long range antiferromagnetism in the dope superconducting compound it is still possible for local antiferromagnetic fluctua tions to arise. Remember too that the normal state electrical behaviour is anom alous indicating other than purely phonon scattering. Another mechanism that ha therefore been proposed is the formation, within the Cu-O planes, of Cooper pair via spin fluctuations. This may lead to d wave pairing, the pair state possessin angular momentum appropriate to a quantum number $l = 2$. The pair state has a two-dimensional order parameter that consists of four lobes alternating in sign and separated by lines of zero amplitude (just like an ordinary atomic d wave function). The gap parameter $\Delta$ is proportional to the order parameter and should therefore vary markedly in k space having the value zero in four specific directions; it must also have different sign in the alternate lobes. The energy gap that arises at the Fermi surface depends only on the magnitude of the energy gap parameter and not on its sign, but d wave pairing will cause an inherent pair state energy gap anisot-ropy different from, but in addition to, any crystalline anisotropy. Furthermore the sign of the order parameter is reflected in the phase of the pair state and this is observable directly in the interferences that arise in the two junction SQUID.

It would therefore seem that, although there is no theoretical consensus regarding the origins of high temperature superconductivity, there is every reason to look forward to new experiments that will allow the proposed models to be tested.

Interest in the cuprate superconductors is enhanced by the prospect of widespread technological application: power transmission cables, energy storage systems, magnetic bearings, powerful magnets, high Q resonators and computer circuitry are a few examples. If these hopes are to be realized then operation at the boiling point of liquid nitrogen, 77 K, is the first stage; this demands that the transition temperature be well over 100 K to ensure good performance. A figure of merit in this context is the critical current density, $J_c$, required to destroy superconductivity; this is zero at $T_c$ and a maximum (corresponding to $B_{cO}$) at 0 K. The superconducting device must, at its operating temperature, have a certain minimum $J_c$ and for power transmission cables it is considered that this minimum value should be $10^6$ A cm$^{-2}$. The critical currents in the bulk cuprates lie at least an order of magnitude below this figure. The crystalline anisotropy, the misoriented granular structure and probable presence of lattice defects cause the occurrence of weak links that degrade the performance. Furthermore it is doubtful whether those compounds with the highest transition temperatures that are based on Tl or Hg would be sanctioned for general use. The most promising candidate appears to be the bismuth based family, particularly (Bi-2223) with $T_c = 120$ K. The latter substance can be compacted and calcined in silver sheaths and thereafter rolled or extruded into wire. The production process causes the orientation of the Cu-O planes to be parallel to the length of the wire which is advantageous for the performance. At present the wire can be produced in lengths of just over 100 m; the critical current density at 77 K is of order $10^4$ A cm$^{-2}$. Although certain applications, such as liquid nitrogen cooled superconducting magnets or intermediate current leads between conventional superconducting magnets and external circuitry, are feasible, much further development is needed before large scale application is technically and economically practical.

The best thin film samples of the cuprate superconductors have a critical current density above $10^6$ A cm$^{-2}$ at 77 K. The implication is that application of these films may find rapid use in microwave circuitry in the form of filters, resonators, antennas and similar devices. Ceramic SQUIDS might also replace the semiconductor in certain computer applications but as yet this appears a rather remote possibility, particularly considering the present accent on optoelectronic development.

Before closing this section we present a brief description of another family of high temperature superconductors based on the fullerene $C_{60}$ (Section 3.7.2).

Crystalline $C_{60}$ is a semiconductor with a bandgap of 1.9 eV, but it dissolves metal atoms (especially the alkali metals) which increase the conductivity by several orders of magnitude. Superconductivity is found with the following transition temperatures: 19 K ($K_3C_{60}$), 28 K ($Rb_3C_{60}$), 33 K ($RbCs_2C_{60}$) and 42 K ($Rb_{2.7}Tl_{2.2}C_{60}$). These temperatures place the doped fullerides in the category of high temperature superconductors. They are type II superconductors with short coherence lengths, $<30$ Å, similar to those found in the cuprates. The metal ions in these compounds are found in rows outside the $C_{60}$ cages and occupy either octahedral or tetrahedral interstices (see Problem 2.3) in the fcc arrangement of the $C_{60}$ molecules. At higher metal concentrations, e.g. $M_6C_{60}$, the alloys become insulators again. The superconductivity of $C_{60}$ alloys can be accounted for within the BCS theory.

## References

CHAKRAVARTY, S., SUDBØ, S., ANDERSON, P. W. and STRONG, S. (1993) *Science* **261**, 337.

CORAK, W. S., GOODMAN, B. B., SATTERTHWAITE, C. B. and WEXLER, A. (1956) *Phys. Rev.* **102**, 656.

DE HAAS, W. J. and VOOGDT, J. (1931) *Commun. Phys. Lab. Univ. Leiden* No. 214c.

FEYNMAN, R. P., LEIGHTON, R. B. and SANDS, M. (1965) *The Feynman Lectures on Physics*, Vol. III. Addison-Wesley, Reading, Massachusetts.

FULDE, P. and HORSCH, P. (1993) *Europhys. News* **24**, 73.

GIAEVER, I. and MEGERLE, K. (1961) *Phys. Rev.* **122**, 1101.

KEESOM, W. H. and VAN LAER, P. H. (1938) *Physica* **5**, 193.

LAGUES, M. *et al.* (1993) *Science* **262**, 1850.

OLSEN, J. (1952) *Proc. Phys. Soc.* **A65**, 518.

THOLENCE, J. L. *et al.* (1994) *Phys. Lett.* **A184**, 215.

## Further Reading

TILLEY, D. R. and TILLEY, J., *Superconductivity*, 3rd Edition (1990) Adam Hilger, Bristol, UK.

RAVENAU, B., Defects and superconductivity in layered cuprates, in *Physics Today* (1992) **45**, 10, pp 53.

BEYERS, R. and SHAW, T. M., The structure of $Y_1Ba_2Cu_3O_4$ and its derivatives, in Ehrenreich, H. and Turnbull, D. (Editors), *Solid State Physics* (1989) **42**, 135, Academic Press Inc., New York.

## Problems

**13.1** Estimate the strength of the effective electron–phonon interaction, the parameter $V$ in equation (13.28) for Al and Pb. All the information needed is available in the text.

**13.2** Calculate the London penetration depth in Al at 0 K under the assumptions (a) that all the valence electrons are responsible for the superconducting property, and (b) that only electrons in an energy interval $\Delta$ in the neighbourhood of the Fermi energy contribute to the supercurrent. To what fraction of the valence electron concentration does the latter case correspond?

**13.3** The metals Cu and Pb have diamagnetic susceptibilities $\kappa = -9.63 \times 10^{-6}$ and $-1.58 \times 10^{-5}$ respectively. Furthermore, Pb is a superconductor with $T_c = 7.2$ K and $B_{c0} = 800$ G. Compare the magnetic energies of Cu and Pb in an external field of 500 G at 7.5 K and 0.1 K; assume the long-wire geometry with the field parallel to the wire. What is the principal difference in the free energy of the Pb sample at the two temperatures and in the presence of the field?

**13.4** As described in Fig. 13.5, we may write

$$B_c = B_{c0}[1 - (T/T_c)^2].$$

Obtain an expression for $B_{c0}$ in terms of $T_c$ and $(c_s - c_n)_{T_c}$. Given that the latter quantities are 1.2 K and 1.6 J K$^{-1}$ kmol$^{-1}$, estimate $B_{c0}$ for Al.

**13.5** Give concise definitions or descriptions of the following: Meissner effect, Cooper pair, coherence length and condensate.

**13.6** Compare the absolute size of the maximum diamagnetic magnetization in Pb with that of the spontaneous magnetization in Ni at 0 K.

**13.7** In a tunnelling experiment using a junction of Sn and Al at 2 K, the following $I(V)$ characteristic was obtained. What can be deduced from this result?

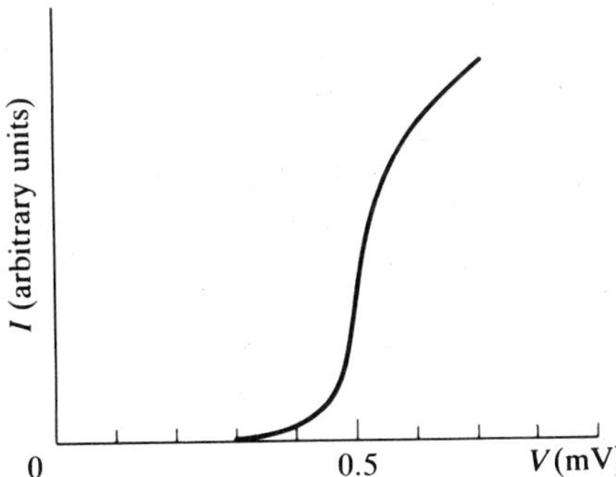

# 14

# Aspects of Surface Physics

## 14.1 Introduction

Previously we have considered the surface of a solid as something to be avoided. Solids were thought of as infinitely large single crystals or the device of periodic boundary conditions was used to simulate infinite size. In real life the surface is unavoidable and it is through the surface that almost all solid state reactions arise – those involving other condensed phases as well as those with gases. The practical importance of surface properties for friction, adhesion, wear, corrosion and catalysis needs no further comment.

On the other hand, the calculation of the physical properties of surfaces, e.g. the surface energy, work function, vibrational spectrum and the electronic structure, is a much more difficult task than the calculation of bulk properties. In the ideal three-dimensional solid the perfect translational symmetry of the lattice is imposed on the electronic wave function, and this allows considerable simplification of a difficult problem. The surface has lateral symmetry, but in the direction normal to the surface the discontinuity in atomic structure severely complicates the issue, making calculation laborious and expensive in computer time. However, significant advances have been made in recent years and comparisons with experiment are encouraging. In what follows we concentrate primarily on surface structures, and surface excitations as those aspects of surface science most relevant in a general course of solid state physics (see e.g. Prutton 1983). For many features of surface science, especially in surface chemistry and catalysis, questions concerning surface composition and the character of the electronic structure of the surface are of great importance. We treat these briefly in Section 14.8; they are considered in depth in a number of books specifically devoted to surface science (see e.g. Woodruff and Delchar 1986).

In the limited discussion that is offered here we must concentrate on the basic characteristics of low-index surfaces. Furthermore, whereas we can readily comprehend the surface as being where the three-dimensional solid ends, there is in fact a transitional sheath that may extend over several atomic layers.

When studying the physics of surfaces as described here, interest is directed towards atomically smooth, clean low-index crystal planes or such surfaces that have been partially or wholly covered by another atomic species in a controlled fashion. The sample is oriented so that the plane of the surface is parallel to within 0.5° of the appropriate crystal plane and carefully polished. It must be further cleaned under ultra-high vacuum conditions prior to measurements being made. If the sample is an elemental metal, this procedure usually involves ion bombardment and annealing procedures.

The need for ultra-high vacuum conditions (of order $10^{-10}$ torr) is readily appreciated. A (001) surface of a Cu crystal contains $1.5 \times 10^{15}$ atoms cm$^{-2}$. It is easily calculated from the kinetic theory of gases that such a surface, when in a gas at a pressure $P$ (in torr), experiences molecular impacts at a rate

$$r \approx 3.5 \times 10^{22} \frac{P}{(MT)^{1/2}} \text{ cm}^{-2} \text{ s}^{-1}, \tag{14.1}$$

$M$ being the molecular weight of the gas and $T$ the absolute temperature. If we exclude hydrogen, most residual gases ($M \approx 16$) in a vacuum system have, at room temperature $r \approx 3.5 \times 10^{20} P$ cm$^{-2}$ s$^{-1}$, with $P$ measured in torr.

If every gas molecule striking the Cu surface has a probability $S$ of sticking then the surface becomes completely covered in a time $t$ given by

$$rSt = 1.5 \times 10^{15},$$

or

$$t \approx \frac{4 \times 10^{-6}}{SP} \text{ s}. \tag{14.2}$$

For $S = 1$ and $P = 10^{-10}$ torr this time to achieve complete coverage is $4 \times 10^{4}$ s or about 10 h. The assumption that $S = 1$ is an exaggeration, but on the other hand an experiment on a purportedly clean surface would be invalidated by the formation of less than $10^{-2}$ of a monolayer of contaminant; so our estimate of $t$ may stand. At a pressure $10^{-8}$ torr the time is reduced to about 5 min. This estimate of contamination time is perhaps better considered as an upper limit, since the evaporation of metals or the operation of equipment involving thermionic filaments usually causes some increase in the ambient pressure. Neither should one forget that bulk materials always contain impurities, which may, and often do, diffuse to the surface in continuous fashion. This is certainly true of the transition metals, for which carbon and sulphur are particularly troublesome.

## 14.2 Two-Dimensional Bravais Lattices

We first consider the somewhat, but not completely, artificial situation of the free-standing two-dimensional periodic array of atoms, which is convenient for a discussion of structure. In two dimensions there are only five ways of arranging geometrical points so that each point has identical surroundings; in other words, there are only five two-dimensional Bravais lattices (Fig. 14.1). Just as in our earlier discussion of crystal structure, each lattice point may be associated with a group of points which itself has symmetry elements about the lattice point. In planar lattices

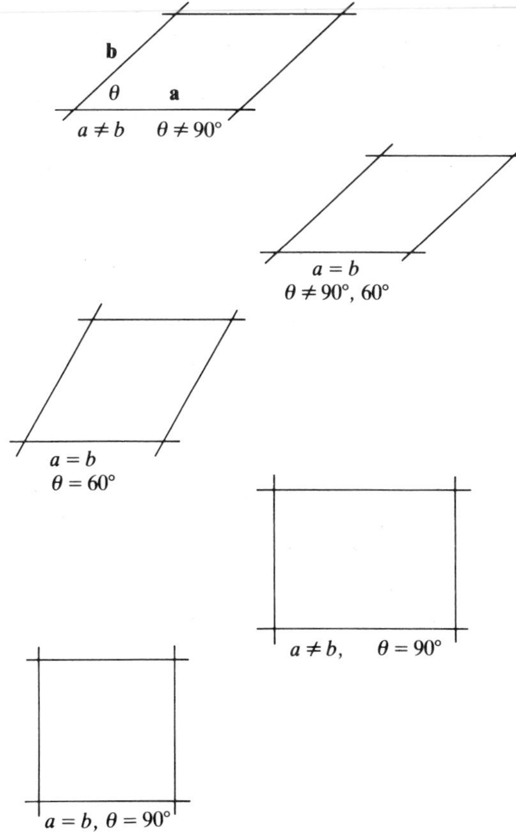


**Figure 14.1** Primitive cells for the five two-dimensional Bravais lattices.

there are ten such point groups. All the concepts developed in our earlier discussion of crystal structure are applicable. We may define primitive cells, unit cells, a reciprocal lattice and Miller indices exactly as for the three-dimensional case.

## 14.3 Determination of Surface Structure

If we imagine a crystal to be divided into two parts along a perfectly planar surface and parallel to a crystallographic plane of low Miller index then we should expect to find the atoms in the newly formed surfaces arranged in, say, square or hexagonal arrays. Laterally, there is no change in the periodic arrangement, but owing to the asymmetry perpendicular to the surface we might expect changes in the separation of the outermost layers owing to the loss of neighbouring atoms and changes in the effective force constants for the surface atoms. If our crystal were a simple metal like Al then, because of the small ion cores and almost uniform density of the electron gas, any readjustment of the atoms in the outermost layers should be readily achieved. Furthermore, because there is no strong directional bonding, only slight atomic rearrangement is to be expected.

A semiconductor like Si, on the other hand, has strongly directional bonds and a much lower packing fraction than Al. The creation of a surface produces uncompensated or dangling bonds associated with a large surface energy. This energy may be reduced by a reorganization or reconstruction of the outermost atoms (extending perhaps three or more layers below the outermost layer). In this way, dangling bonds may be partially or wholly eliminated, causing the surface to display new symmetry features. Such reconstruction is expected to be the rule in covalently bonded semiconductors and the exception for highly coordinated metals: W, Mo, Au, Pt and Ir are among these exceptions.

The structure of a surface is studied using reflection high-energy electron diffraction (RHEED), low-energy electron diffraction (LEED) or atomic-beam scattering experiments. Electrons with energies of about 50 keV are used in RHEED. The angle of incidence must be very large, the electrons merely grazing the surface of the crystal. The diffraction pattern is recorded photographically. The method is used primarily to assess the smoothness of a surface and the forms of overgrowths, i.e. whether atoms, either different from or similar to those of the crystal surface, are added uniformly or form islands projecting up out of the base crystal.

An atomic beam of He (sometimes Ne) may be arranged to have a de Broglie wavelength of order 1 Å. Such a beam may be diffracted only by the outermost layer of a crystal. The technique has been used for the determination of reconstructed surface structures; later we shall describe its application to the study of surface vibrations.

The most widely used method for the study of surface structure is LEED. A schematic arrangement of the apparatus is shown in Fig. 14.2. An electron gun comprising a thermionic emitter and various accelerating and focusing electrodes produces a collimated beam of electrons with energy in the range 50–250 eV, corresponding to a de Broglie wavelength close to 1 Å. These electrons strike the sample

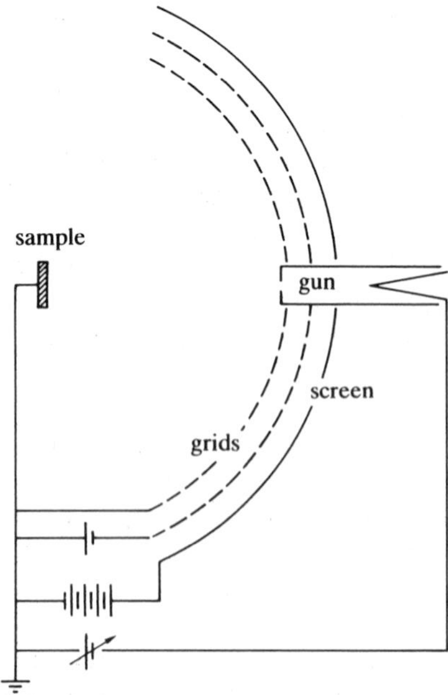


**Figure 14.2** A schematic diagram of apparatus for low-energy electron diffraction (LEED).

in the form of a clean single-crystal surface and are diffracted. The 'back-reflected' electrons travel in field-free space to a system of electrodes or grids, which serve to eliminate inelastically scattered electrons and accelerate those elastically scattered onto a fluorescent screen. The visible pattern on the screen provides a picture of the reciprocal lattice points active in diffraction.

The electrons, on striking the crystal, are scattered by the electric potential associated with the ion cores. The scattering cross-section for the incident electrons is very large, which is advantageous for producing intense diffracted beams. Nevertheless, it is also disadvantageous because the current in a given diffracted beam is not made up of electrons each of which has been coherently scattered by one ion core, but contains electrons that have undergone multiple scattering. Thus, although the diffraction geometry remains unaffected, the electron is scattered by several ions before leaving the sample. This is in direct contrast with the diffraction of X rays by crystals, but the small scattering factors for X rays and their relatively large penetrating power make them of limited use in surface studies. Because of multiple scattering, the intensities of the diffracted electron beams in LEED cannot be given a simple interpretation.† The determination of the details of a reconstructed surface therefore demands (a) a proposed trial structure, (b) an assumed ion core potential and (c) a quantum-mechanical multiple-scattering calculation of the variation of the intensity of the diffracted beams as a function of incident electron energy. However, the situation is further complicated by the strong inelastic scattering that arises when electrons interact with solids, as through the creation of plasmons in metals; but on the other hand, this also means that the diffraction process is restricted to the first three or four atomic layers, because if the electrons penetrate further they are unlikely to get out again. This implies that LEED is surface-sensitive and that complications from multiple scattering are restricted. *If we are only interested in the lateral symmetry of the surface then this is immediately reflected in the geometry of the diffraction pattern.*

We shall return to the geometrical details of LEED patterns when we describe simple overlayer structures. First we need to establish the appropriate reciprocal lattice. The two-dimensional reciprocal lattice may be thought of in the following way. Take the three-dimensional crystal, say Al, and let the base vector $\mathbf{c} \rightarrow \infty$, the vectors $\mathbf{a}$ and $\mathbf{b}$ remaining unchanged. As $\mathbf{c} \rightarrow \infty$, the corresponding reciprocal lattice vector $\mathbf{C} \rightarrow 0$, which means that the discrete points parallel to $\mathbf{C}$ in the reciprocal lattice become closer and closer, finally coalescing to form lines, or rods as they are usually called. The reciprocal lattice associated with the (001) plane of Al becomes a square perpendicular array of parallel and infinitely long rods. We may choose our origin in the central plane of this array and form the Ewald sphere to illustrate the geometry of the diffraction pattern for a particular electron wavelength (Fig. 14.3).

Recently, a new and more direct method for the geometrical characterization of clean metal surfaces and the positions of adsorbed atoms has been developed. It uses the principle of vacuum tunnelling. If two electrodes, one of which is an exceedingly fine metal tip, are separated by a distance $d$ ($\approx 5$ Å) then, when a potential difference exists between the electrodes, a current $j$ may flow through the vacuum

† Multiple scattering causes the atomic form factor to be dependent on the crystallographic environment, this means that the geometrical structure factor (3.15) cannot be used in a straightforward manner as for X-ray scattering.

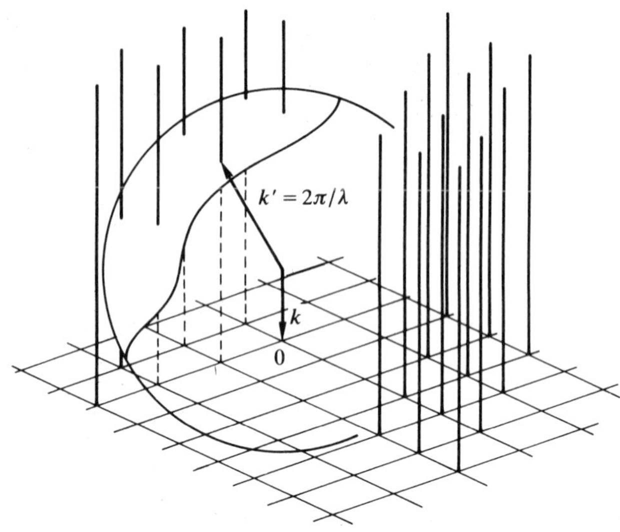


**Figure 14.3** An attempt to illustrate the Ewald sphere within the two-dimensional reciprocal lattice; the latter is represented by an assembly of 'rods'. Only back-reflected beams are observed in LEED.

gap, and

$$j \propto \exp(-A\phi^{1/2}d), \tag{14.3}$$

where $\phi$ is the barrier height, i.e. the average work function of the electrodes, and $A$ is a constant of order 1 Å$^{-1}$ eV$^{-1/2}$. The current $j$ therefore depends strongly on $d$, and changes in $d$ of about 1 Å can produce an order-of-magnitude change in the current. Thus if the fine metal tip is initially positioned at a fixed distance, say 5 Å, from the second electrode (a carefully prepared single-crystal surface) and the separation adjusted to maintain a constant tunnelling current as the point traverses the surface then its position traces the topography of the crystal surface. The point electrode may be moved in regular way to scan a limited area of the sample, and the instrument is therefore known as the Scanning Tunnelling Microscope (STM) (Fig. 14.4). The success of the method depends upon the ability to form a suitably sharp point electrode and control its position to less than 0.1 Å. Sample and point electrode are controlled by piezoelectric crystals, and those positioning the tip have a sensitivity of about 2 Å V$^{-1}$. The practical realization of the instrument lies in the elimination of external sources of vibration.

The STM is able to display individual atom positions, Fig. 14.5, and it has found wide use in surface physics to determine the topography of crystal surfaces and the structure of adsorbed layers. Atomic writing, using the controlled positioning of adsorbed foreign atoms to form letters, has been demonstrated as well as the preparation of artificial surface structures, such as ring enclosures, that trap electrons occurring in the surface states of the substrate crystal; see Further Reading (*Physics Today*, 1993). In addition to its use as a microscope the STM can also probe the local electronic structure at particular surface sites. When the tip is held stationary, a current–voltage characteristic may be determined; $dI/dV$ plotted against $V$ is then an approximate measure of the distribution of the density of electron states at the chosen site; the arrangement is used as an atomic vacuum tunnelling junction.

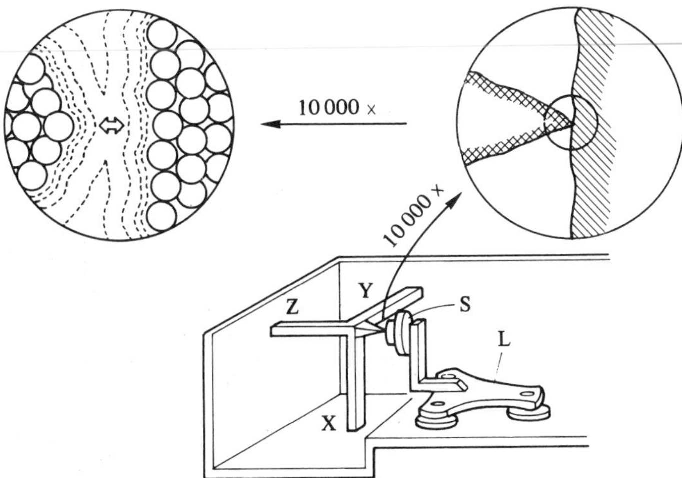


**Figure 14.4** Schematic arrangement of an early version of the tunnelling microscope. The sample S is mounted and fixed in position on the 'louse' L, which has three piezoelectric 'feet', its position is adjustable to within 100 Å. The point electrode is controlled in the *x*, *y* and *z* directions to within 0.1 Å, also by piezoelectric crystals. The insets show the tunnelling gap on a highly magnified scale. Tunnelling arises as a consequence of the overlap of the wave functions of occupied states in the point electrode and those of empty states of equal energy in the sample. The success of the method depends upon the exponential variation of the tunnel current on electrode-sample separation. (Reproduced with permission from *Europhysics News*, Vol. 15, No. 5, p. 15 (1984).)

The STM requires a conducting sample and ultra-high vacuum conditions. Other probing techniques of more general applicability have therefore been developed and information on their function and use is available in special treatises by Chen (1993) and Wiesendanger (1994). By far the most popular is the atomic force probe used in Scanning Force Microscopy (SFM). The force between the pointed probe and the sample is measured by the deflection of a very small rectangular cantilever, typical size 100 $\mu$m × 20 $\mu$m × 1.5 $\mu$m, or if greater lateral stability is required it may be V-shaped. The cantilever is often made of Si, $SiO_2$ or $Si_3N_4$. The probe, usually in the form of a pyramid or cone, is integral with the cantilever. Silicon processing technology allows the batch production of cantilever-probe units with reproducible characteristics.

A major advantage of the SFM is that it may be used with all solid substances and in almost any atmosphere, gaseous, liquid or vacuum. The function is based on the atomic force between the probe tip and the sample, which lies in the range $10^{-6}$ to $10^{-12}$ N. This force causes the cantilever to bend and its displacement is, in many instruments, monitored by a laser beam reflected from the back of the cantilever. After reflection the change in the light beam is determined interferometrically or by a space sensitive detector (e.g. a sectored photon sensitive device). The optical lever allows cantilever displacements of fractions of an Ångström to be detected.

The atomic force between probe and sample varies with their separation and may in fact change sign. This occurs because at large separations the force originates from attractive van der Waals interaction, whereas at small separations the repulsive ion core interaction dominates. In addition to microscopy it is possible to pursue atomic force spectroscopy. The interested reader is referred to the above

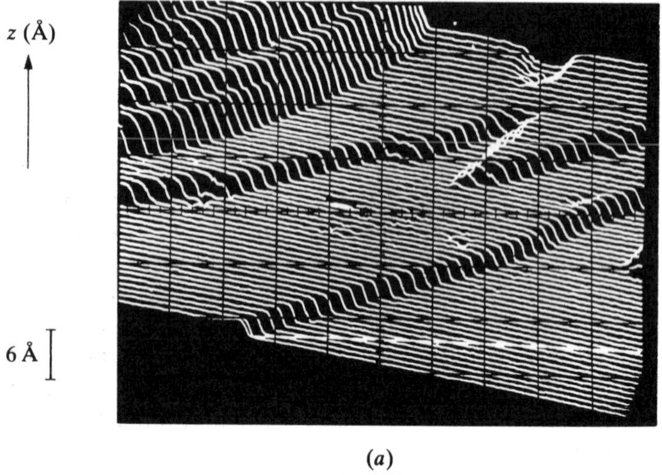


(*a*)

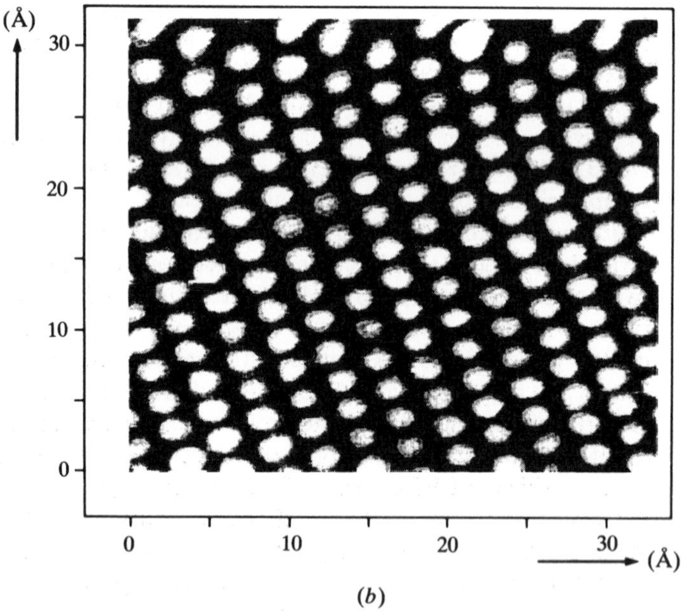


(*b*)

**Figure 14.5** (*a*) Image of an Al(111) surface taken with a modern version of the STM. The white lines represent successive scans of the point electrode across the crystal surface. Flat terraces and mono-atomic steps are clearly seen. (*b*) Similar experiments allow the individual atoms to be resolved. Thus even in a nearly free electron metal like Al there must arise corrugations in the electron density that allow the details of the surface lattice to be resolved. (STM photographs by courtesy of R. J. Behm.)

mentioned sources for further details of this and other probing techniques and their uses.

## 14.4 Overlayer Structures

A clean, atomically smooth, single-crystal surface may be used to support one or more two-dimensional overlayers of another atomic species. The latter might be a gas, e.g. CO, or an element like S or another metal, e.g. Na. We shall confine attention to one overlayer, a so-called monolayer. Just as a metal may not necessarily dissolve another metal (we say that they are immiscible), so one metal when deposited on top of another, usually by evaporation from a molten source, can coagulate and form islands that have no structural relationship to the substrate crystal; however, in a surprising number of instances the deposited metal forms a crystallographically ordered overlayer with a symmetry related to that of the underlying crystal surface.

Consider the (001) surface of an fcc crystal. This has its atoms arranged on a square net. We can, simply as an exercise in geometry, try to find those overlayer nets that can be symmetrically overlaid on this (001) net. Some examples are shown in Fig. 14.6. These overlayer nets may be described in terms of their cell dimensions and orientation with regard to the substrate net, and the terms primitive (p) and centred (c) have obvious meanings. The different nets have different coverages $\theta$, the latter being defined as the ratio of the number of overlayer atoms to the number of atoms in the substrate net. Another way to describe an overlayer is to express the base vectors of the primitive or unit overlayer cell in terms of those of the substrate net. The appropriate transformation matrix is then a complete description of the overlayer.

It will be seen from Fig. 14.6 that certain overlayer structures may have two equivalent orientations (e.g. parallel to the **a** and **b** axes); this leads to domain structures and pseudosymmetries. Thus the hexagonal net of Fig. 14.6 may be aligned parallel to the **a** or the **b** axes. Both versions may arise together and in approximately equal proportions, causing a LEED pattern to indicate twelve-fold symmetry. We emphasize that Fig. 14.6 shows the relationship between overlayer and substrate nets. The exact positions of the overlayer atoms relative to the substrate (both in and perpendicular to the surface) can only be decided by a calculation of the intensities of LEED beams. The hexagonal overlayer of Fig. 14.6 is an example of a coincidence net. Another such net is shown in Fig. 14.7; we see that only certain atoms in the overlayer have positions that coincide with those of the substrate net. If the dimensions of the coincidence net are large then it may have different contents of overlayer atoms and still produce the same coincidence geometry. The content of the overlayer net becomes somewhat arbitrary and one must resort to an intensity analysis or other data to establish the exact structure.

We can now better appreciate the simpler forms of surface reconstruction as an overlayer structure containing the same atom species as the substrate crystal. In the case of the (001) surface of Au, one expects in LEED to see a simple diffraction pattern (Fig. 14.8*a*). Instead, that of Fig. 14.8(*b*) is seen. This pattern may be understood as an overlayer structure arising in domains parallel to **a** and **b**. The domain parallel to the **a** axis produces a pattern (Fig. 14.8*c*) typical of a reciprocal lattice cell with dimensions ($\frac{1}{5} \times 1$) corresponding to a direct lattice cell ($5 \times 1$) in terms of the

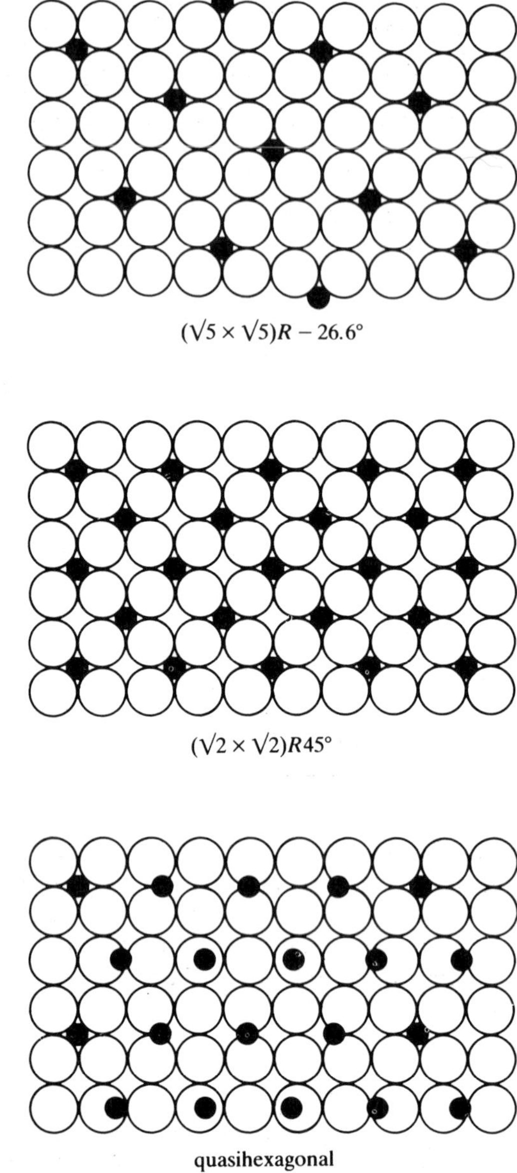


**Figure 14.6** Certain overlayer structures on a square net. The notation has the following interpretation: $(\sqrt{5} \times \sqrt{5})R - 26.6°$ means that the overlayer net is square with side $\sqrt{5}\,a$, where $a$ is the parameter of the substrate square net; the overlayer axes are rotated $-26.6°$ with regard to the substrate net axes. A corresponding interpretation may be given to the $(\sqrt{2} \times \sqrt{2})R\ 45°$ overlayer net. Both these overlays have primitive cells, but the prefix 'p' is often omitted. The term 'quasihexagonal' is used in the third figure because the overlayer does not have perfect hexagonal symmetry; one can establish that an angle that should be 30° is in fact 30.26°, but this difference would not be detectable in a LEED experiment.

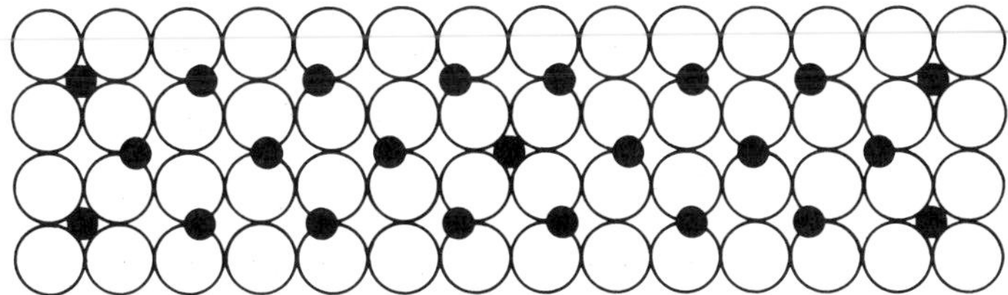

**Figure 14.7** A c(12 × 2) coincidence structure. The overlayer unit cell contains an integral number of substrate unit cells and is defined by the sites that are coincident with sites in the substrate.

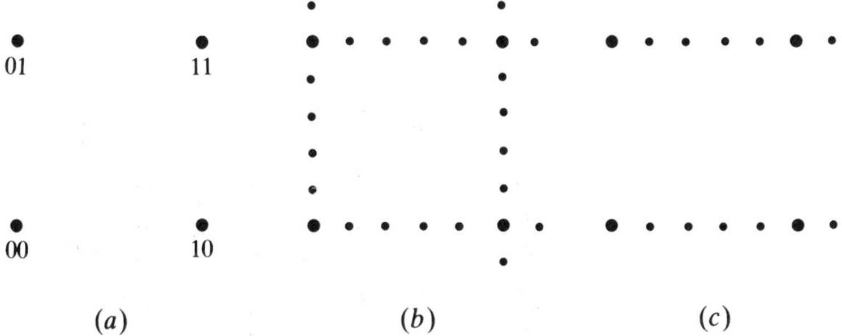


**Figure 14.8** LEED patterns as observed with an Au(001) surface. One expects to see pattern (a), but pattern (b) occurs, showing that the surface is reconstructed. Pattern (b) may be resolved into subpatterns, one of which is shown in (c).

basic parameters of the substrate net. This new structure has been associated with a hexagonal overlayer of Au atoms, but it is necessary to assume approximately 4% compression so that six rows of the hexagonal overlayer fit onto five rows of the underlying square net (Fig. 14.9). If the Au atom maintains a constant diameter, this must involve a certain rippling or buckling of the overlayer. A similar (1 × 5) overlayer also occurs.

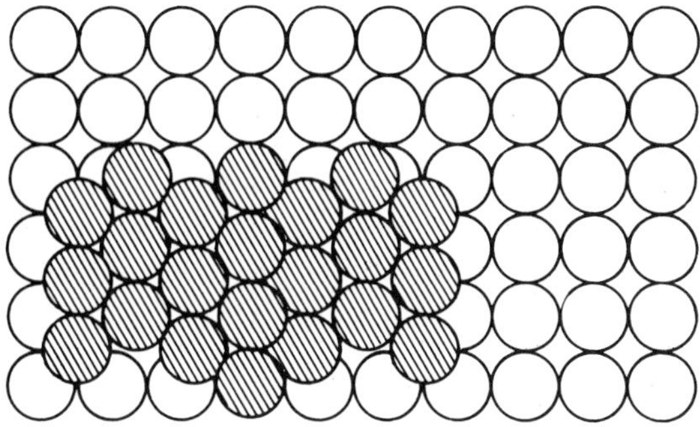

**Figure 14.9** A possible arrangement for the reconstructed Au(001) surface. (After Palmberg and Rhodin 1967.)

## 14.5 Surface Vibrations

Earlier we have described the motion of atoms in a solid using the concepts c vibrational mode, phonons and the first Brillouin zone. We used the notion o periodic boundary conditions to avoid the effect of the surface that always arises i real solids. The vibratory motions of surface atoms are very complicated, but it i still possible to present a useful description in qualitative terms. First of all we mus remember that the total number of vibratory modes is fixed by the number of atom in the sample. Thus when a surface exists we do not create extra modes, but certai of the modes in the perfect crystal change character and become localized in th surface atoms. How should we define a surface mode? One definition says that in a pure element a surface mode is one that is not found in the bulk bands of the perfec crystal. By bulk bands we mean the integrated vibrational frequency spectrum. Thi definition will be found too exclusive.

Suppose first that we could make an imaginary surface without changing th vibrational frequencies of the solid. Within our imaginary surface we have definit periodicities in the arrangement of the atoms, and we can construct a two dimensional Wigner-Seitz cell in reciprocal space that is the two-dimensional Bril louin zone shown in Fig. 14.10. For the fcc (001) plane this zone is related to that ir three dimensions as shown in Fig. 14.11. We first attempt to illustrate the bulk modes in this two-dimensional zone; in other words, we shall illustrate how th modes in the bulk crystal are distributed according to their $q_x$ and $q_y$ components o wave vector. Every point in the two-dimensional zone may be associated with a range of values for $q_z$; this range is determined by the length of intercept in the $q$ direction in the three-dimensional zone. Thus when mapping the bulk behaviour i two dimensions each point $(q_x, q_y)$ becomes multivalued. When we plot the disper sion curves (i.e. $\omega(q)$ curves) along particular symmetry directions in the two dimensional zone we do not find curves as we are used to, but areas as in Fig. 14.12 This figure shows the bulk bands as they would appear on an imaginary surfac within a perfect crystal. It is the starting point for a discussion of the real surface. I our imaginary surface now becomes a real surface then new vibration patterns tha are confined to the atoms in the surface layers appear. These are called localize surface modes and are 'stolen' or 'peeled' from the bulk bands.

Calculation shows that in a surface mode an atom vibrates so that it moves in a elliptical path. The general motion is very complicated, but becomes simpler th

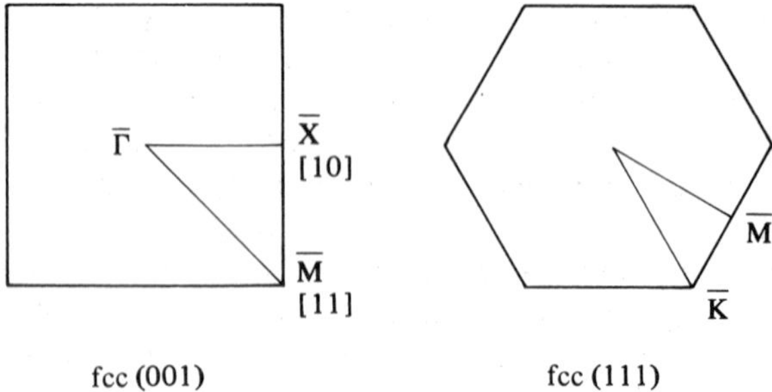


**Figure 14.10** The two-dimensional Brillouin zones corresponding to the two-dimensional nets appropriate to the (001) and (111) planes of the fcc lattice. The symmetry points are given the conventional symbols and the bar over each letter signifies the two-dimensional character.

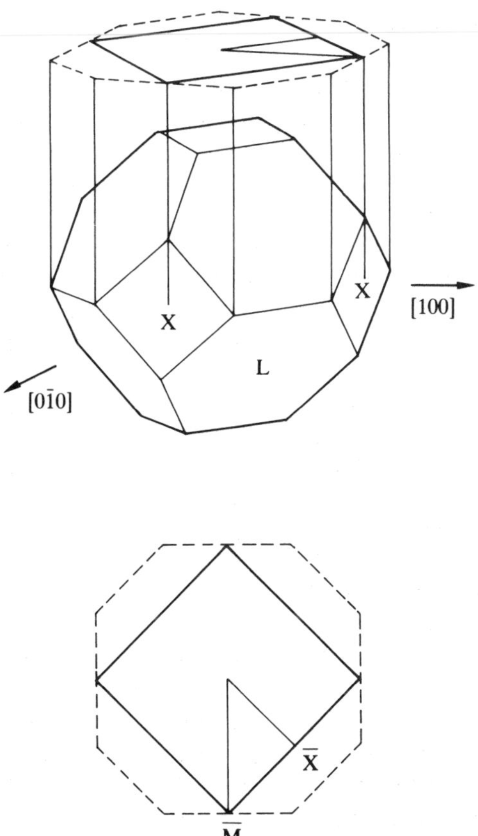


**Figure 14.11** Illustrating the relation between the three- and two-dimensional zone structures for the fcc lattice. Note that it is the $\bar{M}$ point that lies above the X point.

more symmetrical the surface structure. If there is complete mirror symmetry with respect to a plane normal to the surface then there are two principal classes of vibration: (a) those with their displacements in the so-called sagittal plane (that plane containing **q** and the surface normal); and (b) those with their displacements strictly perpendicular to the sagittal plane, i.e. surface shear modes. The former vibrational mode is also known as a Rayleigh wave because Lord Rayleigh in 1885 had shown its existence in a continuous medium; we may liken it to a propagating ripple on the solid surface – it is a motion involved in earthquakes. In a discrete lattice such a waveform is confined to the first 3–4 surface layers and the wavelength may approach the interatomic separation.

The occurrence of the surface modes is a direct result of the discontinuity and the alterations in the forces between atoms in the surface layers. The lack of restraint on the 'empty' side of the surface allows a larger amplitude of vibration and produces associated softer restoring forces, leading to a lower frequency of vibration. Thus the atomic Rayleigh wave is predicted to lie below the bulk bands, and, according to the definition mentioned earlier, it is a true surface mode. The larger the change in force constants in the surface layers the more likely localized surface modes are to be found in the second and third layers below the surface.

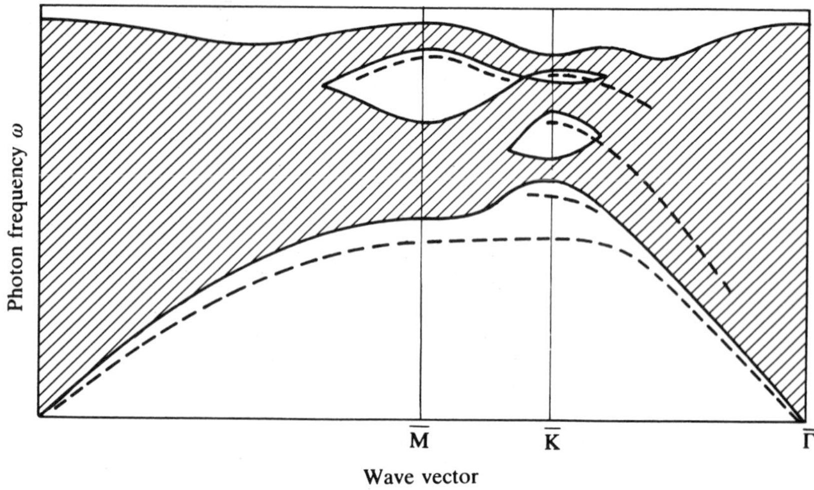


**Figure 14.12** An illustration of the bulk phonon modes described in the two-dimensional zone. The phonon wave vector has three-dimensional character, but on this diagram the modes are shown in terms of a two-dimensional wave vector. The multiple values of frequency that arise for each such wave vector stem from the third component. The shaded area indicates the allowed frequencies that arise in the ideal three-dimensional solid for the **q** values available in the two-dimensional Brillouin zone. In a real substance we find a surface and surface vibrational modes. These surface modes as they are expected in the (111) plane of a fcc lattice are indicated schematically by the broken lines in this diagram.

It will be noticed that gaps arise in the bulk bands (near and between the symmetry points $\bar{M}$ and $\bar{K}$). True surface modes again arise within these gaps, but they may also continue into the bulk bands; in the latter event there are two possibilities If the mode is orthogonal to the bulk modes (i.e. the atomic displacements are in a plane normal to those of the bulk modes) then the mode remains a true surface mode (thus our earlier definition is too restrictive). On the other hand, if the surface mode has similar symmetry to the bulk modes then it will resonate or mix with them when it enters the bulk bands and quickly become indistinguishable from bulk modes. The modes expected to arise on the (111) surface of a face-centred cubic structure are shown schematically in Fig. 14.12.

The vibratory motions of surface atoms contribute to the free energy and other thermodynamic properties of the surface; they influence the intensities of LEED beams and can be significant for the electron transport properties of thin films through the electron-phonon interaction. Furthermore, surface or interface modes arise on any boundary between two media, and at present much interest is directed at the physical properties of multilayer films. There is therefore every reason to obtain experimental information about the surface modes of crystals.

In Section 5.11 we described how the dispersive properties of bulk modes may be determined by the inelastic scattering of neutrons. We need to do the same for the surface modes, but neutrons are completely unsuitable as a probe in this case because they are surface-insensitive. Because of their low absorption coefficient, neutrons sample the behaviour of all the atoms in the crystal equally well. Electrons with energy of order 1 eV have been used and probe the behaviour in say the first five layers, but the most suitable radiation is a beam of inert gas atoms, usually He

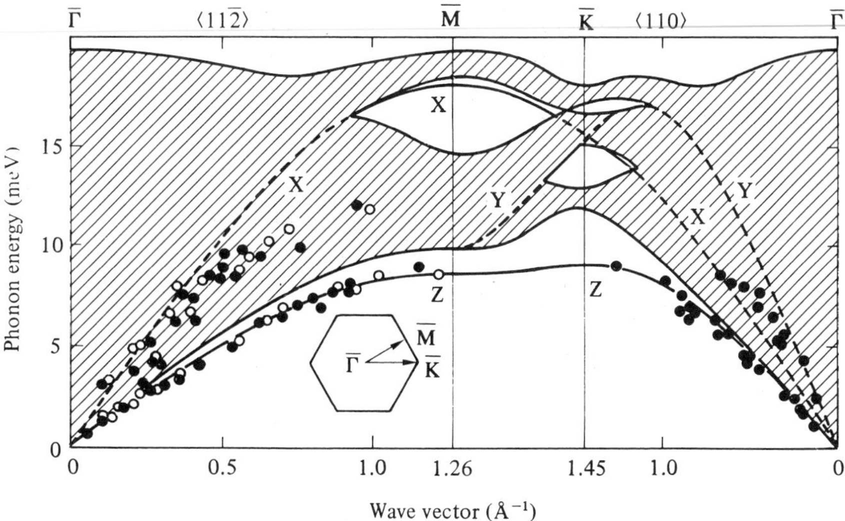


**Figure 14.13** Surface modes in the (111) surface of an Ag crystal as determined by the inelastic scattering of He atoms. (After Doak *et al.* 1983.)

but occasionally Ne. Such a beam can be used to determine the energy and crystal momentum of phonons confined to the first surface layer. By expanding cold (about 80 K) He gas through a nozzle and a series of suitable apertures, a collimated monochromatic beam of He atoms with velocity of order 1000 m s$^{-1}$ and spread in velocity $\leqslant 0.5\%$ may be obtained. In such a beam the He atom has energy about 20 meV and a de Broglie wavelength about 1 Å. Using differential pumping, the beam may be directed onto a crystal surface maintained under ultra-high vacuum. The incident beam is 'chopped' (regularly interrupted by a rotating slit) so that sharp pulses of He atoms are obtained. The He atoms exchange energy and momentum with the sample, exciting or annihilating surface phonons. The scattered atoms are detected in a mass spectrometer. The majority (about 99.9%) of the He atoms are elastically scattered, but those inelastically scattered are well resolved. The momentum exchange is determined by the predetermined scattering geometry, and the energy exchange by the associated change in velocity of the He atoms, which results in different flight times to the detector after scattering. Experimental data for Ag are shown in Fig. 14.13.

## 14.6 The Work Function

The work function is the least energy required to remove an electron from a metal. We have seen that the cohesive energy of a metal derives from the electrons, particularly those in the outermost orbitals, having a lower total energy than in the free atoms. We therefore expect the most energetic electrons, those near the Fermi level, to be more tightly bound the greater the cohesive energy of the metal. For the simpler sp metals this volume contribution to the work function is made evident when we plot the work function against the cohesive energy (Fig. 14.14) (although certain metals, particularly Zn, Cd and Hg, appear to be anomalous). The transition metals, both d and f series, show a similar but less marked trend.

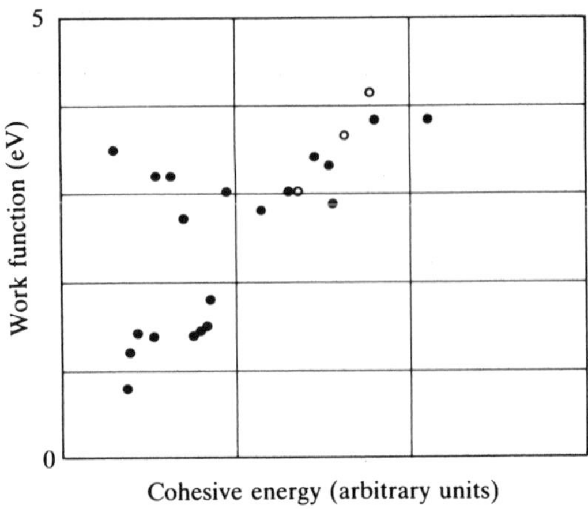


**Figure 14.14** The work function of certain simple metals and its approximately linear dependence on the cohesive energy. The deviations in the upper-left part of the figure arise in Zn, Cd and Hg.

Nevertheless, an electron can only be extracted through the surface of the metal and the existence of the surface introduces another contribution to the work func tion. We have seen that the surface discontinuity may cause an atomic reconstruc tion; a redistribution of electronic charge is more easily envisaged. Two extreme situations that may arise are (a) the electrons spill out of the surface, producing a negatively directed dipole layer (with regard to the outwardly directed surface normal), and (b) the electron charge clouds contract, producing a positive dipole layer. The first possibility increases the work function, the second decreases it. We also expect the work function to vary with the crystallography of the surface. The calculation of the work function is difficult, but values have been obtained for the alkali metals that are in good agreement with experiment. The surface contribution to the work function is markedly affected by the presence of adsorbed atoms, partic ularly those that are strongly electropositive or electronegative with regard to the substrate. The electropositive alkali metals in the form of adsorbed layers are partic ularly effective in lowering the work function. In the bulk form the alkali metals have the lowest work functions among the metals (from 1.8 eV for Cs to 2.4 eV for Li). As free atoms they have small ionization energies. Although very reactive, they may, under ultra-high vacuum conditions, be studied in the form of ordered overlay ers on Cu, Au, Ni, W or similar substrates. The work function is normally deter mined experimentally through the photoelectric effect or by measuring the contac potential difference between the sample and a tungsten filament (often that of the LEED electron gun), which form a simple diode assembly; in the latter case one measures the change in work function as the coverage is increased. Such data are shown for the Cs/Cu(111) system in Fig. 14.15. We see that the presence of the alkal metal adsorbate leads to a rapid initial decrease in $\phi$, which, after passing through a minimum, stabilizes at a value appropriate to the bulk alkali metal. The monolaye structure in this case corresponds to a hexagonal net with parameter twice that o the underlying Cu(111) net; we describe it as $(2 \times 2)$ or $p(2 \times 2)$.

We attribute the initial decrease of the work function to charge transfer from th Cs atoms to the substrate; the Cs atoms become positively charged, they repel on another and therefore spread uniformly over the Cu(111) surface. Each Cs aton

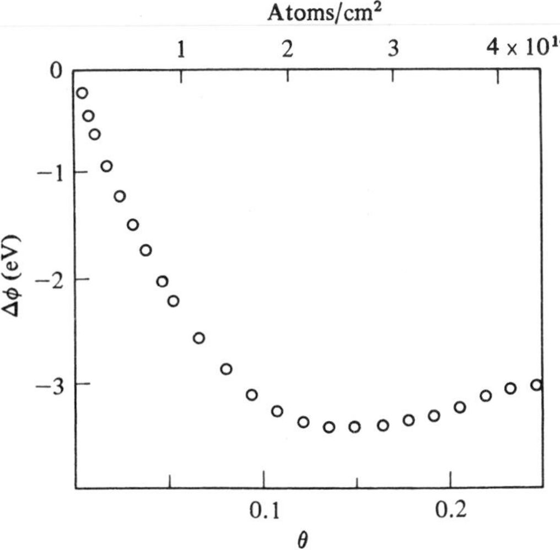


**Figure 14.15** The marked change in the work function of the Cu(111) surface in the presence of adsorbed Cs atoms. (Reprinted with permission from S. Å. Lindgren and L. Walldén, *Solid State Commun.*, **25**, 1978. Pergamon Press, Oxford.)

becomes an electric dipole directed so as to decrease the work function. As more Cs is adsorbed, the atoms and their associated dipoles necessarily become more closely packed, causing the dipole fields to interact and leading to a depolarization effect. The dipoles lose strength owing to this interaction, negative charge flows back from the Cu to the Cs and the work function increases. When the complete overlayer is formed little ionicity remains and the work function of the surface is close to that of Cs metal. It is readily appreciated that an adsorbed electronegative atom (e.g. Cl or I) has the opposite effect.

A qualitative understanding of the mechanism for the above behaviour may be obtained with the aid of Fig. 14.16. In the free atom the valence electron of the alkali metal atom is in a sharp atomic level. On a potential-energy diagram this level lies well above the Fermi level of the substrate because the ionization energy of the alkali atom is less than the work function of Cu. When adsorbed, the valence state of the alkali metal resonates strongly with the conduction states of the metal

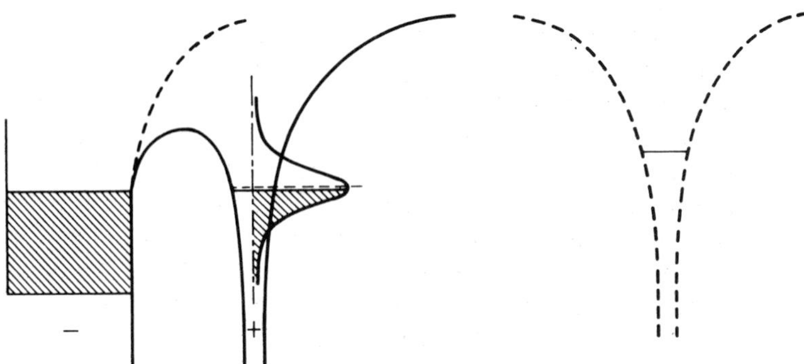


**Figure 14.16** The resonant valence state arising in the adsorbed alkali metal atom. The diagram on the right shows the valence electron in the free atom.

and undergoes both a shift and a marked broadening. At low coverages the resonant valence electron level is less than half filled, charge having been transferred to the Cu crystal. This process continues until the depolarization effects cause the alkali metal atom to become essentially neutral again.

## 14.7 Surface Plasmons

Earlier we described how the conduction electron gas, particularly in a nearly free electron metal like Al, may support charge density or plasma oscillations. It is also found that similar surface charge density waves propagate along the boundary between a metal and vacuum or the interface between two metals. If we have such an interface between two metals with dielectric constants $\varepsilon_1$ and $\varepsilon_2$ respectively then it is found that the characteristic frequency of the surface plasmons is determined by

$$\varepsilon_1 + \varepsilon_2 = 0. \qquad (14.4$$

In the case of the metal-vacuum interface this becomes

$$\varepsilon + 1 = 0,$$

and, assuming $\varepsilon$ to be of the simple Drude form (6.46), we find that the surface plasmon has frequency

$$\omega_s = 2^{-1/2}\omega_p. \qquad (14.5$$

The surface plasmon is readily observed as an inelastic energy loss when electron beams are reflected from a metal surface (Fig. 14.17); it also appears in less marked fashion as a satellite loss peak in core electron photoemission spectroscopy.

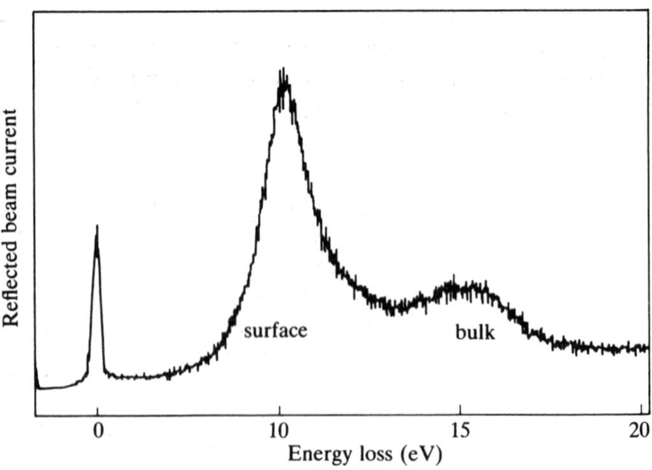


**Figure 14.17** Inelastic plasmon losses arising in electrons reflected from an Al(001) surface. The large angle of incidence (about 80°) for the primary beam accentuates the surface plasmon loss. The energies observed satisfy the relationship $\omega_s = 2^{-1/2}\omega_p$.

## 14.8 Surface Electronic Structures and Adsorbed Molecules

Even for an atomically clean surface, the electronic structure at the surface will differ from that of the bulk in both spatial distribution and distribution in energy. There will inevitably be a spatial redistribution consequent on changes in the atomic structural arrangement, but there will also be changes associated with the abrupt end to the periodic lattice potential. For metals we therefore expect a modification of the wave functions of the electron states, and theoretical calculations can be made of the local density of states (as a function of energy) for atoms in the surface layer or those a little way below it. For semiconductors, as has been recognized for many decades, the dangling bonds left at the surface give rise to both filled and empty states (see Section 10.7), which play an important role in determining the electron barriers that arise at interfaces with vacuum, metals or other semiconductors.

A great impetus has been given to theoretical efforts in this field by the increasing possibility of comparing theoretical results with the experimental data obtainable with the wide range of photoemission electron spectroscopy techniques that are now available. The growth in the number of synchrotron light sources suitable for such work has been of great help. In its simplest form the technique involves irradiating the surface with ultraviolet or X-ray photons and measuring the intensity of the emitted electrons as a function of their energy (see Sections 6.3.2 and 9.4). However, the possibilities of varying the polarization of the radiation, using angular resolved detection and, for magnetic materials, resolving the spin of the emitted electrons have allowed the surface electronic structure to be studied in very great detail. Data for the surface electronic band structure of the aluminium (001) surface are given in Fig. 14.18, where the band gap arising from the two-dimensional character can be seen, but in addition there are truly localized surface states with energies lying within the surface band gap.

The techniques of surface science find extensive application in the study of semiconductor surface states and the electronic states of molecules adsorbed onto the surfaces of metals, especially in connection with attempts to understand catalytic processes. Many gaseous reactions are catalysed on surfaces, and much fundamental research on metal-gas behaviour has been encouraged by the technological importance of heterogeneous catalysis (when the catalyst and the reacting substances have different forms, e.g. solid-gas, solid-liquid). We have earlier described the use of experimental techniques such as XPS, UPS and LEED, other methods are Auger

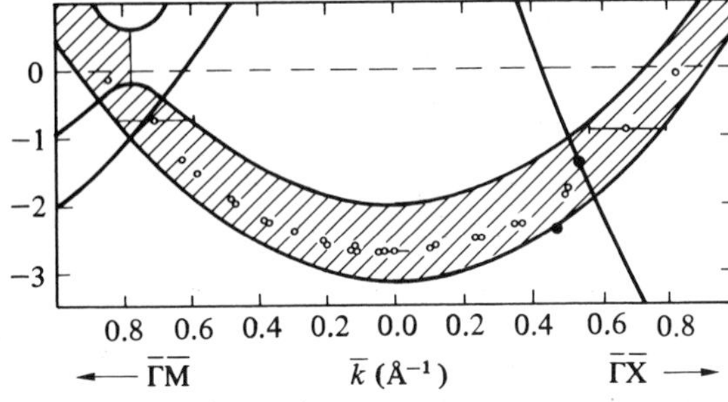


**Figure 14.18** The surface electron state in Al(001). Note that in this figure the region of the energy gap is shown shaded. (After Hansson and Flodström 1978.)

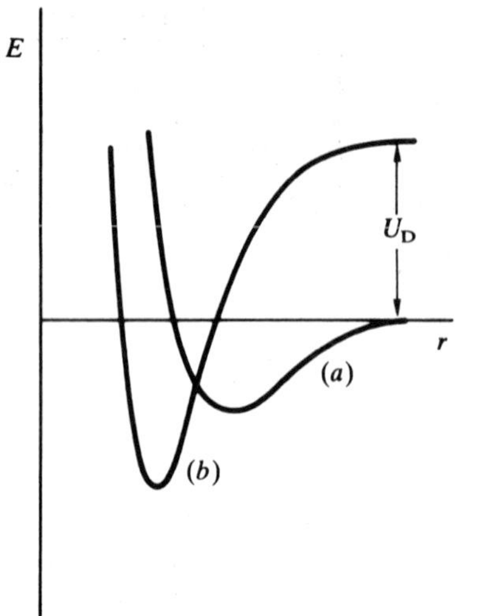


Figure 14.19 The form of the potential-energy curve for the physisorbed molecule (a) and the chemisorbed dissociated molecule (b).

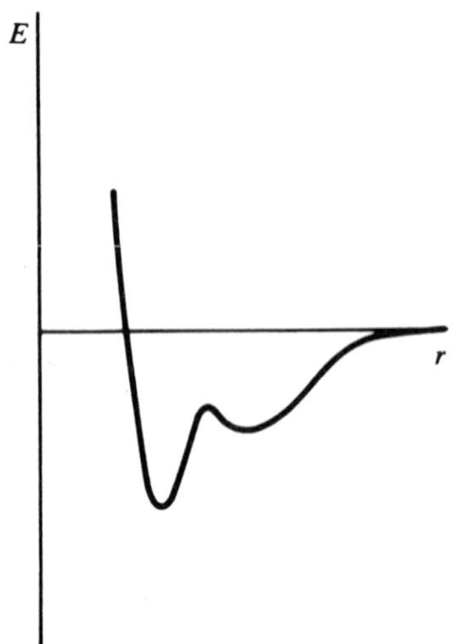


Figure 14.20 The complete-absorption potential-energy curve. However, it is now believed that dissociation and chemisorption are preceded by molecular-ion formation.

spectroscopy, infrared spectroscopy and electron energy-loss spectroscopy (EELS) (see Woodruff and Delchar 1986); the latter is a particularly powerful method for the study of adsorbed molecules on metals and we shall describe certain features of its use.

Schematically, the potential energy of any molecule as it approaches a metal surface along the surface normal may be drawn as in Fig. 14.19*a*. If a minimum occurs in the potential energy then the molecule will be adsorbed onto the surface. However, it often happens that a molecule dissociates after adsorption. We can assess the likelihood of dissociation if we compare the first potential-energy diagram with that for the separate atoms of the molecule. The molecule is first dissociated, costing an energy $U_D$, and the atoms approach the surface along the potential-energy curve of Fig. 14.19*b*, where there is a much deeper minimum. The molecular and atomic absorption curves may be combined to produce a curve of least potential energy as drawn in Fig. 14.20. Whether or not dissociation occurs then depends on the height of the potential barrier separating the two minima compared with the thermal energy $k_B T$.

An electron spectrometer suitable for the study of adsorbed molecules is illustrated in Fig. 14.21. It uses semicylindrical electrostatic mirrors to monochromatize the incident electrons and to analyse those scattered by the sample. The incident electron energy is 1–5 eV and the energy spread in the beam is less than 10 meV.

Diatomic molecules have vibrational modes with quantized energies in the range 50–500 meV; if a molecule dissociates then the atoms become bound to the surface at certain sites, where they also vibrate with characteristic frequencies; in the case of hydrogen the vibrational energies lie in the range 50–100 meV. Incident electrons may suffer inelastic energy exchange with vibrating atoms and molecules; an energy analysis of the reflected electron beam therefore allows the measurement of the vibrational energies of adsorbed species.

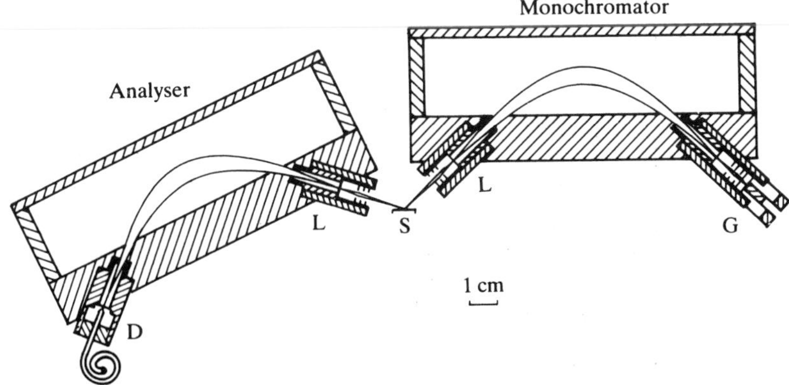


**Figure 14.21** An electron spectrometer for surface physics studies. Both monochromator and analyser assemblies utilize semicylindrical electrostatic mirrors to disperse and focus the electron beam. G, electron gun; L, electron lens; S, sample; D, channeltron detector. Both monochromator and analyser may be rotated in relation to the sample. (Courtesy of C. Nyberg and C. G. Tengstål.)

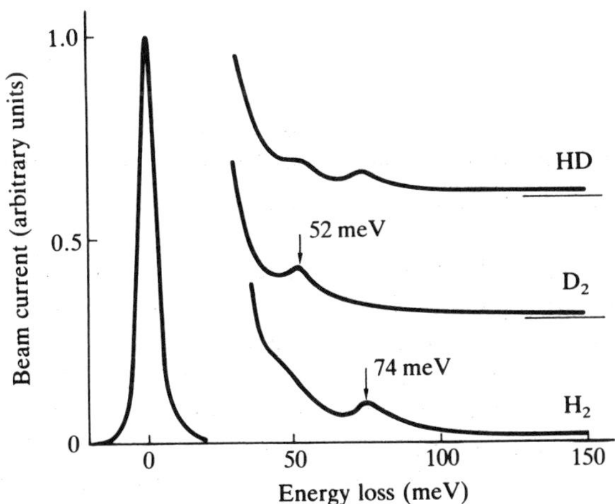


**Figure 14.22** Electron energy-loss spectra for $H_2$, $D_2$ and HD adsorbed onto Ni(001). The data confirm that these molecules are dissociated on the Ni surface. (After Andersson 1978.)

In a study of $H_2$ on Ni(001) the energy-loss curve was obtained, showing a loss at 74 eV (Fig. 14.22). In similar measurements with $D_2$ the loss was at 52 meV (Fig. 14.22). The question is whether these losses are characteristic of the molecules or the atoms of hydrogen and deuterium, but this question is resolved by the spectrum for HD, which shows both the loss at 52 meV and that at 74 meV, indicating that the atoms are responsible for the losses and that the molecules are dissociated on the Ni(001) surface (Fig. 14.22). With the aid of theoretical estimates of the vibrational frequency of the hydrogen atom in various sites of the Ni(001) surface, the adsorption site is associated with the square hollows between the Ni atoms in the surface layer.

## References

ANDERSON, S. (1978) *Chem. Phys. Lett.* **55**, 185.

DOAK, R. B., HARTEN, U. and TOENNIES, J. P. (1983) *Phys. Rev. Lett.* **51**, 578.

HANSSON, G. V. and FLODSTRÖM, S. A. (1978) *Phys. Rev.* **B18**, 1562.

LINDGREN, S.Å. and WALLDÉN, L. (1978) *Solid State Commun.* **25**, 13.

PALMBERG, P. W. and RHODIN, T. N. (1967) *Phys. Rev.* **161**, 586.

PRUTTON, M. (1983) *Surface Physics*, 2nd edn. Oxford University Press, Oxford.

WOODRUFF, D. P. and DELCHAR, T. A. (1986) *Modern Techniques of Surface Science.* Cambridge University Press, Cambridge.

## Further Reading

CHEN, C. J. *Introduction to Scanning Tunnelling Microscopy*, (1993) Oxford University Press, Oxford.

*Physics Today*, (1993) **46**, no. 11 (Search & Discovery), 17.

WIESENDANGER, R. *Scanning Probe Microscopy and Spectroscopy*, (1994) Cambridge University Press, Cambridge.

15

# The Nucleus and Solid State Physics

So far we have considered the nuclei in a solid solely as the seats of the mass and the positive charge, but an increasing number of experimental techniques for the study of solids use nuclear properties as probes. In particular, many nuclei possess angular momentum, measured by the quantum number $I$, and an associated magnetic moment $\mu$; the latter allows interaction of the nucleus with the outer electron clouds. Certain nuclear properties may therefore reflect changes in the electronic structure. Furthermore, the instability of some nuclei leads to the emission of soft $\gamma$ radiation (10 keV $\leqslant h\nu \leqslant$ 100 keV), and when such nuclei form part of a crystalline solid the emission and absorption of the $\gamma$ rays may occur in a recoilless fashion. This is the Mössbauer effect, which is the basis for a spectroscopy with higher resolution than any other form of solid state spectroscopy. It is particularly applicable to the study of magnetic materials.

## 15.1 The Nucleus and Magnetic and Quadrupole Effects

The nucleus is an assembly of neutrons and protons. Each nucleon, i.e. each neutron or proton, has an effective radius of about $1.4 \times 10^{-15}$ m and an intrinsic angular momentum corresponding to a quantum number $\frac{1}{2}$. In nuclei the nucleons also possess orbital angular momentum, which, when combined with their intrinsic spin, produces a resultant angular momentum, which is by tradition called the nuclear spin and denoted by **I** so that

$$|\mathbf{I}| = \hbar[I(I+1)]^{1/2},$$

where $I$ is the appropriate quantum number. The resultant nuclear spin may take quite large values; thus $^{209}$Bi has $I = \frac{9}{2}$.

The nuclear magneton is defined by

$$\mu_N = \frac{e\hbar}{2M} = 5.05 \times 10^{-27}\ \mathrm{J\ T^{-1}},$$

$M$ being the mass of the proton. Even so, the proton does not have a magnetic moment equal to one nuclear magneton, but rather $2.79\mu_N$. Similarly, although the neutron has no net charge, it has magnetic moment $-1.9\mu_N$ (i.e. the magnetic moment is antiparallel to $\mathbf{I}$). The nuclear magnetic moment is expressed as

$$\boldsymbol{\mu} = \gamma \mathbf{I}, \tag{15.1}$$

$\gamma$ being the nuclear magnetomechanical ratio; it varies from nucleus to nucleus. Although this ratio is usually positive in sign, there are nuclei with negative values (e.g. $^{107}$Ag, $I = \frac{1}{2}$, $\mu = -0.133\mu_N$). Owing to spatial quantization, a nucleus with quantum number $I$ possesses $2I + 1$ levels, which, in the absence of an external field, are degenerate. An external field leads to the occurrence of $2I + 1$ discrete nuclear Zeeman levels, each of which is separated from its neighbours by an energy $\hbar\gamma B_0$. The frequency $\omega_0$, also known as the Larmor frequency, defined by

$$\omega_0 = \gamma B_0, \tag{15.2}$$

is that with which the nuclear moment $\mu$ precesses around the direction of the applied field $B_0$, assumed to be directed along the $z$ axis (Fig. 15.1). For the proton in a magnetic field of 1 T, the value of $\hbar\gamma B_0$ is $1.76 \times 10^{-7}$ eV, which, expressed as a frequency, corresponds to 42.58 MHz, an ordinary radio frequency. In addition to Zeeman splitting, nuclear spin levels are also perturbed by electric quadrupole moments. The electrostatic potential along say the $z$ axis of an axially symmetrical distribution of electric charge $\rho(r)$ (Fig. 15.2) may be expressed as a power series in $r/z$ ($r \ll z$), leading to the following expression for the potential $\phi_P$ at the point $P$:

$$4\pi\varepsilon_0 \phi_P = \frac{1}{z}\int \rho \, dv + \frac{1}{z^2}\int \rho r \cos\theta \, dv + \frac{1}{z^3}\int \rho r^2 (3\cos^2\theta - 1)\, dv + \cdots . \tag{15.3}$$

The coefficient of the first term is often called the monopole contribution, being a measure of the net charge density; in a neutral charge distribution this is zero. The second term is the dipole-moment contribution to the potential arising when the positive and negative charge distributions do not have a common centre of gravity, in a symmetrical charge distribution this is also zero. The third term arises when the charge distribution has a centre of symmetry, but is not spherically symmetrical. Nuclei with spin $I \geqslant 1$ often have an ellipsoidal form. The nuclear charge distribution adopts an oblate (flattened at the poles) or a prolate (elongated at the poles) form, leading to nuclear quadrupole moments of negative and positive sign respectively. The quadrupole moment is $Q$, where $eQ$ is the coefficient of the term in $z^{-3}$. In the presence of an axially symmetrical electric field gradient $eq$ the quadrupole moment causes a precession of the angular momentum and a splitting of the nuclear spin levels, even in the absence of a magnetic field. In the presence of a magnetic field the Zeeman splittings are distorted in a manner to be described later. In crystals with cubic symmetry the electric field gradient at a nucleus is zero and quadrupole effects are absent, but if the cubic symmetry is disturbed by mechanical strain, by the presence of dislocations or by alloying, quadrupole effects are again introduced.

## 15.2 Nuclear Levels and Thermal Equilibrium

When discussing the electronic structure of metals or the paramagnetism of a transition metal compound, we tacitly assume that the temperature of the electrons or

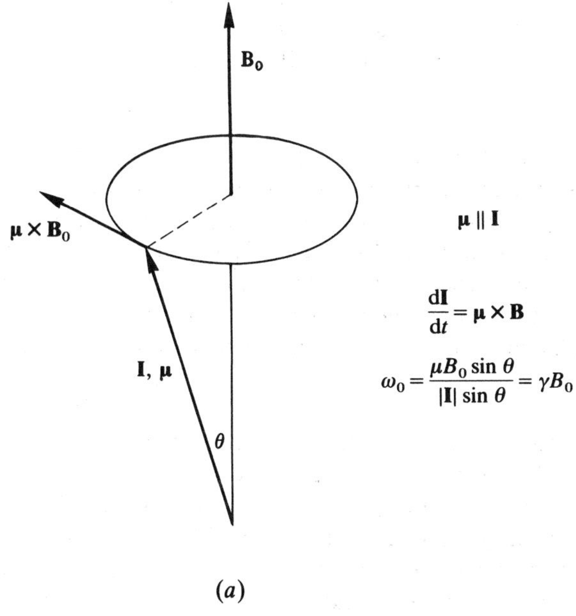


(a)

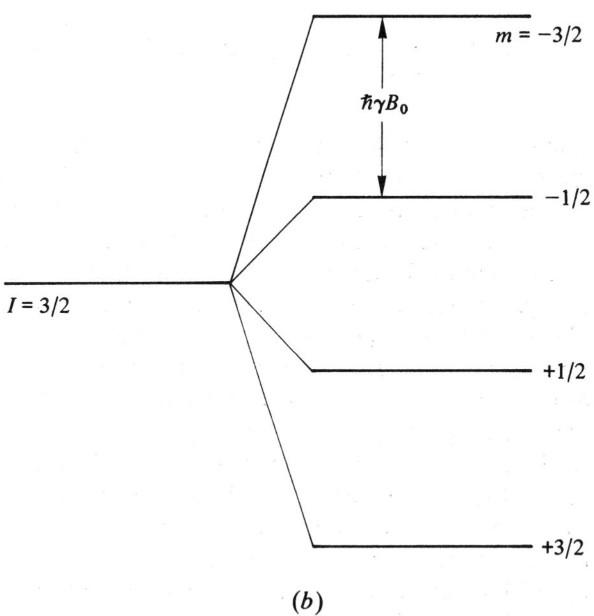


(b)

**Figure 15.1** (a) The precession of the nuclear spin **I** in an external magnetic field $\mathbf{B}_0$. The rate of change of angular momentum ($\omega_0 I \sin\theta$) is equated to the torque ($\boldsymbol{\mu} \times \mathbf{B}_0$) produced by the magnetic field. (b) The Zeeman splitting of nuclear levels by an external magnetic field $\mathbf{B}_0$.

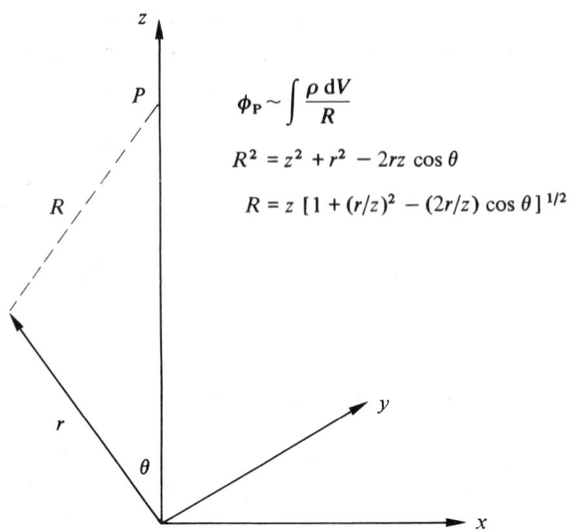


**Figure 15.2** The geometry for the derivation of (15.3). We imagine an axially symmetric distribution of charge and need to calculate the potential at a point $P$ on the $z$ axis.

that of the atomic moments is the same as that of the lattice. The temperature of a substance is associated with the vibratory motion of the atoms. However, a solid is not always a single homogeneous arrangement of atoms; we often need to divide it into 'the phonon system', 'the electron system', 'the atomic-magnet system' and 'the nuclear-magnet system'. Each of these systems has its own characteristic energy domain. Thus the phonons have mean energy of order $10^{-2}$ eV, electrons in metals may have excitation energies anywhere from 0 eV to as high as we can arrange. The Zeeman splitting of the levels associated with an atomic magnetic moment is $g\mu_B B_0$, which, for a field of 1 T, corresponds to $1.16 \times 10^{-4}$ eV when $g = 2$ (or an equivalent frequency 28.0 GHz), whereas we have seen that the nuclear Zeeman levels have separation of order $10^{-7}$ eV. These different systems are only in thermal equilibrium with one another if energy exchange between them can occur rapidly – just how rapidly depends upon the timescale of the measurement being made. The phonon spectrum extends over a vast range of frequency 0–$10^{14}$ Hz. The conduction electrons in a metal experience the ionic potential modulated by the ionic motion – this electron-phonon interaction provides good thermal contact between these systems. Classically, we should say that the phonon and electron gases are in good thermal contact via collisions. The same is true for the phonons and the atomic magnetic moments; energy transfer here requires that transitions between the electronic Zeeman levels (separation $<10^{-4}$ eV) can be mediated by the phonons, but the latter always possess components of frequency that can match the electronic Larmor frequency, and the ion potential modulated at this frequency allows exchange of quanta between the two systems, ensuring thermal equilibrium.

The situation is similar for the system of nuclear spins in a metal or a magnetic insulator. We require fluctuating magnetic fields at the nucleus that can populate or depopulate the nuclear Zeeman levels. These fields, at the appropriate frequencies, are always available in the dipolar fields arising from conduction electrons or the atomic magnetic moments. For metals there is always good thermal equilibrium between the different systems. Insulators, however, particularly diamagnetic ones, have poor coupling to the nuclear spin system because the only source of fluctuating magnetic field is the dipolar interaction between nuclei, which, as we shall see later, is always very weak. In a diamagnetic insulator, and the more so the purer or more stoichiometric the composition, the nuclear spin system may be at a very different temperature from that of the lattice, especially at very low temperatures. Because of this, small amounts of paramagnetic impurity are sometimes added to a sample to promote better thermal contact with the nuclear spin system. In the following we shall usually assume that thermal equilibrium is always achieved.

## 15.3 Nuclear Magnetic Resonance, NMR

If transitions may be induced between the nuclear Zeeman levels by electromagnetic radiation then, using (15.2), we may determine $\omega_0$ and thereby $\gamma$. If $I$ is known, we obtain the nuclear magnetic moment $\mu$, but a knowledge of $\omega_0$ has wider significance because the resonant frequency is found to depend on the nuclear environment.

Nuclear absorption spectroscopy, in the case of the proton for example, requires that $\Delta m = -1$, where $m$ is the nuclear magnetic quantum number. However, under conditions of thermal equilibrium at 300 K and in a magnetic field of 1 T, the populations of the two levels are in the ratio

$$N_{-1/2}/N_{+1/2} = 0.9999932.$$

Furthermore, the ratio of the probability for stimulated absorption and emission events to that for spontaneous events is, for the assumed conditions, of order $10^5$. Spontaneous emission is negligible and stimulated emission occurs to almost the same extent as stimulated absorption, leading to an extremely low absorptivity that can only be measured using sensitive resonance techniques. Various procedures are available, but all require an electromagnet producing a highly homogeneous polarizing field $B_0$ and a radio-frequency tuned circuit that includes a coil surrounding the specimen and operating at a fixed frequency near the expected resonant frequency. The field of the electromagnet is augmented slightly by that of another coil carrying a low-frequency current producing the sweep field $\Delta B$. The polarizing field repeatedly varies over the range $B_0 \pm \Delta B$ within which resonance is fulfilled. At resonance, the quality factor of the resonant circuit containing the sample changes by about 1%, which is sufficient to make measurements practicable.

The major applications of NMR arise because, for a given $B_0$ and for a particular nucleus, the resonant frequency is found to be dependent upon the chemical environment. The reason is that the absolute value of the magnetic field at a particular nucleus comprises not only contributions from the electrons of the atom containing the nucleus but also from the nuclei and electrons of surrounding atoms. In non-conducting substances, particularly organic liquids, this leads to a chemical shift of the resonance, which may be as large as 1 G, but as small as or smaller than 1 mG. Structural and analytical organic chemists have therefore made extensive use of high-resolution NMR; $^{1}H$, $^{19}F$ and $^{31}P$ followed by $^{13}C$, $^{29}Si$ and $^{15}N$ are important nuclei in this context (Fig. 15.3). It is also the case that molecular groupings may be recognized by their NMR pattern (Fig. 15.3).

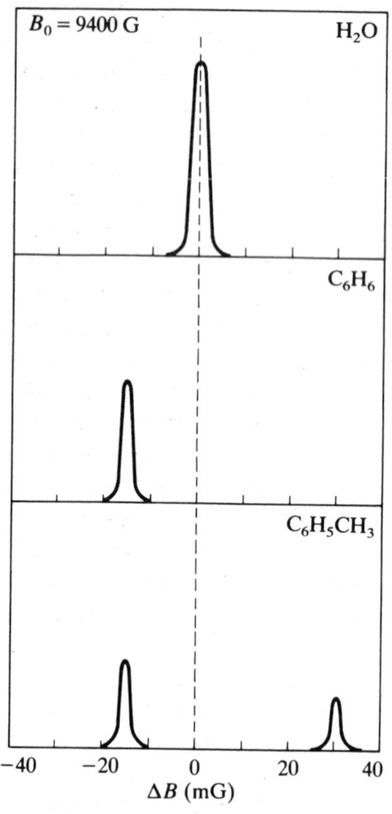


**Figure 15.3** Proton magnetic resonance and the chemical shift in benzene and toluene relative to water. (Reprinted with permission from E. H. Rogers, *NMR and EPR Spectroscopy*, 1960. Pergamon Press, Oxford.)

From the physical point of view, significant features of NMR concern the conditions for its experimental observance. Consider the proton again. To observe the resonance in water, we populate the lower spin level ($m = +\frac{1}{2}$) in slight preference to the upper level by applying the field $B_0$. This requires that we remove energy from the nuclear spin system and transfer it to the lattice; we have implied earlier that this may be a time-consuming process. We cannot magnetize the nuclei more rapidly than the energy can be removed.

Phenomenologically, we introduce a relaxation time $T_1$ and write

$$\frac{dM}{dt} = \frac{M_\infty - M}{T_1}, \tag{15.4}$$

$$M = M_\infty(1 - e^{-t/T_1}), \tag{15.5}$$

where $M_\infty$ is the magnetization appropriate to $t \to \infty$, and $M$ is the magnetization at time $t$. $T_1$ is called the longitudinal spin-lattice relaxation time. The adjective 'longitudinal' is used because we consider magnetization changes parallel to the field $B_0$. For an insulator $T_1$ becomes larger the purer the material and the lower the temperature. At low temperatures, of order $10^{-3}$ K, it may approach several days. Similarly, even if equilibrium magnetization is achieved, the resonant absorption, in the presence of a large $T_1$, may lead to a rapid equalization of the populations of the Zeeman levels, causing the resonant absorption to disappear because stimulated emission and stimulated absorption events occur in equal amounts. Observation of NMR therefore demands that the exchange of energy between the nuclear spins and the lattice occur on a timescale less than that used in passing through the resonance. In distilled water at 20°C $T_1 \approx 2$ s, but the addition of paramagnetic impurities (usually $O_2$ from dissolved air) leads to a significant reduction. In metallic samples, which are usually studied as filings because of the radio-frequency skin effect, $T_1 \approx 1$ ms since the conduction electrons are a good coupling medium.

The absorption resonance is characterized by its shape, e.g. its symmetry or asymmetry and its width. The width is quoted in terms of magnetic field, and may vary from the order of milligauss in organic liquids to the order of gauss in metals. Neglecting experimental features like the inhomogeneity of $B_0$, one source of broadening lies in the lifetime of the excited level; another occurs because the nuclei experience a range of magnetic fields arising from their immediate environments. Even in an elemental sample, there arises a varying magnetic field owing to the motion of the nuclei. Thus, for given $B_0$, the resonant frequency varies among the nuclei; or a given nucleus experiences a different magnetic field at different times. If there is a distribution of resonant frequencies, this means that the nuclear magnetic moments do not precess about the field axis with the same frequency, so they do not remain in phase with one another. If there were perfect phase coherence in the precession, it would lead to a magnetization transverse to $B_0$ and rotating with frequency $\omega_0$. The transverse relaxation time $T_2$ is a measure of the rate at which the nuclei lose their phase coherence and their transverse magnetization. The resonance line has a width proportional to $1/T_2$. In crystalline solids $T_2 \approx 0.1$–1 ms. The narrowest lines are observed in pure organic liquids.

Liquids produce narrow resonance lines owing to the rapid and random motion of the molecules – rapid on a timescale relative to the precession time. This rapid and random motion causes the local magnetic field produced by nearby atoms to have value zero when averaged over a time of order $2\pi/\omega_0$. The effect is called motional narrowing. It is also observable in certain metallic samples; as the temperature is increased, leading to diffusive motion, the resonance line width narrows in marked and continuous fashion. Diffusion coefficients for certain low-melting-point metals (Li, Na, Al) have been measured in this way.

In metals and alloys the resonant frequency of a particular nucleus is different from that observed when it forms part of an insulator. The resonance arises at a higher frequency in the metal, and the change is considerably larger than the conventional chemical shift in insulators, being 0.1–1% $\omega_0$. It is called the Knight shift and occurs because the conduction electron gas becomes polarized by the field $B_0$, thereby creating an additional magnetic field at the nucleus. The effect is produced by the presence of electrons with s symmetry in the conduction electron gas. All electron radial wave functions of symmetry other than s have zero amplitude at the centre of the atom, but s electrons have a maximum amplitude at the nucleus. If the occupation of the spin-up and spin-down states is out of balance owing to an external field, this causes a net polarization of the electron spin density; the contact of s electrons with the nucleus then produces a significant enhancement of the magnetic field at the nucleus and thereby a large shift in $\omega_0$ relative to an insulating medium. The Knight shift is proportional to the product of the susceptibility $\chi_s$ of the s electrons and the square of their wave-function amplitude at the nucleus. Only in exceptional circumstances (e.g. Li and Na) is $\chi_s$ known independently. Knight-shift data must therefore be subject to theoretical analysis. The shift has been studied as a function of temperature, pressure and alloying, and is a valuable monitor of the s component of conduction electrons under different conditions.

Although not a nuclear phenomenon, it is clear that a paramagnetic ion in the presence of a magnetic field also produces a system of Zeeman levels (Section 10.4) but the Larmor frequency lies in the microwave region of the spectrum ($\lambda \approx 1$ cm). Electron paramagnetic resonance, EPR (or electron spin resonance, ESR) is observed in a manner analogous to NMR. The technique has been used to study the effects of the crystal fields (Section 11.4.2) and the hyperfine interaction with the nuclear spin in paramagnetic ions. The method is particularly well adapted to the study of paramagnetic impurity centres as may arise in semiconductors, e.g. the electron bound to a donor atom, the electron in an $F$ centre or similar defect, because the s character of the bound state leads to a strong contact interaction with the nucleus (see Bleaney and Bleaney 1976, Chap. 24).

### 15.3.1 Quadrupole effects

In substances with $I \geqslant 1$ and with symmetry other than cubic, e.g. $^{69}$Ga, pure quadrupole resonance may be observed in the absence of a magnetic field, allowing the determination of the product of the quadrupole moment and the electric field gradient, but these quantities are not readily separated.

Quadrupole effects may be introduced in a cubic substance like Cu by the presence of mechanical strain or alloying elements. Natural Cu contains 66.7% $^{63}$Cu and 33.3% $^{65}$Cu – both isotopes have $I = \frac{3}{2}$, but with different magnetic moments, so their nuclear magnetic resonances do not coincide. The Zeeman levels arising in a field $B_0$ are perturbed by the presence of the quadrupole moment (Fig. 15.4). Instead of one frequency interval, we now find $\omega_0$ and two satellites one on each side of the main line. If these satellites are clearly separated from the unperturbed $\omega_0$ then the latter is reduced in intensity and by an amount dependent on the spin $I$. For $^{63}$Cu the intensity at $\omega_0$ is reduced to 40% of its former value; nevertheless, each $^{63}$Cu nucleus contributes to the signal. If, however, Cu is alloyed with polyvalent metals like Zn, Ga, Ge or As then, even for small concentrations of order 1%, the satellite resonances fall far from $\omega_0$. Furthermore, the intensity of the main resonance at $\omega_0$ drops markedly, without change in width, with concentration of alloying element – increasingly with larger valence.

The reasons are two-fold. The alloying atom has a different size from Cu and therefore causes mechanical strain, which disturbs the cubic symmetry. More important is the fact that each impurity atom perturbs the density of the conduction electrons. In the first approximation we assume that the extra valence electrons introduced with the impurity atom remain in its immediate vicinity to maintain electrostatic neutrality. Even so, the extra 1, 2, 3 or 4 electrons associated with the Zn, Ga, Ge and As atoms represent a large perturbation, the effects of which extend out to the region of the 7th distant neighbours. The electric field gradient is so strong in the region surrounding the impurity atom that no magnetic resonance is observable. The decay of the perturbation is proportional to $r^{-3}$ and is not monotonic but of oscillatory form (Fig. 15.5). In the case of Cu alloys it is found that each Zn, Ga, Ge and As atom destroys the resonance in 18, 38, 64 and 82 $^{63}$Cu atoms

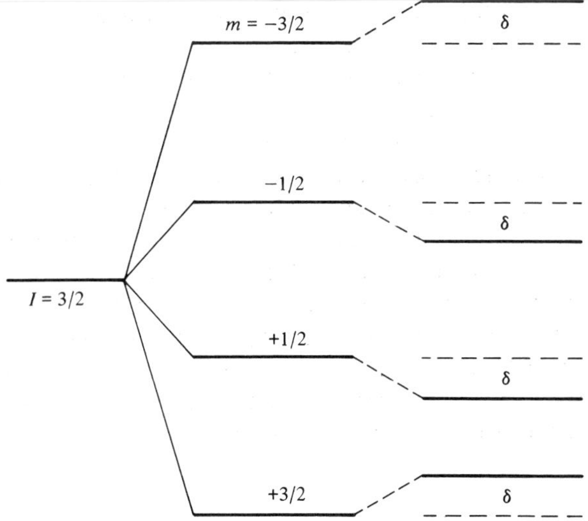


$$\Delta = -\hbar\gamma B_0 + (-1)^{|m|+1/2}\frac{e^2qQ}{4}\left(\frac{3\cos^2\theta - 1}{2}\right)$$

$\delta = \frac{1}{4}e^2qQ$ **when** $\theta = 0$, **i.e. when I is parallel to the field gradient** $e\mathbf{q}$

**Figure 15.4** The perturbation of the Zeeman splitting by a nuclear quadrupole moment $Q$. $\Delta$ is the complete energy shift and $\delta$ the quadrupole component.

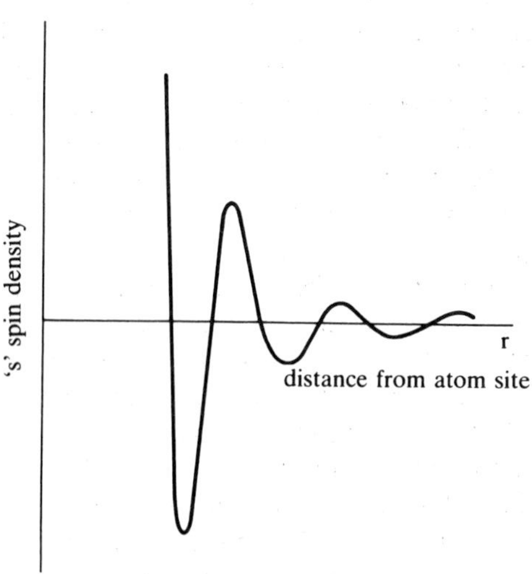


**Figure 15.5** Oscillations in the conduction electron spin density arising in the presence of a nuclear magnetic moment; these oscillations are important for the Knight shift and the coupling of the nuclear magnetic moments. A similar variation of electron charge density is associated with an impurity element of different valence to the host matrix.

respectively. This is a particularly striking demonstration of the long range ove which an impurity atom may disturb the host matrix.

## 15.4 The Nuclear Heat Capacity

In substances possessing an electronic magnetic moment, e.g. Sm, there is always large magnetic field present at the nucleus, which produces an intrinsic nuclea Zeeman splitting: this is the hyperfine field. Similarly, in a non-magnetic substanc of low symmetry, but containing nuclei with a quadrupole moment, the nuclea levels are split. At all normal temperatures, say $>1$ K, these nuclear levels are esser tially equally populated and do not contribute to the heat capacity, but at very lo temperatures we expect to find the nuclei in the lowest possible energy state, and, a the temperature increases, energy will be needed to populate the higher levels. W therefore find a nuclear contribution to the heat capacity.

Consider a simple two-level system as shown in Fig. 15.6. This might well rep resent the quadrupole splitting for a nucleus with $I = \frac{3}{2}$ in the ground state. Th energy separation is $\Delta$ and in thermal equilibrium

$$\frac{N_2}{N_1} = \mathrm{e}^{-\Delta/k_B T},$$

$$N = N_1 + N_2,$$

whence

$$N_2 = \frac{N}{1 + \mathrm{e}^{\Delta/k_B T}},$$

and the energy of the system is $N_2 \Delta$. We readily find that the heat capacity is

$$C = \Delta \frac{\mathrm{d}N_2}{\mathrm{d}T} = Nk_B\left(\frac{\Delta}{k_B T}\right)^2 \frac{\mathrm{e}^{\Delta/k_B T}}{(1 + \mathrm{e}^{\Delta/k_B T})^2} \qquad (15.$$

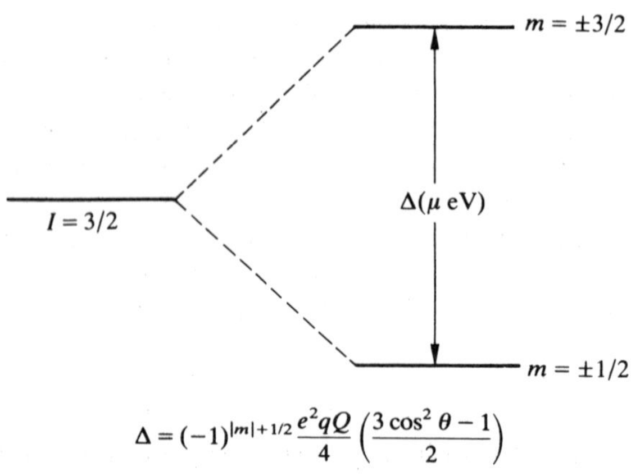


**Figure 15.6** Pure quadrupole splitting of a nuclear level.

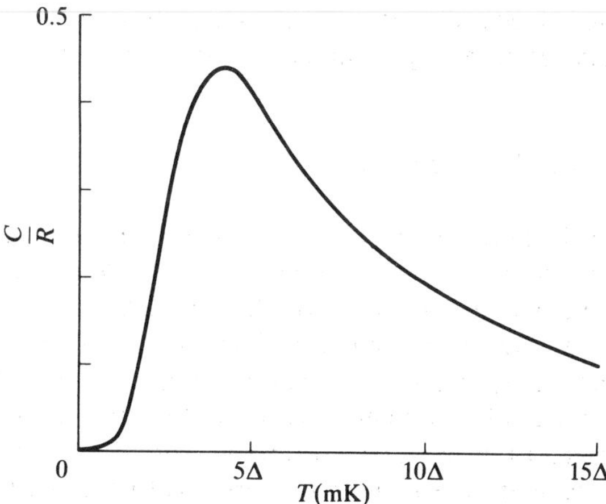


**Figure 15.7** The nuclear quadrupole heat capacity; the shape is typical for any two-level system. $\Delta$ is the numerical separation of the energy levels in $\mu$eV.

and has the form shown in Fig. 15.7:

$$C \propto \begin{cases} e^{-\Delta/k_B T} & (k_B T \ll \Delta), \\ T^{-2} & (k_B T \gg \Delta). \end{cases}$$

For our two-level system the heat capacity attains a maximum value of $0.44Nk_B$ at a temperature 4.9 $\Delta$ mK when $\Delta$ is measured in units of $\mu$eV. At a temperature of 1 mK the lattice and conduction electron components of the heat capacity are extremely small and the value $0.44Nk_B$ represents an enormous contribution. A more complicated system of nuclear levels produces a similar effect, we just need to sum over a larger number of levels using the appropriate Boltzmann factors.

It will also be appreciated that any system of closely spaced levels produces such a contribution; thus if an ion has an excited configuration close to the ground state, such an anomalous heat capacity will appear. The behaviour is known as a Schottky anomaly.

A pure quadrupole nuclear heat capacity arises in the elements As, Sb and Bi. The latter are somewhat exceptional substances; they are normally called 'semimetals', although 'poor metals' might be a better description. In the liquid state they have much better electrical conductivity than in the crystalline form. The reason is that these elements crystallize in a slightly distorted simple cubic structure. The primitive cell is rhombohedral and the basis contains two atoms. They are pentavalent, so there are ten electrons to be accommodated in the zone structure. Five zones are almost, but not quite, filled. There are pockets of electrons in the sixth zone and pockets of holes in the fifth, but these are so small that there are only of order $10^{-3}$ charge carriers per atom; that is why their Hall coefficients are so very large (Section 6.6). The electronic heat capacities of these metals are therefore exceptionally small.

The nuclear spins of the semimetals are large and produce a more complicated system of energy levels than the simple doublet, but the behaviour is similar in principle, and at high temperatures ($>10$ mK) we expect a $T^{-2}$ dependence of the heat capacity. Experimental data are shown in Fig. 15.8. For As and Sb the

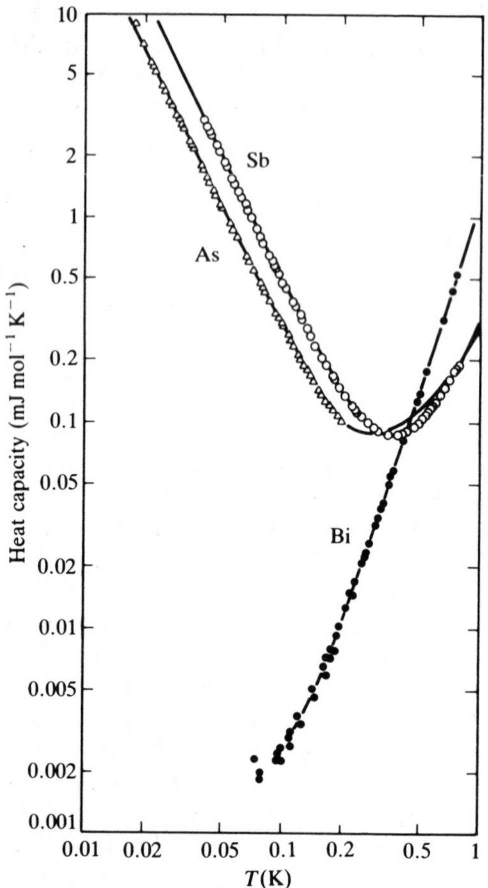


**Figure 15.8** The nuclear contribution to the heat capacity in the semimetals As, Sb and Bi. (After Krusius 1971. See also Collan *et al.* 1970.)

expected nuclear heat is observed and is in excellent agreement with theory and NMR data. Bi appears anomalous, but this may be understood if the nuclear spin system is in poor thermal contact with the lattice. The spin-lattice relaxation time is so long for Bi that thermal contact cannot be made to the nuclear spins on the timescale of the measurement. A probable reason for the large $T_1$ of Bi is its very low charge carrier concentration, which is only about $10^{-2}$ that of As and Sb. When this carrier concentration is increased by the addition of small amounts of Te the nuclear contribution becomes observable. Similar large nuclear heats arise from magnetic hyperfine splittings, particularly among the rare earth metals, but also in Fe and Co for example.

## 15.5 Nuclear Antiferromagnetism

If the nuclei carry magnetic moments, the question of cooperative interactions leading to ferromagnetic or antiferromagnetic arrangements arises. The nucleus has diameter of order $10^{-15}$ m and the internuclear separation in solids is seldom less than $10^{-10}$ m, so clearly there can be no direct nuclear coupling of the spins. In a metal like Cu an indirect interaction of the nuclear moments may arise via the conduction electrons with s character. Because these electrons are in contact with the nucleus, they experience a slight magnetic polarization, which, owing to their approximately uniform density, is felt by adjacent nuclei. In this manner the nuclear moments may be coupled into ferromagnetic or antiferromagnetic arrangements. The only interaction that can arise in a non-magnetic insulator, however, is the classical dipole-dipole interaction – that which we rejected in our discussion of conventional magnetism (Section 10.2). In an insulator, given atoms with a moment of one nuclear magneton, we may readily estimate the magnetic field at a distance of $10^{-10}$ m:

$$\frac{\mu_0 \mu_N}{4\pi r^3} \approx 5 \text{ G.}$$

The energy of another nuclear magneton in this field is $1.5 \times 10^{-12}$ eV, which, expressed as a temperature, is $2 \times 10^{-7}$ K. Taking account of the coordination, we might expect critical temperatures for cooperative nuclear magnetic behaviour to be of order $10^{-6}$ K. If the only way to observe such behaviour were by the cooling of the nuclei, the phonon gas, then it would be extremely difficult. At the present time, using $^3$He–$^4$He refrigerators, it is only possible to reach and *regularly maintain* lattice temperatures $\geqslant 5$ mK. But this is where the spin-lattice relaxation time comes to our aid, because it is possible to cool the nuclear spins to about $10^{-6}$ K while maintaining the lattice at about 10 mK or in certain cases even at 0.3 K.

Before the advent of the $^3$He–$^4$He refrigerator, the only method of obtaining temperatures in the millikelvin range was by the adiabatic demagnetization of a paramagnetic salt, usually cerium magnesium nitrate. The specimen was arranged to be in good thermal contact with the salt and both were in contact with a bath of liquid $^3$He at low pressure maintaining a temperature 0.3 K. The contact to this low-temperature bath was via a thermal switch, enabling good contact for the removal of heat from the sample and magnetic salt or poor contact when these parts were to be thermally isolated. The thermal switch was usually a metal wire that, by the aid of an auxiliary coil, could be put into the normal and good thermal conductivity state or the superconducting state when thermal isolation was required. The salt was magnetized by a large external field, cooled by the low-temperature bath and then thermally isolated. The loss of magnetization in a decreasing external field resulted in an increase in entropy, which, since the change was adiabatic, could only be achieved by the transfer of heat from the lattice to the electron system. The lattice of the magnetic salt and that of the sample were therefore cooled. The cooling could proceed in several demagnetizing stages and extend over many hours. The same principles may be applied to obtain cooling in Cu by adiabatic nuclear demagnetization. Spin temperatures approaching 1 nK have been obtained in this way and the nuclear paramagnetic susceptibility measured (Fig. 15.9). Recently the occurrence of nuclear antiferromagnetism in $^{65}$Cu below 60 nK was demonstrated by neutron diffraction.

In a diamagnetic insulator the nuclear spin system may be more readily isolated from the lattice. The sample is doped with a small concentration of paramagnetic ions (approximately 100 ppm) and a large static magnetic field of several tesla applied. This field polarizes the electronic magnets, but leaves the nuclear spins of the sample essentially unaffected. The nuclear spins are then polarized by a reso-

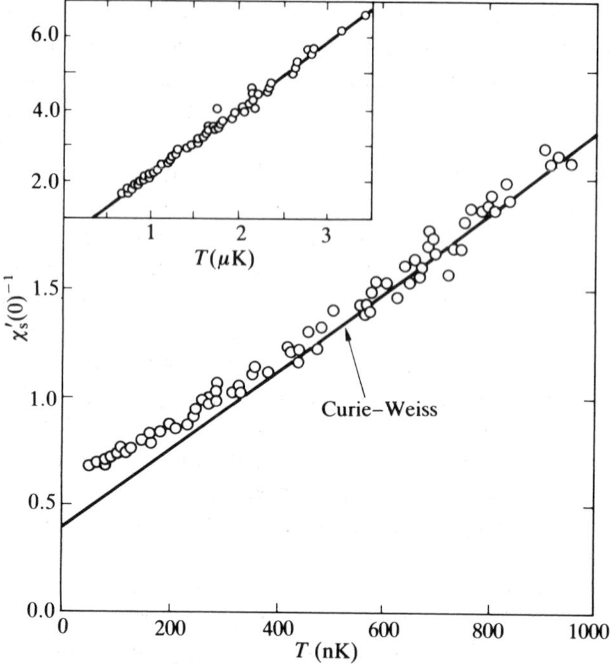


**Figure 15.9** The Curie-Weiss plot for the nuclear magnetism of Cu; it is believed that three anti-ferromagnetic phases exist below a critical temperature of about 70 nK. $\chi_s'(0)$ is the low frequency susceptibility determined using a SQUID technique. The Curie-Weiss behaviour may be interpreted in terms of a combination of dipolar and RKKY couplings between the nuclear magnets. (After Huiku *et al.* 1986.)

nance technique. A microwave magnetic field is applied in addition to the static field, its frequency is $\omega_{0e}-\omega_{0n}$, the difference between the electronic and nuclear Larmor frequencies. The quantum energy corresponds to that required to reverse the direction of an impurity electron spin and at the same time induce a nuclear spin flip in the opposite direction – the situation is depicted schematically in Fig. 15.10.

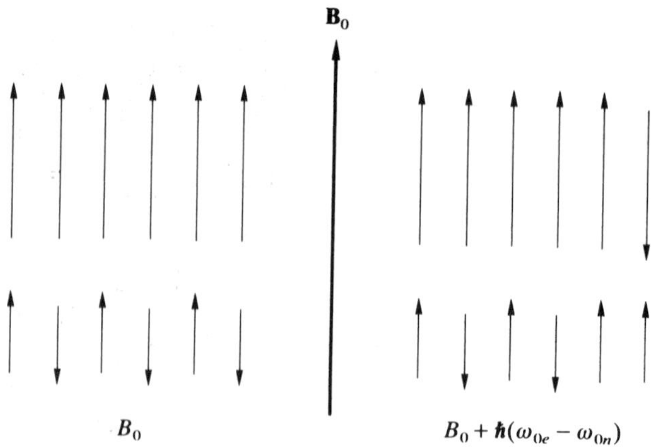


**Figure 15.10** A schematic illustration of the resonant polarization of nuclear magnetic moments. (After Lounasmaa 1974.)

For every adsorbed quantum $\hbar(\omega_{0e} - \omega_{0n})$ an electron moment is caused to become antiparallel to the static field and a nuclear spin parallel to the field. The impurity electronic moments are in good thermal contact with the lattice, so they quickly become parallel to the static field again, but the sample's nuclear spin system is isolated because of the large spin-lattice relaxation time, and its polarization grows. In this manner a system of polarized nuclear spins in a static field and at a high temperature (that of the appropriate reservoir, say $10^{-2} - 10^{-1}$ K) is obtained. It now remains to cool the nuclear spins alone. This is achieved by adiabatically depolarizing the nuclear spins, in the presence of the external static field, again using a resonance technique. A circularly polarized radio-frequency magnetic field of small amplitude and transverse to the static field is swept through the nuclear Larmor frequency. Analysis of the above situation in a rotating frame of coordinates shows that, in spite of the presence of a large static field, demagnetization occurs by nuclear spin flip in an effective field that is that of the transverse radiofrequency field $<1$ G (see Lounasmaa 1974). Entropy is squeezed out of the nuclear spin system, leading to a decrease in its temperature.

In this way nuclear spin temperatures in the range 0.1–1 $\mu$K have been achieved over sufficient time to allow neutron diffraction studies of the nuclear spin arrangements at these temperatures. Such experiments on $^7Li^1H$ have clearly demonstrated the occurrence of an antiferromagnetic arrangement of nuclear magnetic moments (Fig. 15.11). The observed structures are in agreement with a Weiss-type molecular field model. Studies of nuclear cooperative magnetism are confined to a handful of laboratories; on the other hand, the nuclear magnetism of $^3He$ has received wider attention, but this falls outside the scope of our treatment.

## 15.6 Mössbauer Spectroscopy

The Mössbauer effect is the recoilless emission and absorption of $\gamma$ rays with energy in the range 10–100 keV. The significance of these recoilless processes is that they allow radiation with a natural linewidth of order $10^{-8}$ eV to be used in resonant absorption spectroscopy, thereby producing the highest resolution available in any

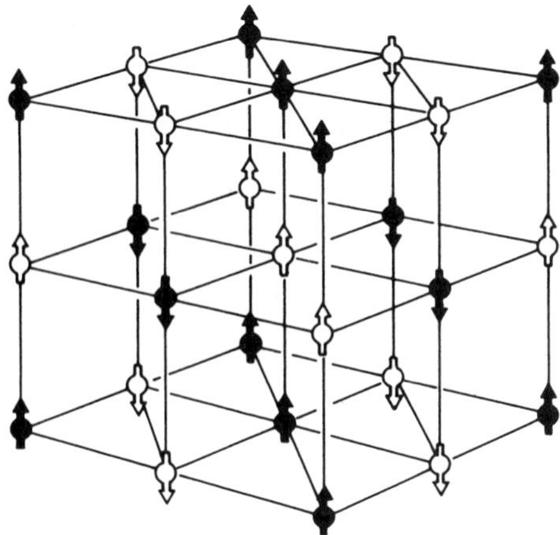

**Figure 15.11** Antiferromagnetic nuclear order as observed by neutron diffraction in $^7Li^1H$. (After Abragam *et al.* 1978.)

spectroscopic technique. For a Mössbauer isotope the ratio of the photon energy to the linewidth is of order $10^{12}$, which may be compared with the same quantity for laser light, $10^{10}$, or the quality factor of a good quartz crystal oscillator, $10^5$. This extremely high resolution has allowed the testing of important aspects of general relativity theory, but has found its widest application in the study of magnetic compounds and alloys.

### 15.6.1 *Recoil and Doppler shifts*

Consider the emission of radiation of frequency $\nu_0$ from an atom in a gas. The atom has thermal energy and at the moment of emission possesses momentum $\mathbf{P}_0$ with a component $P$ in the direction of the emitted photon (Fig. 15.12). Conservation of momentum in the direction of emission gives the recoil momentum

$$R = \frac{h\nu_0}{c}.$$

Prior to emission of the photon, the atom is in an excited state with energy $E'$, and afterwards it has energy $E_0$. The motion of the atom transverse to the direction of the emitted photon remains unchanged, so the energy balance is as follows:

$$E' + \frac{P^2}{2M} = E_0 + \frac{(P-R)^2}{2M} + h\nu_0 .$$

The shift in the energy of the radiation by the atomic recoil and the motion of the source is

$$\begin{aligned}\Delta(h\nu) &= (E' - E_0) - h\nu_0 \\ &= \frac{h^2\nu_0^2}{2Mc^2} - \frac{Ph\nu_0}{Mc} \\ &= E_R - (\pm E_D), \qquad (15.7)\end{aligned}$$

$E_R$ and $E_D$ being the shifts from recoil and from the Doppler effect respectively. The latter is written $\pm E_D$ because the source may be moving parallel or antiparallel to the photon. If the photon is absorbed by a similar atom then recoil and Doppler shifts arise to a similar extent, but with a change of sign.

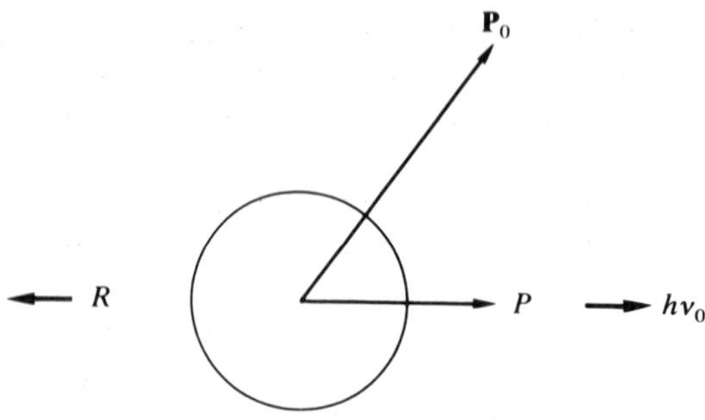


**Figure 15.12** The atomic recoil associated with the emission of a $\gamma$-ray photon.

**Table 15.1** Recoil and Doppler shifts in electronic transitions

| | | |
|---|---|---|
| $h\nu_0$ (eV) | 1 | $10^4$ |
| $E_R$ (eV) | $10^{-11}$ | $10^{-3}$ |
| $E_D$ (eV) | $\pm 10^{-6}$ | $\pm 10^{-2}$ |
| Natural linewidth (eV) | $10^{-7}$ | 1 |

We first estimate the effects of these shifts for photons near the visible and X-ray regions of the spectrum, e.g. $h\nu_0 = 1$ eV and 10 keV, arising from electronic transitions within the atom. At room temperature the average thermal velocity of an atom is about $3 \times 10^2$ m s$^{-1}$. Suppose our atom has mass 50; then we readily find the values given in Table 15.1.

For photons of electronic origin the recoil energy is truly insignificant with regard to both the Doppler shift and especially the linewidth of the radiation. The energies of absorption and emission overlap perfectly, thereby ensuring that resonant absorption always arises. If this were not the case, chemical analysis by atomic absorption spectroscopy would not be possible. The situation is very different for soft $\gamma$-ray photons emitted by certain nuclei where the excited state has a lifetime of $10^{-7}$ s, leading to a linewidth of $10^{-9}$ eV. A 10 keV $\gamma$ ray from a nucleus of mass 50 then experiences a recoil shift some $10^6$ times greater than the line width. The only chance for resonance arises if the atomic motion produces a Doppler overlap of the emission and absorption energies, but in fact the chances are very small.

### 15.6.2 Recoilless processes

The previous discussion concerned the free atom. As first demonstrated by Mössbauer, the emission and absorption of certain nuclear $\gamma$ rays by atoms embedded in a crystal is a very different matter. The atom is no longer free, neither can the recoil energy cause it to become free; we have seen earlier that displacement energies for atoms in solids are about 25 eV. If then the atom is considered to be rigidly fixed in the sample, the recoil must be adsorbed by the sample as a whole. We therefore replace the masses of the emitting and absorbing atoms by those of the source and absorber; since they may contain anywhere between say $10^{15}$ and $10^{18}$ atoms, the recoil energy is reduced to at least $10^{-18}$ eV, which is $10^9$ times less than the inherent line width. Furthermore, the source and adsorber remain essentially stationary, so even the Doppler shift would be vanishingly small. If this were the case then resonant absorption would be the rule. On the other hand, we know that atoms in crystals have a vibratory motion and the previous assumption of complete rigidity is a very coarse approximation.

Even so, the assumption that the recoil energy is absorbed by the crystal lattice as a whole is implicit in our earlier discussion of elastic X-ray or neutron scattering that produces Laue diffraction. The incident photons or neutrons possess energy and momentum, so the diffracting atoms should recoil and thereby degrade the energy of the scattered photon or neutron. In general both elastic (lattice recoil) and inelastic (atomic recoil) scattering occur: the former produces Laue diffraction, the latter produces background radiation. We find a similar situation in the scattering of low-energy $\gamma$ rays. Only the elastically scattered rays produce the resonant absorption that characterizes the Mössbauer effect.

The probability that recoilless emission or absorption of $\gamma$ rays occurs is given by

$$f = \exp\left(-\frac{4\pi^2}{\lambda^2}\langle s^2\rangle\right), \tag{15.8}$$

where $\lambda$ is the wavelength of the $\gamma$ ray and $\langle s^2\rangle$ the mean-square thermal displacement of the appropriate atom in the direction of photon emission (see Greenwood and Gibb 1971). We may rewrite (15.8) as

$$f = \exp\left(-\frac{E^2}{\hbar^2 c^2}\langle s^2\rangle\right), \tag{15.9}$$

$E$ being the $\gamma$-photon energy $h\nu_0$. When $\langle s^2\rangle$ is calculated on the basis of the Debye model, $f$ is known as the Debye-Waller factor. We note that $f$ becomes larger the smaller the photon energy and the smaller $\langle s^2\rangle$ – the latter implies the lower the temperature and the larger the Debye or Einstein temperature. Because $f$ decreases as $E$ increases, there is an upper limit to $h\nu_0$ of 150 keV, but in practice most Mössbauer $\gamma$-ray sources have energies $<100$ keV. On the other hand, $\gamma$ photons with $E < 10$ keV are strongly absorbed by matter, so this energy forms an approximate lower limit. Clearly a proportion $1-f$ of emission or absorption events are subject to recoil and cause a change in the frequency of the radiation. This produces a background of absorption, but it is of no consequence for the resonant behaviour.

The resonant absorption of $\gamma$ rays is observed in the following manner. We assume the source and absorber to be similar (not just the emitting and absorbing nuclei but also their chemical environments). If the absorber is stationary and the source moved at a constant velocity, a Doppler shift of the emitted photon occurs; a source velocity $<1$ mm s$^{-1}$ is often sufficient to destroy the conditions for resonant absorption (Fig. 15.13).

A Mössbauer spectrometer therefore normally comprises a source that is regularly subjected to a constant acceleration so that the velocity follows a sawtooth pattern.

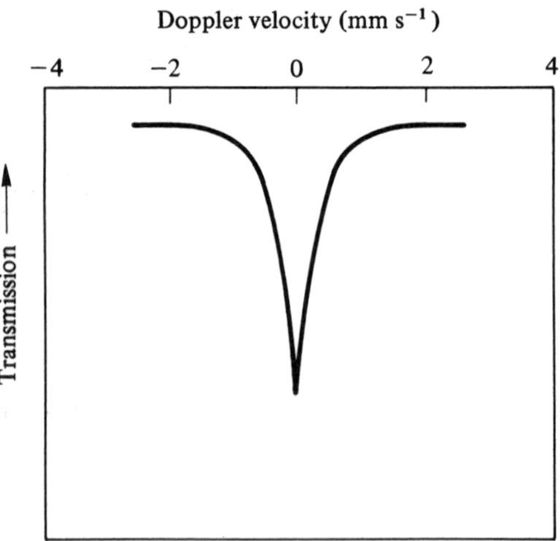


**Figure 15.13** Resonant absorption in Doppler spectroscopy, assuming identical source and absorber.

**Table 15.2** Mössbauer sources

| | $h\nu_0$ (keV) | $E_R$ (meV) | Linewidth (mm $s^{-1}$) |
|---|---|---|---|
| $^{57}$Fe | 14.412 | 1.95 | 0.192 |
| $^{119}$Sn | 23.875 | 2.58 | 0.626 |
| $^{121}$Sb | 37.15 | 6.12 | 2.1 |
| $^{124}$Te | 35.48 | 1.54 | 5.0 |
| $^{127}$I | 57.60 | 14.0 | 2.54 |
| $^{149}$Sm | 22.5 | 1.82 | 1.6 |
| $^{153}$Eu | 97.43 | 33.2 | 10.7 |
| $^{156}$Gd | 88.97 | 27.2 | 1.4 |

The absorber, in the form of a thin foil or powder, is stationary. The transmitted radiation may be detected in a proportional counter. The repeat frequency for the constant acceleration is usually 20 Hz and the velocity limits 0–$10^2$ mm $s^{-1}$. As shown in Fig. 15.13, it is customary to plot the spectrum as a transmission spectrum. Facilities for cooling and heating the absorber are required, particularly in the study of magnetic behaviour.

Some 70 isotopes of about 40 chemical elements provide $\gamma$ rays suitable for Mössbauer spectroscopy, and a selection of these is given in Table 15.2. We shall confine attention to the isotope $^{57}$Fe, which is by far the most important example. It is ideally suited from both its nuclear aspects and its importance in the physics and chemistry of solids. It is obtained through the decay of $^{57}$Co according to the scheme of Fig. 15.14. $^{57}Co_{27}$ is produced by a (d, n) reaction in $^{56}$Fe, which is the major constituent (91.65%) of ordinary iron. The Co is then diffused into a host with cubic symmetry (e.g. Pd) to avoid the complications associated with the quadrupole moment of the excited nuclear state (see Table 15.3). Ordinary iron contains 2.19% $^{57}$Fe and usually it is not necessary to enrich the absorbing sample.

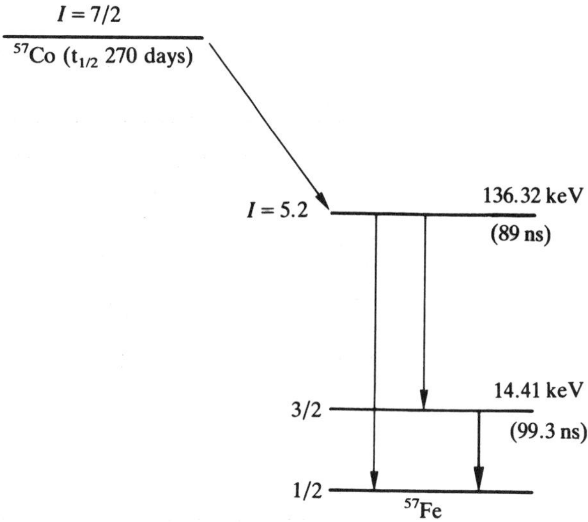


**Figure 15.14** The nuclear decay scheme of $^{57}Co_{27}$. The important transition for Mössbauer spectroscopy is that producing the 14.41 keV photon.

The Mössbauer spectrum of a sample absorber is influenced by:

(a) *The chemical environment of the absorber.* The excited and groundstate nuclei have different radii and their energies are dependent upon the electron density at the nucleus; as we have already seen, the latter can only be of s character. The s-electron density varies with the chemical environment of the atom and the occupation of the outer electron orbitals. Normally we should call such an effect a chemical shift, but in Mössbauer spectroscopy it is termed the isomer shift. Since the source nuclei are also subject to a similar perturbation, a comparison of isomer shifts observed in different samples is only significant if they are obtained using the same source. An example of an isomer shift is shown in Fig. 15.15*a*.

(b) *Quadrupole effects.* The first excited state of $^{57}Fe$ has a quadrupole moment, so, in samples with other than cubic symmetry, we find a two-line spectrum. This quadrupole splitting is superposed on the isomer shift (Fig. 15.16). As in NMR,

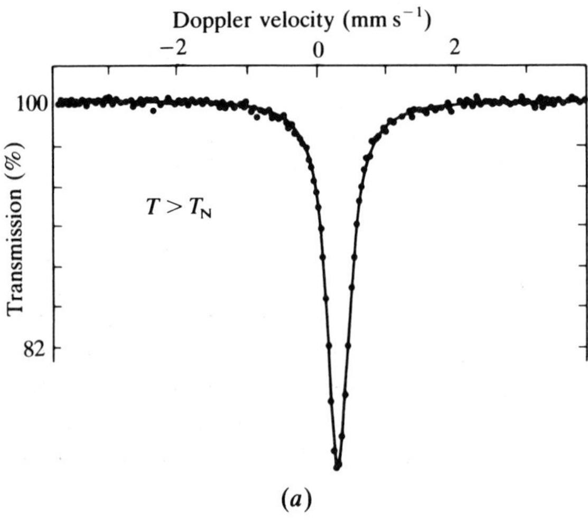


(*a*)

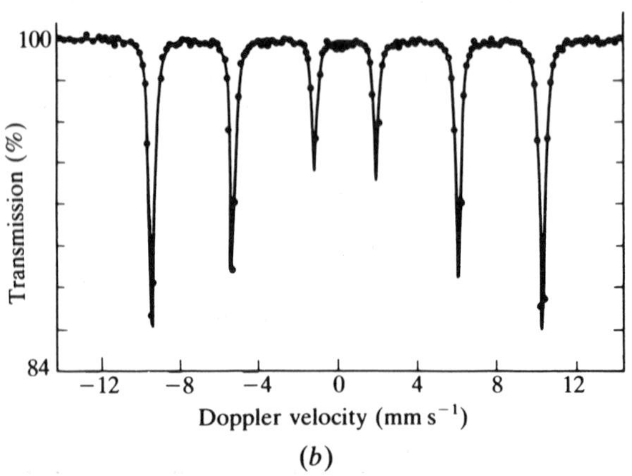


(*b*)

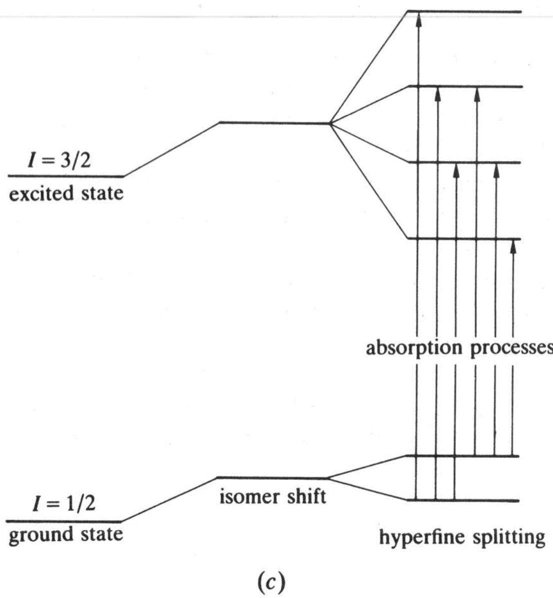


**Figure 15.15** (a) The isomer shift in paramagnetic $FeF_3$. (b) The hyperfine splitting in the Mössbauer spectrum of antiferromagnetic $FeF_3$ at 4.2 K. (c) The origin of the characteristic six-line spectrum is shown. (After Wertheim *et al.* 1968.)

this provides indirect information regarding the electron configuration of the matrix.

(c) *The hyperfine field.* If the Mössbauer nucleus resides in an atom possessing an electronic magnetic moment, and the host matrix is ferromagnetic or antiferromagnetic, then even in the absence of an external magnetic field there arises a large internal magnetic field at the nucleus, the so-called hyperfine field, which causes an intrinsic splitting of both the ground and the excited states of the nucleus.

In the case of $^{57}Fe$ the hyperfine field produces a characteristic six-line spectrum that is slightly offset from the origin because of the isomer shift with respect to the source. This spectrum is evident in pure iron metal below its Curie temperature, but

**Table 15.3** Some properties of $^{57}Fe$

| | Ground state | First excited state |
|---|---|---|
| Energy (keV) | 0 | 14.41 |
| Half life (ns) | $\infty$ | 99.3 |
| Magnetic moment $\mu_N$ | +0.0902 | −0.1547 |
| Quadrupole moment ($10^{-28}$ $m^2$) | 0 | 0.2 |
| $E_R$ (meV) | 1.95 | 1.95 |
| Linewidth (meV) | 0 | 9.2 |
| Linewidth (mm $s^{-1}$) | 0 | 0.192 |

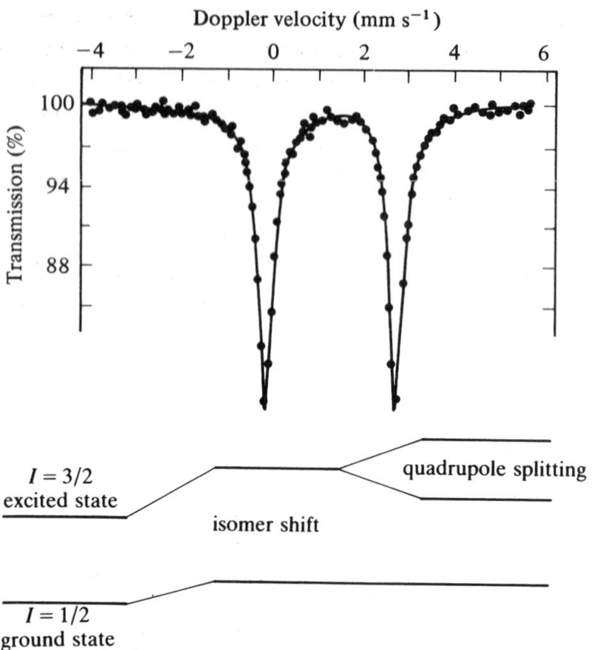


**Figure 15.16** Pure quadrupole splitting in $FeF_2$ at 78.21 K, just above its Néel temperature. (After Wertheim and Buchanan 1967.)

perhaps seen at its best in a cubic antiferromagnetic salt like $FeF_3$ (Fig. 15.15*b*). Above the Néel temperature this pattern is replaced by a single line. Quadrupole effects in $FeF_2$ above the Néel temperature are shown in Fig. 15.16. The dependence of the magnetic state on temperature, pressure or composition may be studied directly through the Mössbauer spectrum.

Although the atomic magnetic moment persists above the Néel temperature, its presence is not observed in Mössbauer spectroscopy. Within the lifetime of the excited nuclear state, $10^{-7}$ s in the case of $^{57}Fe$, the moment changes direction so often that the average hyperfine field at the nucleus is zero. Mössbauer spectroscopy can only illustrate phenomena that arise on a timescale longer than the half-life of the excited state.

The atomic magnetic moments reside in the 3d or 4f shells: these magnetic dipoles make an insignificant contribution to the hyperfine field at the nucleus. For substances with cubic symmetry their contribution is in fact zero – for the same reason that the near electrostatic field in the Lorentz model of a cubic dielectric is zero (Section 11.2).

The major source of the hyperfine field arises from the polarization of electrons in the 1s, 2s, 3s and 4s shells (for the 3d magnetic elements). In the presence of the atomic moment these s states become polarized and the spin-up and spin-down electron wave functions have different amplitudes at the nucleus. The effect is pronounced owing to their contact interaction with the nucleus (just as in the Knight shift), and in the case of Fe metal the resultant effect produces a hyperfine field of 33 T. It must be understood that this is a net field because the 3d electron moment polarizes the 1s and 2s shells in opposite manner to the 3s and 4s shells.

In chemical complexes of iron where the valence electron distributions are better defined than in the metallic phase, the hyperfine field, at low temperatures, corresponds to about 11 T per unpaired d electron; thus the $Fe^{3+}$ ion has a hyperfine field of 55 T. The sign of the hyperfine field may be determined with the aid of an external field because this either augments or diminishes the effective field at the nucleus. It is found that for iron the sign of the hyperfine field is negative. This is because the largest contribution arises from the 2s shell, which is polarized antiparallel to the d-shell atomic moment.

Since the discovery of the Mössbauer effect in 1957, a vast literature regarding its applications to physics and chemistry has evolved (see e.g. Greenwood and Gibb 1971).

## References

ABRAGAM, A., BOUFFARD, V., GOLDMAN, M. and ROINEL, Y. (1978) *J. Phys. (Paris)* **39**, Colloq. C6, Suppl. No. 8, 1436.

BLEANEY, B. I. and BLEANEY, B. (1976) *Electricity and Magnetism*, 3rd edn. Oxford University Press, Oxford.

COLLAN, H. K., KRUSIUS, M. and PICKETT, G. R. (1970) *Phys. Rev.* **B1**, 2888.

GREENWOOD, N. N. and GIBB, T. C. (1971)*Mössbauer Spectroscopy*. Chapman and Hall, London.

HUIKU, M. T., JYRKKIÖ, T. A., KYYNÄRÄINEN, J. M., OJA, A. S. J., CLAUSEN, K. N. and STEINER, M. (1986) *Physica Scripta* **T13**, 114.

KRUSIUS, M. (1971) Thesis, Helsinki University of Technology, Espoo, Finland.

LOUNASMAA, O. V. (1974) *Experimental Principles and Methods Below 1 K*, Academic Press, London.

ROGERS, E. H. (1960) *NMR and EPR Spectroscopy; Varian 3rd Annual Workshop*, Pergamon Press, New York.

WERTHEIM, G. K., GUGGENHEIM, H. J. and BUCHANAN, D. N. E. (1978) *Phys. Rev.* **169**, 465.

WERTHEIM, G. K. and BUCHANAN, D. N. E. (1967) *Phys. Rev.* **161**, 478.

## Further Reading

FREEMAN, A. J. and SCHNEIDER, K. A. (Editors), *Magnetism in the Nineties* (1991) North Holland – Elsevier Science Publishers, Amsterdam.

# Answers to Problems

## Chapter 2

**2.1**
$$\text{Atomic diameter} = \begin{cases} 3^{1/2}a/2 & \text{(bcc)}, \\ a/2^{1/2} & \text{(fcc)}. \end{cases}$$

**2.2** $0.291R$, where $R$ is the hard-sphere radius of Problem 2.1. The sites are at the centres of irregular tetrahedra, one such position being $(0\frac{1}{2}\frac{1}{4})$; there are also sites with radius $0.154R$ at irregular octahedral positions, e.g. at $(0\frac{1}{2}\frac{1}{2})$.

**2.3** Octahedral sites at $(\frac{1}{2}00)$, $(0\frac{1}{2}0)$ etc. Tetrahedral sites located on the cube diagonal and distant $\frac{3}{4}d_{111}$ from the cube corner.

**2.4** Expand bcc along [001] so that $c\ (=b=a)$ becomes $2^{1/2}a[001]$. In terms of the bcc of side $a$, the fcc cell becomes $a[110]$, $a[\bar{1}10]$ and $2^{1/2}\,a[001]$.

| | | | |
|---|---|---|---|
| bcc | (100) | (110) | (111) |
| fcc | $(\bar{1}10)$ | $(\bar{2}00)$ | $(\bar{2}0\bar{1})$ |

**2.6**

Fig. 2.13a NaCl — fcc lattice,
basis Na(000)
Cl$(\frac{1}{2}00)$

Fig. 2.13b CsCl — sc lattice,
basis Cs(100)
Cl$(\frac{1}{2}\frac{1}{2}\frac{1}{2})$

Fig. 2.13d $BaTiO_3$ — sc lattice,
basis Ba(000)
Ti$(\frac{1}{2}\frac{1}{2}\frac{1}{2})$
O$(\frac{1}{2}\frac{1}{2}0)$
$(\frac{1}{2}0\frac{1}{2})$
$(0\frac{1}{2}\frac{1}{2})$

Fig. 2.13f $Nb_3Sn$ — sc lattice,
basis Sn(000), $(\frac{1}{2}\frac{1}{2}\frac{1}{2})$
Nb$(\frac{1}{2}\frac{1}{4}0)$, $(\frac{1}{2}\frac{3}{4}0)$
$(\frac{1}{4}\frac{1}{2}0)$, $(\frac{3}{4}\frac{1}{2}0)$
$(0\frac{1}{2}\frac{1}{4})$, $(0\frac{1}{2}\frac{3}{4})$

**2.8** Two examples: $(01\bar{1}0)$ becomes (010); $(1\bar{1}00)$ becomes $(1\bar{1}0)$.

**2.9** (100) becomes $(\frac{1}{2}\frac{1}{2}0) \rightarrow (110)$; (110) becomes $(\frac{1}{2}1\frac{1}{2}) \rightarrow (121)$; (111) becomes (111).

**2.10** See Chapter 14, Section 14.1.

**2.12** $H = \frac{1}{2}(h - k),$

$K = \frac{1}{2}(h + k).$

**2.13** ... *ABABCBCAC* ...

## Chapter 3

**3.1** $6.23 \times 10^{23}$ (the correct value is $6.02 \times 10^{23}$).

**3.2** (a) Lattice parameter 4.103 Å.
(b) Planes $\{100\}, \{110\}, \{111\}, \{200\}, \{210\}$.
(c) $h^2 + k^2 + l^2 = 27$, $\{511\}, \{333\}$, CsCl.

**3.3** Simple hexagonal (0001), $(10\bar{1}0)$, (0002).
Close-packed hexagonal $(10\bar{1}0)$, (0002).

**3.4** The atomic form factors of the K and Br ions are quite different, but those of the K and Cl ions are similar because these ions have identical electron content.

**3.5** (a) 2.46 Å; (b) 2.80 Å.

**3.6** In the random alloy every lattice site is associated with the same form factor $f = \frac{1}{2}(f_{Cu} + f_{Zn})$, and we obtain the bcc diffraction pattern. In the ordered alloy we may associate cube corners with Cu atoms and the body-centre positions with Zn atoms. We have the CsCl arrangement, and the diffraction pattern has all the lines of the sc lattice but with varying intensities.

**3.7** fcc, $a = 3.919$ Å.

**3.8** Tetragonal, $a = 3.905$ Å, $c/a = 1.0026$.

**3.9** Differentiate Bragg's law to obtain

$$\Delta\lambda = 2(\Delta d \sin\theta + d \cos\theta\, \Delta d),$$

$$\frac{\Delta a}{a} = \frac{\Delta d}{d} = -\cot\theta\, \Delta\theta \quad \text{for constant } \lambda$$

**3.10**

| | $(hkl)$ | $(hhl)$ | $(hhh)$ | $(0kl)$ | $(0kk)$ | $(00l)$ |
|---|---|---|---|---|---|---|
| Cubic | 48 | 24 | 8 | 24 | 12 | 6 |
| Orthorhombic | 8 | 8 | 8 | 4 | 4 | 2 |

**3.13**

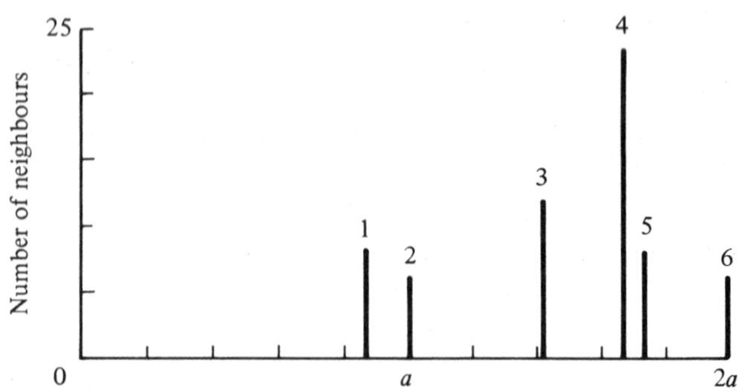


The number of first, second, third, etc. nearest neighbours is shown.

**3.15** $S_{hkl} = f + f e^{i\pi(h+k+l)/n}$. $S = 0$ when $(h + k + l)/n$ is an odd number; thus, for $n = 2$, when $h + k + l = 2, 6, 10, 14$ etc.

**3.16** $S$ contains components with arbitrary phase differences; the intensity is controlled by $SS^*$.

**3.17** See Fig. 3.5 and standard tables for the energies involved. $\lambda$ (in Å) is as follows:

| | Cu | Ag | Au |
|---|---|---|---|
| $\alpha_1$ | 1.5404 | 0.558 | 0.180 |
| $\alpha_2$ | 1.5444 | 0.563 | 0.185 |

**3.18** All integer values of $H$ and $K$ are allowed, but reflections are absent when $h + k$ is an odd number. For a primitive cell all $(HK)$ are permitted; but if a unit cell and basis are used, the basis leads to destructive interferences, causing certain lines to be absent – check the behaviour of the bcc and fcc structures.

**3.19** Factorize the structure factor for the diamond cubic cell into lattice and basis terms as follows:

$$S_{hkl}(\mathrm{Si}) = f_{\mathrm{Si}}(1 + e^{i\pi(h+k+l)/2})(1 + e^{i\pi(h+k)} + \ldots),$$

$$S_{hkl}(\mathrm{GaAs}) = (f_{\mathrm{Ga}} + f_{\mathrm{Ga}}\, e^{i\pi(h+k+l)/2})(1 + e^{i\pi(h+k)} + \ldots).$$

For GaAs, the basis term is finite for all $hkl$, so the structure factor zeros are determined by the lattice term and all the lines of the fcc structure appear. For Si, however, the basis term is zero whenever $h + k + l = 2n$, unless $2n$ is a multiple of 4. So we then obtain the diamond cubic diffraction pattern (see Fig. 3.4).

**3.20** Primitive tetragonal, basis ○ at (000), (1/2 1/2 0);
● at (1/2 0 1/4), (0 1/2 3/4)

Diffraction lines are absent whenever $(h + k)$ is odd and $l$ is even.

## Chapter 4

**4.3** To the right.

**4.5** 0.18 MPa (0001), $[2\bar{1}\bar{1}0]$.

**4.10** $E_D = 1.7$ eV, giving $E_v = 0.9$ eV. $n/N$ (700 K) $= 3.3 \times 10^{-7}$ and $n/N$ (1000 K) $= 2.9 \times 10^{-5}$.

**4.11** $\Delta\rho = 1.93 \times 10^5 \exp(-0.9\ \mathrm{eV}/k_B T)$ nΩ cm and $E_v = 0.9$ eV. 1% vacancies cause incremental resistance 1.93 μΩ cm.

**4.12** $t(727°\mathrm{C}) = 1.6 \times 10^6$ s $\approx$ 18 d; $t(927°\mathrm{C}) = 3.0 \times 10^4$ s $\approx$ 0.3 d.

**4.13** (a) $4.26 \times 10^{-12}$ cm$^2$ s$^{-1}$; (b) $2.49 \times 10^{-6}$ cm$^2$ s$^{-1}$; (c) $8.52 \times 10^{-7}$ cm$^2$ and $4.98 \times 10^{-1}$ cm$^2$; (d) $1.7 \times 10^{-7}$ cm$^2$ s$^{-1}$ (γ-Fe).

**4.14** $D \geqslant 1.9 \times 10^{-13}$ cm$^2$ s$^{-1}$, 750 K. $t_{1223}/t_{750} = 1.50 \times 10^{-5}$.

**4.15**
$$n = (NN')^{1/2} \exp\left(\frac{S}{2k_B}\right) \exp\left(-\frac{U}{2k_B T}\right).$$

## Chapter 5

**5.1** Highest: C(diamond) (2230 K), Be (1460 K), Cr (630 K); lowest: Cs (38 K), Rb (56 K), Hg (69 K).

**5.2**
$$C_D = 9R\left(\frac{T}{\theta_D}\right)^3 \int_0^{\theta_D/T} \frac{z^4 e^z}{(e^z - 1)^2}\,dz;$$
$$C_D = \begin{cases} 3R & (T \gg \theta_D), \\ \frac{12}{5}\pi^4 R\left(\frac{T}{\theta_D}\right)^3 & (T < 0.1\theta_D). \end{cases}$$

**5.5**
$$N(\omega) = \frac{2N}{\pi(\omega_0^2 - \omega^2)^{1/2}}, \qquad \omega_0 = 2\left(\frac{c}{M}\right)^{1/2}.$$

**5.10**
$$\frac{M\omega^2}{2c} = 3 - \cos \mathbf{k}\cdot\mathbf{a}_1 - \cos \mathbf{k}\cdot\mathbf{a}_2 - \cos \mathbf{k}\cdot(\mathbf{a}_1 - \mathbf{a}_2);$$
$$\omega^2 = \begin{cases} 8c/M & \text{(middle of a Brillouin-zone side),} \\ 9c/M & \text{(corner of the Brillouin zone)} \end{cases}$$

**5.13** They have equal size.

## Chapter 6

**6.3** $\Delta E_F = -2\alpha E_F\,\Delta T$; $\Delta E_F = \frac{2}{3}\kappa E_F\,\Delta P$.
(a) Na 0.014% $K^{-1}$, Cu 0.0034% $K^{-1}$;
(b) Na $1.02 \times 10^4$ Pa, Cu $2.05 \times 10^5$ Pa.

**6.6** (a) Al 7.8 K; (b) Pb 1.3 K.

**6.7** 280 T, $7.5 \times 10^{-3}\ \mu_B$ per atom.

**6.8** $\gamma = 2$ mJ mol$^{-1}$ K$^{-2}$; $\theta_D = 87$ K: potassium.

**6.9** $r/k_F = 1.14$.

**6.10** (a) $2.34 \times 10^{23}$ eV$^{-1}$; (b) $2.34 \times 10^{11}$ eV$^{-1}$; (c) 2.34 eV$^{-1}$.

**6.11** 787 Å.

**6.12** 0.999 993.

**6.13** 1.4%.

**6.14** 366 Å.

**6.15** $\hbar\omega_p = 12.7$ eV; $m^* = 1.49$; $\hbar/\tau \approx 0.1$ eV; $\tau \approx 6.4 \times 10^{-15}$ s.

**6.16** 1168 Å.

**6.18** $2.95 \times 10^8\ \Omega^{-1}$ m$^{-1}$; 556 W m$^{-1}$ K$^{-1}$.

**6.19**
$$C_{mol} = 3R + 0.00135T + \frac{6.03 \times 10^{23}}{k_B T^2}(0.78 \times 1.6 \times 10^{-19})^2 \exp\left(-\frac{0.78 \times 1.6 \times 10^{-19}}{k_B T}\right).$$

**6.20** $\theta_c = 88.57°$.

## Chapter 7

**7.1** bcc 1.48; fcc 1.36.

**7.3** Use the geometry of Fig. 7.16 as an aid, in addition to the illustration of the fcc Brillouin zone (Fig. 7.15*b*). The various symmetry points have the following values of $k$; the energies are the free electron values for $a = 5.58$ Å:

| | Γ | X | L | U | W |
|---|---|---|---|---|---|
| $k$ | 0 | $2\pi/a$ | $1.73\pi/a$ | $6.68/a$ | $7.05/a$ |
| $E$ (eV) | 0 | 4.84 | 3.63 | 5.45 | 6.09 |

The minimum energy gaps on the zone boundaries are therefore 1.25 eV at the X point and 2.46 eV at the L point.

**7.5**
$$N(E) = \frac{2L}{\pi a} \frac{1}{[B^2 - (A - E)^2]^{1/2}}.$$

## Chapter 9

**9.6** $\hbar\omega = 0.64E_F$.

## Chapter 10

**10.2** $E_g = 0.69$ eV.

**10.3** $\sigma_i(\text{RT}) = 0.31\ \Omega^{-1}\ \text{m}^{-1}$; $\sigma = 2.68 \times 10^3\ \Omega^{-1}\ \text{m}^{-1}$; approx. 900 K.

**10.4** $\sigma = 4.1 \times 10^{-5}\ \Omega^{-1}\ \text{m}^{-1}$; $p = 2 \times 10^{10}\ \text{m}^{-3}$; $n = 10^{20}\ \text{m}^{-3}$; $\mu = 0.87$ eV.

**10.5** $\sigma = 2.27 \times 10^4\ (\Omega\ \text{m})^{-1}$; $\mu = 0.152$ eV; (a) 0.0008 eV; (b) 545 Å; (c) $1.47 \times 10^{21}\ \text{m}^{-3}$; (d) $n = 1.82 \times 10^{22}\ \text{m}^{-3}$; $p = 1.67 \times 10^{22}\ \text{m}^{-3}$; $\mu = 0.153$ eV.

**10.6** $k_e = k_h = k_0$;

$$\frac{\hbar^2 k_0^2}{2}\left(\frac{1}{m_e} + \frac{1}{m_h}\right) = \hbar\omega - E_g;$$

$k_0 = 0.58 \times 10^9\ \text{m}^{-1}$; $E_h = 0.0187$ eV; $E_e = 0.1813$ eV.

**10.7**
$$N_a\left[\exp\left(\frac{E_a - \mu}{k_B T}\right) + 1\right]^{-1} = p_0 \exp\left(-\frac{\mu}{k_B T}\right).$$

When both donor and acceptor impurities are present, electrostatic neutrality demands that $n + N_a^- = p + N_d^+$. The Fermi functions complicate matters and the equation must be solved numerically; usually it is the Fermi level that is sought.

**10.8** $N_a^- = 4.77 \times 10^{22} = 95\%\ N_a$; $\mu(300\ \text{K}) = 0.13$ eV; $\mu(100\ \text{K}) = 0.045$ eV; $\sigma(300\ \text{K})/\sigma(100\ \text{K}) = 5.19$.

**10.9** $E_g = 0.70$ eV; $p = 10^{22}\ \text{m}^{-3}$; $n \approx 10^{15}\ \text{m}^{-3}$; $\rho = 0.0035\ \Omega$ m; $\mu = 0.158$ eV.

**10.10** Electrons; $n = 9.6 \times 10^{20}\ \text{m}^{-3}$; $\mu_e = 0.313\ \text{m}^2\ \text{V}^{-1}\ \text{s}^{-1}$.

**10.11** (a) n doping.
(b) Saturation value of $R_H \approx -3000$; $n \approx 2 \times 10^{15}\ \text{cm}^{-3}$.

## Chapter 11

**11.1**

| | Fe | Co | Ni |
|---|---|---|---|
| $\mu/\mu_B$ | 2.22 | 1.72 | 0.60 |
| cgs units: | | | |
| $M = N\mu$ (erg $G^{-1}$ $g^{-1}$) | 222 | 162.8 | 57.1 |
| $I = M\rho$ (erg $G^{-1}$ $cm^{-3}$) | 1745 | 1449 | 509 |
| SI units: | | | |
| $M = N\mu$ (J $T^{-1}$ $kg^{-1}$) | 222 | 162.8 | 57.1 |
| $M = N\mu$ (J $T^{-1}$ $m^{-3}$) | $1.745 \times 10^6$ | $1.45 \times 10^6$ | $0.509 \times 10^6$ |
| $\mu_0 M$ (T) | 2.19 | 1.82 | 0.64 |
| $C_{SI} = 4\pi \times 10^{-6}\ C_{cgs}$ | | | |

**11.2** $\lambda = 6.3 \times 10^{-4}$; $B_i = 1.09 \times 10^3$ T $\approx 10^7$ G.

**11.3**

| | $\kappa$ | $\chi_{mol}$ | $\chi$ |
|---|---|---|---|
| SI | $4.31 \times 10^{-2}/T$ | $4.71 \times 10^{-6}/T$ ($m^3$ $mol^{-1}$) | $1.89 \times 10^{-5}/T$ ($m^3$ $kg^{-1}$) |
| cgs | $3.43 \times 10^{-3}/T$ | $3.75 \times 10^{-1}/T$ ($cm^3$ $mol^{-1}$) | $1.5 \times 10^{-3}/T$ ($cm^3$ $g^{-1}$). |

**11.4** $\kappa$(core) $= -2.07 \times 10^{-5}$.

**11.5** Calculated $\gamma = 16.8$ mJ $mol^{-1}$ $K^{-2}$; measured $\gamma = 7$ mJ $mol^{-1}$ $K^{-2}$.

**11.6** (a) $\kappa_\perp = 2\mu_0/v$.

(b) $M_A = \frac{1}{2}MB(J, \alpha_A), \qquad \alpha_A = \frac{\mu}{k_B T}(B_0 + vM_B);$

$M_B = \frac{1}{2}MB(J, \alpha_B), \qquad \alpha_B = \frac{\mu}{k_B T}(-B_0 + vM_A);$

$$\kappa_\parallel = \frac{\mu_0(M_A - M_B)}{B_0}.$$

(c) The axis of antiferromagnetism rotates to become perpendicular to the applied magnetic field, this being the state of minimum magnetostatic energy (the susceptibility has its largest value in this direction).

**11.7** Domain boundary thickness = 345 atom diameters.

## Chapter 12

**12.2** $\mathbf{E}_{loc} = \mathbf{E}$ gives

$$\alpha = \frac{e^2}{\varepsilon_0 m} \frac{1}{(\omega^2 - \omega_i^2 - i\gamma\omega)};$$

$\mathbf{E}_{\mathrm{loc}} = \mathbf{E}_{\mathrm{L}}$ gives

$$\varepsilon_r - 1 = \frac{N\alpha}{1 - \frac{1}{3}N\alpha}, \quad \alpha \text{ as above.}$$

**12.3** $\alpha_{\mathrm{Ag}} = 29.78 \times 10^{-30}$ and $30.66 \times 10^{-30}$ $\mathrm{m}^3$ respectively. Each ion is displaced $1.8 \times 10^{-7}$ Å.

## Chapter 13

**13.1**

| | $\frac{1}{2}N(E_{\mathrm{F}})V$ | $V$ (eV atom) |
|---|---|---|
| Al | 0.17 | 0.6 |
| Pb | 0.37 | 0.59 |

$N(E_{\mathrm{F}})$ is determined from the experimental values of $\gamma$.

**13.2** $\lambda = 12.7$ nm and 2.2 $\mu$m; $3.2 \times 10^{-5}$.

**13.3** At 7.5 K, $E_{\mathrm{Cu}} = 1.86 \times 10^{-2}$ J m$^{-3}$ and $E_{\mathrm{Pb}} = 3.14 \times 10^{-2}$ J m$^{-3}$. At 0.1 K, $E_{\mathrm{Cu}}$ is unchanged and $E_{\mathrm{Pb}} = 2 \times 10^3$ J m$^{-3}$. $\Delta F = 3 \times 10^3$ J m$^{-3}$.

**13.4** $c_s - c_n = 4B_{c0}^2/\mu_0 T_c$ ; $B_{c0} = 78$ G.

**13.6** $M_{\mathrm{Pb}} = -6.4 \times 10^4$ A m$^{-1}$; $M_{\mathrm{Ni}} = +5.1 \times 10^5$ A m$^{-1}$.

**13.7** The threshold voltage gives a measure of $\Delta$(Sn).

# Index